CONTROL AND DYNAMIC SYSTEMS

Advances in Theory and Applications

Volume 51

CONTRIBUTORS TO THIS VOLUME

S. P. BHATTACHARYYA
F. BLANCHINI
BEN M. CHEN
DA-WEI GU
MASAO IKEDA
L. H. KEEL
UY-LOI LY
MOHAMED MANSOUR
IAN POSTLETHWAITE
ALI SABERI
DRAGOSLAV D. ŠILJAK
KENNETH M. SOBEL
MI-CHING TSAI
LE YI WANG
WANGLING YU
GEORGE ZAMES

CONTROL AND DYNAMIC SYSTEMS

ADVANCES IN THEORY AND APPLICATIONS

Volume Editor

C. T. LEONDES

School of Engineering and Applied Science
University of California, Los Angeles
Los Angeles, California

VOLUME 51: ROBUST CONTROL SYSTEM TECHNIQUES AND APPLICATIONS
Part 2 of 2

ACADEMIC PRESS, INC.
Harcourt Brace Jovanovich, Publishers
San Diego New York Boston
London Sydney Tokyo Toronto

ACADEMIC PRESS RAPID MANUSCRIPT REPRODUCTION

This book is printed on acid-free paper. ♾

Academic Press, Inc.
1250 Sixth Avenue, San Diego, California 92101-4311

United Kingdom Edition published by
Academic Press Limited
24–28 Oval Road, London NW1 7DX

Library of Congress Catalog Number: 64-8027

International Standard Book Number: 0-12-012751-2

PRINTED IN THE UNITED STATES OF AMERICA
92 93 94 95 96 97 BC 9 8 7 6 5 4 3 2 1

CONTENTS

ROBUST CONTROL SYSTEM TECHNIQUES AND APPLICATIONS

EXTENDED CONTENTS

CONTRIBUTORS

Numbers in parentheses indicate the pages on which the authors' contributions begin.

S. P. Bhattacharyya (31), *Department of Electrical Engineering, Texas A&M University, College Station, Texas 77843*

F. Blanchini (129), *Departimento di Matematica e Informatica, Universita' degli Studi di Udine, 33100 Udine, Italy*

Ben M. Chen (247, 295), *School of Electrical Engineering and Computer Science, Washington State University, Pullman, Washington 99164*

Da-Wei Gu (183), *Department of Engineering, Leicester University, LE1 7RH, United Kingdom*

Masao Ikeda (1), *Faculty of Engineering, Kobe University, Kobe 657, Japan*

L. H. Keel (31), *Center of Excellence in Information Systems, Tennessee State University, Nashville, Tennessee 37203*

Uy-Loi Ly (247, 295), *Department of Aeronautics and Astronautics, University of Washington, Seattle, Washington 98195*

Mohamed Mansour (79), *Automatic Control Laboratory, Swiss Federal Institute of Technology, Zurich, CH-8092 Zurich, Switzerland*

Ian Postlethwaite (183), *Department of Engineering, Leicester University, LE1 7RH, United Kingdom*

Ali Saberi (247, 295), *School of Electrical Engineering and Computer Science, Washington State University, Pullman, Washington 99164*

Dragoslav D. Šiljak (1), *Santa Clara University, Santa Clara, California 95053*

Kenneth M. Sobel (407), *Department of Electrical Engineering, The City College of New York, New York, New York 10031*

Mi-Ching Tsai (183), *Department of Mechanical Engineering, National Cheng Kung University, Taiwan 70101, People's Republic of China*

Le Yi Wang (349), *Department of Electrical and Computer Engineering, Wayne State University, Detroit, Michigan 48202*

Wangling Yu (407), *Department of Electrical Engineering, The City College of New York, New York, New York 10031*

George Zames (349), *Department of Electrical Engineering, McGill University, Montreal, Quebec H3A 2A7, Canada*

PREFACE

In the early days of modern control theory, the techniques developed were relatively simple but, nevertheless, quite effective for the relatively simple systems applications of those times. Basically, the techniques were frequency domain analysis and synthesis techniques. Then, toward the latter part of the 1950s, system state space techniques began to emerge. In parallel with these developments, computer technology was evolving. These two parallel developments (i.e., increasingly effective system analysis and synthesis techniques and increasingly powerful computer technology) have resulted in a requisite powerful capability to deal with the increasingly complex systems of today's world.

In these modern day systems of various levels of complexity, the need to deal with a wider variety of situations, including significant parameter variations, modeling large scale systems with models of lower dimension, fault tolerance, and a rather wide variety of other problems, has resulted in a need for increasingly powerful techniques, that is, system robustness techniques, for dealing with these issues. As a result, this is a particularly appropriate time to treat the issue of robust system techniques in this international series. Thus, this volume is the first volume of a two-volume sequence devoted to the most timely theme of "Robust Control Systems Techniques."

The first contribution to this volume is "Robust Stabilization of Nonlinear Systems via Linear State Feedback," by Masao Ikeda and Dragoslav D. Šiljak. It presents rather highly effective techniques for achieving robustness in nonlinear systems.

The next contribution is "Robust Stability and Control of Linear and Multilinear Interval Systems," by S. P. Bhattacharyya and L. H. Keel. This contribution provides system robustness techniques based on Kharitonov's theorem and its generalization.

The next contribution is "Robust Stability in Systems Described by Rational Functions," by Mohamed Mansour. It offers a comprehensive treatment of a unified approach to dealing with system stability and robustness problems.

The next contribution is "Constrained Control for Systems with Unknown Disturbances," by F. Blanchini. It discusses techniques for effectively dealing with system disturbances in the case of bounds on the control and system state variables.

The next contribution is "H^∞ Super-Optimal Solutions," by Da-Wei Gu, Ian Postlethwaite, and Mi-Ching Tsai. It presents techniques for robust systems design in the event of system modeling approximations, parameter uncertainties, and unpredictable system disturbances.

The next two contributions are "Closed-Loop Transfer Recovery with Observer-Based Controllers—Part 1: Analysis" and "Part 2: Design," by Ben M. Chen, Ali Saberi, and Uy-Loi Ly. These articles set forth analysis and design techniques for LTR (Loop Transfer Recovery), i.e., compensator design to recover a specific open-loop transfer function, with a view toward system performance and robust stability objectives.

The next contribution is "Robust Adaptation in Slowly Time-Varying Systems: Double-Algebra Theory," by Le Yi Wang and George Zames. It introduces notions of local spectrum and global spectrum (which are both defined in this contribution) that make it possible to unify diverse results with application to system stability analysis.

The final contribution to this second volume of this two-volume sequence on the theme of "Robust Control System Techniques" is "Robust Control Techniques for Systems with Structured State Space Uncertainty," by Kenneth M. Sobel and Wangling Yu. It provides a rather comprehensive review of the various approaches to system robustness, following which robust eigenstructure assignment techniques are presented and exemplified.

This volume is a particularly appropriate one as the second of a companion set of two volumes on robust control system analysis and synthesis techniques. The authors are all to be commended for their superb contributions, which will provide a significant reference source for workers on the international scene for years to come.

Robust Stabilization of Nonlinear Systems via Linear State Feedback

Masao Ikeda
Faculty of Engineering
Kobe University
Kobe 657, Japan

Dragoslav D. Šiljak
B&M Swig Professor
Santa Clara University
Santa Clara, California 95053

I. INTRODUCTION

Ever since it was shown by Kalman [1] that a linear plant driven by the optimal LQ control is globally exponentially stable by virtue of a quadratic form as a Liapunov function, there has been a great number of results demonstrating robustness of stability to unstructured nonlinear perturbations of the plant. On the basis of Kalman's results it has been possible to demonstrate additional robustness features of the LQ control to unmodeled uncertainty within the open–loop dynamics and to distortions of the optimal control law. One of the most gratifying aspects of this development has been the fact that the robustness features have been expressed in terms of classical robustness measures of the gain and phase margins (Anderson and Moore [2], Safonov and Athans [3], Safonov [4]).

The existence of unstructured perturbations in the plant destroys the optimality of the LQ control. If stability is preserved, then it can be shown that the LQ control is suboptimal, and the robustness in terms of gain and phase margins holds albeit at reduced levels determined by the degree of suboptimality (Sezer and Šiljak [5]). A way to recapture the original magnitudes of the stability margins is to restore optimality by modifying the control law, or the performance index, or both (Ikeda and Šiljak [6]). This result is, in fact, a solution to one of a number of inverse optimal control problems which have been considered over the years for a wide variety of plants (Moylan and Anderson [7], Furasov [8], Özgüner [9], Šiljak [10], Ikeda *et al.* [11, 12], Zheng [13]. Saberi [14]). Almost exclusively, all these results have addressed the decentralized control of interconnected systems. The reason is that, in modeling of systems composed of interconnected systems, it is natural to assume that the models of subsystems are available to the control designer with a high degree of accuracy, while the interconnections are sources of uncertainty and are represented as unstructured perturbations of subsystem dynamics.

Recently, Petersen [15] showed how the standard Riccati control design can be modified to robustify a linear system with respect to a structured nonlinear and time–varying uncertainty. This opened up a possibility to include a structured perturbation in the nominal system and still have a quadratic Liapunov function for stabilization of the nominal system by state feedback. This quadratic Liapunov function can again be used as in [6] to establish global exponential stability of the closed–loop system perturbed by a non-structured uncertainty. Our ability to handle both structured and unstructured uncertainty in this way has a special significance in the decentralized control of complex systems [16], because we can consider more realistic models for subsystems by including structured uncertainty in their descriptions. The purpose of this paper is to use the results of [15] and show how the structured uncertainty can be handled in the context of the Riccati LQ design proposed in [6] and retain, at the same time, the robustness of the closed–loop system with respect to both the distortions of the local control laws and the unstructured interconnection perturbations of the subsystems dynamics.

II. STABILIZATION

We start our consideration with a nonlinear system which is composed of a linear time–invariant part and two additive nonlinear and time–varying terms representing structured and unstructured uncertainty. To stabilize the system, we apply the quadratic stabilization method of Petersen [15] to the linear part with structured uncertainty, treating this part as a nominal system and the unstructured uncertainty as a perturbation.

Thus our model of a nonlinear system is

$$\mathbf{S}: \; \dot{x} = Ax + Bu + Ek(t, Cx, u) + f(t, x, u), \tag{1}$$

where $x(t) \in \mathbf{R}^n$ is the state and $u(t) \in \mathbf{R}^m$ is the input of $\mathbf{S}$ at time $t \in \mathbf{R}$. In (1), the constant matrices A and B are $n \times n$ and $n \times m$, and constitute a stabilizable pair. The term $Ek(t, Cx, u)$ represents a structured perturbation, where E and C are $n \times q$ and $p \times n$ constant matrices, and $k : \mathbf{R} \times \mathbf{R}^p \times \mathbf{R}^m \to \mathbf{R}^q$ is a sufficiently smooth function satisfying the bound

$$\|k(t, Cx, u)\| \leq \|Cx\| \text{ for all } (t, x, u) \in \mathbf{R} \times \mathbf{R}^n \times \mathbf{R}^m, \tag{2}$$

where $\|\cdot\|$ denotes the Euclidean norm. The function $f : \mathbf{R} \times \mathbf{R}^n \times \mathbf{R}^m \to \mathbf{R}^n$ represents the unstructured uncertainty and is also sufficiently smooth so that (1) has the unique solution $x(t) = x(t; t_0, x_0, u)$ for any initial time t_0, any initial state x_0, and any fixed piecewise continuous input $u(\cdot)$. Furthermore, we assume that $k(t, 0, 0) = 0$ and $f(t, 0, 0) = 0$ for all $t \in \mathbf{R}$, and $x = 0$ is the unique equilibrium of $\mathbf{S}$ when $u(t) \equiv 0$.

We take the linear part with structured uncertainty,

$$\mathbf{S}_N: \; \dot{x} = Ax + Bu + Ek(t, Cx, u) \tag{3}$$

as a nominal system and assume the existence of a quadratically stabilizing state feedback [15],

$$u_N = -R^{-1}B^T Px \quad , \tag{4}$$

for the system $\mathbf{S}_N$, where P is the positive definite solution of the Riccati equation

$$A^T P + PA - PBR^{-1}B^T P + \mu PEE^T P + \mu^{-1}CC^T + Q = 0, \tag{5}$$

Q and R are constant positive definite matrices of proper dimensions, and μ is an appropriate positive number. We prove the following:

THEOREM 1. If there exist positive numbers α and β such that

$$x^T Qx - 2x^T Pf(t,x,u) + u^T Ru \geq \alpha x^T Qx + \beta u^T Ru$$
$$\text{for all } (t,x,u) \in \mathbf{R} \times \mathbf{R}^n \times \mathbf{R}^m, \tag{6}$$

then the control law u_N stabilizes the system $\mathbf{S}$.

PROOF. The proof is a straightforward application of the Liapunov theory. We consider the overall closed-loop system

$$\hat{\mathbf{S}}: \ \dot{x} = (A - BR^{-1}B^T P)x + Ek(t, Cx, -R^{-1}B^T Px)$$
$$+f(t,x,-R^{-1}B^T Px) \tag{7}$$

and use the quadratic form

$$v(x) = x^T Px \tag{8}$$

as a candidate for Liapunov function. Computing the total time derivative of $v(x)$ with respect to (7), we get

$$\begin{aligned}
\dot{v}(x)_{(7)} =& x^T(A^T P + PA - 2PBR^{-1}B^T P)x \\
& + 2x^T PEk(t, Cx, -R^{-1}B^T Px) \\
& + 2x^T Pf(t, x, -R^{-1}B^T Px) \\
=& - [x^T Qx - 2x^T Pf(t,x,u_N) + u_N^T Ru_N] \\
& - \mu x^T PEE^T Px + 2x^T PEk(t, Cx, u_N) \\
& - \mu^{-1} x^T C^T Cx \\
\leq& - (\alpha x^T Qx + \beta u_N^T Ru_N) - \mu(\|E^T Px\|^2 \\
& - 2\mu^{-1}\|E^T Px\| \, \|Cx\| + \mu^{-2}\|Cx\|^2) \\
\leq& - x^T(\alpha Q + \beta PBR^{-1}B^T P)x \quad \text{for all } x \in \mathbf{R}^n \quad ,
\end{aligned} \tag{9}$$

where u_N is defined in (4). This inequality implies global exponential stability of the equilibrium $x = 0$ of $\hat{\mathbf{S}}$. Q.E.D.

REMARK 1. It is obvious that condition (6) of Theorem 1 is equivalent to the existence of positive numbers α' and β' such that

$$x^T Qx - 2x^T Pf(t,x,u) + u^T Ru \geq \alpha' x^T x + \beta' u^T u$$
$$\text{for all } (t,x,u) \in \mathbf{R} \times \mathbf{R}^n \times \mathbf{R}^m. \tag{10}$$

This inequality is simpler to establish than (6), but may turn out to be inferior in the robustness analysis of $\hat{\mathbf{S}}$.

In applications, a knowledge about the nonlinearity $f(t, x, u)$ is often imprecise, and only sector bounds are available in the form

$$\|f(t,x,u)\| \leq \xi\|x\| + \eta\|u\|$$
$$\text{for all } (t,x,u) \in \mathbf{R} \times \mathbf{R}^n \times \mathbf{R}^m, \tag{11}$$

where ξ and η are nonnegative numbers. We denote by $\lambda_m(\cdot)$ and $\lambda_M(\cdot)$ the minimum and maximum eigenvalue of a given matrix, and prove the following:

COROLLARY 1. If the nonlinearity $f(t,x,u)$ satisfies condition (11) and the inequality

$$2\xi\lambda_m(R) + \eta^2\lambda_M(P) < \frac{\lambda_m(Q)\lambda_m(R)}{\lambda_M(P)} \quad , \tag{12}$$

holds, then the control law u_N stabilizes the system $\mathbf{S}$.

PROOF. Inequality (12) follows from the inequality

$$x^T Qx - 2x^T Pf(t,x,u) + u^T Ru \geq [\lambda_m(Q) - 2\xi\lambda_M(P)]\|x\|^2$$
$$-2\eta\lambda_M(P)\|x\| \ \|u\| + \lambda_m(R)\|u\|^2. \tag{13}$$

It is easy to see that (12) implies (10) for some positive numbers α' and β' such that

$$[\lambda_m(Q) - 2\xi\lambda_M(P) - \alpha'] \ [\lambda_m(R) - \beta'] = \eta^2\lambda_M^2(P). \tag{14}$$

Q.E.D.

A way to enhance stabilizability of $\mathbf{S}$ is to increase the feedback gain as proposed in [12]. For a gain increase to work, the function $f(t,x,u)$ should be separable as

$$f(t,x,u) = Bg(t,x,u) + h(t,x), \tag{15}$$

where the functions $g:\ \mathbf{R}\times\mathbf{R}^n\times\mathbf{R}^m \to \mathbf{R}^m$ and $h:\ \mathbf{R}\times\mathbf{R}^n \to \mathbf{R}^n$ are sufficiently smooth, and $g(t,x,u)$ satisfies the sector condition

$$\|g(t,x,u)\| \le \xi'\|x\| + \eta'\|u\|$$

$$\text{for all } (t,x,u)\in\mathbf{R}\times\mathbf{R}^n\times\mathbf{R}^m \tag{16}$$

with positive numbers ξ' and $\eta' < \lambda_m^{1/2}(R)/\lambda_M^{1/2}(R)$. Then, to reflect the increase of feedback gain, we introduce the number $\rho > 1$ into the control law of (4) to get

$$u_\rho = -\rho R^{-1}B^T Px. \tag{17}$$

THEOREM 2. If there exists a positive number ν such that

$$x^TQx - 2x^TPh(t,x) \ge \nu x^Tx \quad \text{for all } (t,x)\in\mathbf{R}\times\mathbf{R}^n, \tag{18}$$

then a positive number ρ can be chosen so that the control law u_ρ stabilizes the system $\mathbf{S}$.

PROOF. We use again the function $v(x)$ of (8) and compute $\dot v(x)$ with respect to the closed-loop system

$$\hat{\mathbf{S}}_\rho:\quad \dot x = (A-\rho BR^{-1}B^TP)x + Ek(t,Cx,-\rho R^{-1}B^TPx)$$

$$+f(t,x,-\rho R^{-1}B^TPx) \tag{19}$$

to get

$$
\begin{aligned}
\dot{v}(x)_{(19)} = & -[x^T Q x - 2x^T P h(t,x)] \\
& + (\rho - 1)^{-1} g^T(t,x,u_\rho) R g(t,x,u_\rho) \\
& - \rho^{-1} u_\rho^T R u_\rho \\
& - (\rho - 1)[\rho^{-1} u_\rho - (\rho - 1)^{-1} g(t,x,u_\rho)]^T \\
& \times R[\rho^{-1} u_\rho - (\rho - 1)^{-1} g(t,x,u_\rho)] \\
& - [\mu x^T P E E^T P x - 2x^T P E k(t, Cx, u_\rho) \\
& + \mu^{-1} x^T C^T C x]. \qquad (20)
\end{aligned}
$$

This expression together with conditions (16) and (18) imply that for a sufficiently large positive number ρ, there is a positive number $\epsilon < \nu$ such that

$$\dot{v}(x)_{(19)} \leq -(\nu - \epsilon) x^T x \quad \text{for all } x \in \mathbf{R}^n. \qquad (21)$$

that is, the equilibrium $x = 0$ of $\hat{\mathbf{S}}_\rho$ is globally exponentially stable. Q.E.D.

REMARK 2. Theorem 2 implies that increasing the feedback gain, we can enhance stabilizability if the part of the nonlinearity, which cannot be cancelled by any choice of the input u, is independent of u and is sufficiently small. A realistic special case is the situation when f depends only on t and x, and satisfies the so-called "matching" conditions [16,17],

$$f(t,x) = B g(t,x), \qquad (22)$$

with a bounded gain with respect to x.

REMARK 3. We note that the function $h(t,x)$ in the decomposition (15) of $f(t,x,u)$ is not allowed to depend on the input u, otherwise the proof of Theorem 2 cannot go through. This fact implies that, although we enhanced stabilizability using the control u_ρ, the condition for stabilizability has not been relaxed enough for Theorem 2 to include the results of Theorem 1.

III. STABILITY MARGINS

We recall from the classical LQ theory [1-4] that there are significant robustness implications in driving linear systems by optimal control laws. The use of quadratic Liapunov functions to establish stability in this context, provides for considerable stability margins in terms of gain and phase margins, gain reduction tolerance, as well as linear and nonlinear distortions of the control law [6]. In this section, we show how quadratic Liapunov functions can be used to exchange the optimality requirement for a broader scope of stability margins, which includes systems with structured and unstructured perturbations considered in the preceding section. For this purpose, we consider the insertion of a smooth memoryless time-varying nonlinearity $\phi: \mathbf{R} \times \mathbf{R}^m \to \mathbf{R}^m$, $\phi(t, 0) \equiv 0$, or a linear time-invariant stable element having an $m \times m$ proper transfer function $L(s)$, in the feedback loop of $\hat{\mathbf{S}}$ defined in (7). The corresponding perturbed versions of $\hat{\mathbf{S}}$ are

$$\begin{aligned} \hat{\mathbf{S}}_\phi : \dot{x} =& Ax + B\phi(t, -R^{-1}B^TPx) \\ &+ Ek[t, Cx, \phi(t, -R^{-1}B^TPx)] \\ &+ f[t, x, \phi(t, -R^{-1}B^TPx)] \end{aligned} \tag{23}$$

for the nonlinear distortion of u_N, and

$$\begin{aligned} \hat{\mathbf{S}}_\mathcal{L} : \dot{x} =& Ax + B[\mathcal{L} * (-R^{-1}B^TPx)] \\ &+ Ek[t, Cx, \mathcal{L} * (-R^{-1}B^TPx)] \\ &+ f[t, x, \mathcal{L} * (-R^{-1}B^TPx)], \end{aligned} \tag{24}$$

for the insertion of a linear element, where $*$ denotes the convolution and $\mathcal{L}(t)$ is the inverse Laplace transform of $L(s)$.

By δ we denote a nonnegative number for which

$$\alpha Q - \delta PBR^{-1}B^TP > 0, \tag{25}$$

and prove the following:

THEOREM 3. Under the condition of Theorem 1, the equilibrium $x = 0$ of the system $\hat{\mathbf{S}}_\phi$ is globally exponentially stable for any nonlinearity $\phi(t,u)$ such that

$$[\phi(t,u) - u]^T R[\phi(t,u) - u] \le \beta\phi^T(t,u)R\phi(t,u) + \delta u^T Ru$$

$$\text{for all } (t,x) \in \mathbf{R} \times \mathbf{R}^n. \tag{26}$$

PROOF. The proof is similar to that of Theorem 1. We consider again the quadratic form $v(x)$ and compute $\dot{v}(x)_{(23)}$ using the Riccati equation (5) and inequalities (2) and (6), to get

$$\begin{aligned}\dot{v}(x)_{(23)} \le & -x^T(\alpha Q - \delta PBR^{-1}B^T P)x \\ & + (1-\beta)\phi^T(t,u_N)R\phi(t,u_N) - 2\phi^T(t,u_N)Ru_N \\ & + (1-\delta)u_N^T Ru_N \quad \text{for all } (t,x) \in \mathbf{R} \times \mathbf{R}^n.\end{aligned} \tag{27}$$

Global exponential stability of the equilibrium $x = 0$ of $\hat{\mathbf{S}}_\phi$ follows from $\dot{v}(x)_{(23)} \le -\lambda_m(\alpha Q - \delta PBR^{-1}B^T P)x^T x$ under the condition (26). Q.E.D.

THEOREM 4. Under the condition of Theorem 1, the system $\hat{\mathbf{S}}_\mathcal{L}$ is stable for any stable linear time–invariant distortion having the transfer function $L(s)$ such that

$$\beta L^H(j\omega)RL(j\omega) + \delta R \ge \Big[L^H(j\omega) - I\Big]R\Big[L(j\omega) - I\Big]$$

$$\text{for all } \omega \in \mathbf{R}\ . \tag{28}$$

PROOF. We first note that the superscript H means conjugate transpose. Then, we again consider the quadratic form

$v(x)$ and compute

$$
\begin{aligned}
x^T P x - x_0^T P x_0 &= \int_{t_0}^{t} \dot{v}(x)_{(24)} d\tau \\
&\leq - \int_{t_0}^{t} x^T(\alpha Q - \delta P B R^{-1} B^T P) x d\tau \\
&\quad + \int_{t_0}^{t} \Big[(1-\beta)(\mathcal{L} * u_N)^T R (\mathcal{L} * u_N) \\
&\quad - 2(\mathcal{L} * u_N)^T R u_N + (1-\delta) u_N^T R u_N \Big] d\tau \quad ,
\end{aligned} \tag{29}
$$

using (2), (5) and (6). By representing the second integral in the last inequality of (29) using Parseval's formula, we can show that it is nonpositive provided inequality (28) holds. Then, $x(t) \to 0$ as $t \to \infty$. This fact and stability of $\mathcal{L}$ implies stability of $\hat{\mathbf{S}}_{\mathcal{L}}$. Q.E.D.

The deeper significance of Theorems 3 and 4 is that they imply the classical robustness properties of the optimal control design.

COROLLARY 2. If the condition (6) is satisfied, then each input channel of the closed–loop system $\hat{\mathbf{S}}$ in (7) has:

(i) at least

$$20 \log \frac{1 + \sqrt{\theta(\beta - \theta\beta\delta + \delta)}}{1 - \theta\beta} dB \tag{30}$$

gain margin;

(ii) at least

$$20 \log \frac{1 - \theta\delta}{1 + \sqrt{\theta(\beta - \theta\beta\delta + \delta)}} dB \tag{31}$$

gain reduction tolerance; and

(iii) at least

$$\pm \cos^{-1} \Big[1 - \frac{\theta}{2}(\beta + \delta) \Big] deg, \tag{32}$$

phase margin, where $\theta = \lambda_m(R)/\lambda_M(R)$.

PROOF. To estimate the gain margin and gain reduction tolerance, we consider as usual the nonlinearity $\phi(t,u)$ channel by channel, that is, $\phi(t,u)$ is described as $\phi(t,u) = [\phi_1(t,u_1), \phi_2(t,u_2), \ldots, \phi_m(t,u_m)]^T$ and $\phi_i : \mathbf{R} \times \mathbf{R} \to \mathbf{R}$. From Theorem 3 we conclude stability of $\hat{\mathbf{S}}_\phi$ if each component $\phi_i(t,u_i)$ of $\phi(t,u)$ satisfies the inequality

$$\lambda_M(R)[\phi_i(t,u_i) - u_i]^2 - \beta\lambda_m(R)\phi_i^2(t,u_i) - \delta\lambda_m(R)u_i^2 \le 0$$
$$\text{for all } (t,u_i) \in \mathbf{R} \times \mathbf{R}, \quad i = 1,2,\ldots,m \ . \tag{33}$$

or, equivalently,

$$(1-\theta\beta)\left[\frac{\phi_i(t,u_i)}{u_i}\right]^2 - 2\frac{\phi_i(t,u_i)}{u_i} + (1-\theta\delta) \le 0$$
$$\text{for all } (t,u_i) \in \mathbf{R} \times \mathbf{R}, \quad i = 1,2,\ldots,m \ . \tag{34}$$

Inequality (34) is solved as

$$\frac{1-\sqrt{\theta(\beta-\theta\beta\delta+\delta)}}{1-\theta\beta} \le \frac{\phi_i(t,u_i)}{u_i} \le \frac{1+\sqrt{\theta(\beta-\theta\beta\delta+\delta)}}{1-\theta\beta}$$
$$\text{for all } (t,u_i) \in \mathbf{R} \times \mathbf{R} \ , \quad i = 1,2,\ldots,m \ , \tag{35}$$

so that the left side provides (ii) the gain reduction tolerance, while the right side of the inequality gives (i) the gain margin of the ith channel.

For the linear distortion, we assume that the linear element is described by $L(s) = \text{diag}\,\{L_1(s), L_2(s), \ldots, L_m(s)\}$. Then inequality (28) is implied by

$$\lambda_M(R)|L_i(j\omega) - 1|^2 - \beta\lambda_m(R)|L_i(j\omega)|^2 - \delta\lambda_m(R) \le 0$$
$$\text{for all } \omega \in \mathbf{R}, \quad i = 1,2,\ldots,m \ . \tag{36}$$

To estimate the phase margin, we use $|L_i(j\omega)| = 1$ in (36) and get

$$\text{Re}L_i(j\omega) \ge 1 - \frac{\theta}{2}(\beta+\delta) \quad \forall\omega \in \mathbf{R} \ , \quad i = 1,2,\ldots,m \tag{37}$$

and (iii) follows. Q.E.D.

REMARK 4. If the matrix R is diagonal, the stability conditions of Theorems 3 and 4 can be written channel by channel in the scalar form. Then, we can replace both $\lambda_M(R)$ and $\lambda_m(R)$ by r_i in (33) and (36), where r_i's are positive diagonal elements of R. Consequently, Corollary 2 holds for $\theta = 1$, which results in better estimates (i)–(iii) of the robustness properties.

REMARK 5. An interesting special case occurs when the nonlinearity $f(t, x, u)$ is independent of the input u and satisfies the inequality

$$x^T Qx - 2x^T Pf(t, x) \geq \alpha x^T Qx \quad \text{for all } (t, x) \in \mathbf{R} \times \mathbf{R}^n \quad , \tag{38}$$

for some positive number α. In this case, condition (6) holds for $\beta = 1$, and if R is diagonal, implying $\theta = 1$, $\hat{\mathbf{S}}$ has:

- (i) infinite gain margin,
- (ii) at least 50% gain reduction tolerance,
- (iii) at least $\pm 60°$ phase margin,

which are the standard robustness properties of the optimal LQ control systems.

REMARK 6. It is more or less obvious that the results of this section carry over to the control law modifications of the preceding section. Under the condition (18) of Theorem 2, Theorems 3 and 4 can be rephrased to accommodate the control law modification with $\beta = \rho\beta_\rho$ and $\delta = \rho^{-1}\delta_\rho$, where δ_ρ is a positive number such that the matrix $\alpha_\rho Q - \delta_\rho PBR^{-1}B^TP$ is positive definite, and ρ, α_ρ and β_ρ are positive numbers defined in the proof of Theorem 2.

IV. DECENTRALIZED CONTROL

In decentralized control of complex systems [16], it is standard to assume that local subsystem models are known to a high degree of accuracy, while the interconnections among the subsystems are poorly understood, with only their bounds available for control design. The union of the decoupled subsystems is considered as a nominal system, and the local control laws are chosen to provide satisfactory performance of individual subsystems. At the same time, the control laws are required to be robust so that the overall interconnected system remains stable in the presence of uncertain interconnections.

It is realistic to assume that more often than not, the models of subsystems are not known precisely, which places additional robustness requirements on the decentralized feedback design. By redefining the nominal system to be composed of decoupled subsystems with local structured uncertainty added to their dynamics, we can use the results of the previous sections to capture the broadened scope of robustness requirements which includes the subsystem uncertainty as well.

Let us consider a system

$$\mathbf{S}: \; \dot{x} = A_D x + B_D u + E_D k_D(t, C_D x, u) + f_C(t, x, u), \tag{39}$$

which is an interconnection of N subsystems

$$\mathbf{S}_i: \; \dot{x}_i = A_i x_i + B_i u_i + E_i k_i(t, C_i x_i, u_i) \quad , \quad i = 1, 2, \ldots, N \tag{40}$$

where $x_i(t) \in \mathbf{R}^{n_i}$ is the state and $u_i(t) \in \mathbf{R}^{m_i}$ is the input of $\mathbf{S}_i$ at time $t \in \mathbf{R}$, and A_i and B_i are constant matrices of appropriate dimensions, which constitute stabilizable pairs (A_i, B_i). In (39), $x = (x_1^T, x_2^T, \ldots, x_N^T)^T$ is the state, $u = (u_1^T, u_2^T, \ldots, u_N^T)^T$ is the input of the interconnected system $\mathbf{S}$, and the system matrices are defined as

$$A_D = \text{diag}\{A_1, A_2, \ldots, A_N\},\; C_D = \text{diag}\{C_1, C_2, \ldots, C_N\},$$
$$B_D = \text{diag}\{B_1, B_2, \ldots, B_N\},\; E_D = \text{diag}\{E_1, E_2, \ldots, E_N\}. \tag{41}$$

The term $E_i k_i(t, C_i x_i, u_i)$ represents the structured uncertainty in $\mathbf{S}_i$, where E_i and C_i are $n_i \times q_i$ and $p_i \times m_i$

constant matrices, and $k_i : \mathbf{R} \times \mathbf{R}^{p_i} \times \mathbf{R}^{m_i} \to \mathbf{R}^{q_i}$, which is a vector component of $k_D = (k_1^T, k_2^T, \ldots, k_N^T)^T$, satisfies the inequality

$$\|k_i(t, C_i x_i, u_i)\| \leq \|C_i x_i\|$$
$$\text{for all} \quad (t, x_i, u_i) \in \mathbf{R} \times \mathbf{R}^{n_i} \times \mathbf{R}^{m_i}. \tag{42}$$

Finally, the functions $k_D : \mathbf{R} \times \mathbf{R}^p \times \mathbf{R}^m \to \mathbf{R}^q$ and $f_C : \mathbf{R} \times \mathbf{R}^n \times \mathbf{R}^m \to \mathbf{R}^n$ are sufficiently smooth so that the solutions of (39) exist and are unique for all initial conditions and all fixed inputs $u(\cdot)$. Furthermore, $k_D(t, 0, 0) \equiv 0$, $f_C(t, 0, 0) \equiv 0$, and $x = 0$ is assumed to be the unique equilibrium of $\mathbf{S}$ when $u(t) \equiv 0$.

The special assumption in this section is the decentralized information structure constraint on the system $\mathbf{S}$, which is compatible with the subsystems $\mathbf{S}_i$. The constraint restricts the feedback control law to

$$u = -K_D x, \tag{43}$$

where $K_D = \text{diag}\{K_1, K_2, \ldots, K_N\}$ and the submatrices K_i are the gains of the local subsystem state feedback.

To stabilize the system $\mathbf{S}$ by the control law (43), we apply the quadratic stabilization scheme of Section 2. A stabilizing control law for $\mathbf{S}_i$ is computed (when one exists) as

$$u_i = -R_i^{-1} B_i^T P_i x_i \tag{44}$$

where P_i is the positive definite solution of the Riccati equation

$$A_i^T P_i + P_i A_i - P_i B_i R_i^{-1} B_i^T P_i + \mu_i P_i E_i E_i^T P_i$$
$$+ \mu_i^{-1} C_i^T C_i + Q_i = 0, \tag{45}$$

Q_i, R_i are positive definite matrices with proper dimensions, and μ_i is an appropriate positive number. Then, our decentralized control law is

$$u_D = -R_D^{-1} B_D^T P_D x, \tag{46}$$

where

$$\begin{aligned} P_D &= \mathrm{diag}\{P_1, P_2, \ldots, P_N\} \\ R_D &= \mathrm{diag}\{R_1, R_2, \ldots, R_N\}. \end{aligned} \tag{47}$$

To obtain a stability condition for the closed–loop system

$$\begin{aligned} \hat{\mathbf{S}}: \ \dot{x} =& (A_D - B_D R_D^{-1} B_D^T P_D)x \\ &+ E_D k_D(t, C_D x, -R_D^{-1} B_D^T P_D x) \\ &+ f_C(t, x, -R_D^{-1} B_D^T P_D x) \end{aligned} \tag{48}$$

we consider a function $V: \mathbf{R}^n \to \mathbf{R}_+$ defined as

$$V(x) = \sum_{i=1}^{N} d_i x_i^T P_i x_i = x^T \bar{P}_D x \tag{49}$$

where d_i's are positive numbers to be determined later, and $\bar{P}_D = \mathrm{diag}\{d_1 P_1, d_2 P_2, \ldots, d_N P_N\}$. We compute the time derivative of $V(x)$ with respect to (48) as

$$\begin{aligned} \dot{V}(x)_{(48)} =& \sum_{i=1}^{N} d_i [x_i^T (A_i^T P_i + P_i A_i - 2P_i B_i R_i^{-1} B_i^T P_i) x_i \\ &+ 2x_i^T P_i E_i k_i(t, C_i x_i, -R_i^{-1} B_i^T P_i x_i)] \\ &+ 2x^T \bar{P}_D f_C(t, x, -R_D^{-1} B_D^T P_D x) \\ \leq& \sum_{i=1}^{N} d_i [-(x_i^T Q_i x_i + u_i^T R_i u_i) - \mu_i(\|E_i^T P_i x_i\|^2 \\ &- 2\mu_i^{-1} \|E_i^T P_i x_i\| \, \|C_i x_i\| + \mu_i^{-2} \|C_i x_i\|^2)] \\ &+ 2x^T \bar{P}_D f_C(t, x, -R_D^{-1} B_D^T P_D x) \\ \leq& -[x^T \bar{Q}_D x - 2x^T \bar{P}_D f_C(t, x, u_D) + u_D^T \bar{R}_D u_D] \\ & \qquad \text{for all } x \in \mathbf{R}^n \end{aligned} \tag{50}$$

by using the Riccati equation (45), the inequality (42), and the inputs u_i of (44) and u_D of (46), where $\bar{Q}_D = \mathrm{diag}\{d_1 Q_1,$

$d_2Q_2, \ldots, d_NQ_N\}$ and $\bar{R}_D = \text{diag}\{d_1R_1, d_2R_2, \ldots, d_NR_N\}$. Then, the Liapunov theory implies the following:

THEOREM 5. If there exists positive numbers d_i, $i = 1, 2, \ldots, N$, such that the inequality

$$x^T\bar{Q}_Dx - 2x^T\bar{P}_Df_C(t,x,u) + u^T\bar{R}_Du \geq \bar{\alpha}'x^Tx + \bar{\beta}'u^Tu$$

$$\text{for all } (t,x,u) \in \mathbf{R} \times \mathbf{R}^n \times \mathbf{R}^m, \tag{51}$$

holds for some positive numbers $\bar{\alpha}'$ and $\bar{\beta}'$, then the decentralized control u_D stabilizes the overall system $\mathbf{S}$.

Although the expression $\bar{\alpha}x^T\bar{Q}_Dx + \bar{\beta}u^T\bar{R}_Du$ could have been used in (51), we have chosen a simplified bound because it is easier to test.

In the context of complex systems [16], it is of interest to restate Theorem 5 in terms of the decomposition–aggregation framework using the approach proposed in [6]. We assume that there are nonnegative numbers ξ_{ij} and η_{ij} such that the components $f_i : \mathbf{R} \times \mathbf{R}^n \times \mathbf{R}^m \to \mathbf{R}^{n_i}$ of the interconnection function $f_C = (f_1^T, f_2^T, \ldots, f_N^T)^T$ satisfy the inequality

$$\|f_i(t,x,u)\| \leq \sum_{j=1}^{N}(\xi_{ij}\|x_j\| + \eta_{ij}\|u_j\|)$$

$$\text{for all } (t,x,u) \in \mathbf{R} \times \mathbf{R}^n \times \mathbf{R}^m, \tag{52}$$

where $\|\cdot\|$ denotes the Euclidean norm. Using the numbers ξ_{ij} and η_{ij}, we define three $N \times N$ aggregate matrices $W = (w_{ij})$, $Y = (y_{ij})$, and $Z = \text{diag}\{z_1, z_2, \ldots, z_N\}$ as

$$\begin{aligned} w_{ij} &= \begin{cases} \frac{1}{2}\lambda_m(Q_i) - \lambda_M(P_i)\xi_{ii}, & i = j \\ -\lambda_M(P_i)\xi_{ij}, & i \neq j \end{cases} \\ y_{ij} &= \lambda_M(P_i)\eta_{ij}, \\ z_i &= \lambda_m(R_i). \end{aligned} \tag{53}$$

At this point we need a few facts about the class $\mathcal{M}$ of M–matrices (*e.g.*, [18]). When a constant $N \times N$ matrix $W = (w_{ij})$ has nonpositive off–diagonal elements ($w_{ij} \leq 0, i \neq j$), then one of the conditions (Tartar, [19]; Araki and Kondo, [20]) for $W \in \mathcal{M}$ is that there exists a positive diagonal matrix $D = \text{diag}\{d_1, d_2, \ldots, d_N\}$ such that the matrix $W^T D + DW$ is positive definite. This condition is equivalent to the positivity of all the leading principal minors of W, which is an efficient test for concluding whether or not $W \in \mathcal{M}$ and, thus, for establishing the existence of the matrix D.

We return now to the aggregate matrices W, Y, and Z, and obtain the following inequality

$$x^T \bar{Q}_D x - 2x^T \bar{P}_D f_C(t,x,u) + u^T \bar{R}_D u \geq (\bar{x}^T, \bar{u}^T) W_C (\bar{x}^T, \bar{u}^T)^T$$

$$\text{for all } (t,x,u) \in \mathbf{R} \times \mathbf{R}^n \times \mathbf{R}^m \tag{54}$$

where

$$W_C = \begin{bmatrix} W^T D + DW & -DY \\ -Y^T D & DZ \end{bmatrix}, \tag{55}$$

and $\bar{x} = (\|x_1\|, \|x_2\|, \ldots, \|x_N\|)^T$, $\bar{u} = (\|u_1\|, \|u_2\|, \ldots, \|u_N\|)^T$. Obviously, $\bar{x}^T \bar{x} = x^T x$ and $\bar{u}^T \bar{u} = u^T u$.

An immediate corollary to Theorem 5 is the following:

COROLLARY 3. If there exists a positive diagonal matrix D such that the matrix W_C is positive definite, then the decentralized control law u_D stabilizes the system $\mathbf{S}$.

The necessary and sufficient condition for the matrix W_C to be positive definite is

$$W^T D + DW - DY(DZ)^{-1} Y^T D > 0. \tag{56}$$

For (56) to hold, $W^T D + DW$ has to be positive definite, which in turn implies that $W \in \mathcal{M}$. Therefore, to test (56) we should first check positivity of the leading principal minors of W and, if they are positive, we compute a positive D and proceed to test positive definiteness of W_C in (56). It is more or less obvious from (55) that if the interconnections $f_i(t,x,u)$ are weak and ξ_{ij} and η_{ij} are sufficiently small, then the condition of Corollary 3 is satisfied.

There are several interesting special cases of the interconnection function $f_C(t,x,u)$. An easy case is when the function is independent of the control u, that is, we have $f_C(t,x)$. Then, $Y=0$ and stabilizability property of the control law u_D is guaranteed when $W \in \mathcal{M}$. This case has been considered extensively. Another case occurs when $\mathbf{S}$ is an input decentralized system [18] and the ith component $f_i(t,x,u_i)$ of the interconnection function f_C depends only on the ith component u_i of the control u for all $i=1,2,\ldots,N$. In this case, Y is diagonal and the left side of (56) can be written as

$$W^TD + DW - DY(DZ)^{-1}DY = \left(W - \frac{1}{2}YZ^{-1}Y\right)^T D + D\left(W - \frac{1}{2}YZ^{-1}Y\right). \tag{57}$$

Using this expression, we get:

COROLLARY 4. If $\mathbf{S}$ is an input decentralized system and $W - \frac{1}{2}YZ^{-1}Y \in \mathcal{M}$, then the decentralized control law u_D stabilizes $\mathbf{S}$.

We turn our attention to stability margins of the decentralized control, which were established via optimality in [6]. Unlike [6], we will obtain the margins relying on the quadratic Liapunov functions only. The payoff is the additional robustness to structured perturbations on the subsystem level.

Let us consider the system $\hat{\mathbf{S}}$ of (48) and assume that the nonlinear distortion of u_D has occurred, which is described by a nonlinear function $\phi_D: \mathbf{R} \times \mathbf{R}^m \to \mathbf{R}^m$ defined as $\phi_D(t,u) = [\phi_1^T(t,u_1), \phi_2^T(t,u_2), \ldots, \phi_N^T(t,u_N)]^T$, and $\phi_D(t,0) \equiv 0$. The system $\hat{\mathbf{S}}$ becomes

$$\begin{aligned}\hat{\mathbf{S}}_\phi:\ \dot{x} = {} & A_Dx + B_D\phi_D(t, -R_D^{-1}B_D^TP_Dx) \\ & + E_Dk_D[t, C_Dx, \phi_D(t, -R_D^{-1}B_D^TP_Dx)] \\ & + f_C[t, x, \phi_D(t, -R_D^{-1}B_D^TP_Dx)].\end{aligned} \tag{58}$$

To show robustness, we introduce the diagonal matrices

$$\begin{aligned} D_\alpha &= \text{diag}\{\alpha_1 I_{n_1}, \alpha_2 I_{n_2}, \dots, \alpha_N I_{n_N}\} \\ D_\beta &= \text{diag}\{\beta_1 I_{n_1}, \beta_2 I_{n_2}, \dots, \beta_N I_{n_N}\} \\ D_\delta &= \text{diag}\{\delta_1 I_{n_1}, \delta_2 I_{n_2}, \dots, \delta_N I_{n_N}\} \end{aligned} \tag{59}$$

where α_i's and β_i's are positive numbers and δ_i's are non-negative numbers, and assume that these matrices exist such that

$$\begin{aligned} &x^T \bar{Q}_D x - 2x^T \bar{P}_D f_C(t, x, u) + u^T \bar{R}_D u \\ &\qquad \geq x^T D_\alpha \bar{Q}_D x + u^T D_\beta \bar{R}_D u \\ &\qquad \text{for all } (t, x, u) \in \mathbf{R} \times \mathbf{R}^n \times \mathbf{R}^m \\ &D_\alpha \bar{Q}_D - P_D B_D D_\delta \bar{R}_D^{-1} B_D^T P_D > 0. \end{aligned} \tag{60}$$

It is obvious that the existence of such matrices is assured by (51) of Theorem 5. We have:

THEOREM 6. Under the conditions (60), the system $\hat{\mathbf{S}}_\phi$ is stable for any nonlinearity $\phi_D(t, u)$ such that

$$\begin{aligned} &[\phi_i(t, u_i) - u_i]^T R_i [\phi_i(t, u_i) - u_i] \\ &\qquad \leq \beta_i \phi_i^T(t, u_i) R_i \phi_i(t, u_i) + \delta_i u_i^T R_i u_i \\ &\qquad \text{for all } (t, u_i) \in \mathbf{R} \times \mathbf{R}^{m_i}. \end{aligned} \tag{61}$$

Linear distortions are handled similarly. We introduce the matrix $L_D(s) = \text{diag}[L_1^T(s), L_2^T(s), \dots, L_N^T(s)]^T$, where $L_i(s)$ is an $m_i \times m_i$ stable transfer function, with the inverse Laplace transform $\mathcal{L}_D(t) = [\mathcal{L}_1^T(t), \mathcal{L}_2^T(t), \dots, \mathcal{L}_N^T(t)]^T$. The linear closed-loop system is described by

$$\begin{aligned} \hat{\mathbf{S}}_\mathcal{L} : \ \dot{x} &= A_D x + B_D[\mathcal{L}_D * (-R_D^{-1} B_D^T P_D x)] \\ &\quad + E_D k_D[t, C_D x, \mathcal{L}_D * (-R_D^{-1} B_D^T P_D x)] \\ &\quad + f_C[t, x, \mathcal{L}_D * (-R_D^{-1} B_D^T P_D x)]. \end{aligned} \tag{62}$$

Imitating the proof of Theorem 4, we establish the following:

THEOREM 7. Under the condition (60), the system $\hat{\mathbf{S}}_{\mathcal{L}}$ is stable for any stable linear time–invariant distortion having a transfer function $L_D(s)$ such that

$$\begin{aligned} &\beta_i L_i^H(j\omega) R_i L_i(j\omega) + \delta_i R_i \\ &\geq \left[L_i^H(j\omega) - I_{m_i}\right]^T R\left[L_i(j\omega) - I_{m_i}\right] \quad \text{for all } \omega \in \mathbf{R}. \end{aligned} \tag{63}$$

REMARK 7. With Theorems 6 and 7 in hand, we can calculate directly the gain and phase margin of each input channel using the formulas of Corollary 2 as suggested by Remark 4: the numbers β, δ, and θ should be replaced by β_i, δ_i, and θ_i with β_i and δ_i being defined by (60) and $\theta_i = \lambda_m(R_i)/\lambda_M(R_i)$.

The numbers β_i, and δ_i of Remark 7 can also be computed in the decomposition–aggregation framework, but the results may be conservative. If the computational simplicity of the framework is a deciding factor, then we modify the aggregate matrices W and Z as $\tilde{W} = (\tilde{w}_{ij})$ and $\tilde{Z} = \text{diag}\{\tilde{z}_1, \tilde{z}_2, \dots, \tilde{z}_N\}$ which are defined by

$$\begin{aligned} \tilde{w}_{ij} &= \begin{cases} \frac{1}{2}(1-\alpha_i)\lambda_m(Q_i) - \lambda_M(P_i)\xi_{ii}, & i = j \\ -\lambda_M(P_i)\xi_{ij}, & i \neq j \end{cases}, \\ \tilde{z}_i &= (1-\beta_i)\lambda_m(R_i), \end{aligned} \tag{64}$$

and form the matrix $\tilde{W}_C$ as W_C of (55),

$$\tilde{W}_C = \begin{bmatrix} \tilde{W}^T D + D\tilde{W} & -DY \\ -Y^T D & D\tilde{Z} \end{bmatrix}. \tag{65}$$

Now, stability conditions (61) and (63) are satisfied by the positive numbers α_i, β_i, and nonnegative numbers δ_i such that $\tilde{W}_C$ is positive definite and

$$\alpha_i Q_i - \delta_i P_i B_i R_i^{-1} B_i^T P_i > 0. \tag{66}$$

The extension of these facts to control law modifications is straightforward. We present the results without proofs.

Following the approach of [11], we split the interconnection function as

$$f_C(t,x,u) = B_D g_C(t,x,u) + h_C(t,x), \tag{67}$$

where $g_C = (g_1^T, g_2^T, \ldots, g_N^T)^T$, $g_i : \mathbf{R}\times\mathbf{R}^n\times\mathbf{R}^m \rightarrow \mathbf{R}^{m_i}$, and $h_C = (h_1^T, h_2^T, \ldots, h_N^T)^T$, $h_i : \mathbf{R}\times\mathbf{R}^n \rightarrow \mathbf{R}^{n_i}$. We assume that g_i satisfies the inequality

$$\|g_i(t,x,u)\| \leq \xi_i'\|x\| + \eta_i'\|u\| \quad \text{for all } (t,x,u) \in \mathbf{R}\times\mathbf{R}^n\times\mathbf{R}^m, \tag{68}$$

for some positive numbers ξ_i' and $\eta_i' < \lambda_m^{1/2}(R_i)/\lambda_M^{1/2}(R_i)$. The modification of the decentralized control law is similar in form to that of Section II,

$$u_D^\rho = -\rho_D R_D^{-1} B_D^T P_D x, \tag{69}$$

where $\rho_D = \text{diag}\{\rho_1 I_{m_1}, \rho_2 I_{m_2}, \ldots \rho_N I_{m_N}\}$, and $\rho_i > 1$, $i = 1, 2, \ldots, N$. A decentralized version of Theorem 2 is:

THEOREM 8. If there exist positive numbers d_i, $i = 1, 2, \ldots, N$, and a positive number ν' such that

$$x^T \bar{Q}_D x - 2x^T \bar{P}_D h_C(t,x) \geq \nu' x^T x \qquad \text{for all } (t,x) \in \mathbf{R}\times\mathbf{R}^n, \tag{70}$$

then positive numbers ρ_i can be chosen so that the modified control law u_D^ρ stabilizes $\mathbf{S}$.

The condition (70) can be easily tested by the decomposition–aggregation method. For this purpose, we assume that each component h_i of the function h_C satisfies the inequality

$$\|h_i(t,x)\| \leq \sum_{j=1}^{N} \zeta_{ij}'\|x_j\|, \tag{71}$$

where ζ_{ij}' are nonnegative numbers. We define the aggregate matrix $W' = (w_{ij}')$ by

$$w_{ij}' = \begin{cases} \frac{1}{2}\lambda_m(Q_i) - \lambda_M(P_i)\zeta_{ii}', & i = j \\ -\lambda(P_i)\zeta_{ij}', & i \neq j \end{cases} \tag{72}$$

Then, condition (70) holds if $W' \in \mathcal{M}$, that is, W' is an M–matrix, which by Theorem 8 implies that u_D^ρ stabilizes $\mathbf{S}$ when the gain constant ρ_D is chosen sufficiently large.

5. APPLICATION

To illustrate the proposed robust stabilization scheme we shall consider a system of two inverted penduli coupled by a spring, which is shown in Figure 1. This system was considered in the context of inverse optimal control with decentralized constraints [6], where the subsystems were required to be linear for the optimal LQ theory to go through. In the present illustration, the control problem is reformulated as a stabilization problem with nonlinear structured perturbations allowed to enter into subsystem dynamics, thus trading optimality for a more realistic subsystem models.

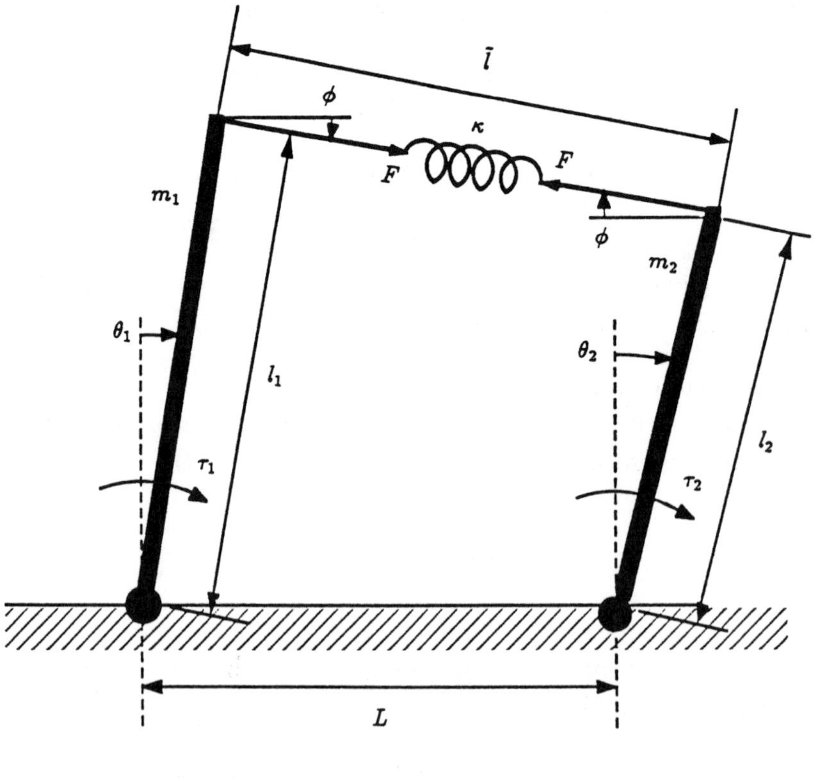

Fig. 1. Inverted penduli.

The variables of the system are:

θ_i — angular displacement of pendulum $i = (i = 1, 2)$

τ_i — torque input generated by the actuator for pendulum $i = (i = 1, 2)$

F — spring force

$\tilde{l}$ — spring length

ϕ — slope of the spring to the earth

and the constants are:

l_i — length of pendulum $i (i = 1, 2)$

m_i — mass of pendulum $i (i = 1, 2)$

L — distance of two penduli

κ — spring constant

The mass of each pendulum is uniformly distributed. The length of the spring is chosen so that $F = 0$ when $\theta_1 = \theta_2 = 0$, which implies that $(\theta_1, \dot{\theta}_1, \theta_2, \dot{\theta}_2)^T = 0$ is an equilibrium of the system if $\tau_1 = \tau_2 = 0$. For simplicity, we assume that the mass of the spring is zero, and restrict the movement of the penduli as $|\theta_i| < \pi/6, i = 1, 2$.

The equations of motion of the coupled penduli are written as

$$\begin{aligned} m_1(l_1/2)\ddot{\theta}_1 &= \tau_1 + m_1 g(l_1/2)\sin\theta_1 + l_1 F\cos(\theta_1 - \phi) \\ m_2(l_2/2)\ddot{\theta}_2 &= \tau_2 + m_2 g(l_2/2)\sin\theta_2 - l_2 F\cos(\theta_2 - \phi) \end{aligned} \tag{73}$$

where $g = 9.8\text{m/sec}^2$ is the constant of gravity, and

$$\begin{aligned} F &= \kappa\left(\tilde{l} - [L^2 + (l_2 - l_1)^2]^{1/2}\right) \\ \tilde{l} &= [(L + l_2\sin\theta_2 - l_1\sin\theta_1)^2 + (l_2\cos\theta_2 - l_1\cos\theta_1)^2]^{1/2} \\ \phi &= \tan^{-1}\frac{l_1\cos\theta_1 - l_2\cos\theta_2}{L + l_2\sin\theta_2 - l_1\sin\theta_1} \end{aligned} \tag{74}$$

We consider each pendulum as one component, and rewrite (73) as an interconnected state equation of (39) with the standard choice of state variables $(\theta_i, \dot{\theta}_i)^T$ for the ith subsystem,

$$\mathbf{S}: \frac{d}{dt}\begin{bmatrix}\theta_1\\ \dot{\theta}_1\\ \theta_2\\ \dot{\theta}_2\end{bmatrix} = \left[\begin{array}{cc|cc} 0 & 1 & 0 & 0\\ g & 0 & 0 & 0\\ \hline 0 & 0 & 0 & 1\\ 0 & 0 & g & 0\end{array}\right]\begin{bmatrix}\theta_1\\ \dot{\theta}_1\\ \theta_2\\ \dot{\theta}_2\end{bmatrix} + \left[\begin{array}{c|c} 0 & 0\\ 2/m_1 l_1 & 0\\ \hline 0 & 0\\ 0 & 2/m_2 l_2\end{array}\right]\begin{bmatrix}\tau_1\\ \tau_2\end{bmatrix} + \left[\begin{array}{c} 0\\ g(\sin\theta_1 - \theta_1)\\ \hline 0\\ g(\sin\theta_2 - \theta_2)\end{array}\right] + \begin{bmatrix} 0\\ 2(1/m_1)F\cos(\theta_1 - \phi)\\ 0\\ -2(1/m_2)F\cos(\theta_2 - \phi)\end{bmatrix}. \tag{75}$$

We note that the nonlinear function in each subsystem $\mathbf{S}_i$ satisfies

$$|g(\sin\theta_i - \theta_i)| \leq 0.25|\theta_i|, \qquad |\theta_i| \leq \pi/6. \tag{76}$$

Then, $g(\sin\theta_i - \theta_i)$ can be described in the form $E_i k_i(C_i x_i)$ using

$$E_i = \begin{bmatrix} 0\\ 0.25\end{bmatrix}, \quad C_i = [1\ 0], \quad k_i(\theta_i) = 4g(\sin\theta_i - \theta_i). \tag{77}$$

We assume the following numerical values:

$$\begin{aligned} l_1 &= 1m & l_2 &= 0.8m\\ m_1 &= 1kg & m_2 &= 0.8kg\\ L &= 1.2m & &\\ \kappa &= 0.04\ N/m \quad \text{and} \quad 4\ N/m & & \end{aligned} \tag{78}$$

Then, with the matrices E_i, C_i in (77), and

$$A_i = \begin{bmatrix} 0 & 1 \\ g & 0 \end{bmatrix}, \qquad B_i = \begin{bmatrix} 0 \\ 2/m_i l_i \end{bmatrix}$$
$$Q_i = \begin{bmatrix} m_i g & 0 \\ 0 & m_i \end{bmatrix}, \quad R_i = 1, \tag{79}$$

the positive definite solutions of the Riccati equations

$$A_i^T P_i + P_i A_i - P_i B_i R_i^{-1} B_i^T P_i + \mu_i P_i E_i E_i^T P_i$$
$$+ \mu^{-1} C_i^T C_i + Q_i = 0, \tag{80}$$

are calculated for $\mu_i = 1$ as

$$P_1 = \begin{bmatrix} 20.513 & 5.478 \\ 5.478 & 1.742 \end{bmatrix}, \quad P_2 = \begin{bmatrix} 10.243 & 2.400 \\ 2.400 & 0.760 \end{bmatrix} \tag{81}$$

and the decentralized control law is obtained as

$$\begin{bmatrix} \tau_1 \\ \\ \tau_2 \end{bmatrix} = - \left[\begin{array}{cc|cc} 10.957 & 3.485 & 0 & 0 \\ \hline 0 & 0 & 7.500 & 2.375 \end{array}\right] \begin{bmatrix} \theta_1 \\ \dot{\theta}_1 \\ \theta_2 \\ \dot{\theta}_2 \end{bmatrix}. \tag{82}$$

The nonlinear interconnection term in (75) is written as

$$f_C(\theta_1, \theta_2) = \begin{bmatrix} 0 \\ 2F\cos(\theta_1 - \phi) \\ 0 \\ -2.5F\cos(\theta_2 - \phi) \end{bmatrix} \tag{83}$$

with

$$F = \kappa\{[3.08 - 2.4\sin\theta_1 + 1.92\sin\theta_2$$
$$- 1.6\cos(\theta_1 - \theta_2)]^{1/2} - 1.217\}$$
$$\phi = \tan^{-1}\frac{\cos\theta_1 - 0.8\cos\theta_2}{1.2 - \sin\theta_1 + 0.8\sin\theta_2}. \tag{84}$$

At this time, we note that, when the nonlinear term f_C in (39) is independent of the input u, the condition (51) of Theorem 5 is reduced to

$$x^T \bar{Q}_D x - 2x^T \bar{P}_D f_C(t,x) \geq \bar{\alpha}' x^T x. \tag{85}$$

We use this fact to show that the decentralized control law of (82) stabilizes the system $\mathbf{S}$ of (75). In case $\kappa = 0.04$, we set $d_1 = d_2 = 1$ and compute

$$\begin{aligned}
&\begin{bmatrix}\theta_1 & \dot{\theta}_1 & \theta_2 & \dot{\theta}_2\end{bmatrix} \begin{bmatrix} 9.8 & 0 & 0 & 0 \\ 0 & 1 & 0 & 0 \\ 0 & 0 & 7.84 & 0 \\ 0 & 0 & 0 & 0.8 \end{bmatrix} \begin{bmatrix}\theta_1 \\ \dot{\theta}_1 \\ \theta_2 \\ \dot{\theta}_2\end{bmatrix} \\
&\quad - 2\begin{bmatrix}\theta_1 & \dot{\theta}_1 & \theta_2 & \dot{\theta}_2\end{bmatrix} \begin{bmatrix} 20.513 & 5.478 & 0 & 0 \\ 5.478 & 1.743 & 0 & 0 \\ 0 & 0 & 10.243 & 2.400 \\ 0 & 0 & 2.349 & 0.760 \end{bmatrix} \\
&\quad \times \begin{bmatrix} 0 \\ 2F\cos(\theta_1 - \phi) \\ 0 \\ -2.5F\cos(\theta_2 - \phi) \end{bmatrix} \\
&\geq \begin{bmatrix}\theta_1 & \dot{\theta}_1 & \theta_2 & \dot{\theta}_2\end{bmatrix} \begin{bmatrix} 9.8 & 0 & 0 & 0 \\ 0 & 1 & 0 & 0 \\ 0 & 0 & 7.84 & 0 \\ 0 & 0 & 0 & 0.8 \end{bmatrix} \begin{bmatrix}\theta_1 \\ \dot{\theta}_1 \\ \theta_2 \\ \dot{\theta}_2\end{bmatrix} \\
&\quad - 2\begin{bmatrix}|\theta_1| & |\dot{\theta}_1| & |\theta_2| & |\dot{\theta}_2|\end{bmatrix} \begin{bmatrix} 10.957|F| \\ 3.485|F| \\ 6.000|F| \\ 1.900|F| \end{bmatrix} \\
&\geq \begin{bmatrix}|\theta_1| & |\dot{\theta}_1| & |\theta_2| & |\dot{\theta}_2|\end{bmatrix} \\
&\quad \times \begin{bmatrix} 5.798 & -0.637 & -2.794 & -0.347 \\ -0.637 & 1 & -0.540 & 0 \\ -2.794 & -0.540 & 5.981 & -0.294 \\ -0.347 & 0 & -0.294 & 0.8 \end{bmatrix} \begin{bmatrix}|\theta_1| \\ |\dot{\theta}_1| \\ |\theta_2| \\ |\dot{\theta}_2|\end{bmatrix} \\
&\geq 0.05(\theta_1^2 + \dot{\theta}_1^2 + \theta_2^2 + \dot{\theta}_2^2),
\end{aligned} \tag{86}$$

where we used the inequality

$$|F| \leq \kappa(4.566|\theta_1| + 3.873|\theta_2|), \tag{87}$$

which holds for $|\theta_i| < \pi/6$. Thus, the condition (85) is satisfied and the decentralized state feedback of (82) stabilizes the overall system $\mathbf{S}$ of (75). In this case, as mentioned in Remark 5, the resultant closed–loop nonlinear system has in each channel (i) infinite gain margin, (ii) at least 50% gain reduction tolerance, and (iii) at least $\pm 60°$ phase margin.

In case $\kappa = 4(N/m)$, it can be readily shown that when $x = (\theta_1, \dot{\theta}_1, \theta_2, \dot{\theta}_2)^T = (-\pi/9, 1.6, \pi/9, -2.2)^T$ the condition (85) does not hold for any $d_1, d_2 > 0$. This implies that we cannot expect that the decentralized control law of (82) stabilizes $\mathbf{S}$ and an increase of the feedback gains is necessary. Fortunately, in this case, the gain increases can eventually stabilize $\mathbf{S}$, because the nonlinearity $f_C(\theta_1, \theta_2)$ of (83) satisfies the matching condition. To see this fact, we write

$$f_C(\theta_1, \theta_2) = \begin{bmatrix} 0 & 0 \\ 2 & 0 \\ 0 & 0 \\ 0 & 3.125 \end{bmatrix} \begin{bmatrix} F\cos(\theta_1 - \phi) \\ -0.8F\cos(\theta_2 - \phi) \end{bmatrix}, \tag{88}$$

and note that the nonlinear functions on the right side of (88) have finite gains with respect of θ_1 and θ_2, which are implied by (87). Then, the condition (70) of Theorem 8 always holds with $h(t, x) = 0$, and with sufficiently large ρ_1, ρ_2 the decentralized control

$$\begin{bmatrix} \tau_1 \\ \tau_2 \end{bmatrix} = -\left[\begin{array}{c|c} \rho_1 & 0 \\ \hline 0 & \rho_2 \end{array}\right] \left[\begin{array}{cc|cc} 10.957 & 3.485 & 0 & 0 \\ \hline 0 & 0 & 7.500 & 2.375 \end{array}\right] \begin{bmatrix} \theta_1 \\ \dot{\theta}_1 \\ \theta_2 \\ \dot{\theta}_2 \end{bmatrix}, \tag{89}$$

stabilizes the system $\mathbf{S}$ of (75).

VI. ACKNOWLEDGMENT

The research reported herein has been supported by the National Science Foundation under the Grant ECS–8813257. Parts of this paper have been published by the authors in the journal *Automatica*, 26 (1990), pp. 499–511 (see reference [6]).

VII. REFERENCES

1. R.E. Kalman, "When is a Linear Control System Optimal?" *Transactions ASME*, **86**, pp. 51–60 (1960).

2. B.D.O. Anderson and J.B. Moore, *Linear Optimal Control*, Prentice–Hall, Englewood Cliffs, NJ. (1971).

3. M.G. Safonov and M. Athans, "Gain and Phase Margin for Multiloop LQG Regulators," *IEEE Transactions on Automatic Control*, **AC–22**, pp. 173–179 (1977).

4. M. G. Safonov, *Stability and Robustness of Multivariable Feedback Systems*, MIT Press, Cambridge, MA (1980).

5. M. E. Sezer and D. D. Šiljak, "Robustness of Suboptimal Control: Gain and Phase Margin," *IEEE Transactions on Automatic Control*, **AC–26**, pp. 907–911 (1981).

6. M. Ikeda and D. D. Šiljak, "Optimality and Robustness of Linear Quadratic Control for Nonlinear Systems," *Automatica*, **26**, pp. 499–511 (1990).

7. P.J. Moylan and B.D.O. Anderson, "Nonlinear Regulator Theory and an Inverse Optimal Control Problem," *IEEE Transactions on Automatic Control*, **AC–18**, pp. 460–465 (1973).

8. V. D. Furasov, "On Vector–Valued Liapunov Functions and Stabilization of Interconnected Systems," *Prikladnaia Matematika i Mekhanika*, **39**, pp. 59–65 (1975).

9. Ü. Özgüner, "Local Optimization in Large Scale Com-

posite Systems," *Proceedings of the 9th Asilomar Conference*, Pacific Grove, CA, pp. 87–91 (1975).

10. D.D. Šiljak, "Reliable Control Using Multiple Control Systems," *International Journal of Control*, **31**, pp. 303–329 (1980).

11. M. Ikeda and D. D. Šiljak, "When is a Linear Decentralized Control Optimal?" *Analysis and Optimization of Systems*, A. Bensoussan and J.L. Lions (eds.), Springer, New York, pp. 419–431 (1982).

12. M. Ikeda, D. D. Šiljak, and K. Yasuda, "Optimality of Decentralized Control for Large–Scale Systems," *Automatica*, **19**, pp. 309–316 (1983).

13. D. Z. Zheng, "Optimalization of Linear–Quadratic Regulator Systems in the Presence of Parameter Perturbations," *IEEE Transactions on Automatic Control*, **AC–31**, pp. 667–670 (1986).

14. A. Saberi, "On Optimality of Decentralized Control for a Class of Nonlinear Interconnected Systems," *Automatica*, **24**, pp. 101–104 (1988).

15. I.R. Petersen, "Disturbance Attenuation and H^∞ Optimization: A Design Method Based on the Algebraic Riccati Equation," *IEEE Transactions on Automatic Control*, **AC–32**, pp. 427–429 (1987).

16. D. D. Šiljak, *Decentralized Control of Complex Systems*. Academic Press, Boston, MA (1991).

17. G. Leitmann, "Guaranteed Asymptotic Stability for Some Linear Systems with Bounded Uncertainties," *Transactions ASME*, **101**, pp. 212–216 (1979).

18. D. D. Šiljak, *Large–Scale Dynamic Systems: Stability and Structure*. North–Holland, New York (1978).

19. L. Tartar, "Une Nouvelle Characterisation des *M* Ma-

trices." *Revue Francaise d'Automatique, Informatique et Richerche Operationelle*, **5**, pp. 127–128 (1971).

20. M. Araki and B. Kondo, "Stability and Transient Behavior of Composite Nonlinear Systems," *IEEE Transactions on Automatic Control*, **AC–17**, pp. 537–541 (1972).

ROBUST STABILITY AND CONTROL OF LINEAR AND MULTILINEAR INTERVAL SYSTEMS[1]

S.P. Bhattacharyya
Department of Electrical Engineering
Texas A&M University
College Station, TX. U.S.A.

L.H. Keel
Center of Excellence in Information Systems
Tennessee State University
Nashville, TN. U.S.A.

ABSTRACT

This paper is a survey of recent results obtained by the authors and their coworkers on the robust stability and control of systems containing parametric uncertainty. The starting point is a generalization of Kharitonov's theorem obtained by Chapellat and Bhattacharyya in 1989. This theorem, called the Linear CB theorem, its generalization to the multilinear case, the singling out of extremal stability subsets and other ramifications now constitutes an extensive and coherent theory of robust parametric stability that is summarized in the results contained here.

I. INTRODUCTION

The stability of a linear time invariant continuous time feedback control system is characterized by the root locations of its characteristic polynomial $\delta(s)$; for stability the polynomial $\delta(s)$ must be Hurwitz i.e. have all its roots in the open left half of the complex plane. Since control systems operate under large uncertainties it is important to determine if stability is robust, that is, preserved under various perturbations. Despite its practical importance the subject of robust parametric stability lay dormant

[1]This research was supported in part by NSF Grant ECS-8914357 and NASA Grant NAG-1-863.

for about 100 years since the Routh-Hurwitz criterion was developed. The field was revived with the advent of Kharitonov's theorem which appeared in 1978. In 1989 Chapellat and Bhattacharyya generalized this result to make it applicable to control systems. This generalization has given rise to a great many useful and insightful results related to stability margin calculations, mixed parametric and unstructured uncertainty, nonlinear perturbations mixed with parametric perturbations, gain and phase margin optimization, development of Bode and Nyquist envelopes and classical design theory for robust systems etc. The present paper a survey of these results without proofs. We expect these results to have an important influence on future developments in this field.

II. LINEAR INTERVAL SYSTEMS

We begin this section by describing the theorem of Kharitonov which deals with the family of real polynomials

$$\delta(s) = \delta_0 + \delta_1 s + \delta_2 s^2 + \cdots + \delta_n s^n \tag{II.1}$$

with coefficient vector $\underline{\delta} = [\delta_0, \delta_1 \cdots \delta_n]$ lying in the box

$$\Delta = [x_0, y_0] \times [x_1, y_1] \times \cdots \times [x_n, y_n]. \tag{II.2}$$

The Kharitonov polynomials associated with the above family of interval polynomials are defined as

$$\begin{aligned} K^1(s) &= x_0 + x_1 s + y_2 s^2 + y_3 s^3 + \cdots \\ K^2(s) &= x_0 + y_1 s + y_2 s^2 + x_3 s^3 + \cdots \\ K^3(s) &= y_0 + x_1 s + x_2 s^2 + y_3 s^3 + \cdots \\ K^4(s) &= y_0 + \beta^1 s + x_2 s^2 + x_3 s^3 + \cdots . \end{aligned} \tag{II.3}$$

Theorem II.1. (Kharitonov's Theorem [1]) *The family Δ contains only Hurwitz polynomials if and only if $K^1(s), K^2(s), K^3(s)$ and $K^4(s)$ are Hurwitz.*

This remarkable theorem unlocked the door leading to the development of a large number of interesting results in the area of real parametric uncertainty. However Kharitonov's theorem itself is of somewhat limited applicability in control problems. To explain this consider the control system shown below in Figure 1.
Let

$$F(s) := \frac{F_1(s)}{F_2(s)} \qquad G(s) := \frac{N(s)}{D(s)} \tag{II.4}$$

and $N(s)$, $D(s)$ and $F_i(s)$, $i = 1, 2$ are polynomials. The characteristic polynomial of the above system is given by

$$\delta(s) = F_1(s)N(s) + F_2(s)D(s). \tag{II.5}$$

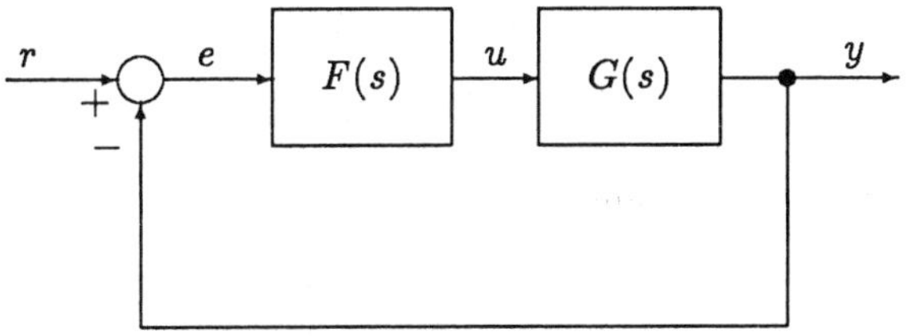

Figure 1. Feedback System

In (II.5) the $F_i(s)$ may denote fixed polynomials corresponding to the controller and $N(s), D(s)$ may be uncertain polynomials corresponding to the plant. Kharitonov's Theorem cannot deal with this situation because of the assumption implicit in it's statement that the coefficients of $\delta(s)$ perturb independently. Motivated by these considerations Chapellat and Bhattacharyya [2] formulated and solved the problem of determining the Hurwitz stability of the family

$$\delta(s) = F_1(s)P_1(s) + F_2(s)P_2(s)\cdots\cdots + F_m(s)P_m(s) \tag{II.6}$$

where the polynomials $P_i(s)$ are interval. This form of the characteristic polynomial (II.6) occurs in m-input (m-output) single output (input) systems. We shall refer to such families of polynomials as *linear interval systems.* The uncertain polynomials may of course correspond to the plant (perturbations) or to the controller (design parameters to be chosen from prescribed intervals). In the following sections we describe various recent results, obtained by the authors and their coworkers on the robust stability of the above types of control systems for both parametric as well as unstructured perturbations. In particular we show the importance of certain segments where the extremal values of various types of stability margins occur. These line segments capture the most important structural information for the analysis and design of robust control systems.

III. THE CB SEGMENTS

We first state the result of of Chapellat and Bhattacharyya [2], [3] for the special case (II.5). $F(s)$ will be assumed to be fixed and $G(s)$ to lie in an uncertainty set described as follows.

Write

$$\begin{aligned} N(s) &:= n_p s^p + n_{p-1}s^{p-1} + \cdots + n_1 s + n_0 \\ D(s) &:= d_q s^q + d_{q-1}s^{q-1} + \cdots + d_1 s + d_0 \end{aligned}$$

and let the coefficients lie in prescribed intervals

$$n_i \in [n_i^-, n_i^+], \quad i \in \{0, 1, \cdots, p\} := \underline{p}$$
$$d_i \in [d_i^-, d_i^+], \quad i \in \{0, 1, \cdots, q\} := \underline{q}.$$

Introduce the interval polynomial sets

$$\begin{aligned} \mathcal{N}(s) &:= \{N(s) = n_p s^p + n_{p-1} s^{p-1} + \cdot + n_0 : n_i \in [n_i^-, n_i^+],\ i \in \underline{p}\} \\ \mathcal{D}(s) &:= \{D(s) = d_q s^q + d_{q-1} s^{q-1} + \cdots + d_0 : d_i \in [d_i^-, d_i^+],\ i \in \underline{q}\} \end{aligned}$$

and the corresponding set of interval transfer functions (or interval systems)

$$\mathbf{G}(s) = \{\frac{N(s)}{D(s)} \ : \ (N(s), D(s)) \in \mathcal{N}(s)\mathrm{x}\mathcal{D}(s)\}. \tag{III.1}$$

The four Kharitonov vertex polynomials associated with $\mathcal{N}(s)$ are

$$\begin{aligned} K_n^1(s) &:= n_0^- + n_1^- s + n_2^+ s^2 + n_3^+ s^3 + n_4^- s^4 + n_5^- s^5 + \cdots \\ K_n^2(s) &:= n_0^- + n_1^+ s + n_2^+ s^2 + n_3^- s^3 + n_4^- s^4 + n_5^+ s^5 + \cdots \\ K_n^3(s) &:= n_0^+ + n_1^- s + n_2^- s^2 + n_3^+ s^3 + n_4^+ s^4 + n_5^- s^5 + \cdots \\ K_n^4(s) &:= n_0^+ + n_1^+ s + n_2^- s^2 + n_3^- s^3 + n_4^+ s^4 + n_5^+ s^5 + \cdots. \end{aligned}$$

and we write

$$\mathcal{K}_{\mathcal{N}}(s) := \{K_n^1(s), K_n^2(s), K_n^3(s), K_n^4(s)\}. \tag{III.2}$$

Similarly the four Kharitonov polynomials associated with $\mathcal{D}(s)$ are denoted $K_d^i(s)$, $i = 1, 2, 3, 4$ and

$$\mathcal{K}_{\mathcal{D}}(s) := \{K_d^1(s), K_d^2(s), K_d^3(s), K_d^4(s)\} \tag{III.3}$$

The four Kharitonov polynomial segments associated with $\mathcal{N}(s)$ are defined as follows:

$$\mathcal{S}_{\mathcal{N}}(s) :=$$
$$[\lambda K_n^i(s) + (1-\lambda) K_n^j(s) \ : \ \lambda \in [0,1],\ (i,j) \in \{(1,2),(1,3),(2,4),(3,4)\}] \tag{III.4}$$

and the four Kharitonov polynomial segments associated with $\mathcal{D}(s)$ are denoted

$$\mathcal{S}_{\mathcal{D}}(s) :=$$
$$[\mu K_d^i(s) + (1-\mu) K_d^j(s) \ : \ \mu \in [0,1],\ (i,j) \in \{(1,2),(1,3),(2,4),(3,4)\}] \tag{III.5}$$

Following Chapellat and Bhattacharyya [2], [3] we introduce the CB subset as follows:

$$\begin{aligned} (\mathcal{N}(s)\mathrm{x}\mathcal{D}(s))_{\mathrm{CB}} &= \{(N(s), D(s)) \ : \ N(s) \in \mathcal{K}_{\mathcal{N}}(s),\ D(s) \in \mathcal{S}_{\mathcal{D}}(s) \\ &\quad \text{or } N(s) \in \mathcal{S}_{\mathcal{N}}(s),\ D(s) \in \mathcal{K}_{\mathcal{D}}(s)\}. \end{aligned} \tag{III.6}$$

The CB subset of the family of interval systems $\mathbf{G}(s)$ is naturally defined as:

$$\mathbf{G}_{\mathrm{CB}}(s) := \{\frac{N(s)}{D(s)} \ : \ (N(s), D(s)) \in (\mathcal{N}(s)\mathrm{x}\mathcal{D}(s))_{\mathrm{CB}}\}. \qquad \text{(III.7)}$$

Theorem III.1. **(CB Theorem [2],[3])** *The control system of Figure 1 is stable for all $G(s) \in \mathbf{G}(s)$ if and only if it is stable for all $G(s) \in \mathbf{G}_{\mathrm{CB}}(s)$.*

We note that each element of $\mathbf{G}_{\mathrm{CB}}(s)$ is a one parameter family of transfer functions and there are at most 32 such distinct elements. The above Theorem therefore gives a constructive solution to the problem of checking robust stability by reducing it to a set of (at most) 32 root locus problems. An alternative way to check the stability of linear interval systems would be to use the Edge Theorem [4] which requires that the exposed edges of the polytope of uncertain polynomials in the coefficient space be checked for stability. While this result also leads to a set of line segments to be checked the number of segments (exposed edges) increases exponentially with the dimension of the uncertain parameter. In the CB theorem the number of segments to be checked is independent of the dimension of the uncertainty set $\mathcal{N}(s)\mathrm{x}\mathcal{D}(s)$. More importantly perhaps these CB segments enjoy many extremal properties that are critical in determining stability margins. We now give an example to illustrate the CB Theorem.

Example III.1. Let us consider the following single input single output plant

$$G(s) = \frac{n(s)}{d(s)} = \frac{s^3 + \alpha s^2 - 2s + \beta}{s^4 + 2s^3 - s^2 + \gamma s + 1},$$

with

$$\alpha \in [-1, -2], \quad \beta \in [0.5, 1], \quad \gamma \in [0, 1].$$

There are two Kharitonov polynomials associated with $n(s)$, namely

$$0.5 - 2s - s^2 + s^3 \text{and } 1 - 2s - 2s^2 + s^3$$

and also two Kharitonov polynomials associated with $d(s)$

$$1 - s^2 + 2s^3 + s^4 \text{and } 1 + s - s^2 + 2s^3 + s^4.$$

In order to check that a given controller $F(s)$ stabilizes the entire family of plants, we only need to check that the controller stabilizes the following four plant segments making up the set $\mathbf{G}_{CB}(s)$:

$$\frac{0.5(1+\lambda) - 2s - (1+2\lambda)s^2 + s^3}{1 - s^2 + 2s^3 + s^4} \qquad \lambda \in [0, 1]$$

$$\frac{0.5(1+\lambda)-2s+(1+2\lambda)s^2+s^3}{1+s-s^2+2s^3+s^4} \quad \lambda \in [0,1]$$

$$\frac{0.5-2s-s^2+s^3}{1+\lambda s-s^2+2s^3+s^4} \quad \lambda \in [0,1]$$

$$\frac{1-2s-2s^2+s^3}{1+\lambda s-s^2+2s^3+s^4} \quad \lambda \in [0,1].$$

On the other hand if one uses the Edge theorem, it is necessary to check the 12 plant segments corresponding to

$$\begin{array}{lll} \alpha=-2, & \beta=0.5, & \gamma\in[0,1] \\ \alpha=-2, & \beta=1, & \gamma\in[0,1] \\ \alpha=-1, & \beta=0.5, & \gamma\in[0,1] \\ \alpha=-1, & \beta=1, & \gamma\in[0,1] \end{array}$$

$$\begin{array}{lll} \alpha=-2, & \beta\in[0.5,1], & \gamma=0 \\ \alpha=-2, & \beta\in[0.5,1], & \gamma=1 \\ \alpha=-1, & \beta\in[0.5,1], & \gamma=0 \\ \alpha=-1, & \beta\in[0.5,1], & \gamma=1 \end{array}$$

$$\begin{array}{lll} \alpha\in[-2,-1], & \beta=0.5, & \gamma=0 \\ \alpha\in[-2,-1], & \beta=0.5, & \gamma=1 \\ \alpha\in[-2,-1], & \beta=1, & \gamma=0 \\ \alpha\in[-2,-1], & \beta=1, & \gamma=1. \end{array}$$

The problem of checking the stability of a line segment of polynomials was solved by Chapellat and Bhattacharyya in [5]; this result called the Segment Lemma is described next. The lemma basically checks for the occurrence of a $j\omega$ root along a line segment of polynomials.

A. Stability of Segments

Let $\delta_1(s)$, and $\delta_2(s)$ be two polynomials of degree n and let

$$\delta_\lambda(s)=\lambda\delta_1(s)+(1-\lambda)\delta_2(s).$$

Denote the segment

$$\{\delta_\lambda(s):\ \lambda\in[0,1]\}:=[\delta_1(s),\delta_2(s)]. \tag{III.8}$$

Also write

$$\begin{aligned} \delta_1(j\omega) &= \delta_1^e(\omega)+j\omega\delta_1^o(\omega) \\ \delta_2(j\omega) &= \delta_2^e(\omega)+j\omega\delta_2^o(\omega). \end{aligned}$$

where $\delta_i^e(\omega)$ and $\delta_i^o(\omega)$ are real.

Lemma III.1. (Segment Lemma [5]) *Suppose that $\delta_1(s)$ and $\delta_2(s)$ are Hurwitz polynomials of degree n with positive coefficients. Then there exists an unstable polynomial on the line segment $[\delta_1(s), \delta_2(s)]$ iff there exists $\omega_0 > 0$, such that*

$$\delta_1^e(\omega_0)\delta_2^o(\omega_0) - \delta_2^e(\omega_0)\delta_1^o(\omega_0) = 0$$

and

$$\delta_1^e(\omega_0)\delta_2^e(\omega_0) \leq 0$$
$$\delta_1^o(\omega_0)\delta_2^o(\omega_0) \leq 0.$$

The Segment Lemma completely solves the problem of checking the Hurwitz stability of a line segment of polynomials. The idea of checking for a root on the stability boundary can of course be extended to other stability regions besides the left half plane.

In general it is known that the stability of the endpoints (vertices) of a line segment does not guarantee that of the entire segment. It is therefore useful to know if there exist some simple additional conditions (simpler than the Segment Lemma) under which the Hurwitz stability of a segment could be guaranteed. It turns out that by restricting the form of the difference polynomial

$$\delta_0(s) = \delta_2(s) - \delta_1(s) \qquad \text{(III.9)}$$

it is possible to conclude segment stability from vertex stability. There exist several known results on this problem. Peterson [6] derived the case when the difference polynomial is antiHurwitz (all roots in the closed right half plane), Chapellat and Bhattacharyya [2] dealt with the case when it is even or odd, Hollot and Yang [7] and Mansour and Kraus [8] proved the result for the difference polynomial being of the form $s^t(as+b)P(s)$ where $P(s)$ is even or odd. The general result given below encompasses all the above cases; the proof may be found in Bhattacharyya [9].

Lemma III.2. (Vertex Lemma) [9]

a) *Let $\delta_1(s)$ and $\delta_2(s)$ be polynomials with positive coefficients and let*

$$\delta_0(s) = A(s)s^t(as+b)P(s)$$

where $A(s)$ is antiHurwitz $t \geq 0$ is an integer, a, b are arbitrary real numbers, and $P(s)$ is even or odd. Then stability of the segment $[\delta_1(s), \delta_2(s)]$ is implied by that of the endpoints $\delta_1(s)$, $\delta_2(s)$.

b) *When $\delta_0(s)$ is not of the form specified in a), stability of the endpoints is not sufficient to guarantee that of the segment.*

Using the above Lemma in conjunction with the Linear CB Theorem it is possible to show that if $F_i(s)$ are of the same form as $\delta_0(s)$ the stability of the CB segments (and therefore the robust stability of a linear interval system) can be ascertained from the Kharitonov vertex polynomials of the system. To state this result let

$$\mathbf{G}_K(s) := \left\{ \frac{N(s)}{D(s)} : (N(s), D(s)) \in (\mathcal{K}_{\mathcal{N}}(s) \mathrm{x} \mathcal{K}_{\mathcal{D}}(s)) \right\}. \qquad \text{(III.10)}$$

Corollary III.1. **(Theorem 2)[9],[3]** *Under the conditions of Theorem III.1, if*

$$F_i(s) = A_i(s)s^{t_i}(a_i s + b_i)Q_i(s), i = 1, 2$$

where $A_i(s)$ is antiHurwitz, t_i is an arbitrary nonnegative integer, a_i, b_i are arbitrary real numbers and $Q_i(s)$ is either an odd or an even polynomial, then the system of Figure 1 is stable for all $G(s) \in \mathbf{G}(s)$ if and only if it is stable for all $G(s) \in \mathbf{G}_K(s)$. Moreover if the $F_i(s)$ do not satisfy the above conditions stability of the system can not be guaranteed by verifying stability for $G(s) \in \mathbf{G}_K(s)$.

This result is obviously useful in control problems where $F(s)$ the compensator consists of integrators, first order lags and leads and unstable and nonminimum phase elements. For instance in Example III.1 if we consider the problem of stabilizing the interval family given, with a controller of the form

$$F(s) = \frac{as + b}{s^t(cs + d)}$$

it is only necessary to check that the controller simultaneously stabilizes the set of plants $\mathbf{G}_K(s)$ shown below:

$$\frac{0.5 - 2s - s^2 + s^3}{1 - s^2 + 2s^3 + s^4} \qquad \frac{0.5 - 2s + s^2 + s^3}{1 + s - s^2 + 2s^3 + s^4}$$

$$\frac{0.5 - 2s - s^2 + s^3}{1 - s^2 + 2s^3 + s^4} \qquad \frac{1 - 2s - 2s^2 + s^3}{1 - s^2 + 2s^3 + s^4}.$$

$$\frac{1 - 2s - 3s^2 + s^3}{1 - s^2 + 2s^3 + s^4} \qquad \frac{1 - 2s + 3s^2 + s^3}{1 + s - s^2 + 2s^3 + s^4}$$

$$\frac{0.5 - 2s - s^2 + s^3}{1 + s - s^2 + 2s^3 + s^4} \qquad \frac{1 - 2s - 2s^2 + s^3}{1 + s - s^2 + 2s^3 + s^4}.$$

To state the CB Theorem for the general case (II.6) we introduce some notation. For any positive integer n let $\underline{n}$ denote the set of integers $\{1, 2, \cdots n\}$. Referring to (II.6) let $d^o(P_i)$ denote the degree of $P_i(s)$ and let p_i^l denote the coefficient of s^l in $P_i(s)$:

$$P_i(s) = p_i^0 + p_i^1 s + \cdots + p_i^l s^l + \cdots + p_i^{d^0(P_i)} s^{d^0(P_i)}. \qquad \text{(III.11)}$$

Write $\underline{p_i} := (p_i^0, p_i^1, \cdots p_i^{d^0(P_i)})$ for each $i \in \underline{m}$, and let

$$\mathbf{p} := [\underline{p}_1, \underline{p}_2 \cdots \cdots \underline{p}_m] \tag{III.12}$$

denote the vector of coefficients of the polynomials $P_i(s)$, $i \in \underline{m}$. Each such coefficient belongs to a given interval:

$$p_i^l \in [\alpha_i^l, \beta_i^l] \quad l = 0, \cdots, d^o(P_i), i \in \underline{m}. \tag{III.13}$$

We let $\mathcal{P}_i(s)$ denote the set of interval polynomials to which $P_i(s)$ belongs and introduce the uncertainty set

$$\mathbf{\Pi} := \{(\mathcal{P}_1(s) \mathrm{x} \mathcal{P}_2(s) \mathrm{x} \cdots, \mathrm{x} \mathcal{P}_m(s))\}$$

Alternatively the uncertainty set can be described in the space of the polynomial coefficients. With mild abuse of notation we use $\mathbf{\Pi}$ to also denote the box of uncertain parameters:

$$\mathbf{\Pi} := \{\mathbf{p} \mid p_i^l \in [\alpha_i^i, \beta_i^l], l = 0, \cdots, d^o(P_i), i \in \underline{m}\}. \tag{III.14}$$

Each point $\mathbf{p} \in \mathbf{\Pi}$ corresponds to a particular choice of the ordered set of polynomials $P_i(s), i \in \underline{m}$. We write $\delta(s, \mathbf{p})$ for the polynomial family (II.6) to display the explicit dependence of $\delta(s)$ on $\mathbf{p}$. For a given fixed set of polynomials $[F_1(s), F_2(s), \cdots, F_m(s)] := \underline{F}$ let $\mathbf{\Delta}$ denote the family of polynomials generated by the map $\underline{F} : \mathbf{\Pi} \Longrightarrow \delta$ as in (II.6) and obtained by letting the parameter vector $\mathbf{p}$, (equivalently, the polynomials $P_i(s)$), range over the box $\mathbf{\Pi}$ described in (III.14). In other words

$$\mathbf{\Delta} := \{\delta(s, \mathbf{p}) | \mathbf{p} \in \mathbf{\Pi}\}. \tag{III.15}$$

Following the previous notation the four Kharitonov polynomials and polynomial segments associated with $\mathcal{P}_i(s)$ are denoted $\mathcal{K}_i(s)$ and $\mathcal{S}_i(s)$ respectively and these definitions hold for each $i \in \underline{m}$ We now introduce some special subsets of $\mathbf{\Pi}$. The linear manifolds $\mathbf{\Pi}_l, l \in \underline{m}$ are defined:

$$\mathbf{\Pi}_l :=$$

$$\{(\mathcal{K}_1(s) \mathrm{x} \cdots \mathrm{x} \mathcal{K}_{l-1}(s) \mathrm{x} \mathcal{S}_l(s) \mathrm{x} \mathcal{K}_{l+1}(s) \mathrm{x} \cdots, \mathrm{x} \mathcal{K}_m(s))\} \tag{III.16}$$

Also

$$\mathbf{\Delta}_l := \{\delta(s, \mathbf{p}) | \mathbf{p} \in \mathbf{\Pi}_l\}. \tag{III.17}$$

Finally, let

$$\mathbf{\Pi}_{CB} := \bigcup_{l=1}^{m} \mathbf{\Pi}_l \tag{III.18}$$

and

$$\mathbf{\Delta}_{CB} := \bigcup_{l=1}^{m} \mathbf{\Delta}_l = \{\delta(s, \mathbf{p}) | \mathbf{p} \in \mathbf{\Pi}_{CB}\}. \tag{III.19}$$

Since there is a one to one correspondence between the elements of $\mathbf{\Pi}_{CB}$ and of $\mathbf{\Delta}_{CB}$ we refer to both sets as CB segments. We also define the *Kharitonov vertices* $\mathbf{K}(\mathbf{\Pi})$, of $\mathbf{\Pi}$ to be the subset of all vertices of $\mathbf{\Pi}$ corresponding to the Kharitonov polynomials of the $P_i(s)$. It is not difficult to see that the number of distinct segments in $\mathbf{\Pi}_{CB}$ in the most general case when all the Kharitonov polynomials associated with each polynomial $P_i(s)$ are distinct, is $m4^m$. With these preliminaries we are ready to state the main result of this section.

Theorem III.2. **(Linear CB Theorem [2],[3])** *$\underline{F}$ stabilizes $\mathbf{\Pi}$ if and only if $\underline{F}$ stabilizes $\mathbf{\Pi}_{CB}$.*

In the next section we show that these CB segments are useful in determining how close one is to instability in the parameter space over a stable set of parameters $\mathbf{\Pi}$.

IV. EXTREMAL PARAMETRIC STABILITY PROPERTIES: LINEAR CASE

We now turn to the question of relative stability of the family $\mathbf{\Pi}$. In other words given a family of polynomials $\mathbf{\Pi}$ which is stable, we wish to know the "distance" to the closest unstable polynomial as the point $\mathbf{p}$ (equivalently the set of polynomials $P_i(s)$) varies over the box $\mathbf{\Pi}$. Before discussing the case of a control system we deal with the special case of a single interval polynomial. In this case we first establish an important extremal property of the Kharitonov polynomials, namely that the closest point to instability over a stable box in coefficient space lies at one of the Kharitonov vertices. The proof of this result was first given in [10].

A. Extremal Property of the Kharitonov Polynomials

Suppose that we have proved the stability of the family of polynomials

$$\delta(s) = \delta_0 + \delta_1 s + \delta_2 s^2 + \cdots + \delta_n s^n, \qquad \text{(IV.1)}$$

with coefficients in the box

$$\mathbf{\Delta} = [x_0, y_0] \times [x_1, y_1] \times \cdots \times [x_n, y_n]. \qquad \text{(IV.2)}$$

Write $\underline{\delta} = [\delta_0, \delta_1, \cdots \delta_n]$, and regard $\underline{\delta}$ as a point in R^{n+1}. Let $||\underline{\delta}||_p$ denote the p norm in R^{n+1} and let this be associated with $\delta(s)$. The set of polynomials which are unstable of degree n or of degree less than n is denoted by $\mathcal{U}$. Then the radius of the stability ball centered at δ is

$$\rho(\delta) = \inf_{u \in \mathcal{U}} ||\underline{\delta} - \mathbf{u}||_p. \qquad \text{(IV.3)}$$

If the polynomial family is stable it is possible to associate with each element the largest stability hypersphere around it. We thus define a mapping from $\mathbf{\Delta}$ to the set of all positive real numbers:

$$\begin{array}{ccc} \mathbf{\Delta} & \xrightarrow{\rho} \mathcal{R}^+\backslash\{0\} \\ \delta(s) & \longrightarrow \rho(\delta) \end{array}$$

and ask the question: Is there a point in $\mathbf{\Delta}$ which is the nearest to instability? Or stated in terms of functions: Has the function ρ a minimum and is there a precise point in $\mathbf{\Delta}$ where it is reached? The answer to that question is given in the following theorem proved in [10]. In the discussion to follow we drop the subscript p from the norm since the result holds for any norm chosen.

Theorem IV.1. (Extremality property of the Kharitonov polynomials [10]). *The function*

$$\begin{array}{ccc} \mathbf{\Delta} & \xrightarrow{\rho} \mathcal{R}^+\backslash\{0\} \\ \delta(s) & \longrightarrow \rho(\delta) \end{array}$$

has a minimum which is reached at one of the four Kharitonov polynomials associated with $\mathbf{\Delta}$.

The above optimal property of Kharitonov polynomials is extremely important and has many uses. Using it it is possible to prove the extremal property of the stability segments occurring in the linear CB Theorem of the previous section.

B. Extremal Parametric Property of the CB Segments

Consider the family $\mathbf{\Delta}$ and the segments $\mathbf{\Pi}_{CB}$ and $\mathbf{\Delta}_{CB}$ which occur in the linear CB Theorem of the previous section. As before consider the family of polynomials lying in the uncertainty set $\mathbf{\Pi}$ and let the coefficients of the polynomials $P_i(s)$, $\mathbf{p} \in R^n$ vary in the prescribed box $\mathbf{\Pi}$. Let $||\cdot||$ denote any norm in R^n and let $\mathcal{P}_u$ denote the set of points $\mathbf{u}$ in R^n for which $\delta(s,\mathbf{u})$ is unstable or loses degree (relative to its generic degree over $\mathbf{\Pi}$. Let

$$\rho(\mathbf{p}) = \inf_{\mathbf{u}\in\mathcal{P}_u} ||\mathbf{p}-\mathbf{u}||_p \tag{IV.4}$$

denote the radius of the stability ball (measured in the norm $||\cdot||$) and centered at the point $\mathbf{p}$. This number serves as the stability margin associated with the point $\mathbf{p}$. If the box $\mathbf{\Pi}$ is stable we can associate a stability margin with each point in $\mathbf{\Pi}$. A natural conterpart of the question posed in the previous section is: Is there a point in $\mathbf{\Pi}$ which is closest to instability in the norm $||\cdot||$ and where is it? The answer to that question is provided in the following theorem first proved in [11].

As before we define a mapping from $\mathbf{\Pi}$ to the set of all positive real numbers:

$$\begin{array}{ll} \mathbf{\Pi} & \xrightarrow{\rho} \mathcal{R}^{+}\backslash\{0\} \\ \mathbf{p} & \longrightarrow \rho(\mathbf{p}). \end{array} \qquad \text{(IV.5)}$$

Theorem IV.2. (Extremal property of the CB Segments [11]) *The function*

$$\begin{array}{ll} \mathbf{\Pi} & \xrightarrow{\rho} \mathcal{R}^{+}\backslash\{0\} \\ \mathbf{p} & \longrightarrow \rho(\mathbf{p}) \end{array}$$

has a minimum which is reached at a point on the CB manifolds $\mathbf{\Pi}_{CB}$.

We remark that this optimality property enjoyed by the CB segments is very useful to find the worst case parametric robustness margin associated with a given controller.

V. PARAMETRIC AND UNSTRUCTURED PERTURBATIONS: LINEAR CASE

We now turn our attention to problems where parametric as well as unstructured uncertainty is simultaneously in operation. The main results in this section will deal with the calculation of the H_∞ stability margin for systems containing parameter uncertainty as defined above. In the following we will use the standard notation: $\mathbf{C}_+ := \{s \in C : \; Re(s) \geq 0\}$, and $H_\infty(\mathbf{C}_+)$ will represent the space of functions $f(s)$ that are bounded and analytic in $\mathbf{C}_+$ with the standard H_∞ norm,

$$\|f\|_\infty = \sup_{\omega \in R} |f(j\omega)|. \qquad \text{(V.1)}$$

Consider first the feedback system below where the fixed stable system with transfer function $G(s)$ is perturbed by H_∞ norm bounded feedback perturbations ΔG. According to the small gain theorem the perturbed system remains stable as long as $\|\Delta G\|_\infty < \frac{1}{\|G\|_\infty}$. We note that $\frac{1}{\|G\|_\infty}$ can be regarded as a "complex gain margin" for the system. The obvious first step is to generalize this by letting $G(s)$ lie in an uncertainty set. We consider the case when $G(s) = \frac{N(s)}{D(s)}$ belongs to an interval family $\mathbf{G}(s)$ as in Section III. We adopt the notation of Section III and let $\mathbf{G}_K(s)$ be the set of 16 *Kharitonov systems* associated with $\mathbf{G}(s)$. Our robust version of the small gain theorem can be stated as follows.

Theorem V.1. (Robust Small Gain Theorem [12]) *Given the interval family* $\mathbf{G}(s)$ *of stable proper systems, the closed-loop system in Figure 2 remains stable for all stable perturbation* ΔP *such that* $\|\Delta P\|_\infty < \alpha$

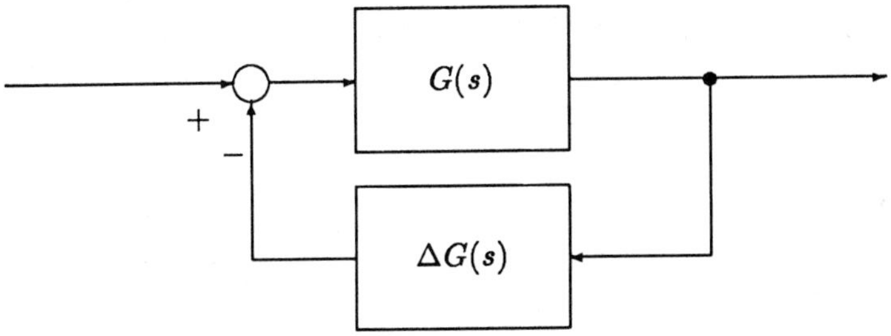

Figure 2.

if and only if,

$$\alpha \leq \frac{1}{\max_{G(s) \in \mathbf{G}_K(s)} \|G\|_\infty}. \tag{V.2}$$

The proof of this theorem given in [12] was based on the following fundamental characterization of H_∞ norms in terms of Hurwitz stability of polynomials. This lemma was also proved in [12].

Lemma V.1. *[12] Let $h(s) = n(s)/d(s)$ be a proper (real or complex) rational function in $H_\infty(\mathbf{C}_+)$, with $deg(d(s)) = q$. Then $\|h\|_\infty < 1$ if and only if*

a1) $|n_q| < |d_q|$,

b1) $d(s) + e^{j\theta}n(s)$ *is Hurwitz for all* θ *in* $[0, 2\pi)$.

We now give an example to illustrate the above theorem

Example V.1. Consider the following stable family $\mathbf{G}(s)$ of interval systems whose generic element is given by,

$$g(s) = \frac{n_0 + n_1 s + n_2 s^2 + n_3 s^3}{d_0 + d_1 s + d_2 s^2 + d_3 s^3}$$

with

$$n_0 \in [1,2],\ n_1 \in [-3,1],\ n_2 \in [2,4],\ n_3 \in [1,3]$$

and

$$d_0 \in [1,3],\ d_1 \in [2,4],\ d_2 \in [6,7],\ d_3 \in [1,2].$$

$\mathbf{G}_K(s)$ consists of the following 16 rational functions,

$$g_1(s) = \frac{1 - 3s + 4s^2 + 3s^3}{1 + 2s + 7s^2 + 2s^3}, \qquad g_2(s) = \frac{1 - 3s + 4s^2 + 3s^3}{1 + 4s + 7s^2 + s^3},$$

$$g_3(s) = \frac{1 - 3s + 4s^2 + 3s^3}{3 + 2s + 6s^2 + 2s^3}, \qquad g_4(s) = \frac{1 - 3s + 4s^2 + 3s^3}{3 + 4s + 6s^2 + s^3},$$
$$g_5(s) = \frac{1 + s + 4s^2 + s^3}{1 + 2s + 7s^2 + 2s^3}, \qquad g_6(s) = \frac{1 + s + 4s^2 + s^3}{1 + 4s + 7s^2 + s^3},$$
$$g_7(s) = \frac{1 + s + 4s^2 + s^3}{3 + 2s + 6s^2 + 2s^3}, \qquad g_8(s) = \frac{1 + s + 4s^2 + s^3}{3 + 4s + 6s^2 + s^3},$$
$$g_9(s) = \frac{2 - 3s + 2s^2 + 3s^3}{1 + 2s + 7s^2 + 2s^3}, \qquad g_{10}(s) = \frac{2 - 3s + 2s^2 + 3s^3}{1 + 4s + 7s^2 + s^3},$$
$$g_{11}(s) = \frac{2 - 3s + 2s^2 + 3s^3}{3 + 2s + 6s^2 + 2s^3}, \qquad g_{12}(s) = \frac{2 - 3s + 2s^2 + 3s^3}{3 + 4s + 6s^2 + s^3},$$
$$g_{13}(s) = \frac{2 + s + 2s^2 + s^3}{1 + 2s + 7s^2 + 2s^3}, \qquad g_{14}(s) = \frac{2 + s + 2s^2 + s^3}{1 + 4s + 7s^2 + s^3},$$
$$g_{15}(s) = \frac{2 + s + 2s^2 + s^3}{3 + 2s + 6s^2 + 2s^3}, \qquad g_{16}(s) = \frac{2 + s + 2s^2 + s^3}{3 + 4s + 6s^2 + s^3}.$$

The H_∞ norms of the above functions are given by,

$$\begin{array}{llll} \|g_1\|_\infty = 2.112, & \|g_2\|_\infty = 3.0, & \|g_3\|_\infty = 5.002, & \|g_4\|_\infty = 3.0, \\ \|g_5\|_\infty = 1.074, & \|g_6\|_\infty = 1.0, & \|g_7\|_\infty = 1.710, & \|g_8\|_\infty = 1.0, \\ \|g_9\|_\infty = 3.356, & \|g_{10}\|_\infty = 3.0, & \|g_{11}\|_\infty = 4.908, & \|g_{12}\|_\infty = 3.0, \\ \|g_{13}\|_\infty = 2.848, & \|g_{14}\|_\infty = 2.0, & \|g_{15}\|_\infty = 1.509, & \|g_{16}\|_\infty = 1.0. \end{array}$$

Therefore, by the above theorem the entire family of systems remains stable under any unstructured feedback perturbations of H_∞ norm less than $\alpha = \frac{1}{5.002} = 0.19992$ which is the smallest "complex stability margin" over the given box of parameters.

A. Computation of the Structured Margin

The converse problem is: given a prescribed bound on the level of unstructured perturbations that are to be tolerated determine the amount of parameter perturbations permissible.

In this case one starts with a nominal stable system

$$g^o(s) = \frac{n_0^o + n_1^o s + \ldots + n_p^o s^p}{d_0^o + d_1^o s + \ldots + d_q^o s^q} \tag{V.3}$$

which satisfies $\|g^o\|_\infty = \alpha$. A bound $\frac{1}{\beta} < \frac{1}{\alpha}$ is then set on the desired level of unstructured perturbations. It is then possible to fix the structure of the parametric perturbations and to maximize a weighted l_∞ ball around the parameters of $g^o(s)$. More precisely, one can allow the parameters n_i, d_j of the plant to vary in intervals of the form

$$n_i \in [n_i^o - \epsilon\nu_i, n_i^o + \epsilon\nu_i], \quad d_j \in [d_j^o - \epsilon\mu_j, d_j^o + \epsilon\mu_j],$$

where the weights ν_i, μ_j are fixed and non negative. For each ϵ we get a family of interval systems $\mathbf{G}(\epsilon)$ and its associated Kharitonov systems $\mathbf{G}_K(\epsilon)$. The structured stability margin is then given by the largest ϵ, say ϵ_{max}, for which every system $g(s)$ in the corresponding interval family $\mathbf{G}(\epsilon_{max})$ satisfies $||g||_\infty \leq \beta$.

An upper bound ϵ_1 for ϵ_{max} is easily found by letting ϵ_1 be the smallest number such that the interval family,

$$\{d(s) = d_0 + \ldots + d_q s^q : \ d_j \in [d_j^o - \epsilon_1 \mu_j, d_j^o + \epsilon_1 \mu_j]\},$$

contains an unstable polynomial. This upper bound is easily calculated, using for example the method proposed in [12] which is straightforward. This can be used to initiate a bisection algorithm. The reader may consult [12] for the details and an example.

We now consider the case where the fixed transfer function $F(s)$ in Figure 1 is not necessarily unity. Let $\mathbf{G}(s)$ be a family of strictly proper interval transfer functions, Assume also that we have found a stabilizing controller $F(s)$ for the entire family. We therefore have a family of stable closed-loop systems and we consider unstructured additive perturbations as shown in Figure 3. Here also we want to determine the amount of unstructured perturbations that can be tolerated by this family of interval plants. In order to do so we have to find the maximum of the H_∞ norm of the closed-loop transfer function $F(s)\big(1 + G(s)F(s)\big)^{-1}$ over all elements $G(s)$ in $\mathbf{G}(s)$. This result is also reported from [12].

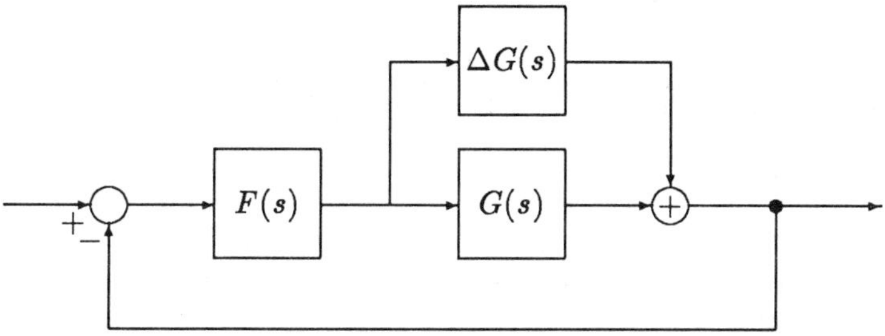

Figure 3.

Theorem V.2. [12] *Given an interval family* $\mathbf{G}(s)$ *of strictly proper plants and a stabilizing controller* $F(s)$ *for* $\mathbf{G}(s)$, *the closed loop system in Figure 3 remains stable for all stable perturbations* ΔG *such that* $||\Delta G||_\infty < \alpha$ *if and only if,*

$$\alpha \leq \frac{1}{\max_{G \in \mathbf{G}_{CB}} ||F(s)\big(1 + G(s)F(s)\big)^{-1}||_\infty}.$$

We illustrate this result with an example.

Example V.2. Consider the following family of interval plants,

$$g_{\beta,\gamma}(s) = \frac{n^p(s)}{d^p(s)} = \frac{\beta s}{1 - s + \gamma s^2 + s^3}, \ \beta \in [1,2], \ \gamma \in [3.4, 5].$$

Using the CB theorem one can easily check that the controller $F(s) = \frac{3}{s+1}$ stabilizes the entire family. The transfer function of interest is given by,

$$F(s)\big(1 + G_{\beta,\gamma}(s)F(s)\big)^{-1} = \frac{3(1 - s + \gamma s^2 + s^3)}{1 + 3\beta s + (\gamma - 1)s^2 + (\gamma + 1)s^3 + s^4}.$$

Following Theorem V.2, we have to find the maximum H_∞ norm of four one-parameter families of rational functions, namely,

$$\begin{aligned}
r_\lambda(s) &= \frac{3(1 - s + \lambda s^2 + s^3)}{1 + 3s + (\lambda - 1)s^2 + (\lambda + 1)s^3 + s^4}, \quad \lambda \in [3.4, 5], \\
r_\mu(s) &= \frac{3(1 - s + \mu s^2 + s^3)}{1 + 6s + (\mu - 1)s^2 + (\mu + 1)s^3 + s^4}, \quad \mu \in [3.4, 5], \\
r_\nu(s) &= \frac{3(1 - s + 3.4s^2 + s^3)}{1 + 3\nu s + 2.4s^2 + 4.4s^3 + s^4}, \quad \nu \in [1, 2], \\
r_\xi(s) &= \frac{3(1 - s + 5s^2 + s^3)}{1 + 3\xi s + 4s^2 + 6s^3 + s^4}, \quad \xi \in [1, 2].
\end{aligned}$$

Consider for example the case of $r_\lambda(s)$. We have,

$$|r_\lambda(j\omega)|^2 = \frac{9\big((1 - \lambda\omega^2)^2 + \omega^2(1 + \omega^2)^2\big)}{\big(1 - (\lambda - 1)\omega^2 + \omega^4\big)^2 + \omega^2\big(3 - (\lambda + 1)\omega^2\big)^2}.$$

Letting $t = \omega^2$ we have to find,

$$\sup_{t \geq 0, \lambda \in [3.4,5]} f(t, \lambda) = \frac{9\big((1 - \lambda t)^2 + t(1 + t)^2\big)}{\big(1 - (\lambda - 1)t + t^2\big)^2 + t\big(3 - (1 + \lambda)t\big)^2}.$$

Differentiating with respect to λ we get a supremum at,

$$\lambda_1(t) = \frac{-2t + 3 + \sqrt{4t^3 + 12t^2 + 1}}{2t}$$

or

$$\lambda_2(t) = \frac{-2t + 3 - \sqrt{4t^3 + 12t^2 + 1}}{2t}.$$

It is then easy to see that $\lambda_1(t) \in [3.4, 5]$ iff $t \in [t_1, t_2] \cup [t_3, t_4]$, where,

$$t_1 \simeq 0.39796, \quad t_2 \simeq 0.64139, \quad t_3 \simeq 15.51766, \quad t_4 \simeq 32.44715\ ,$$

whereas, $\lambda_2(t) \in [3.4, 5]$ iff $t \in [t_5, t_6]$ where,

$$t_5 \simeq 0.15488, \quad t_6 \simeq 0.20095 \ .$$

As a result, the maximum H_∞ norm for $r_\lambda(s)$ is given by,

$$\max\left(\|r_{3.4}\|_\infty, \|r_5\|_\infty, \sqrt{\sup_{t\in[t_1,t_2]\cup[t_3,t_4]} f(t,\lambda_1(t))}, \sqrt{\sup_{t\in[t_5,t_6]} f(t,\lambda_2(t))} \right),$$

where one can at once verify that,

$$f(t,\lambda_1(t)) = \frac{9\left(2t - 1 - \sqrt{4t^3 + 12t^2 + 1}\right)}{2t^2 + 7t - 1 - (t+1)\sqrt{4t^3 + 12t^2 + 1}},$$

and,

$$f(t,\lambda_2(t)) = \frac{9\left(2t - 1 + \sqrt{4t^3 + 12t^2 + 1}\right)}{2t^2 + 7t - 1 + (t+1)\sqrt{4t^3 + 12t^2 + 1}},$$

This maximum is then easily found to be equal to,

$$\max(34.14944, 7.55235, 27.68284, 1.7028) = 34.14944 \ .$$

Proceeding in the same way we finally get the following result,

$$\max_{\beta\in[1,2],\gamma\in[3.4,5]} \|c(s)\left(1 + g_{\beta,\gamma}(s)c(s)\right)^{-1}\|_\infty = 34.14944 \ ,$$

where the maximum is in fact achieved for $\beta = 1, \ \gamma = 3.4$.

The results given above lead to the following theorem.

Theorem V.3. *Let $\mathbf{G}(s)$ be a family of interval plants of fixed degree and let $\alpha > 0$ be given. There exists a linear time invariant controller $F(s)$ that stabilizes $\mathbf{G}(s)$ and that satisfies,*

$$\sup_{G\in\mathbf{G}} \|F(s)\left(1 + G(s)F(s)\right)^{-1}\|_\infty \leq \alpha,$$

if and only if such a controller exists for $\mathbf{G}_{CB}$.

We believe that the above theorem sets the stage for further investigation into the synthesis problem by precisely specifying the role of the controller in the robust stability of a family of interval systems.

In the next section we describe the connection between parametric perturbations and nonlinear feedback perturbations.

VI. PARAMETRIC AND NONLINEAR PERTURBATIONS: LINEAR CASE

An important stability robustness problem that involves unstructured perturbations is the classical Lur'e problem of nonlinear control theory. This problem considers a fixed linear time invariant system subjected to perturbations in the form of nonlinear feedback gains contained in a prescribed sector. In [13] a robust version of the Lur'e problem was treated where structured and unstructured perturbations are simultaneously present. In this formulation the fixed linear system is replaced by the more realistic model of a parametrized family of plants. The "nonlinear stability margin" of the system can be determined by finding the infimum, over the parametrized family, of such stability sectors. From standard results on the Lur'e problem, the size of such a sector can be determined by finding the infimum of the real part of $G(j\omega)$ as $G(s)$ ranges over a parametrized family $\mathbf{G}(s)$. In [13] it was shown how the strict positive realness (SPR) property for a stable family of interval systems can be determined from the set $\mathbf{G}_K(s)$ of the sixteen *Kharitonov systems.* In addition, in the presence of a fixed controller that stabilizes an entire family of interval systems, the SPR property for the family of transfer functions $F(s)(1+G(s)F(s))^{-1}$ is determined from a the CB subset of systems. These results are described in this section and the reader is referred to [13] for proofs.

We begin by giving a stability characterization of the SPR property proved in [13]. Let $G(s) = \frac{n(s)}{d(s)}$ be a real proper transfer function with no poles in the closed right-half plane.

Theorem VI.1. *$G(s)$ is SPR if and only if the following three conditions are satisfied:*

a) Re $G(0) > 0$,

b) $n(s)$ is Hurwitz stable,

c) $d(s) + j\alpha n(s)$ is Hurwitz stable for all α in R.

Based on this result it is possible to formulate a robust SPR result as follows. Let $G(s)$ now belong to an interval family $\mathbf{G}(s)$ as in Section II. Given a real number γ we ask: Under what conditions is $G(s)+\gamma$ SPR for all $G(s)$ in $\mathbf{G}(s)$? The answer is given in the following result from [13].

Lemma VI.1. [13] *$G(s)+\gamma$ is SPR for every element in $\mathbf{G}(s)$ if and only if it is SPR for the 16 Kharitonov systems in $\mathbf{G}_K(s)$.*

This result leads to:

Theorem VI.2. [13] *Given a proper stable family $\mathbf{G}(s)$ of interval plants, the minimum of $Re(G(j\omega))$ over all ω and over all $G(s)$ in $\mathbf{G}(s)$ is achieved at one of the 16 Kharitonov systems in $\mathbf{G}_K(s)$*

We illustrate this result with an example.

Example VI.1. Consider the following stable family $\mathbf{G}(s)$ of interval systems whose generic element is given by

$$G(s) = \frac{1 + \alpha s + \beta s^2 + s^3}{\gamma + \delta s + \epsilon s^2 + s^3}$$

where

$$\alpha \in [1,2],\ \beta \in [3,4],\ \gamma \in [1,2],\ \delta \in [5,6],\ \epsilon \in [3,4].$$

$\mathbf{G}_K(s)$ consists of the following 16 rational functions

$$r_1(s) = \frac{1 + s + 3s^2 + s^3}{1 + 5s + 4s^2 + s^3}, \qquad r_2(s) = \frac{1 + s + 3s^2 + s^3}{1 + 6s + 4s^2 + s^3},$$
$$r_3(s) = \frac{1 + s + 3s^2 + s^3}{2 + 5s + 3s^2 + s^3}, \qquad r_4(s) = \frac{1 + s + 3s^2 + s^3}{2 + 6s + 3s^2 + s^3},$$
$$r_5(s) = \frac{1 + s + 4s^2 + s^3}{1 + 5s + 4s^2 + s^3}, \qquad r_6(s) = \frac{1 + s + 4s^2 + s^3}{1 + 6s + 4s^2 + s^3},$$
$$r_7(s) = \frac{1 + s + 4s^2 + s^3}{2 + 5s + 3s^2 + s^3}, \qquad r_8(s) = \frac{1 + s + 4s^2 + s^3}{2 + 6s + 3s^2 + s^3},$$
$$r_9(s) = \frac{1 + 2s + 3s^2 + s^3}{1 + 5s + 4s^2 + s^3}, \qquad r_{10}(s) = \frac{1 + 2s + 3s^2 + s^3}{1 + 6s + 4s^2 + s^3},$$
$$r_{11}(s) = \frac{1 + 2s + 3s^2 + s^3}{2 + 5s + 3s^2 + s^3}, \qquad r_{12}(s) = \frac{1 + 2s + 3s^2 + s^3}{2 + 6s + 3s^2 + s^3},$$
$$r_{13}(s) = \frac{1 + 2s + 4s^2 + s^3}{1 + 5s + 4s^2 + s^3}, \qquad r_{14}(s) = \frac{1 + 2s + 4s^2 + s^3}{1 + 6s + 4s^2 + s^3},$$
$$r_{15}(s) = \frac{1 + 2s + 4s^2 + s^3}{2 + 5s + 3s^2 + s^3}, \qquad r_{16}(s) = \frac{1 + 2s + 4s^2 + s^3}{2 + 6s + 3s^2 + s^3}.$$

The corresponding minima of their respective real parts along the imaginary axis are given by,

$$\inf_{\omega \in R} Re\ r_1(j\omega) = 0.1385416, \qquad \inf_{\omega \in R} Re\ r_2(j\omega) = 0.1134093,$$
$$\inf_{\omega \in R} Re\ r_3(j\omega) = 0.0764526, \qquad \inf_{\omega \in R} Re\ r_4(j\omega) = 0.0621581,$$
$$\inf_{\omega \in R} Re\ r_5(j\omega) = 0.1540306, \qquad \inf_{\omega \in R} Re\ r_6(j\omega) = 0.1262789,$$
$$\inf_{\omega \in R} Re\ r_7(j\omega) = 0.0602399, \qquad \inf_{\omega \in R} Re\ r_8(j\omega) = 0.0563546,$$
$$\inf_{\omega \in R} Re\ r_9(j\omega) = 0.3467740, \qquad \inf_{\omega \in R} Re\ r_{10}(j\omega) = 0.2862616,$$
$$\inf_{\omega \in R} Re\ r_{11}(j\omega) = 0.3011472, \qquad \inf_{\omega \in R} Re\ r_{12}(j\omega) = 0.2495148,$$
$$\inf_{\omega \in R} Re\ r_{13}(j\omega) = 0.3655230, \qquad \inf_{\omega \in R} Re\ r_{14}(j\omega) = 0.3010231,$$
$$\inf_{\omega \in R} Re\ r_{15}(j\omega) = 0.2706398, \qquad \inf_{\omega \in R} Re\ r_{16}(j\omega) = 0.2345989.$$

Therefore, the entire family is SPR and the minimum is achieved at $r_8(s)$.

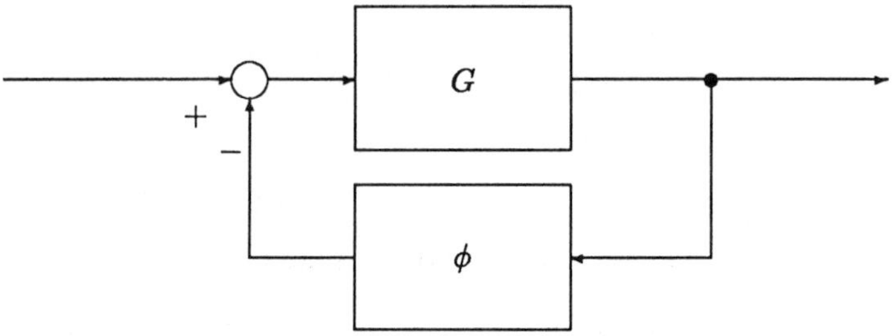

Figure 4.

Turning now to the Lur'e problem let us refer to Figure 4. The class of allowable nonlinearities is described by sector bounded functions. Specifically, the nonlinearity $\phi(t,\sigma)$ satisfies

$$\phi(t,0)=0 \quad \text{for all} \quad t\geq 0$$

$$0\leq \sigma\phi(t,\sigma)\leq k\sigma^2.$$

This implies that $\phi(t,\sigma)$ is bounded by the lines $\phi=0$ and $\phi=k\sigma$. Such nonlinearities are said to belong to a sector $[0,k]$. Referring to Figure 4, we state the following well-known classical result on *absolute stability.*

Theorem VI.3. *If $G(s)$ is a stable transfer function, and ϕ belongs to the sector $[0,k]$, then a sufficient condition for absolute stability is*

$$Re(\frac{1}{k}+g(j\omega))>0, \text{ for all } \omega\in R$$

(i.e $\frac{1}{k}+g(s)$ is SPR).

Combining this with our previous results we have the robust version of the Lur'e problem shown in Figure 5 below.

Theorem VI.4. [13] *Given the interval family $\mathbf{G}(s)$ of stable proper of stable proper systems and the family of sector bounded nonlinearities ϕ belonging to the sector $[0,k]$, a sufficient condition for absolute stability of the closed loop system is that $k>0$ is any number such that*

$$k<\infty, \text{ if } \inf_{\mathbf{G}_K(s)}\inf_{\omega\in R} Re(G(j\omega))\geq 0$$

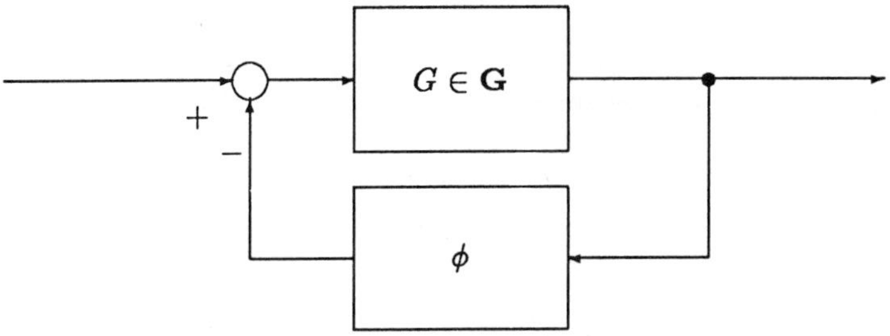

Figure 5.

otherwise

$$k < -\frac{1}{\inf_{\mathbf{G}_K(s)} \inf_{\omega \in R} Re(G(j\omega))}$$

where $\mathbf{G}_K(s)$ *is the set of sixteen Kharitonov systems corresponding to* $\mathbf{G}(s)$.

This theorem may be generalized as follows.

Theorem VI.5. *Given the interval family* $\mathbf{G}(s)$ *of proper systems stabilized by a fixed system* $F(s)$

$$\inf_{\mathbf{G}(s)} \inf_{\omega \in R} Re[F(j\omega)(1 + G(j\omega)F(j\omega))^{-1}] =$$

$$\inf_{\mathbf{G}_K(s)} \inf_{\omega \in R} Re[F(j\omega)(1 + G(j\omega)F(j\omega))^{-1}]$$

In the last section we describe some frequency domain extremal properties of the CB segments.

VII. EXTREMAL FREQUENCY DOMAIN PROPERTIES OF CB SEGMENTS

Consider again the feedback system shown in Figure 1. Since the CB subset characterizes the robust stability of the interval system of Figure 1 it is natural to expect that these subsets also bound the Nyquist and Bode bands of interval systems. This is indeed the case and in this section we present recent results from [14] in this direction. This result was also independently reported in [15]. We expect these results to play a very significant role in synthesis and design issues.

We shall give a quick summary of these results.

A. Nyquist Envelopes

Referring to the control system in Figure 1 we calculate the following transfer functions of interest in analysis and design problems:

$$\frac{y(s)}{u(s)} = G(s) \qquad \frac{u(s)}{e(s)} = F(s) \tag{VII.1}$$

$$\begin{aligned}
T^o(s) &:= \frac{y(s)}{e(s)} = G(s)F(s) \\
T^y(s) &:= \frac{y(s)}{r(s)} = \frac{G(s)F(s)}{1+G(s)F(s)} \\
T^e(s) &:= \frac{e(s)}{r(s)} = \frac{1}{1+G(s)F(s)} \\
T^u(s) &:= \frac{u(s)}{r(s)} = \frac{F(s)}{1+G(s)F(s)}.
\end{aligned} \tag{VII.2}$$

As $G(s)$ ranges over the interval uncertainty set $\mathbf{G}(s)$ (equivalently, $(N(s), D(s))$ ranges over $\mathcal{N}(s)\mathrm{x}\mathcal{D}(s)$) the transfer functions $T^o(s)$, $T^y(s)$, $T^u(s)$, $T^e(s)$ range over corresponding uncertainty sets $\mathbf{T}^o(s)$, $\mathbf{T}^y(s)$, $\mathbf{T}^u(s)$, and $\mathbf{T}^e(s)$, respectively. In other words

$$\begin{aligned}
\mathbf{T}^o(s) &:= \{G(s)F(s) \ : \ G(s) \in \mathbf{G}(s)\} \\
\mathbf{T}^y(s) &:= \{\frac{G(s)F(s)}{1+G(s)F(s)} \ : \ G(s) \in \mathbf{G}(s)\} \\
\mathbf{T}^u(s) &:= \{\frac{F(s)}{1+G(s)F(s)} \ : \ G(s) \in \mathbf{G}(s)\} \\
\mathbf{T}^e(s) &:= \{\frac{1}{1+G(s)F(s)} \ : \ G(s) \in \mathbf{G}(s)\}.
\end{aligned} \tag{VII.3}$$

The CB subsets of the transfer function sets (VII.3) are also introduced:

$$\begin{aligned}
\mathbf{T}^o_{CB}(s) &:= \{G(s)F(s) \ : \ G(s) \in \mathbf{G}_{\mathrm{CB}}(s)\} \\
\mathbf{T}^y_{CB}(s) &:= \{\frac{G(s)F(s)}{1+G(s)F(s)} \ : \ G(s) \in \mathbf{G}_{CB}(s)\} \\
\mathbf{T}^u_{CB}(s) &:= \{\frac{F(s)}{1+G(s)F(s)} \ : \ G(s) \in \mathbf{G}_{\mathrm{CB}}(s)\} \\
\mathbf{T}^e_{CB}(s) &:= \{\frac{1}{1+G(s)F(s)} \ : \ G(s) \in \mathbf{G}_{\mathrm{CB}}(s)\}
\end{aligned} \tag{VII.4}$$

In frequency domain analysis and design problems the complex plane image of each of the above sets evaluated at $s = j\omega$ plays an important role. We denote each of these two dimensional sets in the complex plane by replacing s in the corresponding argument by ω. Thus, for example,

$$\mathbf{T}^{y}_{CB}(\omega) := \{\mathbf{T}^{y}_{CB}(s) \ : \ s = j\omega\} \tag{VII.5}$$

The Nyquist plot of a set of functions (or polynomials) $\mathbf{T}(s)$ is denoted by $\mathbf{T}$:

$$\mathbf{T} := \cup_{0 \leq \omega < \infty} \mathbf{T}(\omega) \tag{VII.6}$$

The boundary of a set S is denoted ∂S.

We shall give the main results here without proof. The proofs are given in Keel, Shaw and Bhattacharyya [16] and also independently by Tesi and Vicino [15].

Theorem VII.1. [14][17] *For every* $\omega \geq 0$,

$$\begin{aligned} \partial\mathbf{G}(\omega) &\subset \mathbf{G}_{CB}(\omega) \\ \partial\mathbf{T}^{o}(\omega) &\subset \mathbf{T}^{o}_{CB}(\omega) \\ \partial\mathbf{T}^{y}(\omega) &\subset \mathbf{T}^{y}_{CB}(\omega) \\ \partial\mathbf{T}^{u}(\omega) &\subset \mathbf{T}^{u}_{CB}(\omega) \\ \partial\mathbf{T}^{e}(\omega) &\subset \mathbf{T}^{e}_{CB}(\omega) \end{aligned}$$

This result shows that at every $\omega \geq 0$ the image set of each transfer function in (VII.4) is bounded by the corresponding image set of the CB segments.

The next result deals with the Nyquist plots of each of the transfer functions in (VII.4).

Theorem VII.2. [14][17] *The Nyquist plots of each of the transfer function sets* $\mathbf{T}^{o}(s)$, $\mathbf{T}^{y}(s)$, $\mathbf{T}^{u}(s)$, *and* $\mathbf{T}^{e}(s)$ *are bounded by their corresponding CB subsets:*

$$\begin{aligned} \partial\mathbf{T}^{o} &\subset \mathbf{T}^{o}_{CB} \\ \partial\mathbf{T}^{y} &\subset \mathbf{T}^{y}_{CB} \\ \partial\mathbf{T}^{u} &\subset \mathbf{T}^{u}_{CB} \\ \partial\mathbf{T}^{e} &\subset \mathbf{T}^{e}_{CB} \end{aligned}$$

B. Bode Envelopes

For any function say, $T(s)$ let $\mu_T(\omega) := |T(j\omega)|$ and $\phi_T(\omega) := \angle T(j\omega)$ denote the magnitude and phase evaluated at $s = j\omega$. If $\mathbf{T}(s)$ denotes a set of functions we let the extremal values of magnitude and phase at a given frequency be defined as follows:

$$\begin{aligned}\bar{\mu}_{\mathbf{T}}(\omega) &:= \sup_{\mathbf{T}(j\omega)} |T(j\omega)| \\ \underline{\mu}_{\mathbf{T}}(\omega) &:= \inf_{\mathbf{T}(j\omega)} |T(\omega)|.\end{aligned} \tag{VII.7}$$

Similarly

$$\begin{aligned}\bar{\phi}_{\mathbf{T}}(\omega) &:= \sup_{\mathbf{T}(j\omega)} \angle T(j\omega) \\ \underline{\phi}_{\mathbf{T}}(\omega) &:= \inf_{\mathbf{T}(j\omega)} \angle T(j\omega).\end{aligned} \tag{VII.8}$$

Suppose that $\mathbf{G}(s)$ is an interval family. To compute

$$\bar{\mu}_{\mathbf{G}}(\omega), \qquad \underline{\mu}_{\mathbf{G}}(\omega) \tag{VII.9}$$

and

$$\bar{\phi}_{\mathbf{G}}(\omega), \qquad \underline{\phi}_{\mathbf{G}}(\omega), \tag{VII.10}$$

the following two lemmas are necessary.

Lemma VII.1. *Let $\mathcal{A}$ be a closed polygon in the complex plane, and "a" be an arbitrary point in $\mathcal{A}$. Let $V_{\mathcal{A}}$ be the set of vertices and $E_{\mathcal{A}}$ be the set of edges of $\mathcal{A}$. Then the following statements are true.*

1) $$\max_{\mathcal{A}} |a| = \max_{V_{\mathcal{A}}} |a|$$

2) $$\min_{\mathcal{A}} |a| = \min_{E_{\mathcal{A}}} |a|$$

Lemma VII.2. *Let $\mathcal{A}$ and $\mathcal{B}$ be disjoint closed polygons in the complex plane, and "a" and "b" be arbitrary points on $\mathcal{A}$ and $\mathcal{B}$, respectively. Let $V_{\mathcal{A}}$ and $V_{\mathcal{B}}$ be the sets of vertices and let $E_{\mathcal{A}}$ and $E_{\mathcal{B}}$ be the sets of edges of $\mathcal{A}$ and $\mathcal{B}$, respectively. Then the following statements are true.*

1) $$\max_{\mathcal{A}\mathrm{x}\mathcal{B}}\{\angle a - \angle b\} = \max_{V_{\mathcal{A}}\mathrm{x}V_{\mathcal{B}}}\{\angle a - \angle b\}$$

2) $$\min_{\mathcal{A}\mathrm{x}\mathcal{B}}\{\angle a - \angle b\} = \min_{V_{\mathcal{A}}\mathrm{x}V_{\mathcal{B}}}\{\angle a - \angle b\}$$

Proofs of the above two lemmas are obvious from geometric considerations illustrated in [17].

Let $\mathcal{N}(\omega)$ denote the complex plane image of the set of polynomials $N(s) \in \mathcal{N}(s)$ evaluated at $s = j\omega$. Similar definitions hold for $\mathcal{D}(\omega)$,

$\mathcal{S}_{\mathcal{N}}(\omega)$ and $\mathcal{S}_{\mathcal{D}}(\omega)$. $\mathcal{N}(\omega)$ is bounded by the set of Kharitonov segments $\mathcal{S}_{\mathcal{N}}(\omega)$. Similarly, $\mathcal{D}(\omega)$ is bounded by the set $\mathcal{S}_{\mathcal{D}}(\omega)$. These facts along with Lemmas VII.1 and VII.2 lead to the following results. Before we state Theorem VII.3, let us define the following sets.

$$\begin{aligned}
\mathbf{G}(\omega) &:= \{G(j\omega) = \frac{N(j\omega)}{D(j\omega)} \mid \\
&\qquad N(j\omega) \in \mathcal{N}(\omega), D(j\omega) \in \mathcal{D}(\omega)\} && \text{(VII.11)} \\
\bar{\mathbf{G}}_{CB}(\omega) &:= \{G(j\omega) = \frac{N(j\omega)}{D(j\omega)} \mid \\
&\qquad N(j\omega) \in \mathcal{K}_{\mathcal{N}}(\omega), D(j\omega) \in \mathcal{S}_{\mathcal{D}}(\omega)\} && \text{(VII.12)} \\
\underline{\mathbf{G}}_{CB}(\omega) &:= \{F(j\omega) = \frac{N(j\omega)}{D(j\omega)} \mid \\
&\qquad N(j\omega) \in \mathcal{S}_{\mathcal{N}}(\omega), D(j\omega) \in \mathcal{K}_{\mathcal{D}}(\omega)\}. && \text{(VII.13)}
\end{aligned}$$

Theorem VII.3. *For every frequency $\omega \geq 0$,*

$$\begin{aligned}
\bar{\mu}_{\mathbf{G}}(\omega) &= \bar{\mu}_{\bar{\mathbf{G}}_{CB}}(\omega) \\
\underline{\mu}_{\mathbf{G}}(\omega) &= \underline{\mu}_{\underline{\mathbf{G}}_{CB}}(\omega)
\end{aligned}$$

Let us also define the set of systems constructed from Kharitonov vertices as follows:

$$\mathbf{G}_{\mathcal{K}}(\omega) := \{F(j\omega) = \frac{N(j\omega)}{D(j\omega)} \mid N(j\omega) \in \mathcal{K}_{\mathcal{N}}(\omega), D(j\omega) \in \mathcal{K}_{\mathcal{D}}(\omega)\}. \qquad \text{(VII.14)}$$

Theorem VII.4. *For every frequency $\omega \geq 0$,*

$$\begin{aligned}
\bar{\phi}_{\mathbf{G}}(\omega) &= \bar{\phi}_{\mathbf{G}_{\mathcal{K}}}(\omega) \\
\underline{\phi}_{\mathbf{G}}(\omega) &= \underline{\phi}_{\mathbf{G}_{\mathcal{K}}}(\omega)
\end{aligned}$$

Using the above extremal properties it is possible to evaluate the Bode magnitude and phase bands of interval transfer functions. Let us consider the family of transfer functions

$$\mathbf{T}^o(s) = \{\ T^o(s) \mid F(s)G(s),\ G(s) \in \mathbf{G}(s)\ \}. \qquad \text{(VII.15)}$$

Since $F(s)$ is fixed,

$$\begin{aligned}
\bar{\mu}_{\mathbf{T}^o}(\omega) &= |F(j\omega)|\, \bar{\mu}_{\mathbf{G}}(\omega) \\
\underline{\mu}_{\mathbf{T}^o}(\omega) &= |F(j\omega)|\, \underline{\mu}_{\mathbf{G}}(\omega). && \text{(VII.16)}
\end{aligned}$$

Similarly,

$$\begin{aligned}
\bar{\phi}_{\mathbf{T}^o}(\omega) &= \angle F(j\omega) + \bar{\phi}_{\mathbf{G}}(\omega) \\
\underline{\phi}_{\mathbf{T}^o}(\omega) &= \angle F(j\omega) + \underline{\phi}_{\mathbf{G}}(\omega). && \text{(VII.17)}
\end{aligned}$$

These relations are sufficient to construct the Bode magnitude and phase envelopes.

The Nyquist and Bode envelopes are important tools for solving analysis and design problems in robust parametric stability. In the next section, we show how the previous theory can be used to develop techniques to improve a given controller, by choosing an controller from a given set of stabilizing interval controllers, that provides optimal gain (or phase) margin to the closed loop system.

VIII. DESIGN OF LINEAR INTERVAL CONTROL SYSTEMS

In this section, we consider a nominal plant connected to an interval controller and give some design techniques for improving the closed loop gain and phase margins using the Nyquist envelope described in Theorem VII.2. From the results of the previous section it is clear that the main computational task is to determine the stability margin over the CB segments. In the next section, we discuss the problem of determining optimal gain and phase margins over a single segment system.

A. Segment System

The typical CB segment is of the form

$$\mathbf{p}(s) := \{\frac{p_0(s)}{p_1(s) + \lambda p_2(s)} \mid \lambda \in [0,1]\} \tag{VIII.1}$$

or

$$\mathbf{p}(s) := \{\frac{p_1(s) + \lambda p_2(s)}{p_0(s)} \mid \lambda \in [0,1]\} \tag{VIII.2}$$

where $p_i(s)$ are fixed polynomials. In this section, we develop simple techniques to compute the extremal gain and phase margins over a segment. We also determine the optimal value λ^*, equivalently $p^*(s)$, that produces the optimal gain (or phase) margin over the family $\mathbf{p}(s)$.

Let us consider the following <u>segment system</u> with

$$\mathrm{p}(j\omega, \lambda) = \frac{p_o(j\omega)}{p_1(j\omega) + \lambda p_2(j\omega)}.$$

The problem of computing the extremal gain and phase margins at the loop breaking point "m" over the single segment system is described as follows. Let us denote

$$(\Lambda \mathrm{x} \Omega) := \{(\lambda, \omega) \mid \angle p(j\omega, \lambda) = 180^o, \lambda \in [0,1]\} \tag{VIII.3}$$

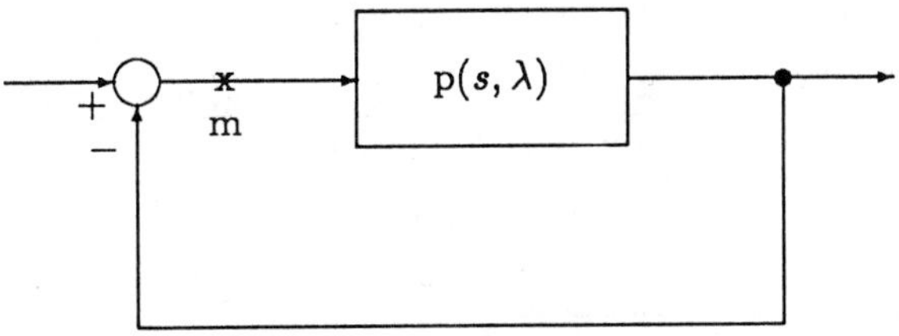

Figure 5. Segment System

and

$$\bar{\mu}_{\mathbf{p}} := \max_{(\Lambda x \Omega)} |p(j\omega, \lambda)| \tag{VIII.4}$$

$$\underline{\mu}_{\mathbf{p}} := \min_{(\Lambda x \Omega)} |p(j\omega, \lambda)| \tag{VIII.5}$$

$$\bar{\phi}_{\mathbf{p}} := \max_{(\Lambda x \Omega)} \angle p(j\omega, \lambda) \tag{VIII.6}$$

$$\underline{\phi}_{\mathbf{p}} := \min_{(\Lambda x \Omega)} \angle p(j\omega, \lambda)\ . \tag{VIII.7}$$

Then

$$\text{maximum gain margin over } \mathbf{p}(s) \ : \ \bar{\rho} := \frac{1}{\underline{\mu}_{\mathbf{p}}} \tag{VIII.8}$$

$$\text{minimum gain margin over } \mathbf{p}(s) \ : \ \underline{\rho} := \frac{1}{\bar{\mu}_{\mathbf{p}}} \tag{VIII.9}$$

$$\text{maximum phase margin over } \mathbf{p}(s) \ : \ \bar{\theta} := \bar{\phi}_{\mathbf{p}} - 180^o \tag{VIII.10}$$

$$\text{minimum phase margin over } \mathbf{p}(s) \ : \ \underline{\theta} := \underline{\phi}_{\mathbf{p}} - 180^o \tag{VIII.11}$$

Similar definitions can be made for the case of gain margins less than 1.

As seen from eqs. (VIII.8) - (VIII.11), the problem of computing the extremal gain or phase margin over the segment system is two parameter optimization problem. This can be reduced to a simple one parameter problem as follows. Write

$$p_i(j\omega) := p_{iR}(\omega) + jp_{iI}(\omega)$$

Then

$$\mathrm{p}(j\omega, \lambda) = \frac{p_o(j\omega)}{p_1(j\omega) + \lambda p_2(j\omega)}$$

$$
\begin{aligned}
&= \frac{p_{0R}(\omega) + jp_{0I}(\omega)}{[p_{1R}(\omega) + \lambda p_{2R}(\omega)] + j[p_{1I}(\omega) + \lambda p_{2I}(\omega)]} \\
&= \underbrace{\frac{p_{0R}(\omega)p_{1R}(\omega) + p_{0I}(\omega)p_{1I}(\omega) + \lambda[p_{0R}(\omega)p_{2R}(\omega) + p_{0I}(\omega)p_{2I}(\omega)]}{[p_{1R}(\omega) + \lambda p_{2R}(\omega)]^2 + [p_{1I}(\omega) + \lambda p_{2I}(\omega)]^2}}_{\mathrm{Re}\{\mathrm{p}(j\omega,\lambda)\}} \\
&+ j\underbrace{\frac{p_{0I}(\omega)p_{1R}(\omega) - p_{0R}(\omega)p_{1I}(\omega) + \lambda[p_{0I}(\omega)p_{2R}(\omega) - p_{0R}(\omega)p_{2I}(\omega)]}{[p_{1R}(\omega) + \lambda p_{2R}(\omega)]^2 + [p_{1I}(\omega) + \lambda p_{2I}(\omega)]^2}}_{\mathrm{Im}\{\mathrm{p}(j\omega,\lambda)\}}
\end{aligned}
\tag{VIII.12}
$$

In order to determine the gain margin, we set

$$\angle \mathrm{p}(j\omega, \lambda) = 180^o \tag{VIII.13}$$

which implies

$$\mathrm{Im}\{\mathrm{p}(j\omega, \lambda)\} = 0. \tag{VIII.14}$$

Note that (VIII.14) will be satisfied when $\angle \mathrm{p}(j\omega, \lambda) = 0^o$ or 180°. We exclude frequencies ω for which $\angle \mathrm{p}(j\omega, \lambda) = 0^o$. From eqs.(VIII.14) and (VIII.12), we have

$$
\begin{aligned}
\mathrm{Im}\{\mathrm{p}(j\omega, \lambda)\} &= [p_{0I}(\omega)p_{1R}(\omega) - p_{0R}(\omega)p_{1I}(\omega)] \\
&\quad + \lambda[p_{0I}(\omega)p_{2R}(\omega) - p_{0R}(\omega)p_{2I}(\omega)] \\
&= 0
\end{aligned}
\tag{VIII.15}
$$

equivalently

$$\lambda(\omega) = \frac{p_{0R}(\omega)p_{1I}(\omega) - p_{0I}(\omega)p_{1R}(\omega)}{p_{0I}(\omega)p_{2R}(\omega) - p_{0R}(\omega)p_{2I}(\omega)}. \tag{VIII.16}$$

From this representation, we can easily conclude that instead of searching both $\omega \in [0, \infty)$ and $\lambda \in [0, 1]$, searching only selected ranges of ω that satisfy $\lambda \in [0, 1]$ is enough. Thus, we let

$$
\begin{aligned}
\lambda(\omega) &= \frac{p_{0R}(\omega)p_{1I}(\omega) - p_{0I}(\omega)p_{1R}(\omega)}{p_{0I}(\omega)p_{2R}(\omega) - p_{0R}(\omega)p_{2I}(\omega)} \\
&= 0 \text{ or } 1.
\end{aligned}
\tag{VIII.17}
$$

Without loss of generality, we have

$$
\begin{aligned}
&\text{for } \lambda = 1 \qquad p_{0R}(\omega)p_{1I}(\omega) - p_{0I}(\omega)p_{1R}(\omega) - p_{0I}(\omega)p_{2R}(\omega) \\
&\qquad\qquad\quad + p_{0R}(\omega)p_{2I}(\omega) = 0 \\
&\text{for } \lambda = 0 \qquad p_{0R}(\omega)p_{1I}(\omega) - p_{0I}(\omega)p_{1R}(\omega) = 0
\end{aligned}
\tag{VIII.18}
$$

The valid ranges of ω with respect to the condition $\lambda \in [0, 1]$ can be easily determined from the roots of the above two equations. Thus, the problem posed in eqs. (VIII.8) and (VIII.9) is reduced to selection of maximum and minimum magnitudes of λ evaluated over the admissible ranges of ω determined from the roots of eq. (VIII.18). Furthermore, the optimal value λ^*, equivalently optimal values of parameters over the segment system, can also be easily determined by substituting ω^* that corresponds to the maximum gain margin into eq. (VIII.16).

If the segment system is of the form in eq. (VIII.2), one can follow a similar procedure to determine the extremal margins and the corresponding optimal systems over the segment system. Similar procedures can also be applied for computing extremal phase margins over a single segment. This is easily derived by replacing the condition (VIII.13) by

$$|p(j\omega, \lambda)| = 1 \tag{VIII.19}$$

B. Optimal Parameter Selection

Applying the procedure described in the previous section to the entire set of segments systems, the extremal margins over the interval plant are determined. Consequently, the optimal system that produces the maximum gain or phase margin over an interval family is also determined. This procedure may be used to solve the following interesting problem.

Suppose that a fixed system $F(s)$ and a family of controllers $\mathbf{G}(s)$ are given, for which the closed loop system is stable. The objective is to select an optimal system $G_{\text{opt}}(s) \in \mathbb{R}^n(s)$ so that the resulting closed loop system has the maximum possible gain margin or phase margin over the family $\mathbf{G}(s)$. Once such an optimal system is found the controller may be reset to the optimal parameter as the new nominal controller. At this point a new family of stabilizing interval controllers can be determined and the previous procedure of selecting the best controller repeated over the new box of parameters. The set of stabilizing interval controllers can be determined by many different methods; for example the locus introduced by Tsypkin and Polyak [18] may be used. This procedure described above can be repeated until 1) improvement of the maximum margin in a given iteration is small or 2) the stability radius in the parameter space is small. Of course there is no guarantee that a globally optimum or even a satisfactory design will be achieved by this method.

In the next section, an illustrative example is given.

C. Illustrative Example

Suppose

$$F(s) := \frac{N_g(s)}{D_g(s)} = \frac{s^2 + 2s + 1}{s^4 + 2s^3 + 2s^2 + s}$$

$$\mathbf{G}(s) \quad := \quad \{\frac{n_1 s + n_0}{d_2 s^2 + d_1 s + d_0}\}$$

where

$$n_0 \in [0.9, 1.1], \quad n_1 \in [0.1, 0.2]$$
$$d_0 \in [1.9, 2.1], \quad d_1 \in [1.8, 2.0], \quad d_2 \in [0.9, 1.0]$$

We first check the stability of the family of closed loop systems with $F(s)$ and $\mathbf{G}(s)$. This can be done by checking the stability of the corresponding CB segment.

We have

$$K_n^1(s) = 0.1s + 0.9 \qquad K_n^2(s) = 0.2s + 0.9$$
$$K_n^3(s) = 0.1s + 1.1 \qquad K_n^2(s) = 0.2s + 0.9$$

$$K_d^1(s) = s^2 + 1.8s + 1.9 \qquad K_d^2(s) = s^2 + 2s + 1.9$$
$$K_d^3(s) = 0.9s^2 + 1.8s + 2.1 \qquad K_d^4(s) = 0.9s^2 + 2s + 2.1$$

and the corresponding segments $\mathbf{G}_{CB}(s)$ are

$$\frac{\lambda K_n^i(s) + (1-\lambda)K_n^j(s)}{K_d^k(s)} \cup \frac{K_n^k(s)}{\lambda K_d^i(s) + (1-\lambda)K_d^j(s)}$$

where $k = 1, 2, 3, 4$

$$(i, j) \in \{(1,2), (1,3), (2,4), (3,4)\}$$

and $\lambda \in [0, 1]$ Using the Segment Lemma [5], we verified that all the above CB segments stabilize the closed loop.

The Bode and Nyquist envelopes associated with $\mathbf{T}^o(s)$, the forward transfer functions $F(s)\mathbf{G}(s)$, are constructed by evaluating the following rational functions over $\omega \in [0, \infty)$ and $\lambda \in [0, 1]$:

$$\frac{N(s)K_n^k(s)}{D(s)[\lambda K_d^i(s) + (1-\lambda)K_d^j(s)]}$$

and

$$\frac{N(s)[\lambda K_n^i(s) + (1-\lambda)K_n^j(s)]}{D(s)K_d^k(s)}.$$

This is shown in Figures 6,7 and 8.

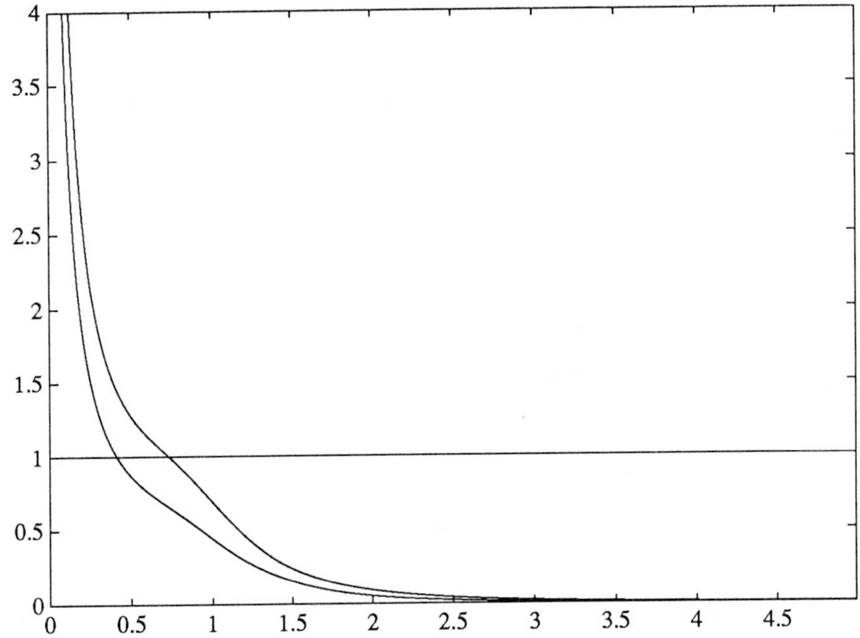

Figure 6. Magintude envelope plot

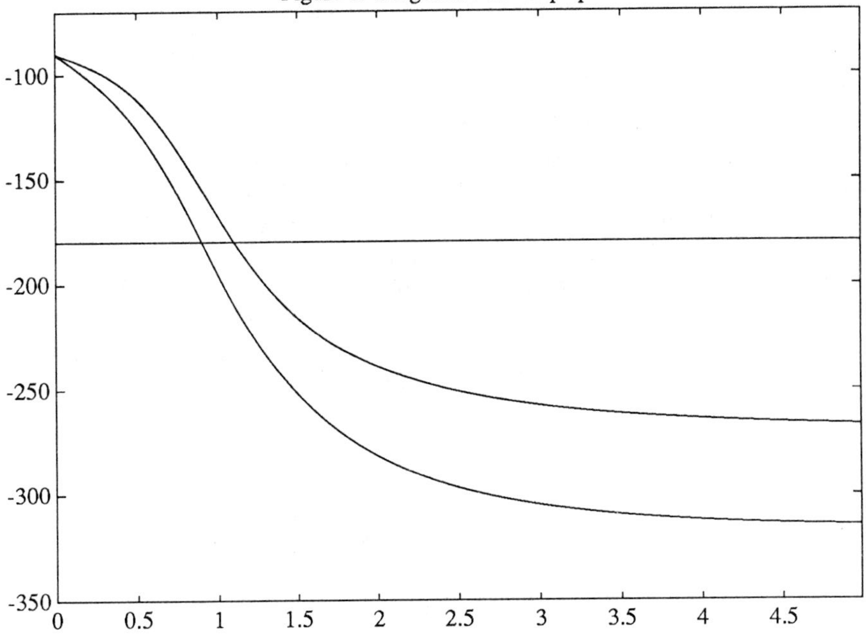

Figure 7. Phase envelope plot

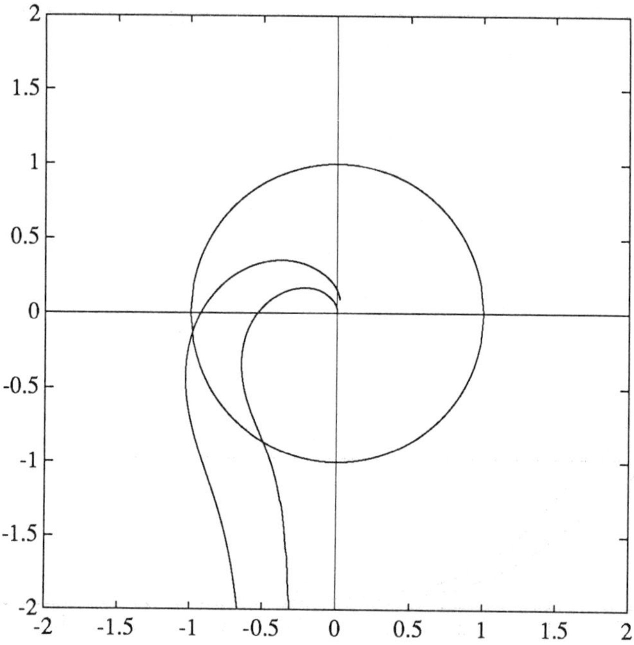

Figure 8. Nyquist envelope plot

From these figures, the minimum and maximum gain margins are found to be:

$$\underline{\rho} = 1.1240. \qquad \bar{\rho} = 1.7582.$$

If we want to improve the gain margin of the system by selecting parameters n_i and d_i beyond its previously given intervals, we can repeat the procedure as follows. With the controller designed in the previous part as the new nominal controller we can again construct an interval family of stabilizing controllers centered at the parameters of $G_{\text{opt}}(s)$. This can be done by several methods. We used the stability locus introduced by Tsypkin and Polyak [18] which is shown in Figure 9. Note that in this case the locus shows the ℓ_2 stability margin.

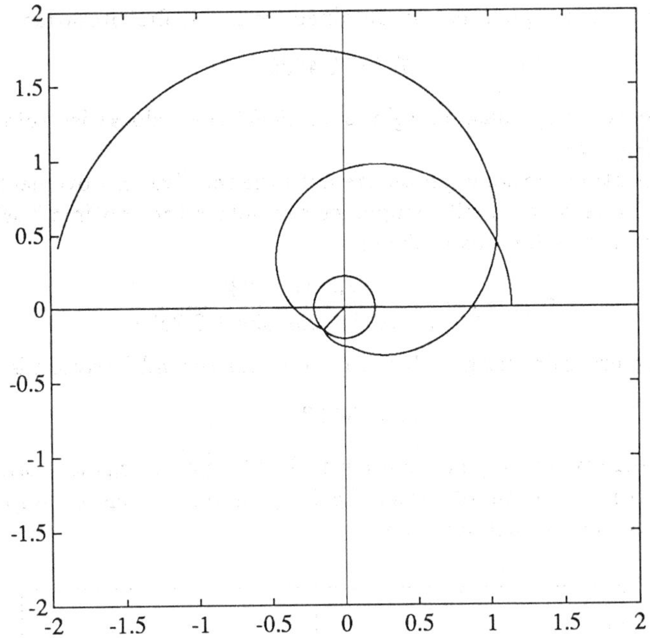

Figure 9. Stability locus (Tsypkin - Polyak)

From this method, we have the parametric stability margin, $\gamma = 0.2128$. Thus we construct the stabilizing intervals around the nominal values n_i and d_i as follows:

$$n_i - \frac{\gamma}{2} \leq n_i \leq n_i + \frac{\gamma}{2} \qquad\qquad d_i - \frac{\gamma}{2} \leq d_i \leq d_i + \frac{\gamma}{2}$$

Consequently, we obtain

$$\mathbf{G}^{(1)}(s) = \{\frac{n_1^{(1)}s + n_0^{(1)}}{d_2^{(1)}s^2 + d_1^{(1)}s + d_0^{(1)}}\}$$

where

$$n_0^{(1)} \in [0.7936, 1.0064], \quad n_1^{(1)} \in [0.0839, 0.2967]$$
$$d_0^{(1)} \in [1.9936, 2.2064], \quad d_1^{(1)} \in [1.8936, 2.1064], \quad d_2^{(1)} \in [0.7936, 1.0064].$$

Now the previous optimization procedure can be applied to this new family of interval controllers. This yields the result:

$$\begin{aligned} G_{\text{opt}}^{(1)}(s) &= \frac{\lambda K_n^1(s) + (1-\lambda)K_n^2(s)}{K_d^4(s)}|_{\lambda=0} \\ &= \frac{0.2967s + 0.7936}{0.7936s^2 + 2.1064s + 2.2064} \end{aligned}$$

and the maximum gain margin obtained from this first iteration is

$$\bar{\rho}^{(1)} = 2.4295.$$

The corresponding phase margin is 71.4506^o (i.e., clockwise rotation of 71.4506 degrees).

By repeating the same procedure until the relative improvement of the gain margin becomes small enough, we can obtain the "optimal" selection of the parameters for this problem.

$$G^*_{\text{opt}}(s) = \frac{.3359s + .0398}{.0398s^2 + 2.8602s + 2.9602}.$$

The maximum gain margin obtained from this second iteration is

$$\bar{\rho}^* = 562.3651$$

and the corresponding phase margin is 95.77^o (i.e., clockwise rotation of 95.77 degrees). Figure 10 shows the Nyquist plot of the corresponding optimal system for each iteration.

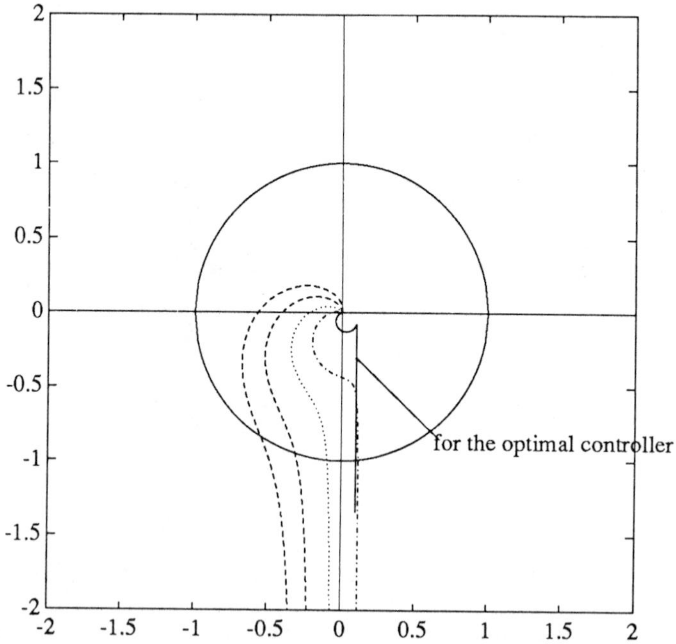

Figure 10. Nyquist Plots for Various Controller Parameters.

IX. MULTILINEAR INTERVAL SYSTEMS

In following sections of the paper, we consider systems in which the uncertain parameters enter the characteristic polynomial affine multilinearly.

As an example consider a feedback control system with a fixed compensator connected to a cascade of two interval plants as in the block diagram below:

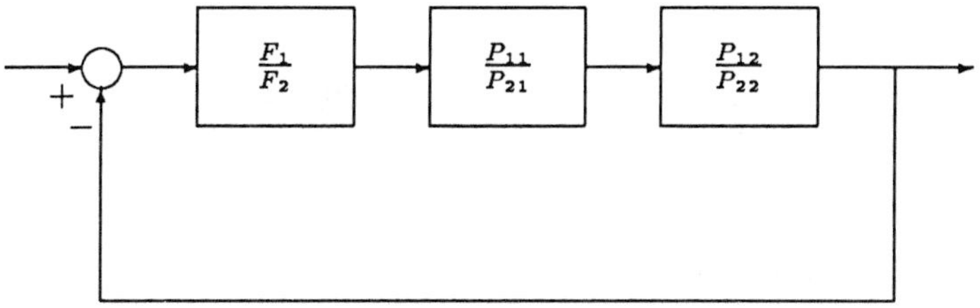

Figure 11. Unity Feedback System.

We have the following expression for the characteristic polynomial

$$\delta(s) = F_1(s)P_{11}(s)P_{12}(s) + F_2(s)P_{21}(s)P_{22}(s) \tag{IX.1}$$

with $P_{ij}(s)$ being interval polynomials $i = 1, 2, j = 1, 2$. The CB Theorem derived in [2] cannot deal with the robust stability of the family (IX.1) because it contains products of interval polynomials. Neither can the stability of the family (IX.1) be checked by using the Edge Theorem of [4] since it is not a polytope.

The above considerations motivate the problem of determining the Hurwitz stability of the family of polynomials

$$\begin{aligned}\delta(s) &= F_1(s)P_{11}(s)P_{12}(s)\cdots P_{1r_1}(s) + \cdots\cdots \\ &\quad + F_m(s)P_{m1}(s)P_{m2}(s)\cdots P_{mr_m}(s)\end{aligned} \tag{IX.2}$$

where $F_i(s)$ are fixed, and the polynomials $P_{ij}(s)$ are interval polynomials. The uncertainty set is therefore a box $\mathbf{\Pi}$ in the space of these coefficients. The family (IX.1) is of this type represents the special case of (IX.2) when products of uncertain polynomials do not occur. Characteristic polynomials of the form (IX.2) always occur in control systems containing several interconnected subsystems with uncertain parameters. We remark here that the vector of uncertain parameters, namely the set of coefficients of the polynomials $P_{ij}(s)$, enters into the characteristic polynomial coefficients affine multilinearly. Since we assume that these parameters vary within prescribed intervals we refer to the family (IX.2) as a *multilinearly parametrized interval family.* This form of the characteristic polynomial occurs in state space descriptions with interval matrices and also in matrix fraction description of multivariable systems when the matrix factors contain interval polynomials.

We first introduce some notation. For any positive integer n let $\underline{n}$ denote the set of integers $\{1, 2, \cdots n\}$. We consider the family of polynomials

$$\begin{aligned} \delta(s) &= F_1(s)P_{11}(s)P_{12}(s)\cdots P_{1r_1}(s) + \cdots\cdots \\ &\quad + F_m(s)P_{m1}(s)P_{m2}(s)\cdots P_{mr_m}(s) \end{aligned} \tag{IX.3}$$

where $F_i(s)$ are fixed, and the polynomials $P_{ij}(s)$ are interval with $i \in \underline{m}, j \in \underline{r}_i$. Let $d^o(P_{ij})$ denote the degree of $P_{ij}(s)$ and let p_{ij}^l denote the coefficient of s^l in $P_{ij}(s)$:

$$P_{ij}(s) = p_{ij}^0 + p_{ij}^1 s + \cdots + p_{ij}^l s^l + \cdots + p_{ij}^{d^0(P_{ij})} s^{d^0(P_{ij})}. \tag{IX.4}$$

Write $\underline{p}_{ij} := (p_{ij}^0, p_{ij}^1, \cdots p_{ij}^{d^0(P_{ij})})$ for each $i \in \underline{m}, j \in \underline{r}_i$ and let

$$\mathbf{p} := [\underline{p}_{11}, \underline{p}_{12} \cdots\cdots \underline{p}_{mr_m}] \tag{IX.5}$$

denote the vector of coefficients of the polynomials $P_{ij}(s)$, $i \in \underline{m}, j \in \underline{r}_i$. Each such coefficient belongs to a given interval:

$$p_{ij}^l \in [\alpha_{ij}^l, \beta_{ij}^l] \quad l = 0, \cdots, d^o(P_{ij}), i \in \underline{m}, j \in \underline{r}_i. \tag{IX.6}$$

In the space of these coefficients we have the box $\mathbf{\Pi}$ of uncertain parameters:

$$\mathbf{\Pi} := \{\mathbf{p} | p_{ij}^l \in [\alpha_{ij}^i, \beta_{ij}^l], l = 0, \cdots, d^o(P_{ij}), i \in \underline{m}, j \in \underline{r}_i\}. \tag{IX.7}$$

Each point $\mathbf{p} \in \mathbf{\Pi}$ corresponds to a particular choice of the ordered set of polynomials $P_{ij}(s), i \in \underline{m}, j \in \underline{r}_i$. We write

$$\begin{aligned} \delta(s, \mathbf{p}) &= F_1(s)P_{11}(s)P_{12}(s)\cdots P_{1r_1}(s) + \cdots\cdots \\ &\quad + F_m(s)P_{m1}(s)P_{m2}(s)\cdots P_{mr_m}(s) \end{aligned} \tag{IX.8}$$

to display the explicit dependence of $\delta(s)$ on $\mathbf{p}$. For a given fixed set of polynomials $[F_1(s), F_2(s)\cdots F_m(s)] := \underline{F}$ let $\mathbf{\Delta}$ denote the family of polynomials generated by the map $\underline{F} : \mathbf{\Pi} \Longrightarrow \delta$ as in (IX.3) and obtained by letting the parameter vector $\mathbf{p}$, (equivalently, the polynomials $P_{ij}(s)$), range over the box $\mathbf{\Pi}$ described in (IX.7). In other words

$$\mathbf{\Delta} := \{\delta(s, \mathbf{p}) | \mathbf{p} \in \mathbf{\Pi}\}. \tag{IX.9}$$

The four Kharitonov polynomials associated with the family of interval polynomials corresponding to $P_{ij}(s)$ are

$$\begin{aligned} K_{ij}^1(s) &= \alpha_{ij}^0 + \alpha_{ij}^1 s + \beta_{ij}^2 s^2 + \beta_{ij}^3 s^3 + \cdots \\ K_{ij}^2(s) &= \alpha_{ij}^0 + \beta_{ij}^1 s + \beta_{ij}^2 s^2 + \alpha_{ij}^3 s^3 + \cdots \\ K_{ij}^3(s) &= \beta_{ij}^0 + \alpha_{ij}^1 s + \alpha_{ij}^2 s^2 + \beta_{ij}^3 s^3 + \cdots \\ K_{ij}^4(s) &= \beta_{ij}^0 + \beta_{ij}^1 s + \alpha_{ij}^2 s^2 + \alpha_{ij}^3 s^3 + \cdots, \end{aligned}$$

and these definitions hold for each $i \in \underline{m}, j \in \underline{r_i}$. Corresponding to each $P_{ij}(s)$ we define the four polynomial segments

$$S_{ij}^1 := [K_{ij}^1(s), K_{ij}^2(s)], \qquad S_{ij}^2 := [K_{ij}^1(s), K_{ij}^3(s)]$$
$$S_{ij}^3 := [K_{ij}^2(s), K_{ij}^4(s)], \qquad S_{ij}^4 := [K_{ij}^3(s), K_{ij}^4(s)],$$

which we call the *Kharitonov segments.* These segments were introduced originally in [2]. A typical element of the segment S_{ij}^1, for example, is a polynomial, denoted by $S_{ij}^1(\lambda, s)$ which is a convex combination of the form called the *Kharitonov segments*

$$(1-\lambda)K_{ij}^1(s) + \lambda K_{ij}^2(s) := S_{ij}^1(\lambda, s), \qquad \lambda \in [0,1].$$

We now need to introduce some special subsets of $\mathbf{\Pi}$ called the *CB manifolds.* Fixing $i = l$ we let $\mathbf{\Pi}_l \subset \mathbf{\Pi}$ denote the union of all the r_l dimensional linear manifolds obtained by letting $P_{ij}(s), i \neq l$ range over the corresponding Kharitonov polynomials $K_{ij}^k(s), k \in \underline{4}, j \in \underline{r_i}, i \in \underline{m}$, and $P_{lj}(s)$, range over the Kharitonov segments $S_{lj}^k, k \in \underline{4}, j \in \underline{r_l}$. The CB manifolds $\mathbf{\Pi}_l, l \in \underline{m}$ map into the corresponding multilinear surfaces $\mathbf{\Delta}_l \subset \mathbf{\Delta}, l \in \underline{m}$ in the space of coefficients of the polynomials $\delta(s)$, under the previously defined mapping. More concretely,

$$\mathbf{\Pi}_l :=$$
$$\{[K_{11}^{i(1,1)}(s)K_{12}^{i(1,2)}(s)\cdots K_{1r_1}^{i(1,r_1)}(s), K_{21}^{i(2,1)}\cdots K_{2r_2}^{i(2,r_2)}(s), \cdots,$$
$$S_{l1}^{i(l,1)}(\lambda_1, s)S_{l2}^{i(l,2)}(\lambda_2, s)\cdots S_{lr_l}^{i(l,r_l)}(\lambda_{r_l}, s), \cdots\cdots,$$
$$K_{m1}^{i(m,1)}(s)K_{m2}^{i(m,2)}(s)\cdots K_{mr_m}^{i(m,r_m)}(s)] \mid$$
$$i(k,n) \in \underline{4}, k \in \underline{m}, n \in \underline{r_k}, \lambda_j \in [0,1], j \in \underline{r_l}.\}$$

and

$$\begin{aligned}\mathbf{\Delta}_l : &= \{\delta(s) = F_1(s).K_{11}^{i(1,1)}(s).K_{12}^{i(1,2)}(s).\cdots.K_{1r_1}^{i(1,r_1)}(s)\\ &+ \cdots + F_l(s).S_{l1}^{i(l,1)}(\lambda_1, s).S_{l2}^{i(l,2)}(\lambda_2, s).\cdots.S_{lr_l}^{i(l,r_l)}(\lambda_{r_l}, s) + \cdot\\ &\cdots + F_m(s).K_{m1}^{i(m,1)}(s).K_{m2}^{i(m,2)}(s)\cdots.K_{mr_m}^{i(m,r_m)}(s)\\ &\mid i(k,n) \in \underline{4}, k \in \underline{m}, n \in \underline{r_k}, \lambda_j \in [0,1], j \in \underline{r_l}.\}\end{aligned}$$

Equivalently

$$\mathbf{\Delta}_l := \{\delta(s, \mathbf{p}) | \mathbf{p} \in \mathbf{\Pi}_l\}.$$

Finally, let

$$\mathbf{\Pi}_{CB} := \bigcup_{l=1}^{m} \mathbf{\Pi}_l$$

denote the set of all linear *CB manifolds* and let the corresponding set of multilinear manifolds

$$\mathbf{\Delta}_{CB} := \bigcup_{l=1}^{m} \mathbf{\Delta}_l = \{\delta(s,\mathbf{p}) | \mathbf{p} \in \mathbf{\Pi}_{CB}\}.$$

Because there is a one to one correspondence between the elements of $\mathbf{\Pi}_{CB}$ and of $\mathbf{\Delta}_{CB}$ we refer to both sets as CB manifolds. Define the *Kharitonov vertices*, $\mathbf{K}(\mathbf{\Pi})$, of $\mathbf{\Pi}$ to be the subset of all vertices of $\mathbf{\Pi}$ corresponding to the Kharitonov polynomials of the $P_{ij}(s)$.

To illustrate the definition of the manifolds consider the special case of the family (IX.1). In this case each manifold in $\mathbf{\Pi}_1$ is the union of polynomial vectors of the form:

$$[S_{11}^i(\lambda_1, s), S_{12}^j(\lambda_2, s), K_{21}^k(s), K_{12}^l(s)], \tag{IX.10}$$

where $\lambda_t \in [0,1], t \in \underline{2}$ and (i,j,k,l) range over $(\underline{4} \times \underline{4} \times \underline{4} \times \underline{4})$. Similarly $\mathbf{\Pi}_2$ consists of the union of the polynomial vectors

$$[K_{11}^i(s), K_{12}^j(s), S_{21}^k(\lambda_1, s), S_{22}^l(\lambda_2, s)] \tag{IX.11}$$

where (i,j,k,l) range over $(\underline{4} \times \underline{4} \times \underline{4} \times \underline{4})$ and $\lambda_t \in [0,1], t \in \underline{2}$. The polynomial manifold $\mathbf{\Delta}_1$ consists of polynomials of the form:

$$F_1(s)S_{11}^i(\lambda_1, s)S_{12}^j(\lambda_2, s) + F_2(s)K_{21}^k(s)K_{12}^l(s), \lambda_t \in [0,1], t \in \underline{2}, \tag{IX.12}$$

and $\mathbf{\Delta}_1$ consists of the union of such manifolds obtained by letting (i, j, k, l) range over $(\underline{4} \times \underline{4} \times \underline{4} \times \underline{4})$. Similarly, the manifolds contained in $\mathbf{\Delta}_2$ are of the form:

$$F_1(s)K_{11}^i(s)K_{12}^j(s) + F_2(s)S_{21}^k(\lambda_1, s)S_{22}^l(\lambda_2, s), \lambda_t \in [0,1], t \in \underline{2}, \tag{IX.13}$$

where (i,j,k,l) range over $(\underline{4} \times \underline{4} \times \underline{4} \times \underline{4})$.

It is not difficult to see that the number of distinct manifolds in $\mathbf{\Pi}_l$ in the most general case when all the Kharitonov polynomials associated with each polynomial $P_{ij}(s)$ are distinct, is $4^{(r_1+r_2+\cdots+r_m)}$. Since this holds true for each $l \in \underline{m}$ the total number of CB manifolds in $\mathbf{\Pi}_{CB}$ or $\mathbf{\Delta}_{CB}$ is $m4^R, R = r_1 + r_2 + \cdots r_m$. With these preliminaries we are ready to state the main result of the next section.

X. STABILITY OF MULTILINEAR MANIFOLDS

A. The Multilinear CB Theorem

In this section we give necessary and sufficient conditions for the Hurwitz stability of the family (IX.2). Using the notation introduced in the last section we shall say that $\underline{F}$ stabilizes the family $\mathbf{\Pi}$ if and only if each polynomial of the family $\boldsymbol{\Delta}$ is Hurwitz stable. Similarly we shall say that $\underline{F}$ stabilizes $\boldsymbol{\Pi}_{CB}$ if and only if every polynomial in $\boldsymbol{\Delta}_{CB}$ is Hurwitz stable.

Theorem X.1. (Multilinear CB Theorem) *$\underline{F}$ stabilizes $\boldsymbol{\Pi}$ if and only if $\underline{F}$ stabilizes $\boldsymbol{\Pi}_{CB}$.*

The proof of this theorem is based on induction and may be found in [11].

Remark X.1. The assumption of independence of the perturbations can be easily relaxed. The reader is is referred to [11] for the details.

B. Simple Determination of Stability of Two Dimensional Multilinear Manifolds

In this section we consider the problem of checking the stability of a CB manifold of dimension 2. This case will arise when $r_i = 2$ in (IX.2) and is interesting because it can be solved analytically. Consider therefore the following two dimensional manifold

$$\begin{aligned} \delta(\lambda_1, \lambda_2, s) &= p_0(s)\lambda_1\lambda_2 + p_1(s)\lambda_1 + p_2(s)\lambda_2 + p_3(s), \\ & \lambda_i \in [0,1], i = 1,2. \end{aligned} \tag{X.1}$$

Assuming the four vertices are stable, the manifold $\delta(\lambda_1, \lambda_2, s)$ is unstable if and only if it has a $j\omega$ root for a set of real values $(\lambda_1, \lambda_2) \in [0,1]\mathrm{x}[0,1]$. To test for this we separate (X.1) into real and imaginary parts after substituting $s = j\omega$, and set them equal to zero. This gives

$$p_{0r}(\omega)\lambda_1\lambda_2 + p_{1r}(\omega)\lambda_1 + p_{2r}(\omega)\lambda_2 + p_{3r}(\omega) = 0 \tag{X.2}$$
$$p_{0i}(\omega)\lambda_1\lambda_2 + p_{1i}(\omega)\lambda_1 + p_{2i}(\omega)\lambda_2 + p_{3i}(\omega) = 0 \tag{X.3}$$

where

$$p_k(s)|_{s=j\omega} := p_{kr}(\omega) + jp_{ki}(\omega), \qquad \text{for } k = 0,1,2,3.$$

From (X.2), we have

$$[p_{0r}(\omega)\lambda_2 + p_{1r}(\omega)]\lambda_1 + p_{2r}(\omega)\lambda_2 + p_{3r}(\omega) = 0 \tag{X.4}$$

and

$$\lambda_1 = -\frac{p_{2r}(\omega)\lambda_2 + p_{3r}(\omega)}{p_{0r}(\omega)\lambda_2 + p_{1r}(\omega)}. \tag{X.5}$$

Similarly, from (X.3),

$$\lambda_1 = -\frac{p_{2i}(\omega)\lambda_2 + p_{3i}(\omega)}{p_{0i}(\omega)\lambda_2 + p_{1i}(\omega)}. \tag{X.6}$$

Since $\lambda_1 = \infty \notin [0,1]$ we can without loss of generality, deal only with the case in which the denominators of (X.5) and (X.6) are nonzero. By equating (X.5) and (X.6) we have,

$$[p_{2i}(\omega)p_{0r}(\omega) - p_{0i}(\omega)p_{2r}(\omega)]\lambda_2^2$$
$$+[p_{3i}(\omega)p_{0r}(\omega) + p_{2i}(\omega)p_{1r}(\omega) - p_{0i}(\omega)p_{3r}(\omega) - p_{1i}(\omega)p_{2r}(\omega)]\lambda_2$$
$$+ [p_{3i}(\omega)p_{1r}(\omega) - p_{1i}(\omega)p_{3r}(\omega)] = 0 \tag{X.7}$$

From (X.7) and (X.5) we can solve for $\lambda_1(\omega)$ and $\lambda_2(\omega)$ and verify if $(\lambda_1(\omega), \lambda_2(\omega))$ intersects the set $[0,1]$x$[0,1]$ for some ω. If the intersection is empty the manifold is Hurwitz stable otherwise it is unstable.

XI. EXTREMAL PARAMETRIC STABILITY PROPERTY: MULTILINEAR CASE

We now consider the family $\mathbf{\Delta}$ and the manifolds $\mathbf{\Pi}_{CB}$ and $\mathbf{\Delta}_{CB}$ which occur in the Multilinear CB Theorem of the last section. As before let

$$\mathbf{p} := [\underline{p}_{11}, \underline{p}_{12} \cdots\cdots \underline{p}_{mr_m}]$$

denote the n dimensional parameter vector consisting of the ordered set of coefficients of the polynomials $P_{ij}(s)$ and let $\mathbf{p} \in R^n$ vary in the prescribed box $\mathbf{\Pi}$ specified by the given upper and lower bounds:

$$p_{ij}^l \in [\alpha_{ij}^l, \beta_{ij}^l] \quad l = 0, \cdots, d^o(P_{ij}), i \in \underline{m}, j \in \underline{r}_i$$

Let $\|\cdot\|$ denote any norm in R^n and let $\mathcal{P}_u$ denote the set of points $\mathbf{u}$ in R^n for which $\delta_(s, \mathbf{u})$ is unstable or loses degree (relative to its generic degree over $\mathbf{\Pi}$. Let

$$\rho(\mathbf{p}) = \inf_{\mathbf{u} \in \mathcal{P}_u} \|\mathbf{p} - \mathbf{u}\|_p$$

denote the radius of the stability ball (measured in the norm $\|\cdot\|$) and centered at the point $\mathbf{p}$. This number serves as the stability margin associated with the point $\mathbf{p}$. If the box $\mathbf{\Pi}$ is stable we can associate a stability margin with each point in $\mathbf{\Pi}$. A natural question to ask then is: Is there a point in $\mathbf{\Pi}$ which is closest to instability in the norm $\|\cdot\|$ and where is it? The answer to that question is provided in the following theorem.

As before we define a mapping from $\mathbf{\Pi}$ to the set of all positive real numbers:

$$\begin{array}{ccc} \mathbf{\Pi} & \xrightarrow{\rho} & \mathcal{R}^+ \backslash \{0\} \\ \mathbf{p} & \longrightarrow & \rho(\mathbf{p}) \end{array}$$

Our question stated in terms of functions is: Has the function $\rho(\mathbf{p})$ a minimum and is there a precise point in $\mathbf{\Pi}$ where it is reached?

Theorem XI.1. (Extremal property of the stability manifolds) *The function*

$$\begin{array}{ccc} \mathbf{\Pi} & \xrightarrow{\rho} & \mathcal{R}^+\backslash\{0\} \\ \mathbf{p} & \longrightarrow & \rho(\mathbf{p}) \end{array}$$

has a minimum which is reached at a point on the CB manifolds $\mathbf{\Pi}_{CB}$.

The proof of this theorem may be found in [11] and omitted here.

XII. PARAMETRIC AND UNSTRUCTURED PERTURBATIONS: MULTILINEAR CASE

In this section we will analyze the problem of robust stability in the presence of both parameter variations and unstructured perturbations modelled in the usual way as norm bounded perturbations. The subject of robust stability under mixed types of perturbations is of current interest (see for example [19], [20], [12], [13] and [21]). We model this situation by considering a multilinear interval plant, namely one whose transfer function is a ratio of polynomials of the type that was introduced in IX.2. To be specific we will consider single-input, single-output, proper, stable systems with transfer function of the form

$$g(s) = \frac{\gamma(s)}{\delta(s)}$$

Here

$$\begin{aligned} \gamma(s) \;&=\; H_1(s)L_{11}(s)L_{12}(s)\cdots L_{1r_1}(s) + \cdots\cdots \\ &\quad + H_m(s)L_{m1}(s)L_{m2}(s)\cdots L_{mr_m}(s) \end{aligned}$$

where the polynomials $H_i(s)$ are fixed and the polynomials $L_{ij}(s)$ are interval polynomials, that is their coefficients vary in a prescribed box $\mathbf{\Lambda}$; the corresponding family of polynomials $\gamma(s)$ is denoted by $\mathbf{\Gamma}$. We suppose as before that

$$\begin{aligned} \delta(s) \;&=\; F_1(s)P_{11}(s)P_{12}(s)\cdots P_{1r_1}(s) + \cdots\cdots \\ &\quad + F_m(s)P_{m1}(s)P_{m2}(s)\cdots P_{mr_m}(s) \end{aligned}$$

where the polynomials $F_i(s)$ are fixed, the polynomials $P_{ij}(s)$ are interval polynomials, with coefficients that vary in the prescribed box $\mathbf{\Pi}$ and the resulting family of polynomials $\delta(s)$ is denoted $\mathbf{\Delta}$. As in Section 2 we let $\mathbf{p}$ denote the vector of coefficients of the polynomials $\{P_{ij}(s)\}$ and we similarly let $\mathbf{l}$ denote the vector of coefficients $\{L_{ij}(s)\}$. We also denote explicitly, the dependence of $\delta(s)$ on $\mathbf{p}$ and of $\nu(s)$ on $\mathbf{l}$ by writing $\delta(s, \mathbf{p})$ and $\nu(s, \mathbf{l})$ whenever necessary. It is assumed that the parameters $\mathbf{p}$ and l perturb independently. From these polynomial families we form the parametrized family of transfer functions

$$\mathcal{G} = \{\frac{\gamma(s,\mathbf{l})}{\delta(s,\mathbf{p})}|\mathbf{p} \in \mathbf{\Pi}, \text{ and } \mathbf{l} \in \mathbf{\Lambda}\}. \quad \text{(XII.1)}$$

To display the dependence of a typical element $g(s)$ of $\mathcal{G}$ on $\mathbf{l}$ and $\mathbf{p}$ we write $g(s,\mathbf{p},\mathbf{l})$. Introduce the Kharitonov polynomials and segments associated respectively with the $P_{ij}(s)$ and $L_{ij}(s)$ respectively. As in Section 2 these are used to generate the extremal subsets $\mathbf{\Pi}_{CB}$ of $\mathbf{\Pi}$ and $\mathbf{\Lambda}_{CB}$ of $\mathbf{\Lambda}$ respectively. The Kharitonov extreme points of $\mathbf{\Pi}$ and $\mathbf{\Lambda}$ are denoted respectively by $\mathbf{K}(\mathbf{\Pi})$ and $\mathbf{K}(\mathbf{\Lambda})$. Finally, we denote the polynomial manifolds resulting from $\mathbf{K}(\mathbf{\Pi})$, $\mathbf{K}(\mathbf{\Lambda})$, $\mathbf{\Lambda}_{CB}$ and $\mathbf{\Pi}_{CB}$ as follows:

$$\mathbf{\Gamma}_{CB} = \{\gamma(s,\mathbf{l})|\mathbf{l} \in \mathbf{\Lambda}_{CB}\}, \mathbf{\Gamma}_K = \{\gamma(s,\mathbf{l})|\mathbf{l} \in \mathbf{K}(\mathbf{\Lambda})\}$$

$$\mathbf{\Delta}_{CB} = \{\delta(s,\mathbf{p})|\mathbf{p} \in \mathbf{\Pi}_{CB}\}, \mathbf{\Delta}_K = \{\delta(s,\mathbf{p})|\mathbf{p} \in \mathbf{K}(\mathbf{\Pi})\}.$$

The main results in this section will deal with the calculation of the H_∞ stability margin for systems containing parameter uncertainty as defined above. In the following we will use the standard notation: $\mathbf{C}_+ := \{s \in C : \ Re(s) \geq 0\}$, and $H_\infty(\mathbf{C}_+)$ will represent the space of functions $f(s)$ that are bounded and analytic in $\mathbf{C}_+$ with the standard H_∞ norm,

$$||f||_\infty = \sup_{\omega \in R} |f(j\omega)|.$$

To determine the unstructured stability margin of the family $\mathcal{G}$ we need to determine the supremum of the H_∞ norm of certain transfer functions over $\mathcal{G}$. Specifically we formulate the following problems: Let $W(s)$ be a scalar stable weight, with a stable inverse, and write $W(s) = \frac{n_w(s)}{d_w(s)}$.

A) Consider the feedback configuration shown in Figure 12, $\mathcal{G}$ is a stable family, and ΔP is any H_∞ perturbation that satisfies $||\Delta P|| < \alpha$.

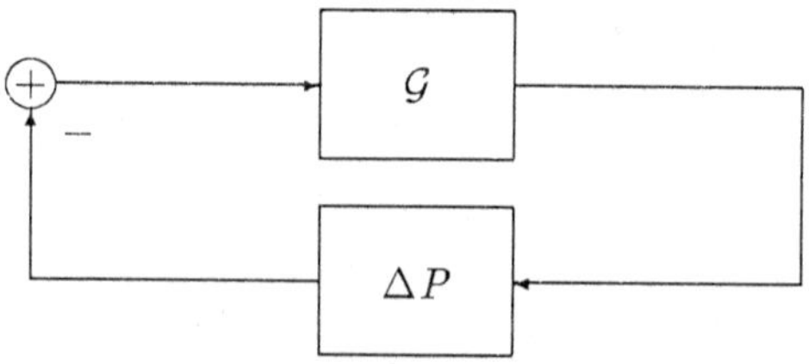

Figure 12. Multiplicative Perturbations

B) Consider the feedback configuration shown in Figure 13, ΔP is any H_∞ perturbation that satisfies $||\Delta P|| < \alpha$, and C is a controller that simultaneously stabilizes every element in the set $\mathcal{G}$.

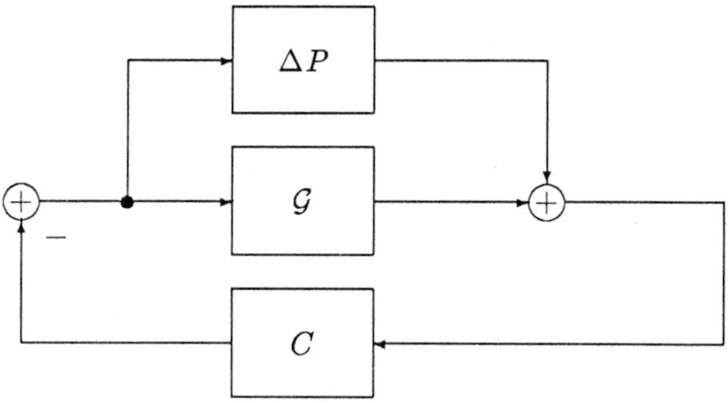

Figure 13. Additive Perturbations

The above problems are generalized versions of standard H_∞ robust stability problems (see [22]) where a fixed plant is considered. The solution is accomplished once again by showing that the H_∞ norms in question attain their supremum value over a certain extremal set of transfer functions $\mathcal{G}_{CB} \subset \mathcal{G}$. This set is defined as follows:

$$\mathcal{G}_{CB} := \{\frac{\gamma(s,\mathbf{l})}{\delta(s,\mathbf{p})} | (\mathbf{l} \in \mathbf{K}(\Lambda), \mathbf{p} \in \Pi_{CB}) \text{ or } (\mathbf{l} \in \Lambda_{CB}, \mathbf{p} \in \mathbf{K}(\Pi)\}.$$

We can now state the main result of this Section.

Theorem XII.1. (Extremal properties)

A) $$\sup_{g\in\mathcal{G}} ||Wg||_\infty = \sup_{g\in\mathcal{G}_{CB}} ||Wg||_\infty,$$

B) $$\sup_{g\in\mathcal{G}} ||WC(1+gC)^{-1}||_\infty = \sup_{g\in\mathcal{G}_{CB}} ||WC(1+gC)^{-1}||_\infty.$$

Corollary XII.1. (Unstructured Margins)

1) The configuration of Figure 12 will be stable if and only if α *satisfies*

$$\alpha \leq \frac{1}{\sup_{g\in\mathcal{G}_{CB}} ||g||_\infty} := \alpha_o^*.$$

2) The configuration of Figure 13 will be stable if and only if α *satisfies*

$$\alpha \leq \frac{1}{\sup_{g\in\mathcal{G}_{CB}} ||C(1+gC)^{-1}||_\infty} := \alpha_c^*.$$

The proof of this theorem is similar to that used in [12] and details are omitted here. The idea behind this approach is to replace the question of finding an upper bound of the H_∞ norm of a transfer function by an equivalent question concerning the stability of a certain parametrized family of polynomials, for which the results of the previous sections apply. For this purpose we need the following lemma [12] which gives a characterization of proper rational functions $g(s)$ which are in $H_\infty(\mathbf{C}_+)$ and which satisfy $||g||_\infty < 1$.

Lemma XII.1. *Let $h(s) = n(s)/d(s)$ be a proper (real or complex) rational function in $H_\infty(\mathbf{C}_+)$, with $deg(d(s)) = q$, then $||h||_\infty < 1$ if and only if*

a1) $|n_q| < |d_q|$,

b1)$d(s) + e^{j\theta} n(s)$ *is Hurwitz for all* θ *in* $[0, 2\pi)$.

Remark XII.1. The quantities α_o^* and α_c^* serve as unstructured H_∞ stability margins for the respective open and closed loop parametrized systems treated in Problems I and II.

XIII. PARAMETRIC AND NONLINEAR PERTURBATIONS: MULTILINEAR CASE

Another stability robustness problem that involves structured and unstructured perturbations is the classical Lur'e problem of nonlinear control theory. This problem considers a fixed linear time invariant system subjected to perturbations in the form of nonlinear feedback gains contained in a prescribed sector. In [13] a robust version of the Lur'e problem was treated. In this formulation the fixed linear system is replaced by the more realistic model of a parametrized family of plants. The "nonlinear stability margin" of the system can be determined by finding the infimum, over the parametrized family, of such stability sectors. From standard results on the Lur'e problem, the size of such a sector can be determined by finding finding the infimum of the real part of $g(j\omega)$ as g ranges over the parametrized family. In [13] it was shown how the strict positive real (SPR) property for a stable family of interval systems can be determined from a set of sixteen plants called the *Kharitonov systems.* In addition, in the presence of a fixed controller that stabilizes an entire family of interval systems, the SPR property for the family of transfer functions $C(1+gC)^{-1}$ is determined from a set of 32 one parameter family of systems. Here we consider the more general situation where the parametrized family considered is the family $\mathcal{G}$ defined in the previous section. Using the extremal properties established in the last section and the proof developed in [13], it is possible to establish the following theorem. The proof is omitted as it is very similar to that of the last section.

Theorem XIII.1. (Extremal properties)

1) *Let $\mathcal{G}$ be the multilinear family defined above, and assume that $\mathcal{G}$ is stable then*

$$\inf_{g\in\mathcal{G}} \inf_{\omega\in R} Re(W(j\omega)g(j\omega)) = \inf_{g\in\mathcal{G}_{CB}} \inf_{\omega\in R} Re(W(j\omega)g(j\omega)).$$

2) *If C is a controller that stabilizes the entire family $\mathcal{G}$, then*

$$\inf_{g\in\mathcal{G}} \inf_{\omega\in R} Re(WC(1+gC)^{-1}(j\omega)) =$$

$$\inf_{g\in\mathcal{G}_{CB}} \inf_{\omega\in R} Re(WC(1+gC)^{-1}(j\omega)).$$

XIV. CONCLUDING REMARKS

The summary of results presented here form the beginnings of a complete theory of interval control systems. We expect such a theory to develop over the next few years. We expect such a theory to impact on the design of control systems filters and communication systems.

XV. REFERENCES

1. V. L. Kharitonov, "Asymptotic stability of an equilibrium position of a family of systems of linear differential equations," *Differential Uravnen*, vol. 14, pp. 2086 – 2088, 1978.
2. H. Chapellat and S. P. Bhattacharyya, "A generalization of Kharitonov's theorem: robust stability of interval plants," *IEEE Transactions on Automatic Control*, vol. AC - 34, pp. 306 – 311, March 1989.
3. S. P. Bhattacharyya, "Robust parametric stability: the role of the CB segments," in *Control of Uncertain Dynamic Systems*, (L. H. Keel and S. P. Bhattacharyya, eds.), Littleton, MA: CRC Press, September 1991.
4. A. C. Bartlett, C. V. Hollot, and H. Lin, "Root location of an entire polytope of polynomials: it suffices to check the edges," *Mathematics of Controls, Signals and Systems*, vol. 1, pp. 61 – 71, 1988.
5. H. Chapellat and S. P. Bhattacharyya, "An alternative proof of Kharitonov's theorem," *IEEE Transactions on Automatic Control*, vol. AC - 34, pp. 448 – 450, April 1989.
6. I. R. Peterson, "A new extension to Kharitonov's theorem," in *Proceedings of IEEE Conference on Decision and Control*, December 1987. Los Angeles, CA.

7. C. V. Hollot and F. Yang, "Robust stabilization of interval plants using lead or lag compensators," in *Proceedings of IEEE Conference on Decision and Control*, December 1989. Tampa, FL.

8. M. Mansour and F. J. Kraus, "Argument conditions for Hurwitz and Schur stable polynomials and the robust stability problem," Tech. Rep., ETH, Zurich, Tech. Report, 1990.

9. S. P. Bhattacharyya, "Vertex results in robust stability," Tech. Rep., TCSP Report, Texas A&M University, May 1991.

10. H. Chapellat and S. P. Bhattacharyya, "Calculation of maximal stability domains using an optimal property of Kharitonov polynomials," in *Analysis and Optimization of Systems, Lecture Notes in Control and Information Sciences*, pp. 22 – 31, Springer - Verlag, 1988.

11. H. Chapellat, M. Dahleh, and S. P. Bhattacharyya, "Extremal manifolds in robust stability," Tech. Rep., TCSP Report, Texas A&M University, July 1990.

12. H. Chapellat, M. Dahleh, and S. P. Bhattacharyya, "Robust stability under structured and unstructured perturbations," *IEEE Transactions on Automatic Control*, vol. AC - 35, pp. 1100 – 1108, October 1990.

13. H. Chapellat, M. Dahleh, and S. P. Bhattacharyya, "On robust nonlinear stability of interval control systems," *IEEE Transactions on Automatic Control*, vol. AC - 36, pp. 59 – 67, January 1991.

14. L. H. Keel, J. Shaw, and S. P. Bhattacharyya, "Robust control of interval systems," in *Proceedings of 1991 Robust Control Workshop*, Tokyo, Japan: Springer-Verlag, June 1991.

15. A. Tesi and A. Vicino, "Kharitonov segments suffice for frequency response analysis of plant-controller families," in *Control of Uncertain Dynamic Systems*, (L. H. Keel and S. P. Bhattacharyya, eds.), Littleton, MA: CRC Press, September 1991.

16. L. H. Keel, J. Shaw, and S. P. Bhattacharyya, "Robust control under parameter variations," Tech. Rep., ISE Tech. Report, Tennessee State University, July 1991. Also in TCSP Tech. Report, Texas A&M University, July, 1991.

17. L. H. Keel and S. P. Bhattacharyya, "Frequency domain design of interval controllers," in *Control of Uncertain Dynamic Systems*, (L. H. Keel and S. P. Bhattacharyya, eds.), Littleton, MA: CRC Press, September 1991.

18. Ya. Z. Tsypkin and B. T. Polyak, "Frequency domain criteria for ℓ_p-robust stability of continuous linear systems,". To appear.

19. M. K. H. Fan and A. L. Tits, "Robustness in the presence of joint parametric uncertainty and unmodeled dynamics," in *Proceedings of 1988 American Control Conference*, June 1988. Altanta, GA.

20. D. Hinrichsen and A. J. Prichard, "Robustness of stability of linear state space systems with respect to time-varying, nonlinear and dynamic perturbations," in *Proceedings of the 28th IEEE Conference on Decision and Control*, pp. 52 – 53, December 1989. Tampa, FL.

21. M. K. H. Fan, A. L. Tits, and J. C. Doyle, "Robustness in the presence of mixed parametric uncertainty and unmodeled dynamics," *IEEE Transactions on Automatic Control*, vol. AC - 36, pp. 25 – 38, January 1991.

22. B. Francis, *A Course in H^∞ Control Thoery, Lecture Notes in Control and Information Sciences.* Springer - Verlag, 1987.

Robust Stability in Systems Described by Rational Functions

Mohamed Mansour

Automatic Control Laboratory
Swiss Federal Institute of Technology Zurich, Switzerland

I. INTRODUCTION

In this chapter the robust stability problem of linear continuous and discrete systems is dealt with in a unified manner beginning with principle of the argument as the source of all derivations. From the principle of the argument, criteria for polynomial stability in the frequency domain are derived from which Hermite-Bieler theorems for continuous and discrete systems are obtained. Monotony of the argument of different functions is a direct consequence of Hermite-Bieler theorems.

It was shown in [1], [2], [3], [4] that the different stability criteria of continuous and discrete systems such as Routh-Hurwitz, continuous Schur-Cohn, discrete Routh-Hurwitz and Schur-Cohn can be obtained directly from Hermite-Bieler theorems. From the polynomial stability in the frequency domain one can get directly a robust stability criterion in the frequency domain using the boundary crossing theorem [2] which makes use of the continuity of the roots of a polynomial as a function of its coefficients. An edge theorem, stability of edges and extreme point results are obtained in a simple way. The robust stability of special polynomials for continuous and discrete systems is obtained using the robust stability in the frequency domain or using the Hermite-Bieler theorems. The robust stability of con-

CONTROL AND DYNAMIC SYSTEMS, VOL. 51

trol systems can also be obtained in the frequency domain in the same way as the previous results.

II. PRINCIPLE OF THE ARGUMENT AND STABILITY

A. Principle of the argument

Consider a function $f(z)$ which is regular inside a closed contour C and is not zero at any point on the contour and let $\Delta_C\{\arg f(z)\}$ denote the variation of $\arg f(z)$ round the contour C, then

Theorem 1 [5]

$$\Delta_C\{\arg f(z)\} = 2\pi \mathrm{N} \tag{1}$$

where N *is the number of zeros of* $f(z)$ *inside* C.

B. Polynomial stability in the frequency domain

Here we apply the principle of the argument to the Hurwitz and Schur stability of polynomials. Nyquist criterion for the Hurwitz and Schur stability of control system is obtained in the same way.

1. Continuous systems

Consider the characteristic polynomial of a continuous system

$$f(s) = a_0 s^n + a_1 s^{n-1} + \ldots + a_{n-1} s + a_n \tag{2}$$

Theorem 2 *$f(s)$ has all its roots in the left half of the s-plane if and only if $f(j\omega)$ has a change of argument of $n\frac{\pi}{2}$ when ω changes from 0 to ∞.*

The proof of this criterion which is known in the literature as Cremer-Leonhard-Michailov criterion is based on the fact that a negative real root contributes an angle of $\frac{\pi}{2}$ to the change of argument, and a pair of complex roots in the left half plane contributes an angle of π to the change of argument as ω changes from 0 to ∞. It is easy to show geometrically that the argument is monotonically increasing.

$$f(s) \text{ can be written as } f(s) = h(s^2) + s g(s^2) \tag{3}$$

where $h(s^2)$ and $sg(s^2)$ are the even and odd parts of $f(s)$ respectively.

$$f(j\omega) = h(-\omega^2) + j\omega g(-\omega^2) \tag{4}$$

2. Discrete systems

Consider the characteristic polynomial of a discrete system

$$f(z) = a_0 z^n + a_1 z^{n-1} + \ldots + a_{n-1} z + a_n \tag{5}$$

Theorem 3 *$f(z)$ has all its roots inside the unit circle of the z-plane if and only if $f(e^{j\theta})$ has a change of argument of $n\pi$ when θ changes from 0 to π.*

The proof of this criterion is based on the fact that a real root inside the unit circle contributes an angle of π to the change of argument, and a pair of complex roots inside the unit circle contributes an angle of 2π to the change of argument as θ changes from 0 to π. It is easy to show geometrically that the argument is monotonically increasing

$$f(z) \text{ can be written as } f(z) = h(z) + g(z) \tag{6}$$

where

$$h(z) = \frac{1}{2}\left[f(z) + z^n f\left(\frac{1}{z}\right)\right] = \alpha_0 z^n + \alpha_1 z^{n-1} + \ldots + \alpha_1 z + \alpha_0 \tag{7}$$

and

$$g(z) = \frac{1}{2}\left[f(z) - z^n f\left(\frac{1}{z}\right)\right] = \beta_0 z^n + \beta_1 z^{n-1} + \ldots - \beta_1 z - \beta_0 \tag{8}$$

are the symmetric and antisymmetric parts of $f(z)$ respectively. Also

$$f(e^{j\theta}) = 2e^{jn\theta/2}[h^* + jg^*] \tag{9}$$

where for n even, $n = 2\nu$

$$\begin{aligned} h^*(\theta) &= \alpha_0 \cos\nu\theta + \alpha_1 \cos(\nu-1)\theta + \ldots + \alpha_{\nu-1}\cos\theta + \frac{\alpha_\nu}{2} \\ g^*(\theta) &= \beta_0 \sin\nu\theta + \beta_1 \sin(\nu-1)\theta + \ldots + \beta_{\nu-1}\sin\theta \end{aligned} \tag{10}$$

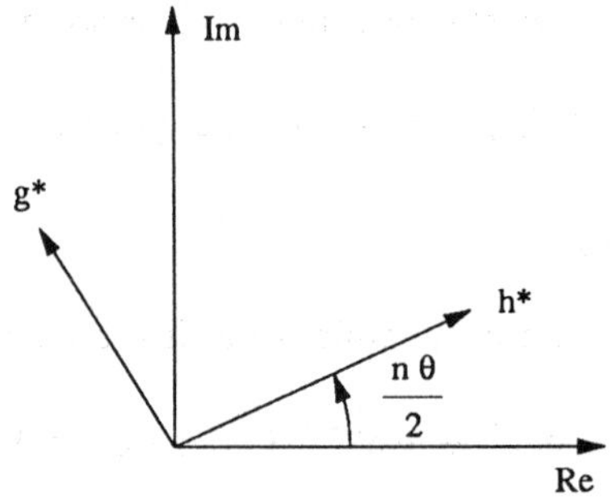

Fig. 1. Change of coordinates

and for n odd, i.e., $n = 2\nu - 1$, we get

$$
\begin{aligned}
h^*(\theta) &= \alpha_0 \cos\left(\nu - \frac{1}{2}\right)\theta + \alpha_1 \cos\left(\nu - \frac{3}{2}\right)\theta + ... + \alpha_{\nu-1}\cos\frac{\theta}{2} \\
g^*(\theta) &= \beta_0 \sin\left(\nu - \frac{1}{2}\right)\theta + \beta_1 \sin\left(\nu - \frac{3}{2}\right)\theta + ... + \beta_{\nu-1}\sin\frac{\theta}{2}
\end{aligned}
\tag{11}
$$

From theorem 3 and Eq. (9) we get

Theorem 4 *$f(z)$ has all its roots inside the unit circle of the z-plane if and only if $f^* = h^* + jg^*$ has a change of argument of $n\frac{\pi}{2}$ when θ changes from 0 to π.*

f^* is a complex function w.r.t. a rotating axis. It can be shown geometrically that the argument of f^* increases monotonically. Fig. 1 shows the change of coordinates.

C. Hermite-Bieler theorem

Here we derive Hermite-Bieler theorem and its equivalent for discrete systems. In [1], [2], [3], [4] Hermite-Bieler theorems were used to derive in a simple way the classical stability criteria like Routh-Hurwitz, continuous Schur-Cohn, discrete Routh-Hurwitz and Schur-Cohn criteria.

1. Continuous systems

Consider the polynomial in Eq. (3) and replace s^2 by λ.

Theorem 5 *(Hermite-Bieler Theorem) $f(s)$ has all its roots in the left half of the s-plane if and only if $h(\lambda)$ and $g(\lambda)$ have simple real negative alternating roots and $\frac{a_0}{a_1} > 0$. The first root next to zero is of $h(\lambda)$.*

The proof of this theorem follows directly from theorem 1; [2]. If theorem 5 is satisfied then $h(\lambda)$ and $g(\lambda)$ are called a positive pair.

2. Discrete systems

Consider the polynomial in Eq. (5).

Theorem 6 *(Discrete Hermite-Bieler Theorem) $f(z)$ has all its roots inside the unit circle of the z-plane if and only if $h(z)$ and $g(z)$ have simple alternating roots on the unit circle and $\left|\frac{a_n}{a_0}\right| < 1$; [6], [3].*

The proof can be sketched as follows:
Due to theorem 4, h^* and g^* and hence h and g will go through 0 alternatively n times as θ changes from 0 to π. The necessary condition $\left|\frac{a_n}{a_0}\right| < 1$ distinguishes between $f(z)$ and its inverse which has all roots outside the unit circle.

D. Monotony of the argument

In the following it is shown that the argument of a Hurwitz polynomial $f(s) = h(s^2)+sg(s^2)$ in the frequency domain $f(j\omega) = h(-\omega^2)+j\omega g(-\omega^2)$ as well as the argument of the modified functions $f_1^*(\omega) = h(-\omega^2)+jg(-\omega^2)$ and $f_2^*(\omega) = h(-\omega^2) + j\omega^2 g(-\omega^2)$ are monotonically increasing functions of ω, whereby ω varies between 0 and ∞. Similarly it is shown that given a Schur polynomial $f(z) = h(z) + g(z)$ where $h(z)$ and $g(z)$ are the symmetric and the antisymmetric parts of $f(z)$ respectively, the same monotony property can be obtained for an auxiliary function $f^*(\theta)$ defined by $f(e^{j\theta}) = 2e^{jn\theta/2} f^*(\theta)$. The argument of $f^*(\theta) = h^*(\theta) + jg^*(\theta)$ as well as the argument of the modified functions

$$f_3^*(\theta) = \frac{h^*(\theta)}{\cos\frac{\theta}{2}} + j\frac{g^*(\theta)}{\sin\frac{\theta}{2}}$$

and

$$f_4^*(\theta) = \frac{h^*(\theta)}{\sin\frac{\theta}{2}} + j\frac{g^*(\theta)}{\cos\frac{\theta}{2}}$$

are monotonically increasing functions of θ whereby θ varies between 0 and π.

These results were proved in [7] using mathematical induction. However, a simpler proof can be obtained using network-theoretical results. The proof given here is based on discussions with Prof. N. Bose and [8], [9]. The results obtained can be directly applied to the robust stability problem [7].

1. Continuous systems

Let

$$f(s) = s^n + a_1 s^{n-1} + a_2 s^{n-2} + \dots + a_{n-1}s + a_n = h(s^2) + sg(s^2) \quad (12)$$

where h and g are the even and odd parts of $f(s)$ respectively.

Theorem 7 *For a Hurwitz stable polynomial $f(s)$ the arguments of*

$$\begin{aligned} f(j\omega) &= h(-\omega^2) + j\omega g(-\omega^2) \\ f_1^*(\omega) &= h(-\omega^2) + jg(-\omega^2) \\ f_2^*(\omega) &= h(-\omega^2) + j\omega^2 g(-\omega^2) \end{aligned}$$

are monotonically increasing functions of ω whereby ω varies between 0 and ∞.

Proof: As shown earlier by Hermite-Bieler theorem, the roots of $h(s^2)$ and $g(s^2)$ for a stable polynomial are simple on the imaginary axis and interlacing. Therefore for n even

$$\frac{sg}{h} = \frac{sa_1(s^2+\omega_2^2)(s^2+\omega_4^2)\dots(s^2+\omega_{n-2}^2)}{(s^2+\omega_1^2)(s^2+\omega_3^2)\dots(s^2+\omega_{n-1}^2)} \quad (13)$$

where

$$\omega_1 < \omega_2 < \omega_3 < \dots < \omega_{n-1} \quad (14)$$

With partial fraction expansion we get

$$\frac{sg}{h} = sa_1 \sum_{i=1,3,\dots,n-1} \frac{K_i}{s^2+\omega_i^2} \quad (15)$$

where

$$K_i = \frac{\prod_{j=2,4,\ldots,n-2}(\omega_j^2 - \omega_i^2)}{\prod_{j=1,3,\ldots,n-1;j\neq i}(\omega_j^2 - \omega_i^2)} \tag{16}$$

K_i is positive because of Eq. (14).

For the monotony of the argument of $f(j\omega)$, $\frac{d}{d\omega}\frac{\omega g(-\omega^2)}{h(-\omega^2)}$ must be positive.

$$\begin{aligned} \frac{d}{d\omega}\frac{\omega g(-\omega^2)}{h(-\omega^2)} &= \frac{d}{d\omega}\omega\, a_1 \sum_{i=1,3,\ldots,n-1} \frac{K_i}{\omega_i^2 - \omega^2} \\ &= a_1 \sum_{i=1,3,\ldots,n-1} K_i \frac{\omega_i^2 + \omega^2}{(\omega_i^2 - \omega^2)^2} > 0 \end{aligned} \tag{17}$$

But

$$\frac{g(-\omega^2)}{h(-\omega^2)} = a_1 \sum_{i=1,3,\ldots,n-1} \frac{K_i}{(\omega_i^2 - \omega^2)} \tag{18}$$

From Eq. (17) and Eq. (18) we get the well known result in network theory [8].

$$\frac{d}{d\omega}\frac{\omega g(-\omega^2)}{h(-\omega^2)} > \left|\frac{g(-\omega^2)}{h(-\omega^2)}\right| \tag{19}$$

For n odd

$$\frac{sq}{h} = \frac{s(s^2+\omega_2^2)(s^2+\omega_4^2)\ldots(s^2+\omega_{n-1}^2)}{a_1(s^2+\omega_1^2)(s^2+\omega_3^2)\ldots(s^2+\omega_{n-2}^2)} \tag{20}$$

where Eq. (14) is valid.

$$\frac{sg}{h} = \frac{s}{a_1}\left[1 + \sum_{i=1,3,\ldots,n-2} \frac{K_i}{s^2+\omega_i^2}\right] \tag{21}$$

where

$$K_i = \frac{\prod_{j=2,4,\ldots,n-1}(\omega_j^2 - \omega_i^2)}{\prod_{j=1,3,\ldots,n-2;j\neq i}(\omega_j^2 - \omega_i^2)} \tag{22}$$

K_i is positive because of Eq. (14)

$$\begin{aligned} \frac{d}{d\omega}\frac{\omega g}{h} &= \frac{d}{d\omega}\frac{\omega}{a_1}\left[1 + \sum_{i=1,3,\ldots,n-2} \frac{K_i}{\omega_i^2 - \omega^2}\right] \\ &= \frac{1}{a_1}\left[1 + \sum_{i=1,3,\ldots,n-2} K_i \frac{\omega_i^2+\omega^2}{(\omega_i^2-\omega^2)^2}\right] > 0 \end{aligned} \tag{23}$$

But

$$\frac{g}{h} = \frac{1}{a_1}\left[1 + \sum_{i=1,3,\ldots,n-2} \frac{K_i}{\omega_i^2 - \omega^2}\right] \tag{24}$$

From Eq. (23) and Eq. (24) we get the same result in Eq. (19).
From Eq. (19)

$$\frac{d}{d\omega}\frac{\omega g}{h} = \omega \frac{d}{d\omega}\frac{g}{h} + \frac{g}{h} > \left|\frac{g}{h}\right|$$

Therefore

$$\omega \frac{d}{d\omega}\frac{g}{h} > \left|\frac{g}{h}\right| - \frac{g}{h} \geq 0$$

Hence

$$\frac{d}{d\omega}\frac{g}{h}0 \tag{25}$$

which means that the function

$$f_1^* = h(-\omega^2) + jg(-\omega^2)$$

has a monotonically increasing argument.
Similarly

$$\frac{d\omega^2 g}{d\omega} = \omega \frac{d}{d\omega}\frac{\omega g}{h} + \frac{\omega g}{h} = \omega\left(\frac{d}{d\omega}\frac{\omega g}{h} + \frac{g}{h}\right) > 0 \tag{26}$$

which means that the function

$$f_2^* = h(-\omega^2) + j\omega^2 g(-\omega^2)$$

has a monotonically increasing argument. □

2. Discrete systems

Let

$$f(z) = z^n + a_1 z^{n-1} + a_2 z^{n-2} + \ldots + a_n = h(z) + g(z) \tag{27}$$

where $h(z)$ and $g(z)$ are the symmetric and antisymmetric parts of $f(z)$ respectively. They are given by Eq. (7) and Eq. (8).

Theorem 8 *For a Schur stable polynomial $f(z)$ the arguments of*

$$\begin{aligned} f^*(\theta) &= h^*(\theta) + jg^*(\theta) \\ f_3^*(\theta) &= \frac{h^*(\theta)}{\cos\frac{\theta}{2}} + j\frac{g^*(\theta)}{\sin\frac{\theta}{2}} \\ f_4^*(\theta) &= \frac{h^*(\theta)}{\sin\frac{\theta}{2}} + j\frac{g^*(\theta)}{\cos\frac{\theta}{2}} \end{aligned}$$

are monotonically increasing functions of θ whereby θ varies between 0 and π.

Proof: From the discrete Hermite-Bieler theorem, the roots of $h(z)$ and $g(z)$ for a Schur-stable polynomial are simple on the unit circle and interlacing.

Therefore, for n even

$$\frac{g(z)}{h(z)} =$$

$$\frac{\beta_0}{\alpha_0}\frac{(z^2-1)(z^2-2\cos\theta_2 z+1)(z^2-2\cos\theta_4 z+1)...(z^2-2\cos\theta_{n-2}z+1)}{(z^2-2\cos\theta_1 z+1)(z^2-2\cos\theta_3 z+1)...(z^2-2\cos\theta_{n-1}z+1)} \tag{28}$$

where

$$\theta_1 < \theta_2 < \theta_3 < ... < \theta_{n-1} \tag{29}$$

Let $z = e^{j\theta}$ then from Eq. (9)

$$\frac{g}{h} = \frac{2e^{jn\theta/2}jg^*}{2e^{jn\theta/2}h^*} = j\frac{g^*}{h^*} \tag{30}$$

From Eq. (28) and Eq. (30) we have

$$\frac{g}{h} = j\frac{g^*}{h^*} =$$

$$\frac{\beta_0}{\alpha_0}j\frac{\sin\theta(\cos\theta-\cos\theta_2)(\cos\theta-\cos\theta_4)...(\cos\theta-\cos\theta_{n-2})}{\cos\theta-\cos\theta_1)(\cos\theta-\cos\theta_3)...(\cos\theta-\cos\theta_{n-1})} \tag{31}$$

Therefore

$$\frac{g^*}{h^*} = \frac{\beta_0}{\alpha_0}\sin\theta \sum_{i=1,3,\ldots,n-1}\frac{K_i}{\cos\theta-\cos\theta_1} \tag{32}$$

where

$$K_i = \frac{\prod_{j=2,4,\ldots,n-2} (\cos\theta_i - \cos\theta_j)}{\prod_{j=1,3,\ldots,n-1; j\neq i} (\cos\theta_i - \cos\theta_j)} \tag{33}$$

As

$$\cos\theta_i - \cos\theta_j > 0 \text{ for } \theta_i < \theta_j \text{ where } \theta_i, \theta_j \in [0,\pi] \text{ then } K_i > 0$$

and

$$\frac{d}{d\theta}\frac{g^*}{h^*} = \frac{\beta_0}{\alpha_0} \sum \frac{K_i(1-\cos\theta\ \cos\theta_i)}{(\cos\theta - \cos\theta_i)^2} > 0 \tag{34}$$

Therefore, $f^* = h^* + jg^*$ has an argument which is monotonically increasing w.r.t. θ. Also

$$\frac{g^*}{h^*} = \frac{\beta_0}{\alpha_0} \sum_{i=1,3,\ldots,n-1} \frac{K_i \sin\theta(\cos\theta - \cos\theta_i)}{(\cos\theta - \cos\theta_i)^2} \tag{35}$$

But

$$1 - \cos\theta\ \cos\theta_i > |\cos\theta - \cos\theta_i|$$

$$([1-\cos\theta\ \cos\theta_i]^2 - [\cos\theta - \cos\theta_i]^2 = [1-\cos^2\theta][1-\cos^2\theta_i] > 0)$$

From Eq. (34) and Eq. (35) we get

$$\frac{d}{d\theta}\frac{g^*}{h^*} > \frac{1}{\sin\theta}\left|\frac{g^*}{h^*}\right| \tag{36}$$

which is the condition corresponding to Eq. (19) for discrete systems. For n odd we have

$$\begin{aligned}\frac{g}{h} &= \frac{\beta_0}{\alpha_0}\frac{(z-1)(z^2-2z\cos\theta_2+1)(z^2-2z\cos\theta_4+1)\ldots(z^2-2z\cos\theta_{n-1}+1)}{(z+1)(z^2-2z\cos\theta_1+1)(z^2-2z\cos\theta_3+1)\ldots(z^2-2z\cos\theta_{n-2}+1)}\\ &= \frac{\beta_0}{\alpha_0}\frac{(z^2-1)(z^2-2z\cos\theta_2+1)(z^2-2z\cos\theta_4+1)\ldots(z^2-2z\cos\theta_{n-1}+1)}{(z^2+2z+1)(z^2-2z\cos\theta_1+1)(z^2-2z\cos\theta_4+1)\ldots(z^2-2z\cos\theta_{n-2}+1)}\end{aligned} \tag{37}$$

From Eq. (30) and Eq. (37) with $z = e^{j\theta}$ we get

$$\frac{g^*}{h^*} = \frac{\beta_0}{\alpha_0}\sin\theta\left[\frac{K_0}{\cos\theta+1} + \sum_{i=1,3,\ldots,n-2}\frac{K_i}{\cos\theta - \cos\theta_i}\right] \tag{38}$$

where

$$K_0 = \frac{\prod_{j=2,4,\ldots,n-1} (-1 - \cos\theta_j)}{\prod_{j=1,3,\ldots,n-2} (-1 - \cos\theta_j)} \tag{39}$$

and

$$K_i = \frac{\prod_{j=2,4,\ldots,n-1} (\cos\theta_i - \cos\theta_j)}{\prod_{j=1,3,\ldots,n-2; j\neq i} (\cos\theta_i + 1)(\cos\theta_i - \cos\theta_j)} \tag{40}$$

Here also

$$\cos\theta_i - \cos\theta_j > 0 \text{ for } \theta_i < \theta_j \text{ and } \theta_i, \theta_j \in [0, \pi]$$

Therefore $K_0, K_i > 0$

$$\frac{d}{d\theta}\frac{g^*}{h^*} = \frac{\beta_0}{\alpha_0}\left[\frac{K_0(1+\cos\theta)}{(\cos\theta+1)^2} + \sum \frac{K_i(1-\cos\theta\ \cos\theta_i)}{(\cos\theta - \cos\theta_i)}\right] > 0 \tag{41}$$

From Eq. (41) it is clear that $h^* + jg^*$ has a monotonically increasing argument. Also from Eq. (38) and Eq. (41) we get the result in Eq. (36). Moreover

$$\begin{aligned}
\frac{d}{d\theta}\frac{\cos\frac{\theta}{2}}{\sin\frac{\theta}{2}}\frac{g^*}{h^*} &= \cot\frac{\theta}{2}\frac{d}{d\theta}\frac{g^*}{h^*} - \frac{1}{2\sin^2\frac{\theta}{2}}\frac{g^*}{h^*} \\
&= \cot\frac{\theta}{2}\left[\frac{d}{d\theta}\frac{g^*}{h^*} - \frac{1}{\sin\theta}\frac{g^*}{h^*}\right] > 0
\end{aligned} \tag{42}$$

and

$$\frac{d}{d\theta}\frac{\sin\frac{\theta}{2}}{\cos\frac{\theta}{2}}\frac{g^*}{h^*} = \tan\frac{\theta}{2}\left[\frac{d}{d\theta}\frac{g^*}{h^*} + \frac{1}{\sin\theta}\frac{g^*}{h^*}\right] > 0 \tag{43}$$

Therefore the functions

$$f_3^* = \frac{h^*}{\cos\frac{\theta}{2}} + j\frac{g^*}{\sin\frac{\theta}{2}} \quad \text{and} \quad f_4^* = \frac{h^*}{\sin\frac{\theta}{2}} + j\frac{g^*}{\cos\frac{\theta}{2}}$$

have monotonically increasing arguments. □

Corollary 1 *For a Schur stable $f(z)$ the function $e^{-j\mu\theta} f\left(e^{j\theta}\right)$ has a monotonically increasing argument for $\mu \le \frac{n}{2}$.*

Proof:

$$\begin{aligned}
e^{-j\mu\theta} f\left(e^{j\theta}\right) &= 2e^{-j\mu\theta}e^{jn\theta/2}(h^* + jg^*) \\
&= 2e^{j(n/2-\mu)\theta}(h^* + jg^*)
\end{aligned}$$

which has a monotonically increasing argument for $\mu \le \frac{n}{2}$. □

E. Hermite-Bieler Theorem and Criteria for Hurwitz and Schur Stability

1. Continuous systems

In this section it is shown how different criteria for Hurwitz stability like Routh-Hurwitz criterion and equivalent of Schur-Cohn criterion for continuous systems are derived from Hermite-Bieler theorem; [1], [2], [4].

a. Routh-Hurwitz Criterion

Without loss of generality consider n even. The first and second rows of the Routh table are given by the coefficients of h and g respectively. The third row is given by the coefficients of $h - \frac{a_0}{a_1}s^2 g$. The following theorem shows that Routh table reduces the stability of a polynomial of degree n to another one of degree $n-1$. Repeating this operation we arrive at a polynomial of degree one whose stability is easily determined.

Theorem 9 [1], [2] *$f(s)$ is Hurwitz stable if and only if $h - \frac{a_0}{a_1}s^2 g + sg$ is Hurwitz stable.*

Proof: Let $f(s) = h + sg$ be Hurwitz stable i.e. $a_0, a_1, ...a_n > 0$ and according to theorem 5, $h(\lambda)$ and $g(\lambda)$ have simple real negative alternating roots. $h - \frac{a_0}{a_1}s^2 g$ and h have the same sign at the roots of g and the number of roots of $h - \frac{a_0}{a_1}s^2 g + sg$ is $n-1$. Moreover $a_1 a_2 - a_0 a_3 > 0$. Therefore, according to theorem 5, $h - \frac{a_0}{a_1}s^2 g + sg$ is Hurwitz stable. The proof of the if part is similar. □

b. Equivalent of Schur-Cohn-Jury Criterion for Continuous Systems

Theorem 10 [4] *$f(s)$ is Hurwitz stable if and only if $f(s) - \frac{f(-1)}{f(1)}f(-s)$ is Hurwitz stable and* $\left|\frac{f(-1)}{f(1)}\right| < 1$

-1 is a root of $f(s) - \frac{f(-1)}{f(1)}f(-s)$ which can be eliminated, so that the stability of a polynomial of degree n is again reduced to the stability of a polynomial of degree $n-1$. This theorem can be proved using Rouché theorem. Here we give a proof based on Hermite-Bieler theorem and on the following Lemma.

Lemma 1 *$h+sg$ is Hurwitz stable if and only if $\gamma h+\delta sg$ is Hurwitz stable for γ, δ positive numbers.*

Proof: The interlacing property does not change by multiplying h and g by positive numbers. Also $\frac{\delta a_1}{\gamma a_0} > 0$. □

Proof: Proof of theorem 10. Let $f(s)$ be Hurwitz stable.
If $f(s) = h + sg$ then $f(-s) = h - sg$

Therefore
$f(s) - \frac{f(-1)}{f(1)} f(-s) = \left(1 - \frac{f(-1)}{f(1)}\right) h + \left(1 + \frac{f(-1)}{f(1)}\right) sg.$

Using Lemma 1 we get $f(s) - \frac{f(-1)}{f(1)} f(-s)$ is Hurwitz stable. The proof of the if part is similar. □

2. Discrete systems

In this section it is shown how different criteria for Schur stability like the equivalent of Routh-Hurwitz criterion for discrete systems and Schur-Cohn-Jury criterion are derived from Hermite-Bieler theorem; [3], [4].
Consider $f(z)$ given by Eq. (5). Without loss of generality we assume $a_0 > 0$ and n even. For n odd a similar treatment can be made. A necessary condition for Schur stability of Eq. (5) is $\left|\frac{a_n}{a_0}\right| < 1$ which results in $\alpha_0 > 0$ and $\beta_0 > 0$.

a. Equivalent of Routh-Hurwitz Criterion for Discrete Systems

If $h(z)$ and $g(z)$ given by Eq. (7) and Eq. (8) fulfill the discrete Hermite-Bieler theorem then their projections on the real axis $h^*(x)$ and $g^*(x)$ have distinct interlacing real roots between -1 and $+1$. In this case $h^*(x)$ and $g^*(x)$ form a discrete positive pair.
From [10] $h^*(x)$ and $g^*(x)$ are given by

$$h^*(x) = \sum_{i=0}^{\nu-1} \alpha_i \quad T_{\nu-i} + \frac{\alpha_\nu}{2}$$

$$g^*(x) = \sum_{i=0}^{\nu-1} \beta_i \quad U_{\nu-i-1} \tag{44}$$

where $n = 2\nu$ and T_k, U_k are Tshebyshev polynomials of the first and second kind respectively.

Let

$$\begin{aligned} h^*(x) &= \gamma_0 x^\nu + \gamma_1 x^{\nu-1} + \ldots \\ g^*(x) &= \delta_0 x^{\nu-1} + \delta_1 x^{\nu-2} + \ldots \end{aligned}$$

Theorem 11 [4] *$f(z)$ is Schur stable if and only if $g^*(x)$ and $h^*(x) - \frac{\gamma_0}{\delta_0}(x-1)g^*(x)$ form a discrete positive pair and $h^*(-1) < 0$ for ν odd and $h^*(-1) > 0$ for ν even.*

Proof: Assume $f(z)$ be Schur stable. $h^*(x) - \frac{\gamma_0}{\delta_0}(x-1)g^*(x)$ has the same value at the roots of $g^*(x)$ as $h^*(x)$. Therefore, the interlacing property is preserved. The condition $h^*(-1) < 0$ for ν odd and $h^*(-1) > 0$ for ν even guarantees that the reduction of the order is not due to a root of $h^*(x)$ between -1 and $-\infty$. Therefore $g^*(x)$ and $h^*(x) - \frac{\gamma_0}{\delta_0}(x-1)g^*(x)$ form a discrete positive pair.

The if part can be proved similarly. □

This theorem can be implemented in table form.

b. Schur-Cohn-Jury Criterion

Theorem 12 [3], [4] *The polynomial in Eq. (5) with $\left|\frac{a_n}{a_0}\right| < 1$ is Schur stable if and only if the polynomial $\frac{1}{z}\left[f(z) - \frac{a_n}{a_0} z^n f\left(\frac{1}{z}\right)\right]$ is Schur stable.*

This was proved in [3] using the following Lemma:

Lemma 2 *$h + g$ is Schur stable if and only if $\gamma h + \delta g$ is Schur stable for γ, δ positive numbers.*

Proof: Proof of Lemma 2: The interlacing property does not change by multiplying h and g by positive numbers. Also $\left|\frac{\gamma\alpha_0 - \delta\beta_0}{\gamma\alpha_0 + \delta\beta_0}\right| < 1$ if $\left|\frac{a_n}{a_0}\right| < 1$. The reverse is true. □

Proof: Proof of theorem 12: $f(z) - \frac{a_n}{a_0} z^n f\left(\frac{1}{z}\right) = h(z) + g(z) - \frac{a_n}{a_0}[h(z) - g(z)] = \left(1 - \frac{a_n}{a_0}\right) h(z) + \left(1 + \frac{a_n}{a_0}\right) g(z)$. The only if part of theorem 12 follows directly from $\left|\frac{a_n}{a_0}\right| < 1$ and Lemma 2.
The if part can be proved similarly. □

III. ROBUST STABILITY IN THE FREQUENCY DOMAIN

In chapter II the foundations for stability investigations were led beginning with the principle of the argument. In this chapter we extend and use the previous results to solve the robust stability problem.
Consider a polynomial family with real coefficients

$$f(s, \underline{a}) = a_0 s^n + a_1 s^{n-1} + ... + a_{n-1} s + a_n \tag{45}$$

whose coefficient vector $\underline{a}$ lies in a simply connected closed set A in the coefficient space, i.e. $\underline{a} \in A$ where $\underline{a}^T \equiv [a_0 \quad a_1 ... a_n]$.
If $\underline{s}^T \equiv \left[s^n \quad s^{n-1} ... 1\right]$ then

$$f(s, \underline{a}) = \underline{a}^T \underline{s} \tag{46}$$

Definition 1 *A polynomial is called D-stable if all its roots lie in the stability domain D where D is an open subset of the complex plane.*

For polynomials with real coefficients, D is symmetrical w.r.t. the real axis.

Definition 2 $f(s^*, A) = \Gamma(s^*)$ *is the value set of the polynomial family for* $s = s^* \in C$. $\partial\Gamma(s^*)$ *is the boundary of* $\Gamma(s^*)$.

A. Robust D-stability

The stability criterion of section II-B can be generalized in the sense that the value set as a whole has a change of argument of $2\pi n$ if s moves once along ∂D the contour of D in the positive direction. However, to get a criterion which can be checked more easily we use the boundary crossing theorem [2] which can be simply explained as follows: If $f(s, \underline{a})$ is D-stable for one value of $\underline{a}$ and $\underline{a}$ continuously changes then instability can occur only when one or more roots cross the boundary ∂D.

Theorem 13 *$f(s, \underline{a})$ with $\underline{a} \in A$ is robust D-stable if and only if the following conditions are satisfied*

(i) $f(s, \underline{a}^*)$ *is D-stable for some* $\underline{a}^* \in A$

(ii) $\Gamma(s^*) \neq 0$ *for some* $s^* \in \partial D$

(iii) $\partial\Gamma(s) \neq 0$ *for all* $s \in \partial D$

The proof of this theorem follows directly from the boundary crossing theorem. See [11].

B. An edge theorem

In [12] an edge theorem was proved which can be used to investigate the position of the roots of a family of polynomials. Here we give a form of this theorem which is adequate for our discussion of the robust stability of continuous and discrete systems [11].
Consider the family of polynomials given by Eq. (45) and let $\underline{a} \in A$ where A is a polytope in the coefficient space. Let D be an open domain in the s-plane symmetric w.r.t. real axis and ∂D its boundary. Let B be the exposed edges of A.

Theorem 14 *All roots of the family $f(s, A)$ lie in D if and only if all the roots of B lie in D.*

Proof: Take two points $\underline{a}_0, \underline{a}_1 \in A$. These are mapped in $f(s^*, \underline{a}_0) = \underline{a}_0^T \underline{s}^* = f_0$ and $f(s^*, \underline{a}_1) = \underline{a}_1^T \underline{s}^* = f_1$ for $s = s^*$.
The straight line joining $\underline{a}_0$ and $\underline{a}_1$ is given by $\lambda\underline{a}_0 + (1-\lambda)\underline{a}_1$ where $\lambda \in [0, 1]$. It is mapped in $(\lambda\underline{a}_0^T + (1-\lambda)\underline{a}_1^T)\underline{s}^* = \lambda f_0 + (1-\lambda) f_1$ which is also a straight line in the value set. Hence the value set $\Gamma(s^*)$ is a convex polygon whose boundary $\partial\Gamma \subset \Gamma(B)$. For real $s^* = \sigma \in \partial D$, the value set $\Gamma(\sigma)$ is a segment of the real axis which is the mapping of the polytope as well as of all the exposed edges. Then all the conditions of theorem 13 are fulfilled. Hence follows the result. □

Actually one has to check the exposed edges of the value set which correspond in general to a part of B. The above theorem can be applied to continuous and discrete systems. For continuous systems the imaginary axis is considered for $\omega = 0$ to ∞. For discrete systems the unit circle $e^{j\theta}$ is considered for $\theta = 0$ to π.

C. Stability of edges and extreme point results

1. Continuous systems

a. Hurwitz stability of an edge

To check the Hurwitz stability of an edge we need the following result

Theorem 15 [13]; *Assume $f_0(s)$ and $f_1(s)$ are Hurwitz stable. Let $f_\lambda(s) = \lambda f_0(s) + (1-\lambda)f_1(s), \lambda \in [0,1]$ then $f_\lambda(s)$ is Hurwitz stable if and only if*

$$\frac{f_1(j\omega^*)}{f_0(j\omega^*)} > 0$$

for every real solution $\omega = \omega^*$ of

$$h_0(-\omega^2)g_1(-\omega^2) - h_1(-\omega^2)g_0(-\omega^2) = 0$$

Proof: $f_0(s) = h_0(s^2) + sg_0(s^2)$ and $f_1(s) = h_1(s^2) + sg_1(s^2)$
The value set is a straight line. Instability will occur when the origin lies on the value set for one $\omega = \omega^*$.
In this case

$$\lambda\left[h_0\left(-\omega^{*2}\right) + j\omega^* g_0\left(-\omega^{*2}\right)\right] + (1-\lambda)\left[h_1\left(-\omega^{*2}\right) + j\omega^* g_1\left(-\omega^{*2}\right)\right] = 0$$

which gives

$$h_0\left(-\omega^{*2}\right)g_1\left(-\omega^{*2}\right) - h_1\left(-\omega^{*2}\right)g_0\left(-\omega^{*2}\right) = 0$$

As $\lambda \in [0,1]$ then $\frac{f_1(j\omega^*)}{f_0(j\omega^*)} = \frac{-\lambda}{1-\lambda} < 0$ □

An equivalent to theorem 15 is given by the following result

Theorem 16 [14]; *Assume $f_0(s)$ and $f_1(s)$ are Hurwitz stable. $f_\lambda(s)$ is Hurwitz stable if and only if $H_1^{-1}H_0$ has no negative eigenvalues where H_0 and H_1 are the Hurwitz matrices of dimension $(n-1)$ corresponding to $f_0(s)$ and $f_1(s)$.*

Proof: $\det H > 0$ is the main critical stability condition. Stability of $f_\lambda(s)$ will be lost if $\det[\lambda H_0 + (1-\lambda)H_1] = 0$ which means that $H_1^{-1}H_0$ has negative eigenvalue.
The other critical stability condition $f(0) > 0$ is also satisfied for $f_\lambda(s)$ if it is satisfied for $f_0(s)$ and $f_1(s)$. □

b. Extreme point results for the Hurwitz stability of edges

[15]. In this section some results are discussed where the stability of an edge is guaranteed by the stability of its vertices. In this case one can say that the edge has the vertex property.

Theorem 17 *If an edge is given by* $f(s,\lambda) = f_0(s) + \lambda(\alpha s + \beta)p^+(s)k(s)$ *where* $\lambda \in [0,1], p^+(s)$ *is an anti-Hurwitz polynomial (all its roots lie in the right half plane),* $k(s)$ *is an even or odd polynomial and* $(\alpha s + \beta)$ *is a first order system, then* $f(s,\lambda)$ *is Hurwitz stable if the two vertex polynomials* $f(s,0)$ *and* $f(s,1)$ *are Hurwitz stable and the degree of* $f(s,\lambda)$ *is* λ*-invariant.*

Proof: Let $f(s,0)$ and $f(s,1)$ be Hurwitz stable. $p^-(s) = p^+(-s)$ is a Hurwitz polynomial. The two extended vertex polynomials $\tilde{f}_0(s) = f_0(s)p^-(s)$ and $\tilde{f}_1(s) = f_1(s)p^-(s)$ are also Hurwitz polynomials. In section II-D-1 it was shown that the use of the mappings $f_1^*(\omega)$ and $f_2^*(\omega)$ does not change the monotony of the argument of any Hurwitz stable polynomial $f(s)$. Using the mapping $f_1^*(\omega)$ for $k(s)$ even and $f_2^*(\omega)$ for $k(s)$ odd we obtain a line with a constant slope. We get the situation as in Fig. 2 which is impossible to occur. □

2. Discrete systems

a. Schur stability of an edge

To check the Schur stability of an edge we need the following result

Theorem 18 [16]. *Assume* $f_0(z)$ *and* $f_1(z)$ *are Schur stable.*
Let $f_\lambda(z) = \lambda f_0(z) + (1-\lambda)f_1(z), \lambda \in [0,1]$ *then* $f_\lambda(z)$ *is Schur stable if*

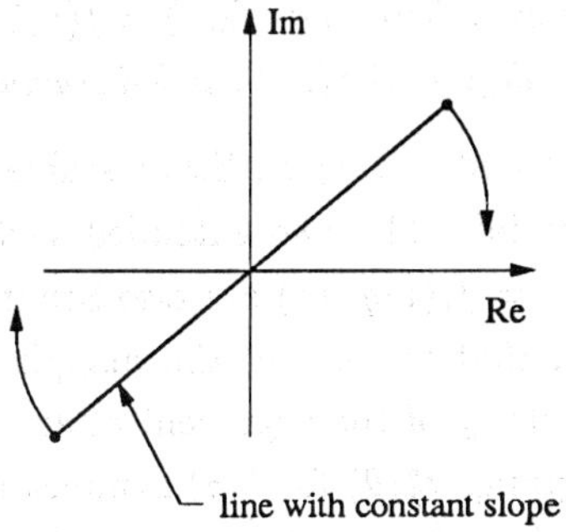

Fig. 2. Impossible situation

and only if

$$\frac{f_1(z^*)}{f_0(z^*)} > 0$$

for every $z = z^*$ *of*

$$z^n\left[f_0(z^{-1})f_1(z) - f_0(z)f_1(z^{-1})\right] = 0 \quad \textit{with} \quad |\, z^* \,| = 1$$

The proof is similar to the proof of theorem 15.
An equivalent to theorem 18 is given by the following result

Theorem 19 [17]. *Assume* $f_0(z)$ *and* $f_1(z)$ *are Schur stable.* $f_\lambda(z)$ *is robust Schur stable if and only if* $\Delta_1^{-1}\Delta_0$ *has no negative real eigenvalues where* Δ_1 *and* Δ_0 *are Jury inner matrices corresponding to* $f_0(z)$ *and* $f_1(z)$.

The proof is similar to that of theorem 16. Here we have three critical stability conditions $\Delta > 0, f(1) > 0$ and $(-1)^n f(-1) > 0$. The last two conditions are linear in the coefficients and hence always satisfied for $f_\lambda(z)$ if $f_0(z)$ and $f_1(z)$ are Schur stable.

b. Extreme point results for the Schur stability of edges

[15]. Similar to Hurwitz stability we consider the following result

Theorem 20 *If an edge is given by*

$$f(z,\lambda) = f_0(z) + \lambda(\alpha z + \beta)q^+(z)k(z) \qquad \lambda \in [0,1]$$

where $q^+(z)$ *is an anti-Schur polynomial of degree* r, $k(z)$ *is symmetric or antisymmetric and* $(\alpha z + \beta)$ *is a first order system, then* $f(s,\lambda)$ *is Schur stable if the two vertex polynomials* $f_0(z) = f(z,0)$ *and* $f_1(z) = f(z,1)$ *are Schur stable and the degree of* $f(z,\lambda)$ *is* λ*-invariant.*

Proof: Let $f(z,0)$ and $f(z,1)$ be Schur stable. $q^-(z) = z^r q^+\left(\frac{1}{z}\right)$ is a Schur stable polynomial. The two extended vertex polynomials $\tilde{f}_0(z) = f_0(z)q^-(z)$ and $\tilde{f}_1(z) = f_1(z)q^-(z)$ are also Schur polynomials. In section II-D-2 it was shown that the use of the mappings $f_3^*(\theta)$ and $f_4^*(\theta)$ does not change the monotony of the argument of any Schur stable polynomial $f(z)$. Using the mapping $f_3^*(\theta)$ for $k(z)$ symmetric and $f_4^*(\theta)$ for $k(z)$ antisymmetric we obtain a line with constant slope. Hence similar to the continuous case the origin cannot lie on the value set of the edge. □

D. Robust stability of special polynomials

One of the most significant results in the stability of uncertain linear systems is due to Kharitonov [18]. Since the publication of his celebrated theorem a large amount of research activity has resulted with far reaching consequences. In this section we give Kharitonov theorem with two simple proofs using the results in II-B and II-C. Also the analog and counterpart results for discrete systems are derived.

1. Continuous systems: robust Hurwitz stability

Kharitonov [18] considers the Hurwitz stability of

$$f(s) = a_0 s^n + a_1 s^{n-1} + \ldots + a_{n-1}s + a_n \tag{47}$$

with

$$\underline{a}_i \leq a_i \leq \overline{a}_i \tag{48}$$

Theorem 21 *(Kharitonov theorem)* $f(s)$ *given by Eq.* (47) *and Eq.* (48) *is Hurwitz stable if and only if the four extreme polynomials*

$$f_1(s) = \overline{a}_n + \underline{a}_{n-1}s + \underline{a}_{n-2}s^2 + \overline{a}_{n-3}s^3 + \overline{a}_{n-4}s^4 + \ldots \tag{49}$$

$$f_2(s) = \overline{a}_n + \overline{a}_{n-1}s + \underline{a}_{n-2}s^2 + \underline{a}_{n-3}s^3 + \overline{a}_{n-4}s^4 + \ldots \tag{50}$$

$$f_3(s) = \underline{a}_n + \overline{a}_{n-1}s + \overline{a}_{n-2}s^2 + \underline{a}_{n-3}s^3 + \underline{a}_{n-4}s^4 + \ldots \tag{51}$$

$$f_4(s) = \underline{a}_n + \underline{a}_{n-1}s + \overline{a}_{n-2}s^2 + \overline{a}_{n-3}s^3 + \underline{a}_{n-4}s^4 + \ldots \tag{52}$$

are Hurwitz stable.

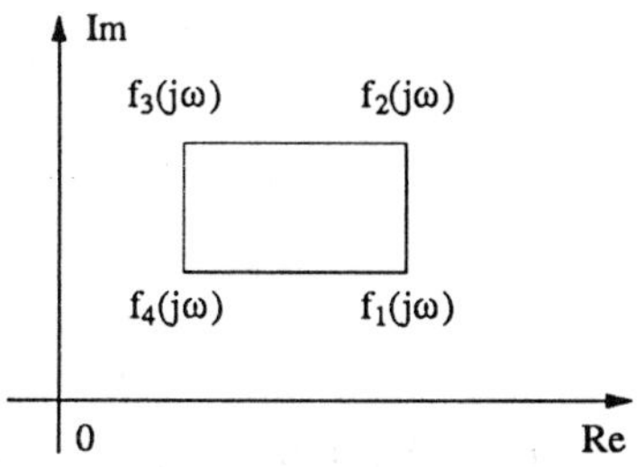

Fig. 3. Value set of Kharitonov box

Theorem 22 [19] *For $n = 3$, $f(s)$ is Hurwitz stable if and only if $f_1(s)$ is Hurwitz stable. For $n = 4$, $f(s)$ is Hurwitz stable if and only if $f_1(s)$ and $f_2(s)$ are Hurwitz stable. For $n = 5$, $f(s)$ is Hurwitz stable if and only if $f_1(s), f_2(s)$ and $f_3(s)$ are Hurwitz stable.*

Proof: Proof of theorems 21 and 22. It was shown in [20] that the value set $\Gamma(\omega)$ of a box parallel to the axis given by Eq. (48) is a rectangle in the complex plane parallel to the axes and whose corners are the mapping of the four Kharitonov polynomials as shown in Fig. 3. Using theorem 2 for f_1, f_2, f_3, f_4 and the monotony of the argument one can easily show in Fig. 4 that the origin cannot come on the boundary of the rectangle for any ω. Hence the conditions of theorem 13 are satisfied and hence the result of theorem 21. Moreover, for $n = 3$ the stability of f_1, for $n = 4$ the stability of f_1 and f_2 and for $n = 5$ the stability of f_1, f_2 and f_3 guarantees the exclusion of the origin; see [21]. □

Theorems 21 and 22 are also proved using Hermite-Bieler theorem as follows: [10]

For $n = 3$ we have

$$f(s) = a_0 s^3 + a_1 s^2 + a_2 s + a_3 = h(s^2) + s g(s^2) \quad (53)$$

where $a_i \in [\underline{a}_i, \overline{a}_i]$

$$h(\lambda) = a_1 \lambda + a_3 \quad (54)$$

$$g(\lambda) = a_0 \lambda + a_2 \quad (55)$$

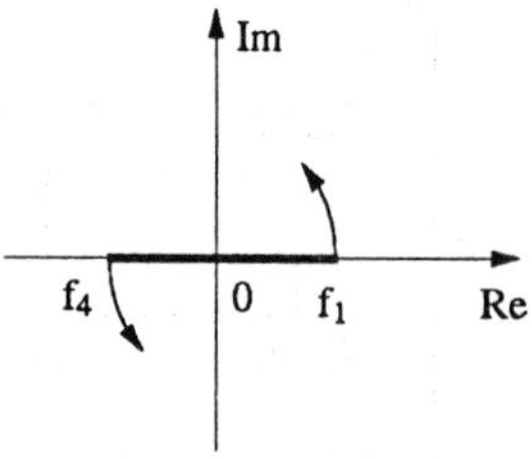

Fig. 4. An impossible situation

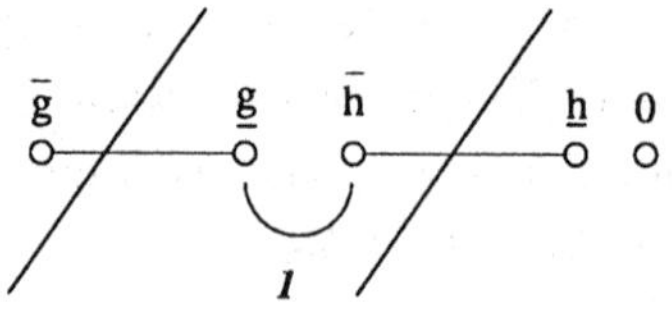

Fig. 5. Transition diagram for $n = 3$

These are straight lines as shown in Fig. 5. In the continuous case we have only one interval which is the negative real axis.

From Fig. 5 we notice that we have only one transition corresponding to the extreme polynomial $f_1(s)$ given by $(\overline{h}, \underline{g})$

$$f_1(s) = \overline{a}_0 s^3 + \underline{a}_1 s^2 + \underline{a}_2 s + \overline{a}_3 \tag{56}$$

If $f_1(s)$ is stable then the interlacing property on the line $[-\infty, 0]$ is satisfied and hence stability.

For $n = 4$ we have

$$f(s) = a_0 s^4 + a_1 s^3 + a_2 s^2 a_3 s + a_4 \tag{57}$$

$$h(\lambda) = a_0 \lambda^2 + a_2 \lambda + a_4 \tag{58}$$

$$g(\lambda) = a_1 \lambda + a_3 \tag{59}$$

Fig. 6 shows the transition diagram in this case. From Fig. 6 we notice that we have two transitions corresponding to the extreme polynomials

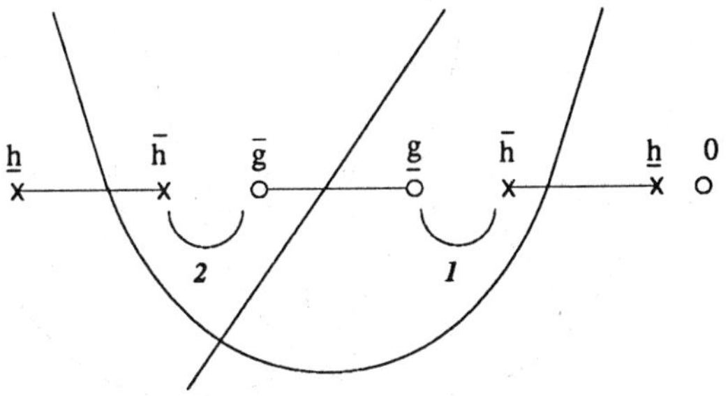

Fig. 6. Transition diagram for $n = 4$

$(\overline{h}, \underline{g})$ and $(\overline{h}, \overline{g})$

$$f_1(s) = \overline{a}_0 s^4 + \overline{a}_1 s^3 + \underline{a}_2 s^2 + \underline{a}_3 s + \overline{a}_4 \tag{60}$$

$$f_2(s) = \overline{a}_0 s^4 + \underline{a}_1 s^3 + \underline{a}_2 + \overline{a}_3 s + \overline{a}_4 \tag{61}$$

If $f_1(s)$ and $f_2(s)$ are stable then the interlacing property on the line $[-\infty, 0]$ is satisfied and hence stability.
For $n = 5$ we have

$$f(s) = a_0 s^5 + a_1 s^4 + a_2 s^3 + a_3 s^2 + a_4 s + a_5 \tag{62}$$

$$h(\lambda) = a_1 \lambda^2 + a_3 \lambda + a_5 \tag{63}$$

$$g(\lambda) = a_0 \lambda^2 + a_2 \lambda + a_4 \tag{64}$$

Fig. 7 shows the transition diagram for $n = 5$. From Fig. 7 we notice that we have three transitions corresponding to the extreme polynomials $(\overline{h}, \underline{g})$, $(\overline{h}, \overline{g})$ and $(\underline{h}, \overline{g})$.

$$f_1(s) = \underline{a}_0 s^5 + \overline{a}_1 s^4 + \overline{a}_2 s^3 + \underline{a}_3 s^2 + \underline{a}_4 s + \overline{a}_5 \tag{65}$$

$$f_2(s) = \overline{a}_0 s^5 + \overline{a}_1 s^4 + \underline{a}_2 s^3 + \underline{a}_3 s^2 + \overline{a}_4 s + \overline{a}_5 \tag{66}$$

$$f_3(s) = \overline{a}_0 s^5 + \underline{a}_1 s^4 + \underline{a}_2 s^3 + \overline{a}_3 s^2 + \overline{a}_4 s + \underline{a}_5 \tag{67}$$

If $f_1(s), f_2(s)$ and $f_3(s)$ are stable then the interlacing property is satisfied and hence stability.

Fig. 8 shows the transition diagram for $n = 6$. As seen from Fig. 8 we have four transitions corresponding to the four extreme polynomials $(\overline{h}, \underline{g})$,

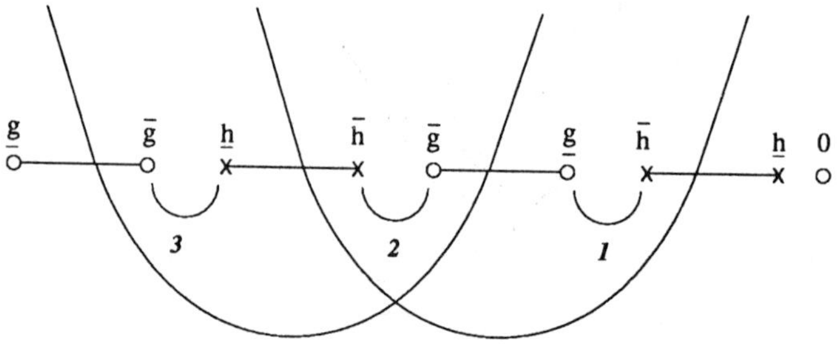

Fig. 7. Transition diagram for $n = 5$

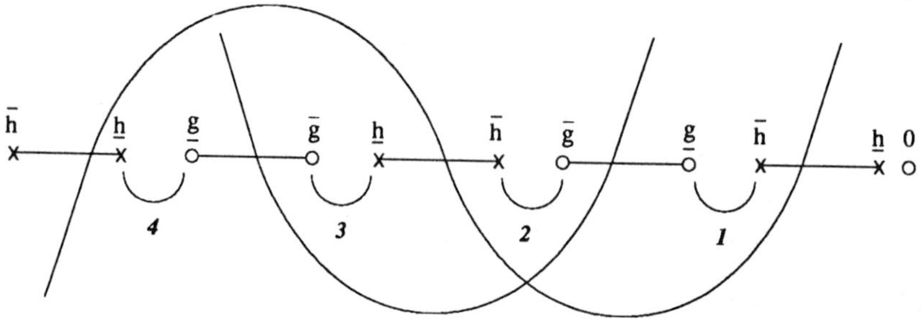

Fig. 8. Transition diagram for $n = 6$

$(\overline{h}, \overline{g})$, $(\underline{h}, \overline{g})$ and $(\underline{h}, \underline{g})$, i.e. we need to check four corners for stability. For $n > 6$ the transition diagram can show only repetitions of one or more of these polynomials, so that the stability of only four corners is necessary and sufficient for stability of the higher order polynomials.

2. Discrete systems: robust Schur stability

It can be easily shown through counter examples [22] that even if all the corners of a box parallel to the axes in the coefficient space are Schur stable, there is no guarantee that the polynomial family given by this box is Schur stable. This statement is correct for monic polynomials with $n \geq 4$ or nonmonic polynomials with $n \geq 3$. In this section we get several results which correspond to results in the continuous case.

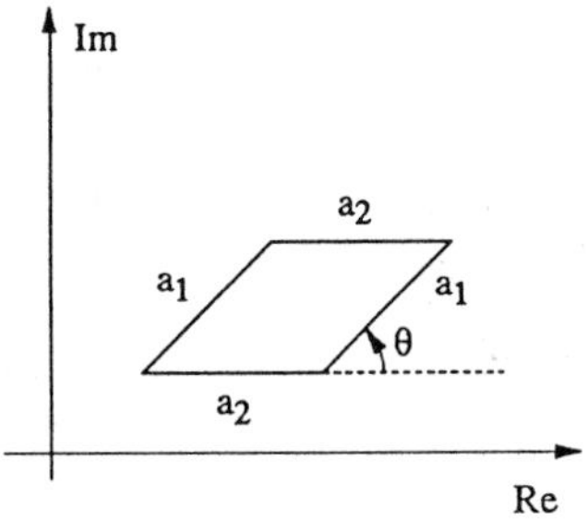

Fig. 9. Value set of a second order system

a. Low order polynomials

[22]. Consider first the monic interval polynomial with $n = 2$.

$$f(z) = z^2 + a_1 z + a_2 \tag{68}$$

where

$$\underline{a}_i \leq a_i \leq \bar{a}_i \tag{69}$$

The value set of this polynomial family with $z = e^{j\theta}$ is given in Fig. 9. From theorem 14 one can conclude that Schur stability is obtained if and only if the a_1-edges and a_2 edges of Fig. 9 are Schur stable. Stability of the a_2-edges can be replaced by the stability of their vertices using the monotony of the argument. Stability of the a_1 edges can be replaced by the stability of their vertices using theorem 8. Hence for $n = 2$ Schur stability of Eqs. (68), (69) is given by the Schur stability of the vertices.
Consider a monic interval polynomial with $n = 3$.

$$f(z) = z^3 + a_1 z^2 + a_2 z + a_3 \tag{70}$$

where

$$\underline{a}_i \leq a_i \leq \bar{a}_i \tag{71}$$

The form of the value set of this polynomial family with $z = e^{j\theta}$ is given in Fig. 10, for $\theta < 90°$ and $\theta > 90°$. The corners from 0 to 7 corresponding to 3 bit binary numeration. 0 is $\underline{a}_1$, $\underline{a}_2$, $\underline{a}_3$.
Applying theorem 14 one can conclude that Schur stability is obtained if

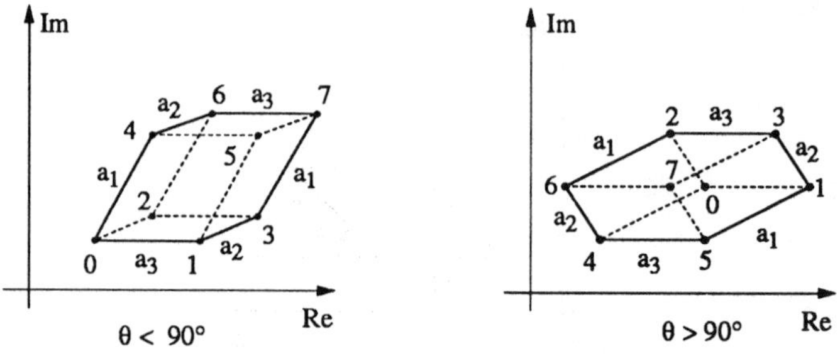

Fig. 10. Value set of a third order system

and only if the a_1, a_2, a_3-edges of Fig. 10 are Schur stable. Stability of these edges can be replaced by the stability of their vertices using theorem 8 and corollary 1. Hence for $n = 3$ Schur stability of Eqs. (70), (71) is given by the Schur stability of the vertices.
For monic interval polynomials with $n \geq 4$, one has to check some of the edges [16].

b. Special case

[23]. Consider the polynomial

$$f(z) = a_0 z^n + a_1 z^{n-1} + \cdots + a_{n-1} z + a_n \tag{72}$$

a_i is fixed for $i = 0, 1, \ldots, \frac{n-1}{2}\rfloor$, and

$$\underline{a}_i \leq a_i \leq \bar{a}_i \qquad \text{otherwise} \tag{73}$$

where $x\rfloor$ is the next lower integer if x is not an integer.
Here we use theorem 14, theorem 8 and corollary 1 as before to get the following result.

Theorem 23 *$f(z)$ given by Eqs. (72), (73) is robust Schur stable if and only if all the vertices are Schur stable.*

c. Counter-part result

[24]. Consider $f(z)$ given by Eq. (72) and let

$$\underline{a}_i \leq a_i \leq \bar{a}_i \qquad \text{for } i = 0, 1, \ldots, n$$

Table I. Angles of the regions in the z-plane

$n=2$	0	$\frac{\pi}{2}$	π			
$n=3$	0	$\frac{\pi}{3}$	$\frac{2\pi}{3}$	π		
$n=4$	0	$\frac{\pi}{4}$	$\frac{\pi}{2}$	$\frac{3\pi}{4}$	π	
$n=5$	0	$\frac{\pi}{5}$	$\frac{2\pi}{5}$	$\frac{3\pi}{5}$	$\frac{4\pi}{5}$	π

Table II. Number of regions as function of n

n	2	3	4	5	6	7	8	9	10
$R(n)$	2	4	6	10	12	18	22	28	32

If we apply the result in the last paragraph we need only to check the a_i-edges, where $i = 0, 1, \ldots, \frac{n-1}{2}\rfloor$.

However it is possible to reduce the number of edges to be checked by considering the exposed edges of the value set. For $z = z^*$ the value set can be obtained through the direction of the edge polynomials. The value set is a parpolygon where two edges of the value set have the same direction. The form of the value set is given by the sequence of the edge polynomials. The set of exposed edges change only when the sequence of edge polynomials changes. This happens when two or more edge polynomials become parallel for a value $z = z^*$. The points in the z plane at which this happens divide the plane into different regions R_j, (see Fig. 11). The set S of the exposed edges to be checked is obtained through the union of the exposed edges in the regions R_j of the z-plane only for $i = 0, 1, \ldots, \frac{n-1}{2}\rfloor$.

The boundaries of the regions in the z-plane are straight lines through the origin whose angles are given by table I. The regions R_j are sectors centered in the region. It is clear that the number of regions in the upper half plane is given by

$$R(n) = R(n-1) + \Phi(n) \tag{74}$$

where $\Phi(n)$ is the Euler function and $R(2) = 2$. See table II. Table III gives the distribution of the exposed edges, where the repeated edges are eliminated. f_i denotes the direction of a_i-edge.

Starting from table III we consider only the exposed edges for the direc-

Table III. Number of exposed edges in each direction

n	f_0	f_1	f_2	f_3	f_4	f_5	f_6	f_7	f_8	f_9	f_{10}
1	2	2									
2	4	2	4								
3	8	4	4	8							
4	12	8	4	8	12						
5	20	12	8	8	12	20					
6	24	20	12	8	12	20	24				
7	36	24	20	12	12	20	24	36			
8	44	36	24	20	12	20	24	36	44		
9	56	44	36	24	20	20	24	36	44	56	
10	64	56	44	36	24	20	24	36	44	56	64

tions $f_i, i = 0, 1, ..., \frac{n-1}{2} \rfloor$.

In this case the minimum number of exposed edges to be checked for Schur stability is given by the recursion formulas

$$K(n) = K(n-1) + 2\left[R(n) - R\left(\frac{n}{2}\right)\right] \qquad \text{for } n \text{ even}$$

and

$$K(n) = K(n-1) + 2R(n) \qquad \text{for } n \text{ odd} \tag{75}$$

where $K(2) = 4$.

Therefore $K(n)$ for different n and in relation to the total number of edges $K_1(n)$ is given by table IV. The reduction factor is about 50 for $n = 10$.

Example $n = 3$: Here the total number of corners is $2^4 = 64$ and from table II the number of regions is 4 (Fig. 11 shows the 4 regions). We numerate the corners from 0 to 15 corresponding to the 4 bit binary coding as in section III-D-2-a. In table V we get the corner sequences for the value set in the four regions R_1 and R_4. On the boundary between R_1 and R_2, f_0 will be parallel to f_3. Therefore going from region R_1 to region R_2, corner sequences in R_1 built by f_0, f_3 will be changed. In R_1 there are two corner

Table IV. Exposed edges and total number of edges

n	2	3	4	5	6	7	8	9	10
$K(n)$	4	12	20	40	56	92	124	180	224
$K_1(n)$	12	32	80	192	448	1024	2304	5120	11264

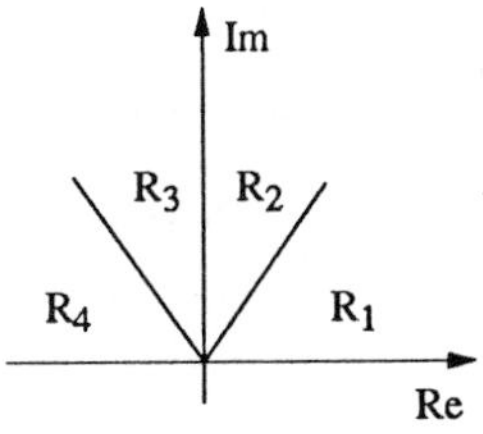

Fig. 11. Regions for $n = 3$

sequences which give edges in direction of f_0, f_3:

8 0 1 and 7 15 14

Fig. 12 shows the geometric setup of the above sequences in region R_1. One

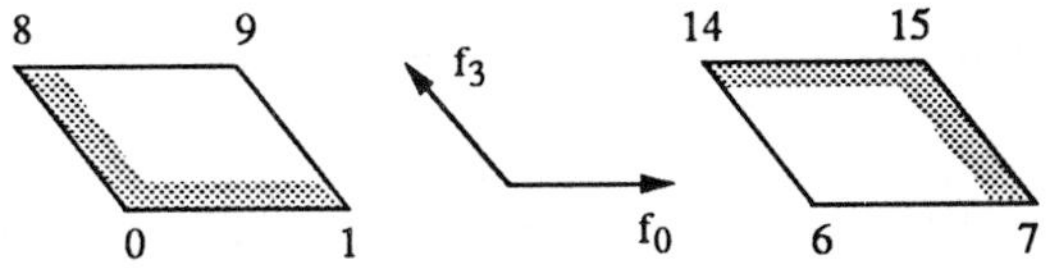

Fig. 12. Directions f_0 and f_3 in region R_1

sees that the corners 0 and 15 go inside the value set and the corners 9 and 6 go on the boundary of the value set in region R_2.

Similarly on the boundary between region R_2 and region R_3, f_0 will be parallel to f_2 and f_1 will be parallel to f_3.

From this the corresponding corner sequences and changes can be generated. The number of exposed edges for each direction f_i can be obtained by counting all the changes in which the edge f_i is involved. This leads to

Table V. Corner sequences in the four regions

R_1	0	1	3	7	15	14	12	8	0
R_2	9	1	3	7	6	14	12	8	9
R_3	9	11	3	2	6	4	12	13	9
R_4	9	11	10	2	6	4	5	13	9

the number of repetition of f_i multiplied by two in the following sequences.

$$\begin{array}{ll} (f_0, f_1, f_2, f_3) & \text{Region } R_1 \\ (f_0, f_3) & \text{change } R_1 \longrightarrow R_2 \\ (f_0, f_2); (f_1, f_3) & \text{change } R_2 \longrightarrow R_3 \\ (f_0, f_3) & \text{change } R_3 \longrightarrow R_4 \end{array}$$

i.e. 8, 4, 4, 8 which is just the row for $n = 3$ in table III. From the sequences in table V one can get immediately the exposed edges in each direction. It is to be noted that before begining the robust stability investigation of Eq. (72) with the coefficients lying in a box parallel to the axes in the coefficients space (Kharitonov box) one should check if the necessary conditions for the coefficients as given in [25] are satisfied. The coefficient ranges build a box which should include the Kharitonov box for monic polynomials.

d. Analog result: strong Kharitonov theorem for discrete systems

[26]. Consider $f(z)$ in Eq. (5) and the corresponding symmetric and anti-symmetric parts $h(z)$ and $g(z)$ given by Eqs. (7) and (8). Assume

$$\underline{\alpha}_i \leq \alpha_i \leq \bar{\alpha}_i \quad \text{and} \quad \underline{\beta}_i \leq \beta_i \leq \bar{\beta}_i$$

which can be represented as in Fig. 13. The corresponding region in the coefficient space is given in Fig. 14. From the discrete Hermite-Bieler theorem [4] $h(z)$ and $g(z)$ must have simple interlacing roots on the unit circle for Schur stability. If we consider their projection on the line $[-1, 1]$ we get for n even, $n = 2\nu$

$$h^*(x) = \sum_{i=0}^{\nu-1} \alpha_i T_{\nu-i} + \frac{\alpha_\nu}{2} \tag{76}$$

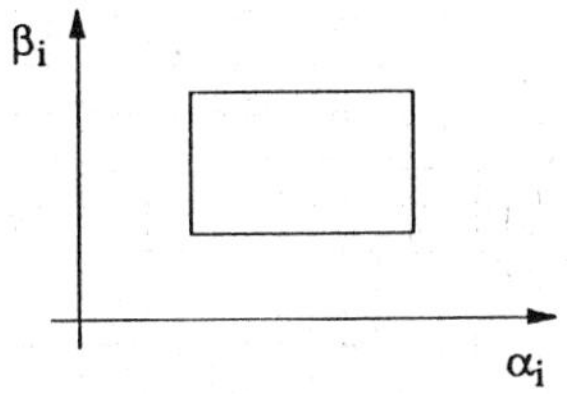

Fig. 13. Parameter region in $\alpha_i - \beta_i$ plane

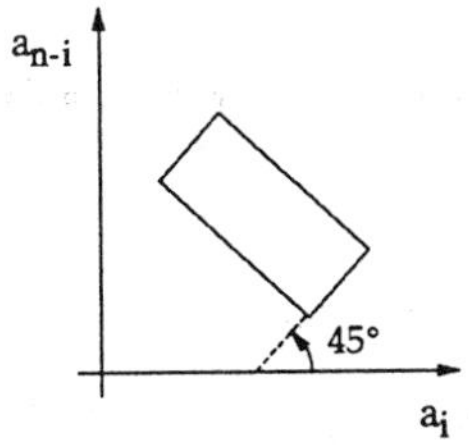

Fig. 14. Parameter region in $a_i - a_{n-i}$ plane

$$g^*(x) = \sum_{i=0}^{\nu-1} \beta_i U_{\nu-i-1} \tag{77}$$

where T_ν and U_ν are Tshebyshev polynomials of the first and second kind respectively.
$h^*(x)$ and $g^*(x)$ must have the interlacing property on the line $[-1, 1]$. The roots of the Tshebyshev polynomials divide this line in intervals whose number is given by

$$I_n = 1 + \sum_{k=4,8,12,\ldots} \Phi(k) + 2 \sum_{k=6,10,14,\ldots} \Phi(k) \qquad n \geq 2 \tag{78}$$

where $\Phi(k)$ is the Euler function.
For n odd, $n = 2\nu + 1$

$$h^*(x) = \sum_{i=0}^{\nu} \alpha_i \tau_{\nu-i} \tag{79}$$

$$g^*(x) = \sum_{i=0}^{\nu} (2\nu - 2i + 1)\beta_i \mu_{\nu-i} \tag{80}$$

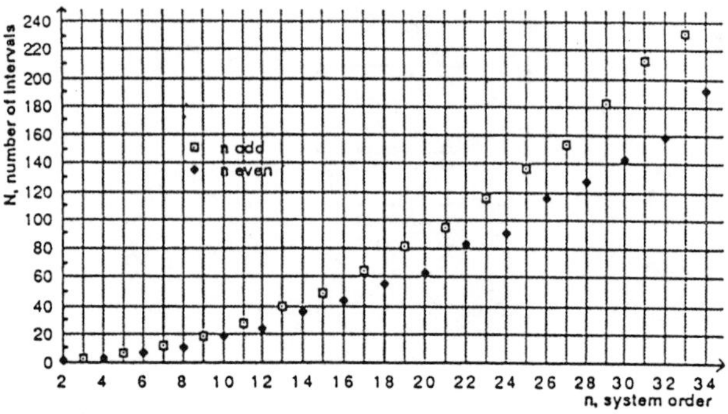

Fig. 15. Numbers of intervals as function of n

where τ_i and μ_i are Jacobi polynomials

$$\tau_i = \frac{\cos((i+0.5)\arccos x)}{\cos(0.5\arccos x)}, \quad \mu_i = \frac{\sin((i+0.5)\arccos x)}{\sin(0.5\arccos x)}$$

The roots of these polynomials divide the line $[-1, 1]$ in intervals whose number is given by

$$I_n = 1 + \sum_{k=3,5,7,\ldots} \Phi(k) \qquad n \geq 3 \tag{80}$$

For Schur stability $h^*(x)$ and $g^*(x)$ must have interlacing zeros on the line $[-1,\ 1]$. This property is satisfied if in every interval on $[-1,\ 1]$ the four Kharitonov polynomials corresponding to $(\bar{h}^*,\ \underline{g}^*)$, $(\bar{h}^*,\ \bar{g}^*)$, $(\underline{h}^*,\ \underline{g}^*)$ and $(\underline{h}^*,\ \underline{g}^*)$ are Schur stable.
Hence the number of corners necessary and sufficient for Schur stability is

$$N = 4I_n \tag{81}$$

For $n = 10$, $N = 4 \times 20 = 80$ corners out of $2^{11} = 2048$ corners and for $n = 30$, $N = 4 \times 144 = 576$ corners out of 2^{31} corners.
Fig. 15 shows the relation between the number of intervals and the system order n. The same result can be obtained in the frequency domain using $h^*(\theta)$ and $g^*(\theta)$ of Eqs. (10) and (11). With respect to rotating axes of Fig. 1, the value set is a rectangle parallel to the axes as shown in Fig. 16. Here we have the same situation as in the continuous case.
Therefore, the Schur stability conditions include only polynomials of the corners points R^* dependent on $\bar{h}^*$, $\underline{h}^*$, $\bar{g}^*$, $\underline{g}^*$. These are in turn dependent

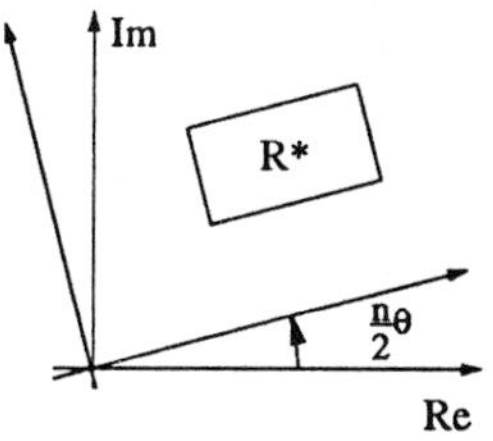

Fig. 16. Value set

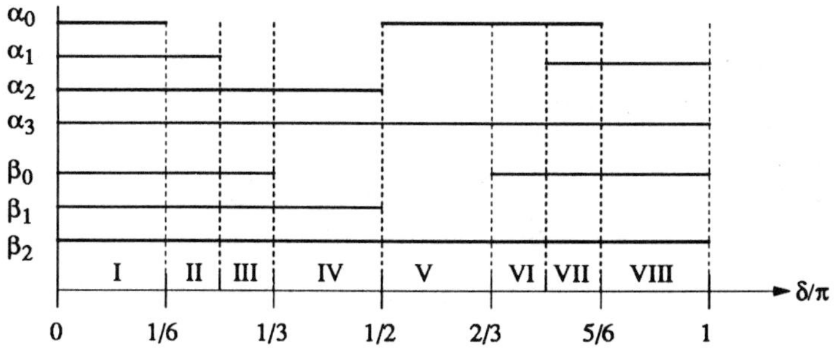

Fig. 17. $\overline{h}^*$ and $\overline{g}^*$ as functions of the angle

on θ. The maxima and minima of h^* and g^* w.r.t. α_k and β_k depend on the sign of the cosine and sine terms in Eqs. (10) and (11). The change of these signs determines the bounds of the intervals where the four corners of the box R^* remain unchanged. Fig. 17 shows the θ-intervals for $n = 6$ where the solid line means maximization of α_k, β_k. For low order polynomials $n = 2, 3, 4, 5$, we need in every interval $0, 1, 2$, and 3 extreme polynomials as in the continuous case. In general the number of corners needed is given by

End conditions + (no. of transitions × no. of intervals)

For $n = 2$ one needs $2 + (0 \times 2) = 2$ corners

For $n = 3$ one needs $1 + (1 \times 3) = 4$ corners

For $n = 4$ one needs $2 + (2 \times 4) = 10$ corners

For $n = 5$ one needs $1 + (3 \times 7) = 22$ corners

For details see [26].

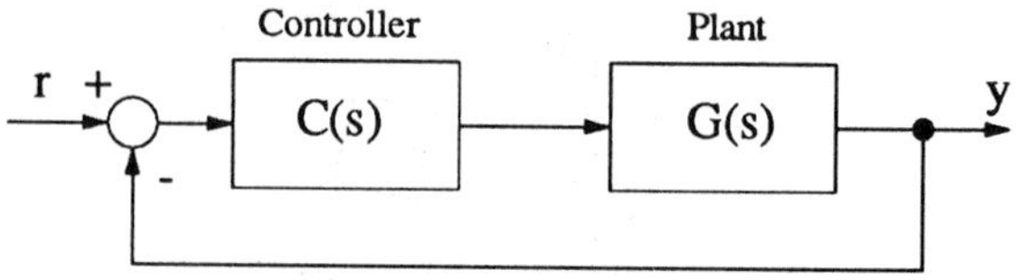

Fig. 18. Control system

E. Robust stability of control systems

1. Continuous case: robust Hurwitz stability of control systems

a. General controller

[27], [28]. Consider the Hurwitz stability of the control system in Fig. 18 with

$$G(s) = \frac{b(s)}{a(s)} \quad ; \quad C(s) = \frac{c(s)}{d(s)}$$

where

$$a(s) = a_0 s^n + a_1 s^{n-1} + ... + a_n$$
$$b(s) = b_0 s^m + b_1 s^{m-1} + ... + b_m$$

$$m \leq n-1; \quad \underline{a}_i \leq a_i \leq \overline{a}_i; \quad \underline{b}_i \leq b_i \leq \overline{b}_i$$

The characteristic equation is given by

$$f(s, \underline{a}, \underline{b}) = a(s)d(s) + b(s)c(s) \tag{83}$$

One can prove the following theorem

Theorem 24 *The interval plant $G(s)$ is Hurwitz stabilizable by a fixed controller $C(s)$ of order m if and only if a maximum of 32 edges (depending on the phase response of the controller) is Hurwitz stable.*

Proof: Fig. 19 shows the value set of the characteristic polynomial of Eq. (83). It is clear that there can be a change of the edges of the boundary of the value set only at $\varphi = 0°, 90°, 180°, 270°$ i.e. there is a maximum of four frequency intervals. As we have 8 exposed edges in each interval then the maximum number of edges to be checked for stability is 4x8=32 edges. Depending on the range of the phase of the controller we may need to check 8, 16 or 24 edges. Fig. 20 shows an example of the phase of the controller where only 24 edges are needed. □

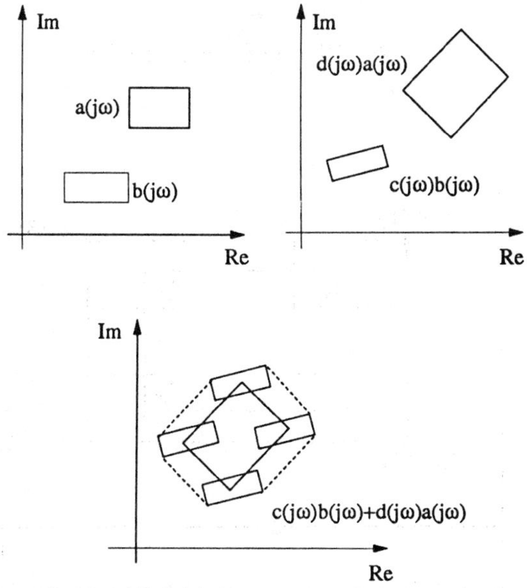

Fig. 19. Value set for one frequency

b. Proportional controller

[29]. If $C(s) = K$ positive gain then the characteristic polynomial is given by

$$f(s, \underline{a}, \underline{b}) = a(s) + Kb(s) \tag{84}$$

and the value set is given by Fig. 21. It is clear that one has to check only four edges which reduce to checking the four vertices $f_{1a} + Kf_{1b}$, $f_{2a}+$

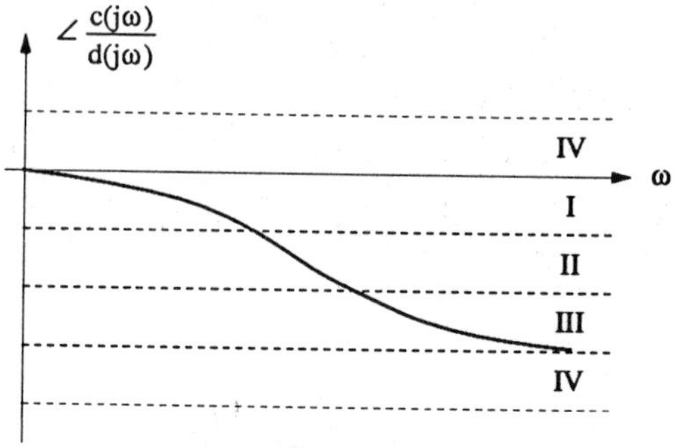

Fig. 20. Phase of the controller

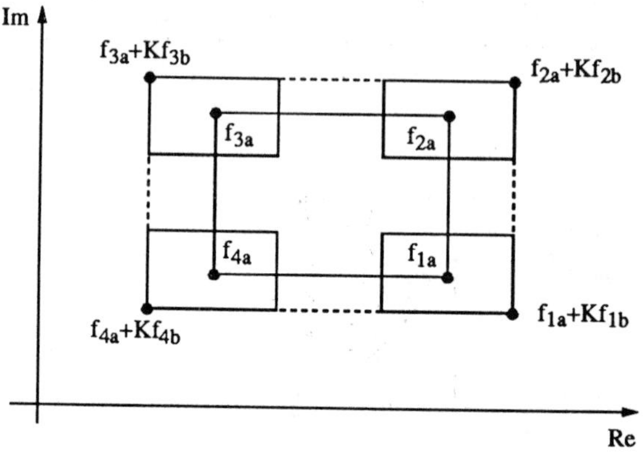

Fig. 21. Value set for control system with proportional controller

$Kf_{2b}, f_{3a} + Kf_{3b}$ and $f_{4a} + Kf_{4b}$ where $f_{1a}, ..., f_{4a}$ and $f_{1b}, ..., f_{4b}$ are the Kharitonov polynomials for $a(s)$ and $b(s)$ respectively.

c. **First order controller**

[30]. The controller is given by

$$C(s) = K\frac{s-c}{s-p} \tag{85}$$

Theorem 25 *The fixed first order controller given by Eq. (85) robustly stabilizes the interval plant if and only if it stabilizes all the sixteen Kharitonov*

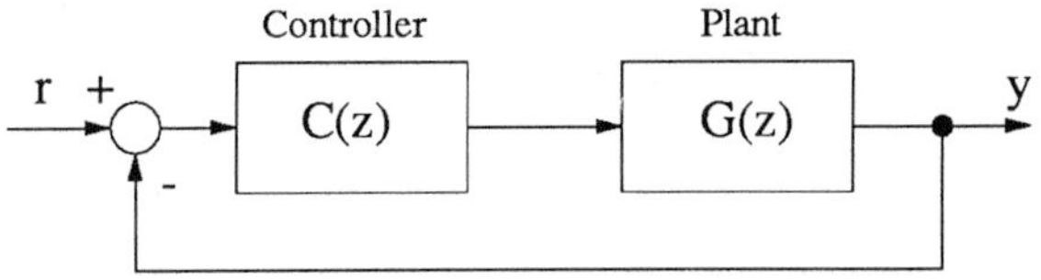

Fig. 22. Discrete system

plants (Kharitonov plants are plants whose numerator and denominator are one of the four Kharitonov polynomials). For lead or lag compensators one needs only to check eight plants.

Proof: According to theorem 14 one needs to check 2x8=16 edges and for lead or lag compensators where the phase changes between zero and 90° or between zero and −90° respectively, one needs to check only eight edges. Each edge can be given by $f(s, \lambda) = f_0(s) + \lambda(as+b)k(s)$ where $k(s)$ is an even or odd function as can be seen from Fig. 19. Using the result of III-C-1-b one has to check the vertices instead of the edges which give directly the result of the theorem. □

2. Discrete systems: robust Schur stability of control systems

a. Counterpart result

[31]. Fig. 22 shows a discrete system with interval plant and fixed controller. With

$$G(z) = \frac{b(z)}{a(z)} \quad ; \quad C(z) = \frac{c(z)}{d(z)}$$

where

$$a(z) = a_0 z^n + a_1 z^{n-1} + ... + a_n$$

$$b(z) = b_0 z^m + b_1 z^{m-1} + ... + b_m$$

$$m \leq n-1 \quad ; \quad \underline{a}_i \leq a_i \leq \overline{a}_i \quad ; \quad \underline{b}_i \leq b_i \leq \overline{b}_i$$

The characteristic equation is given by

$$f(z, \underline{a}, \underline{b}) = a(z)d(z) + b(z)c(z) \tag{86}$$

Fig. 23 shows the value set of the two interval polynomials and of the characteristic polynomial for $n = 2$ and $m = 1$. The value set of $a(z)$

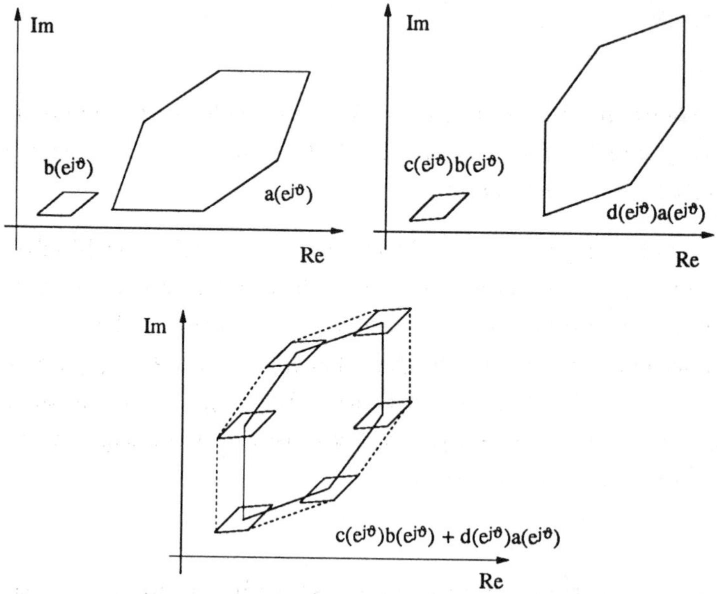

Fig. 23. Value set of a discrete system: $n = 2, m = 1$

is a parpolygon with $2(n + 1)$ edges. The value set stays essentially the same if its boundaries are images of the same set of edges of the parameter polytope. It can change significantly only if at least two edge polynomials z^i, z^j are collinear.

In the z-plane the collinearity boundaries give partitions with essentially the same value sets.

Fig. 24 shows the partitions of the [z]-plane for $n = 3$. The partitions of the [z]-plane are characterized by angles Ψ_i obtained from

$$\{\Psi_i\}_n = \bigcup_{\nu=2}^{n} \{\delta_j\}_\nu \quad 0 \leq \delta_j \leq \pi \tag{87}$$

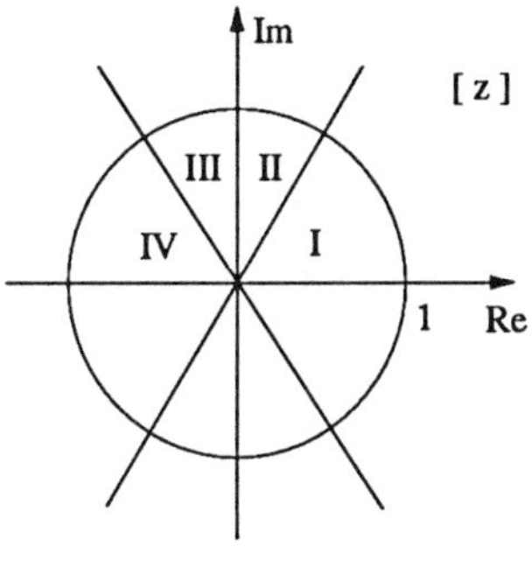

Fig. 24. z-plane partitions

Table VI. δ–angles

ν	$\{\delta_j\}_\nu/\pi$					
2	0	1/2	1			
3	0	1/3	2/3	1		
4	0	1/4	2/4	3/4	1	
5	0	1/5	2/5	3/5	4/5	1

where the angles δ_j for different ν are given by

$$\{\delta_j\}_\nu = \left\{\frac{k}{\nu}\pi; \quad k = 0, 1, ..., \nu\right\}$$

and summarized in table VI. Along the unit circle every partition is associated with an interval of ∂D of essentially the same value set.

For $n = 3$ there are four different δ-intervals. The associated corners and exposed edges are given in table VII, where + denotes the maximum and − the minimum of the associated parameter a_i in the δ-intervals I-IV. The same procedure leads to the value set of $b(z)$.

Now the influence of the controller polynomials $c(z), d(z)$ shall be investigated. The original value sets will be distorted to $a(z)d(z), b(z)c(z)$ where $c(z), d(z)$ are complex values for $|z| = 1$. For the construction of the value set only the relative position of the associated phasors is important.

Because of

$$\frac{f(z, \underline{a}, \underline{b})}{d(z)} = a(z) + \frac{c(z)}{d(z)}b(z)$$

Table VII. Corners and exposed edges for $n = 3$

	I	II	III	IV
a_0	$- - - - + + + +$	$+ + - - - - + +$	$+ + + + - - - -$	$- - + + + + - -$
a_1	$- - - + + + + -$	$- - - - + + + +$	$+ + - - - - + +$	$+ + + - - - - +$
a_2	$- - + + + + - -$	$- - - + + + + -$	$- - - + + + + -$	$- - - - + + + +$
a_3	$- + + + + - - -$	$- + + + + - - -$	$- + + + + - - -$	$- + + + + - - -$

the relative position of the a-phasors and b-phasors depends only on the phase

$$\varphi(z) = \angle \frac{c(z)}{d(z)}$$

of the controller. If the distorted a- and b-phasors become collinear for some $|z| = 1$, then the point of change between two different value sets is reached.

Such collinearity condition can be expressed as

$$\varphi(e^{j\theta}) = (j - k)\theta - \kappa\pi \tag{88}$$

$$\theta \in [0, \pi] \qquad j - k = -m, ..., 0, ..., n \qquad \kappa = 0, \pm 1, \pm 2, ...$$

The graphic interpretation in Fig. 25 for $n = 3$ and $m = 2$ helps to clarify the last formula. Every intersection of the phase of the controller with one of the directions $(-m, ..., 0, ..., n)\pi$ beginning by $..., \pi, 0, -\pi, ...$ gives a border χ_j of an interval of θ such that inside these intervals the relative position of the a- and b-phasors is the same.

The whole range of $\theta \in [0, \pi]$ is partitioned by

$$\{\Psi_i\}_n \cup \{\chi_j\}$$

where Ψ_i are boundaries from changes of the value sets of $a(z)$ and $b(z)$ and χ_j boundaries depending on the phase of the controller.

b. Analog result

[32]. Let $a(z)$ and $b(z)$ be uncertain polynomials with coefficients as given in Fig. 26. In this case we have

$$\underline{\alpha_i} \leq \alpha_i \leq \overline{\alpha_i}$$

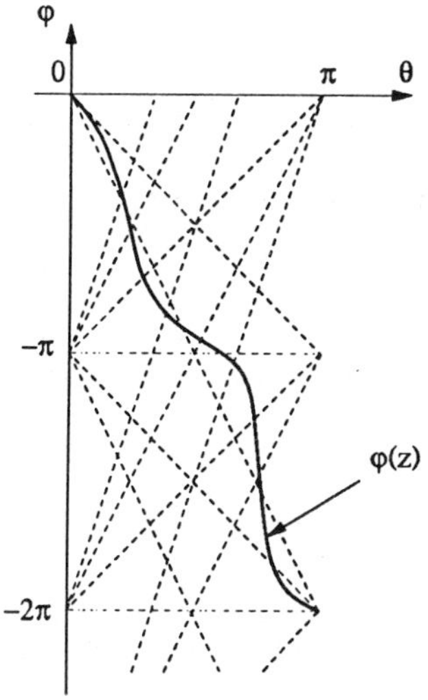

Fig. 25. Construction of the collinearity boundaries

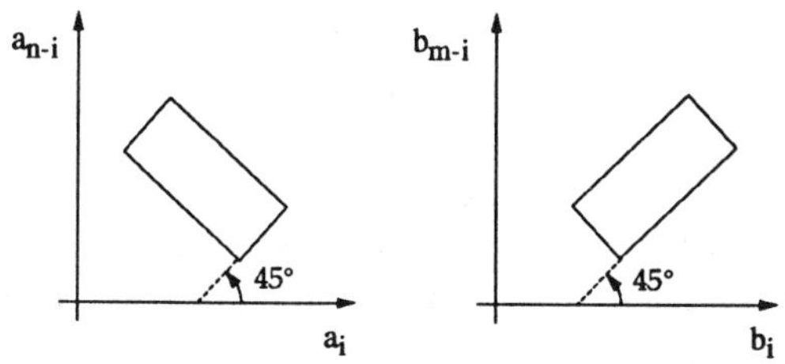

Fig. 26. The domain of the coefficients of $a(z)$ and $b(z)$

Table VIII. δ–angles for n even

ν	$\{\delta_j\}_\nu/\pi$						
2	0	1/2	1				
4	0	1/4	2/4	3/4	1		
6	0	1/6	2/6	3/6	4/6	5/6	1

Table IX. δ–angles for n odd

ν	$\{\delta_j\}_\nu/\pi$							
3	0	1/3	2/3	1				
5	0	1/5	2/5	3/5	4/5	1		
7	0	1/7	2/7	3/7	4/7	5/7	6/7	1

$$\underline{\beta_i} \leq \beta_i \leq \overline{\beta_i}$$

α_i, β_i are the parameters of the symmetrical and antisymmetrical parts of the polynomials $a(z)$ respectively.

It was shown in Fig. 16 that the associated value set for $z = e^{j\theta}$ is a rotated box with corner polynomials depending on θ. The change of the corner polynomials results from the sign change in the coefficients of h^* and g^*. The resulting θ-intervals are

$$\{\Psi_i\}_n = \bigcup_{\nu} \{\delta_j\}_\nu \tag{89}$$

$$\nu = 2, 4, 6, ..., n \qquad \text{or} \qquad \nu = 3, 5, 7, ..., n$$

where the union is over ν even or ν odd depending on n even or odd respectively. Table VIII and table IX give the associated δ-angles for n even or odd respectively. For $n = 3$ the resulting corners and exposed edges of the value set of $a(z)$ are given in table X, where $+$ means the maximum and $-$ denotes the minimum in every interval I, II, III. The same procedure leads to the value set of $b(z)$. Notice that for both m, n odd or even, no new intervals result. For the polynomials $a(z), b(z)$ the exposed edges change simultaneously. For mixed m and n, even and odd, the number of intervals grows and is obtained by mixing of the even and

Table X. Corners and exposed edges for $a(z)$ of degree $n = 3$

	I	II	III
α_0	$++--$	$--++$	$--++$
α_1	$++--$	$++--$	$++--$
β_0	$-++-$	$-++-$	$+--+$
β_1	$-++-$	$-++-$	$-++-$

odd interval sets. We obtain the same intervals $\{\Psi_i\}_n$ as those for the interval polynomials in Eq. (87).

Now the influence of the controller $c(z)/d(z)$ shall be investigated. The resulting value set of the characteristic polynomial can be obtained just by adding the distorted boxes $d(z)a(z), c(z)b(z)$.

For $z = e^{j\theta}$ the a-box is tilted by $\frac{n}{2}\theta$ and the b-box by $\frac{m}{2}\theta$. Because only the relative position of the boxes is important, collinearity and, therefore, also the significant change of the resulting value set occurs if

$$\frac{n}{2}\theta = \frac{m}{2}\theta + \varphi(e^{j\theta}) + \kappa\frac{\pi}{2} \tag{90}$$

with

$$\theta \in [0, \pi] \qquad \kappa = 0, \pm 1, \pm 2, \ldots$$

where

$$\varphi(z) = \angle\frac{c(z)}{d(z)}$$

is the phase of the controller. The interval boundary is obtained by

$$\varphi(e^{j\theta}) = \frac{n-m}{2}\theta + \kappa\frac{\pi}{2} \tag{91}$$

The relative position of the two boxes stays essentially the same as long as φ satisfies the condition

$$\frac{n-m}{2}\theta + (\kappa - 1)\frac{\pi}{2} < \varphi(e^{j\theta}) \leq \frac{n-m}{2}\theta + \kappa\frac{\pi}{2} \qquad \kappa = 0, \pm 1, \pm 2, \ldots$$

This condition can be very easily interpreted as given in Fig. 27. Every intersection of the controller phase with the tilted quadrant boundaries gives a border χ_j of a θ-interval.

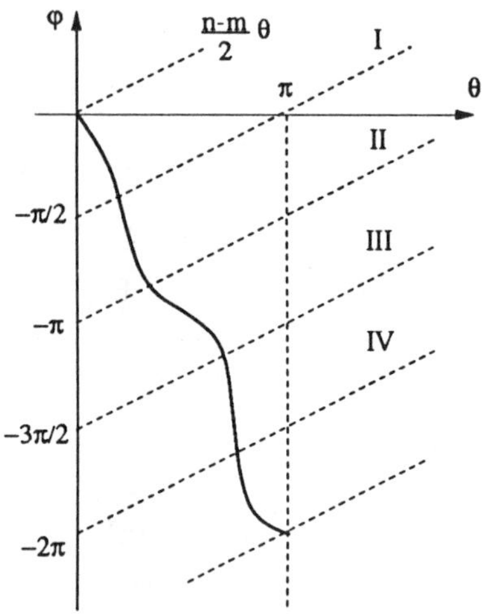

Fig. 27. Construction of the collinearity boundaries

All the θ-intervals are obtained by combining the first and second kind of interval boundaries

$$\{\Psi_i\}_n \cup \{\chi_j\}$$

where $\{\Psi_i\}_n$ results from the changes of the exposed edges of the value sets of $a(z), b(z)$ and $\{\chi_j\}$ depends on the controller phase. For each of these θ-intervals a new set of exposed edges of the value set of $f(z, \underline{a}, \underline{b})$ must be determined.

The total number of exposed edges is not higher than 8λ where λ is the number of intervals resulting from

$$\{\Psi_i\}_n \cup \{\chi_j\}$$

Some of these exposed edges can of course occur repeatedly in different intervals.

c. Proportional controller

If $C(z) = K$ positive gain then the characteristic polynomial is given by

$$f(z, \underline{a}, \underline{b}) = a(z) + Kb(z) \tag{92}$$

and the value set will be similar to Fig. 23 except that there will be no phase shift so that the whole range of $\theta \in [0, \pi]$ is partitioned as given by Eq. (87).

d. First order controller

If we use first order controller together with uncertain plant where the uncertainty is defined by the symmetric and antisymmetric parts as given by Fig. 26 then the stability will be reduced to the stability of edges of the form $f(z, \lambda) = f_0(z) + \lambda(ay + b)k(z)$, where $k(z)$ is symmetric or antisymmetric polynomial.
From section III-C-2-b one can reduce the stability of the edges to the stability of the vertices as in the continuous case.

IV. CONCLUSION

Beginning with the principle of the argument one can derive a stability criterion in the frequency domain from which Hermite-Bieler theorem is derived. From Hermite-Bieler theorem the monotony of the argument can be proved for different functions. Also the robust stability criteria for polynomials and closed loop system can be obtained in a simple way, using the results of the first part. The situation is illustrated in Fig. 28. This gives a unified approach to deal with the stability and robust stability problems. Extension can be made for other stability regions like sectors, circle in the left half plane using δ-transform, MISO and SIMO systems, systems with time delay, two dimensional systems, etc.

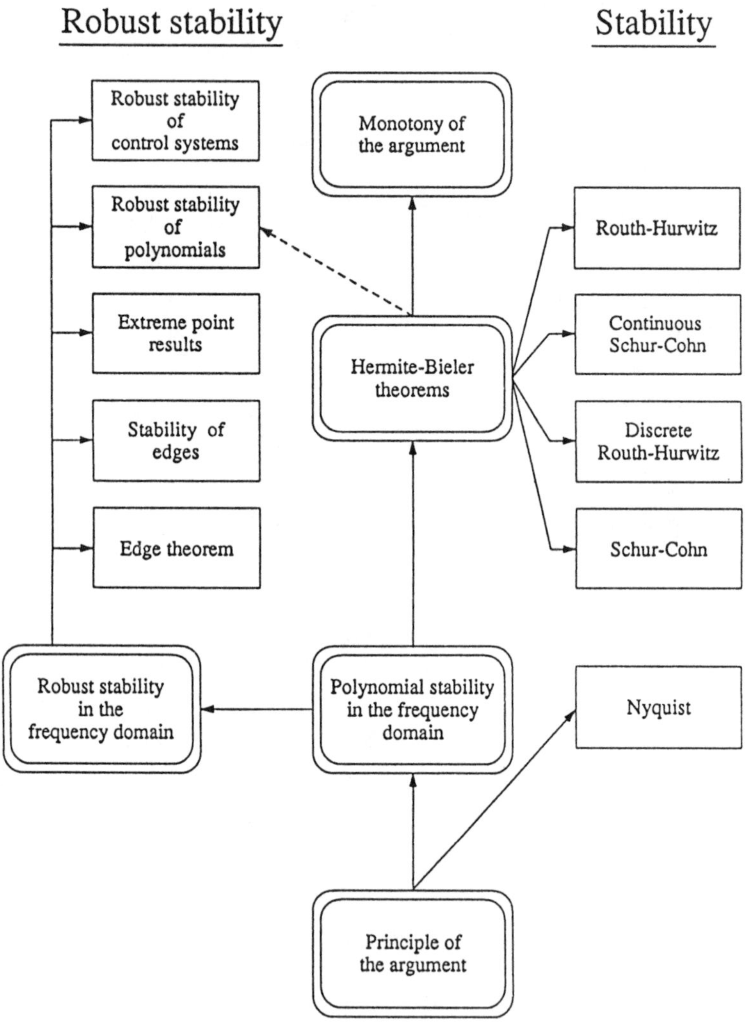

Fig. 28. The principle of the argument as origin of stability and robust stability methods

V. REFERENCES

1. M. Mansour, "A simple proof of the Routh-Hurwitz criterion", *Internal Report 88-04, Automatic Control Laboratory, Swiss Federal Institute of Technology, Zurich* (1988).

2. H. Chapellat, M. Mansour, and S. P. Bhattacharyya, "Elementary proofs of some classical stability criteria", *IEEE Transactions on Education* **33**(3), pp. 232–239 (1990).

3. M. Mansour, "Simple proof of Schur-Cohn-Jury criterium using principle of argument", *Internal Report 90-08, Automatic Control Laboratory, Swiss Federal Institute of Technology, Zurich* (1990).

4. M. Mansour, "Six stability criteria & Hermite-Bieler theorem", *Internal Report 90-09, Automatic Control Laboratory, Swiss Federal Institute of Technology, Zurich* (1990).

5. E.G. Phillips, "Functions of a complex variable with applications", *Oliver and Boyd* (1963).

6. H.W. Schüssler, "A stability theorem for discrete systems", *IEEE Transactions on Acoustics, Speech and Signal Processing* **ASSP-24**, pp. 87–89 (1976).

7. M. Mansour and F.J. Kraus, "Argument conditions for Hurwitz and Schur stable polynomials and the robust stability problem", *Internal Report 90-05, Automatic Control Laboratory, Swiss Federal Institute of Technology, Zurich* (1990).

8. J. F. Delansky and N. K. Bose, "Real and complex polynomial stability and stability domain construction via network realizability theory", *International Journal of Control* **48**(3), pp. 1343–1349 (1988).

9. J. F. Delansky and N. K. Bose, "Schur stability and stability domain construction", *International Journal of Control* **49**(4), pp. 1175–1183 (1989).

10. M. Mansour and F.J. Kraus, "On robust stability of Schur polynomials", *Internal Report 87-05, Automatic Control Laboratory, Swiss Federal Institute of Technology, Zurich* (1987).

11. S. Dasgupta, P. J. Parker, B. D. O. Anderson, F. J. Kraus, and M. Mansour, "Frequency domain conditions for the robust stability of linear and nonlinear systems", *IEEE Transactions on Circuits and Systems* **38**(4), pp. 389–397 (1991).

12. A. C. Bartlett, C. V. Hollot, and H. Lin, "Root locations of an entire polytope of polynomials: it suffices to check the edges", *Mathematics of Control, Signals and Systems* **1**, pp. 61–71 (1988).

13. E. Zeheb, "Necessary and sufficient conditions for root clustering of a polytope of polynomials in a simply connected domain", *IEEE Transactions on Automatic Control* **34**, pp. 986–990 (1989).

14. S. Bialas, "A necessary and sufficient condition for the stability of convex combinations of stable polynomials and matrices", *Bulletin of the Polish Academy of Sciences, Technical Sciences* **33**, pp. 473–480 (1985).

15. F. J. Kraus, M. Mansour, W. Truöl, and B.D.O. Anderson, "Robust stability of control systems: extreme point results for the stability of edges", *Internal Report 91-06, Automatic Control Laboratory, Swiss Federal Institute of Technology, Zurich* (1991).

16. F. J. Kraus, M. Mansour, and E. I. Jury, "Robust Schur-stability of interval polynomials", *Proceedings of the 28th IEEE Conference on Decision and Control* **3**, pp. 1908–1910 (1989).

17. J. E. Ackermann and B. R. Barmish, "Robust Schur stability of a polytope of polynomials", *IEEE Transactions on Automatic Control* **33**, pp. 984–986 (1988).

18. V. L. Kharitonov, "Asymptotic stability of an equilibrium position of a family of systems of linear differential equations", *Differentsial'nye Uravneniya* **14**, pp. 2086–2088 (1978).

19. B. D. O. Anderson, E. I. Jury, and M. Mansour, "On robust Hurwitz polynomials", *IEEE Transactions on Automatic Control* **32**, pp. 909–913 (1987).

20. S. Dasgupta, "Kharitonov's theorem revised", *Systems & Control Letters* **11**, pp. 381–384 (1988).

21. R. J. Minnichelli, J. J. Anagnost, and C. A. Desoer, "An elementary proof of Kharitonov's stability theorem with extensions", *IEEE Transactions on Automatic Control* **34**(9), pp. 995–998 (1989).

22. F. J. Kraus, B. D. O. Anderson, E. I. Jury, and M. Mansour, "On the robustness of low-order Schur polynomials", *IEEE Transactions on Circuits and Systems* **35**, pp. 570–577 (1988).

23. C. V. Hollot and A. C. Bartlett, "Some discrete-time counterparts to Kharitonov's stability criterion for uncertain systems", *IEEE Transactions on Automatic Control* **31**, pp. 355–356 (1986).

24. F. J. Kraus and M. Mansour, "On robust stability of discrete systems", *Proceedings of the 29th IEEE Conference on Decision and Control* **2**, pp. 421–422 (1990).

25. M. Mansour, "Instability criteria of linear discrete systems", *Automatica* **2**, pp. 167–178 (1965).

26. M. Mansour, F. J. Kraus, and B.D.O. Anderson, "Strong Kharitonov theorem for discrete systems", *in Robustness in Identification and Control, Ed. M. Milanese, R. Tempo, A. Vicino, Plenum Publishing Corporation, 1989. Also in Recent Advances in Robust Control, Ed. P. Dorato, R.K. Yedavalli, IEEE Press, 1990.*

27. H. Chapellat and S. P. Bhattacharyya, "A generalization of Kharitonov's theorem: robust stability of interval plants", *IEEE Transactions on Automatic Control* **34**, pp. 306–311 (1989).

28. F. J. Kraus and W. Truöl, "Robust stability of control systems with polytopical uncertainty: a Nyquist approach", *International Journal of Control* **53**(4), pp. 967–983 (1991).

29. B. K. Ghosh, "Some new results on the simultaneouis stabilizability of a family of single input, singel output systems", *Systems & Control Letters* **6**, pp. 39–45 (1985).

30. C. V. Hollot, F.J. Kraus, R. Tempo, and B.R. Barmish, "Extreme point results for robust stabilization of interval plants with first order compensators", *Proceedings of the American Control Conference, San Diego* **3**, pp. 2533–2538 (1990).

31. F. J. Kraus and M. Mansour, "Robust discrete control", *Proceedings of the European Control Conference, Grenoble* (1991).

32. F. J. Kraus and M. Mansour, "Robust Schur stable control systems", *Proceedings of the American Control Conference, Boston* (1991).

CONSTRAINED CONTROL FOR SYSTEMS WITH UNKNOWN DISTURBANCES.

F. Blanchini

Dipartimento di Matematica e Informatica
Universita' degli Studi di Udine
Udine, Italy.

CONTROL AND DYNAMIC SYSTEMS, VOL. 51

I. INTRODUCTION.

One of the main goals a control system is to counteract the effects of the disturbances. With the term "disturbances" (sometimes the term "noise" is used) we mean input signals that affect the system removing it from the desired working conditions.

There are different classifications for the disturbances and each of them is related to the particular perspective we are considering. One important characteristic of the disturbances is their source. From this point of view, we can distinguish essentially two classes of disturbances. The first class is that of the external disturbances, that are due to outside agents, in the sense that they are not intrinsic of the system model. There are many examples of such disturbances: the rain for an interconnected reservoir system, the wind for an airplane.

There are other kinds of disturbances, say the internal ones, which are intrinsic of the system, in the sense that they are due to variations of the system model. Such disturbances are often referred to as parametric variations. Such parameter variations are due to different and unpredictable factors that may not be conveniently assumed as inputs.

Again the examples of such kind of disturbances are very frequent in any field of engineering. In electric networks, for instance, parameters such as the impedances, for which a nominal value is usually assumed, may be subject to fluctuations due to the temperature, air wetness, faults and so on.

Another classification for the disturbances is related to their behavior. From this point of view such classification is essentially based on the kind of model we assume.

In lucky cases the disturbance may be reasonably assumed to be known at least from a qualitative point of view. For example, the load of an elevator may be reasonably supposed to be a positive constant weight and it is natural to assume, in the design stage, the maximal admissible load as representative external input.

Unfortunately, in the major part of cases the disturbances are unknown even from a "qualitative" point of view. Let us think of a system as an airplane subject to wind or of an elastic structure during an earthquake. It is well known that such kinds of disturbances are unpredictable and cannot be reasonably modelled by a fixed input.

There are essentially two approaches to cope with uncertain disturbances. According to the so called stochastic approach, it is assumed that the unknown disturbances follow a statistic law which is supposed to be known. In other terms it is assumed that a probability distribution for the noise is known and statistical objectives are pursued such as the minimization of the output variance.

Even if very elegant results have been derived in this area, in practice difficulties arise since it is not easy to find accurate probability distributions for many real cases. Moreover, many of the results that have been derived, are related to particular kinds of distribution laws, such as the Gaussian one, and such typical distributions are not in many cases satisfactory to have a correct representation for the noise.

The alternative approach is the deterministic one. It is assumed that the disturbances are functions belonging to an assigned family $\mathscr{D}$. The problems arising in this area have the following abstract form: given a property $\mathscr{P}$ of deterministic type (for instance stability or boundedness), assure that such a property holds for every disturbance in $\mathscr{D}$. A typical family $\mathscr{D}$ is the class of all functions having values in an assigned compact set. The property $\mathscr{P}$ is said robust if it is fulfilled for every disturbance of the family $\mathscr{D}$.

Such an approach is often referred to as "worst case analysis" in which the nature, which is seen as a player in competition with the controller, chooses the worst (from the point of view of the property $\mathscr{P}$) disturbing function in $\mathscr{D}$.

The assumption of the deterministic approach seems to be more realistic in the sense that while a probability measure for the noise is not easily determined, a bound for the noise size may be very reasonably fixed.

On the other hand, in many cases, the deterministic approach may be too conservative, in the sense that in general nature is not so adverse as it is assumed by the deterministic approach.

Roughly speaking, the deterministic approach is very reasonable whenever the property $\mathscr{P}$ mentioned above is "crucial". It is clear for instance that no one would like to fly on an airplane which is stable with a (even high) probability less than one. On the other hand, in other kinds of problems, the requirement that a given property $\mathscr{P}$ is satisfied in every case may have a too high price in terms of control action to be realistically applied.

The approach we follow in this contribution is the deterministic one.

In literature different kinds of properties have been considered. We recall among other properties the ultimate boundedness, the quadratic stability, the robust stability. To have a complete overview about these problems, the reader is referred to the other contributions of this book.

In this chapter we consider linear time-invariant systems and we tackle the problem of keeping the output of the system in an assigned set in the presence of an external unknown disturbance which is only assumed to take its values in an assigned compact set.

In view of the real problems it is important to emphasize that there are different kinds of outputs and inputs.

In general there is a set of variables which are the primary concern of the control and constitute what is called the system controlled output. The selection of such variables is however suggested by the context in which the system is operating. Essentially there are two reasons that lead to the selection of the controlled output variables. First the system safety, in the sense that a variable that gets out of its operative range may produce a failure of the system. In a nuclear reactor, a variable such as the core temperature is clearly crucial ad its value must be absolutely kept in a fixed range. Second, the task for which a system is used naturally induces a preference for some variables with respect to other ones. For instance, it is clear that in a servo-position system the angle has a priority if compared with other variables, and that in a power distribution network variables such as tension and frequency deserve a primary attention.

Together with the controlled output, it is important to recall the concept of the measured output, say the variables which are available on-line.

Also for the inputs a distinction may be done between the control input, that is to say the variables that are modified by the compensator in order to perform the control, and the disturbance input. The general structure of this situation is represented in Fig.1.

It is well known [1] that such a situation may be analyzed from the game theoretic point of view, in the sense that the control and the disturbance may be seen as two agents acting each against the other.

The classical control theory, based on the input-output representation, copes with the problem of the disturbance by generically considering that the introduction of a high gain compensator reduces the disturbance effect on the output.

The compensator design is performed by choosing typical forms of disturbance such as steps or impulses and assigning some design requirements such as the steady state error or the maximal overshoot.

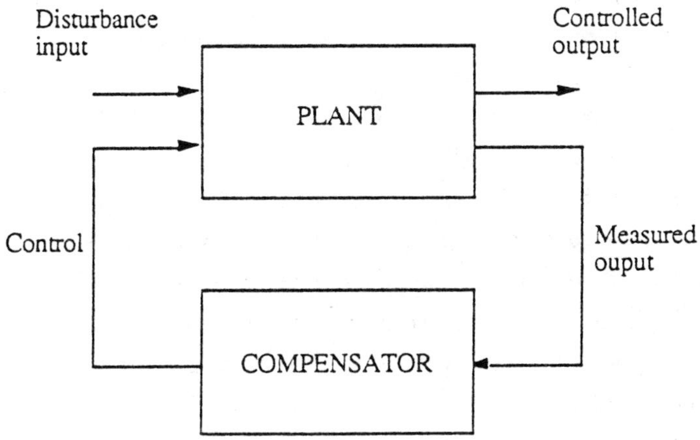

Fig. 1. General feedback structure

The control techniques via pole assignment or linear quadratic optimization, do not consider the problem of the disturbance, since the goal is that of assuring a certain convergence speed of the state to the origin. As it is known, the initial state contains all the information about the effect of a disturbance whose action is confined in the past. For this reason, it has been introduced the concept of "persistent disturbance" to contemplate the cases in which we cannot assume that the disturbance acted before the initial time.

More effective techniques to handle the problem of the disturbance are the results of more recent researches.

The first aspect of the problem of persistent disturbance rejection is to fix a good characterization of the disturbances.

One of the important models of disturbances introduced in literature is obtained by the assumption that the disturbance is the output generated by the free response of a linear system. Such an assumption clearly embodies the disturbances often considered in the classical control such as steps and sinusoidal signals etc. as a particular cases [2][3].

A further generalization of this class of disturbances is the notion of waveform structure disturbance [4]. Essentially a waveform structure disturbance is a function such that the time axis may be partitioned in finite intervals and in each of them the disturbance is of the form above. Such kind of disturbances may be obtained as the output of a linear system having as input a function which is zero almost everywhere with impulses on the interval separation points.

A completely different approach, the so called set-theoretic control, was introduced in [5]. In that paper, the concepts introduced in [1] from a game theoretic point of view, have been reconsidered and a general technique has been presented to solve the problem of the permanence of the state in an assigned constraint set using controls which are subject to constraints, in the presence of disturbances which are unknown but constrained in a compact set. Different cases have been considered: the linear and the nonlinear controller case, the perfect and the imperfect state observation. In the same year very similar results have been presented in [6], even if only nonlinear controllers are considered. In this second paper it was introduced, even if not developed, the idea of using polyhedral sets to confine the state. Here we analyze the properties of the polyhedral sets as candidate state bounding sets in the case of a linear controller.

The set theoretic control problem was analyzed in [7][8]. Rectangular bounds for the state and the control were assumed and a one dimensional bounded disturbance was considered. The problem has been solved by confining the state in an ellipsoidal set. It has been shown how a feedback matrix may be derived by an optimization procedures involving the Lagrange multipliers theory [7] or direct search methods [8]. These results have been more recently applied to the control of a nuclear power plant [9].

The disturbance rejection problem may be approached via H_∞ theory. The goal of such a theory (see for instance [10]) is to synthesize a compensator minimizing the quadratic norm of the output corresponding to the class of inputs of unit quadratic norms. This problem is known to be equivalent to the minimization of the H_∞ norm of the closed-loop transfer function.

Recently the problem of optimal disturbance rejection has been formulated as the minimization of the maximum amplitude of the output corresponding to unit amplitude inputs. Such a minimization is equivalent to the minimization of the L_1

norm of the impulse response and so this theory is known as the L_1 control theory (respectively l_1 in the discrete-time case) [11][12] [13][14][15].
Both L_1 and H_∞ approaches assume zero initial conditions. This is the main distinction between these theories and the set theoretic control approach in which the determination of the set of initial conditions for which the system is kept within its boundaries for every disturbance is a fundamental aspect of the problem. As we shall see in the next chapters, this aspect leads to the notion of positive invariance which is the central point of the theory. A compensator solves the set theoretic control problem if and only if there exists a positively invariant set which is compatible with the constraints. Moreover, if a set with such a property is found, then the problem has a solution for every initial condition in it.
In [7] and [9] (where the notion of PI invariance is not explicitly introduced) the problem is solved by considering ellipsoidal positively invariant bounding sets.
Even if the ellipsoidal sets are convenient since they may be represented by a finite number of parameters which is a function of the state dimension, they may be restrictive for the problem in the sense that in general they do not give a good approximation for the set of all initial states (the maximal positively invariant set) for which a solution exists. Moreover the conditions given in [7] for the positive invariance of a PI set are only sufficient. As a consequence, the derived bounds for the disturbance size that assures that no boundary violations occur may turn out to be too restrictive.
In this contribution we analyze the properties of the polyhedral sets as candidate bounding sets for the state and the control. The polyhedral bounds naturally arise in problems in which the constraints are simply given by an upper and a lower bound on the input and output variables.
One important property of this class of sets is that we can give necessary and sufficient conditions for the positive invariance of a polyhedral set. Moreover under some non-trivial conditions we can arbitrarily closely approximate the maximal positively invariant set in the continuous-time case and we can exactly compute it in the discrete-time case.
If polyhedral constraints are considered, the maximal disturbance size which is compatible with the constraints may be exactly determined. These results are related to the L_1 theory which considers hyper-boxes that are a particular class of polyhedral sets.

II. NOTATION.

In the sequel we use the following notations.

$\mathcal{V}, \mathcal{P}, \mathcal{U}, \mathcal{D}, \mathcal{M}, \ldots$	sets
$\mathcal{R}$	real field
i,j,k,h,l,m,n,p,q,r,s	integers
$\alpha, \beta, \gamma, \ldots$	real or complex numbers
A,B,C, ...	matrices
A_i	i-th row of the matrix A
I	identity matrix
$\sigma(A)$	set of the eigenvalues of A
u,v,w,x,y,z,e	vectors
x_i	i-th component of x
$\lambda \mathcal{S}$, $\lambda \geq 0$	$\{ x: x=\lambda z, \ z \in \mathcal{S} \}$
int$\{\mathcal{S}\}$	interior of $\mathcal{S}$
$\partial \mathcal{S}$	boundary of $\mathcal{S}$
$\mathcal{C}_{\mathcal{S}}(x)$	tangent cone to $\mathcal{S}$ in x
conv$\{\mathcal{S}\}$	convex hull of $\mathcal{S}$
vert$\{\mathcal{S}\}$	(where $\mathcal{S}$ is a polytope) set of vertices of $\mathcal{S}$
C-set	convex and compact set containing the origin in its interior
$\|\cdot\|$	norm (the Euclidean one if not differently specified)
$\phi(\cdot): \mathcal{A} \to \mathcal{R}$	real valued function defined on $\mathcal{A}$
$\mathcal{B}_\phi(\lambda)$	$\{ a \in \mathcal{A} : \phi(a) \leq \lambda \}$
$\wp$	(continuous-time case) derivative operator
$\wp$	(discrete-time case) shift forward operator
$x'(t)$	derivative of x(t)
x^T	transpose of x
$x \leq y$	$x_i \leq y_i$, i=1, 2, ..

III. DISTURBANCE REJECTION PROBLEM.

We consider linear systems represented by the following linear equations:

$$\wp\, x(t) = A\, x(t) + B\, u(t) + E\, v(t) \tag{1a}$$

$$y(t) = C\, x(t) + D\, v(t) \tag{1b}$$

$$e(t) = N\, x(t) + M\, v(t) \tag{1c}$$

where

$x \in \mathcal{R}^n$	is the system state,
$u \in \mathcal{R}^q$	is the control input,
$v \in \mathcal{R}^m$	is the disturbance input,
$y \in \mathcal{R}^p$	is the measured output,
$e \in \mathcal{R}^r$	is the error output.

A,B,E,C,D,N,M are constant matrices of appropriate dimensions. The symbol $\wp$ represents the derivative operator in the continuous-time case and the shift forward operator in the discrete-time case.

The model described by equations (1,a,b,c) represents the general situation in which the disturbance input and the control input enter the system through different channels. Such a model also embodies the case in which there is a distinction between the measured output y and the variables e representing the error, that is to say the deviation from the desired value of the controlled variables.

The structure (1) encompasses the following particular but important cases:

C1) The controlled variables are the state components: N=I, M = O.

C2) The full state measurement is exactly available: C=I, D=O.

C3) The measured output is the controlled output: N=C, D=M.

Our discussion considers the problem with input and output constraints in the presence of disturbance. According to many realistic situations, such disturbances are unknown but they are assumed to take values in a compact and convex set $\mathcal{V}$ containing the origin as an interior point.

The most natural way to impose constraints on inputs and outputs is to assign an upper and a lower bound to each component, say

$$e(t) \in \mathscr{E} = \{e: e^- \le e \le e^+\} \subset \mathscr{R}^r, \ t \ge 0, \tag{2a}$$

$$u(t) \in \mathscr{U} = \{u: u^- \le u \le u^+\} \subset \mathscr{R}^q, \ t \ge 0, \tag{2b}$$

$$v(t) \in \mathscr{V} = \{v: v^- \le v \le v^+\} \subset \mathscr{R}^m, \ t \ge 0, \tag{2c}$$

where e^-, u^-, v^- and e^+, u^+, v^+ are assigned vectors and the inequalities hold componentwise.

The most general form of a linear shift-invariant feedback compensator for (1) is given by the system represented by the following equations

$$\wp\, z(t) = F\, z(t) + G\, y(t) \tag{3a}$$

$$u(t) = H\, z(t) + K\, y(t) \tag{3b}$$

where $z(t) \in \mathscr{R}^l$.

It is known that the dynamic regulator problem may be studied as an equivalent static output feedback problem. The application of the compensator (3) to system (1) produces the same closed-loop system obtained by applying an output feedback compensator of the form

$$u_A(t) = K_A\, y_A(t) \tag{4}$$

with

$$K_A = \begin{bmatrix} K & H \\ G & F \end{bmatrix}$$

to the augmented system

$$\wp\, x_A(t) = A_A x_A(t) + B_A u_A(t) + E_A\, v(t) \tag{5a}$$

$$y_A(t) = C_A\, x_A(t) + D_A\, v(t) \tag{5b}$$

where it has been denoted by

$$x_A(t) = \begin{bmatrix} x(t) \\ z(t) \end{bmatrix},$$

$$A_A = \begin{bmatrix} A & 0 \\ 0 & 0 \end{bmatrix}, \quad B_A = \begin{bmatrix} B & 0 \\ 0 & I \end{bmatrix}, \quad C_A = \begin{bmatrix} C & 0 \\ 0 & I \end{bmatrix},$$

$$E_A = \begin{bmatrix} E \\ 0 \end{bmatrix}, \quad D_A = \begin{bmatrix} D \\ 0 \end{bmatrix}.$$

The resulting closed-loop system is therefore

$$\wp\, x_A(t) = (A_A + B_A K_A C_A)\, x_A(t) + (B_A K_A D_A + E_A)\, v(t) \tag{6a}$$

$$e(t) = [\,N \quad 0\,]\; x_A(t) + M\, v(t) = N_A\, x_A(t) + M\, v(t) \tag{6b}$$

which is what we obtain applying compensator (3) to system (1).

The conversion of the original problem to the static output feedback problem for system (5) introduces a redundancy due to the external equivalence between compensators by means of state transformations. We see that the compensators represented by (F,G,H,K) and $(T^{-1}FT, T^{-1}G, HT, K)$ are strictly equivalent if we consider the input-output behavior of system (6).

However, in many cases the compensator is implemented by a physical device such as an analog electric circuit. So its state variables correspond to physical quantities and may be subject to constraints of the form

$$z(t) \in \mathcal{Z} = \{\, z:\ z^- \leq z \leq z^+ \,\} \subset \mathcal{R}^l,\ t \geq 0. \tag{7}$$

In view of the additional constraints (7), the compensator variables have to be considered as an output of the closed-loop system and so the parametrization (4) is no redundant anymore.

From the considerations above, without loss of generality, we can limit our attention to the case of the linear system (1) with a compensator of the form

$$u(t)=Ky(t).$$

which produces the closed-loop system

$$\wp\, x(t) = (A+BKC)\, x(t) + (E+BKD)\, v(t) \tag{8}$$

We see that the feedback control law modifies the disturbance input matrix E if the measured output is affected by the disturbance, i.e. the matrix D is nonzero.

Let us consider now expression (2a). Such a constraint induces the following constraint for the state x(t)

$$e^- \leq \ N\, x(t) + M\, v(t) \ \leq \ e^+$$

The constraint above encloses the uncertain quantity v(t) that can be eliminated as follows. First observe that at each time, x(t) does not depend on the current value of v(t), and it has been assumed that v(t) is an arbitrary function of time with values in $\mathcal{V}$. So each component of this constraint

$$e_i^- \leq N_i\ x(t) + M_i\ v(t) \leq e_i^+$$

can be replaced by the following constraint for x(t)

$$\mathit{e}_i^- = e_i^- - \min_{v \in \mathcal{V}} (M_i\ v) \leq N_i\ x(t) \leq e_i^+ - \max_{v \in \mathcal{V}} (M_i\ v) = \mathit{e}_i^+$$

The shift terms $\max_{v \in \mathcal{V}} (M_i\ v)$ and $\min_{v \in \mathcal{V}} (M_i\ v)$ may be determined by solving a simple convex programming problem. As a conclusion we can assume that e(t) does not depend directly on v(t), that is to say M=0.

Let us now assume that a compensator u(t)=Ky(t) is given. In view of (1b), the constraint (2b) produces the following linear constraint for the state x(t)

$$u^- \leq KC\ x(t) + KD\ v(t) \leq u^+ .$$

We can eliminate the term KD v(t) if we replace the vectors u^- and u^+ by u^- and u^+ which are respectively the vectors whose components are

$$\mathit{u}_i^- = u_i^- - \min_{v \in \mathcal{V}} ([KD]_i\ v) \ \text{and}\ \mathit{u}_i^+ = u_i^+ - \max_{v \in \mathcal{V}} ([KD]_i\ v)$$

Therefore we have the following constraints for x(t)

$$\mathit{e}^- \leq N\ x(t) \leq \mathit{e}^+ \tag{9a}$$

$$\mathit{u}^- \leq KC\ x(t) \leq \mathit{u}^+ . \tag{9b}$$

It should be noticed that expression (9a) is derived from the only constraints on e(t), while the second one (9b) is a function the compensator parameters.

In the sequel, for the sake of simplicity, we assume D=0 and M=0. This choice is not restrictive since, for any fixed K, we have to cope with a system of the form (8) with the constraints (9) in which the terms related to the disturbance have been eliminated. Moreover, we consider a state feedback given by u=Kx since the output feedback case may be handled by constraining K to be of the form $K=K^*C$.

For any fixed compensator, the linear constraints (9) represent a polyhedral set in which the closed-loop state variables must be included at each time.

We see that although the constraints (2) are hyper-boxes, they naturally lead to general polyhedral constraints and so we generalize our discussion to contemplate this more general case. Thus, in the following, we consider systems having polyhedral constraint sets and we present conditions assuring the permanence of the state and the control in these constraints for any possible action of the disturbance.

IV. INVARIANT SETS AND LYAPUNOV FUNCTIONS

The theory of the invariant sets and of the Lyapunov functions plays an important role in the problem of the constrained control of uncertain systems. In this chapter we analyze in detail these concepts which are also strictly connected to the system stability properties.

We consider now discrete-time systems of the form

$$x(t+1) = A_o\, x(t) + B\, u(t) + E\, v(t) \tag{10a}$$

or continuous-time systems of the form

$$\frac{d}{dt}\, x(t) = A_o\, x(t) + B\, u(t) + E\, v(t)\,. \tag{10b}$$

In the continuous-time case we assume that v(t) is a piecewise continuous function of time (i.e. any finite interval contains at most a finite number of points in which v(t) is not continuous and in each of these points v(t) has finite right and left limits). We assume that the following constraints are given

$$x(t) \in \mathcal{X} \subset \mathcal{R}^n,\ t \geq 0, \tag{11a}$$

$$u(t) \in \mathcal{U} \subset \mathcal{R}^q,\ t \geq 0, \tag{11b}$$

$$v(t) \in \mathcal{V} \subset \mathcal{R}^m,\ t \geq 0. \tag{11c}$$

where $\mathcal{X}, \mathcal{U}, \mathcal{V}$ are closed and convex polyhedral sets containing the origin in their interior, with $\mathcal{U}$ and $\mathcal{V}$ compact.

If we consider a linear state feedback compensator of the form

$$u(t) = K\, x(t). \tag{12}$$

the closed-loop system is

$$x(t+1) = [A_o + BK]\, x(t) + E\, v(t) = A\, x(t) + E\, v(t) \qquad (13a)$$

or

$$\frac{d}{dt} x(t) = [A_o + BK]\, x(t) + E\, v(t) = A\, x(t) + E\, v(t). \qquad (13b)$$

For any assigned compensator, the permanence of the state in its constraint set clearly depends on the initial condition. We can formally introduce the following problem [16].

Problem 1. (Disturbed Linear Constrained Regulator Problem). Find a compensator u=Kx and an initial condition set $\mathcal{S}_0$ such that for each $x(0)\in\mathcal{S}_0$ the constraints (11a) and (11b) are satisfied for each v( ) fulfilling (11c).

Given a regulator K, we say that an initial condition $x(0)\in\mathcal{R}^n$ is feasible if it assures the permanence of the state in its constraint set for all t>0. The feasible initial condition set $\mathcal{S}_0$ required in Problem 1 has an important meaning: it represents a "region of efficacy" for the compensator, in the sense that if the state is contained in it at a given time, then the constraints will be satisfied at all future times.

As it is known, the zero state of a linear model often represents the nominal work point for the system. So it is natural to require that the initial condition set contains the origin.

We may wish to assign the set $\mathcal{S}_0$ and impose the permanence of the state and the control in their constraint sets starting from any state in it. The following problem thus results

Problem 2. (Disturbed Linear Constrained Regulator Problem with Assigned Initial Condition Set). Given the set $\mathcal{S}_0\subset\mathcal{X}$ containing the origin, find a compensator u=Kx such that for each $x(0)\in\mathcal{S}_0$ the constraints (11a) and (11b) are fulfilled for each $v(\cdot)$ fulfilling (11c).

Several works dealing with disturbance rejection [11][12] [13] [14] [15] do not explicitly mention the problem of considering a set of feasible initial conditions since it is assumed x(0)=0. Conversely, here we are mainly interested to the problem of the determination of such a set, in particular we wish to find the set of all feasible initial conditions.

To this aim we introduce the following definition.

Definition 1. The set $\mathcal{S}$ is said to be Positively Invariant (PI) for (13a) (or 13b) if the condition $x(t_0) \in \mathcal{S}$ for some t_0 implies that $x(t) \in \mathcal{S}$ for $t \geq t_0$. We shall also say Positively $\mathcal{V}$- Invariant whenever it may be ambiguous to which set $\mathcal{V}$ we are referring to.

The introduction of the compensator (12) in view of (11b) produces a new constraint for the system state. For this reason we consider the concept of admissibility of a state vector with respect to the control law (12).

Definition 2. The set $\mathcal{S}$ is said to be admissible [17] with respect to the control law (12) if the condition $x(t) \in \mathcal{S}$ implies that $u(t) = Kx(t) \in \mathcal{U}$.

Both concepts of PI and admissibility play a fundamental role in the constrained regulator problem. It is an immediate consequence of the Definitions 1 and 2 that if a set $\mathcal{S}$ contained in the set $\mathcal{X}$ which is positively invariant and admissible with respect to the control law u=Kx is known, then Problem 1 is solved for every subset $\mathcal{S}_0$ of $\mathcal{S}$. Conversely, if there exist a set $\mathcal{S}_0$ and a control K as required by Problems 1 and Problem 2, then we have necessarily that $\mathcal{S}_0 \subset \mathcal{S}$, where $\mathcal{S}$ is a set which is PI and admissible with respect to the control law K [16]. This is immediately seen since the set of all feasible initial conditions is PI and admissible, in fact it is the maximal PI set in the sense that it contains every other PI set (with that control).
We see that the existence of a solution with the control law u=Kx for Problem 1 is equivalent to the existence of a PI set which is admissible (with respect to K) and it is contained in the state constraint set, while Problem 2 is solved if and only if the assigned initial condition set is contained in a set $\mathcal{S} \subset \mathcal{X}$ having these properties.
It should be noted that, if the compensator is given, then the admissibility constraint $Kx(t) \in \mathcal{U}$ is fixed and may be included in the state constraint $\mathcal{X}$, so the problem we have to consider in this case is the simple existence of a PI set contained in $\mathcal{X}$. Conversely, if the compensator is to be found, the admissibility constraints are a function of the matrix K, and this is the reason of the distinction between the natural constraints represented by the set $\mathcal{X}$ which are intrinsic of the system and the admissibility constraints which are due to the compensator.

As we shall see in section VI, we can check if a given control assures that no boundary violations occur for the zero initial condition, without evaluating a PI set. For instance (see [11]-[15]) if the infinity norm is considered (this is equivalent to choose bounding sets of a symmetric rectangular shape) and if it is assumed x(0)=0, the permanence of the output in the unit ball is assured for all unit norm disturbances if and only if the L_1 norm of the impulse response is less than one. If this is the case, a PI set does exist even if it is not explicitly determined.

There exist important reasons that suggest to determine in an explicit way a PI and admissible set. First, such a set is a set of feasible initial conditions and then it represents an effective bound for the system state. Second, a PI set is in some way related to the system stability. Indeed the property of positively invariance of a set is equivalent to the existence of a Lyapunov-like function, and if such a set is a compact and absorbing set (see [18]) then its existence guarantees the system stability.

We consider now some important properties of the PI sets.

Proposition 1. Assume that the set $\mathcal{S}$ is PI for (13a) (or (13b)). Then conv{$\mathcal{S}$} is PI.

Proof. Assume that $x(0) \in \text{conv}\{\mathcal{S}\}$. Then there exist two vectors (non necessarily distinct) in $\mathcal{S}$ we denote by $x_1(0)$ and $x_2(0)$ such that $x(0)= \alpha x_1(0) + (1-\alpha)x_2(0)$ for some α, $0 \leq \alpha \leq 1$. Let $x_1(t)$ and $x_2(t)$ be the solutions of (13) corresponding to v(t). Since $\mathcal{S}$ is PI, $x_1(t)$ and $x_2(t)$ belong to $\mathcal{S}$ for $t \geq 0$. We have that $x(t)=\alpha x_1(t)+(1-\alpha)x_2(t)$ is the solution corresponding to v(t) with initial condition x(0), on the other hand $x(t) \in \text{conv}\{\mathcal{S}\}$ and the proof is completed. □

We see that it is not restrictive to limit the attention to the class of the convex PI sets. In fact by Proposition 1, if $\mathcal{S}$ is PI and it is contained in $\mathcal{X}$, its convex hull is also PI and, since convex constraints are considered, it is contained in $\mathcal{X}$.

It is well known that one of the main properties required for a compensator is to assure the closed-loop stability. In view of this requirement, we give particular attention to the class of convex and compact PI sets containing the origin as an interior point (C-sets). Let us introduce the following definitions.

Definition 3. We say that $\phi(x)$ is a Lyapunov function in the weak sense for the system (13), iff it is continuous and positive definite and there exists an open neighborhood of the origin $\mathcal{Q}$ such that if $x \notin \mathcal{Q}$ and $x(\cdot)$ is a solution of (13) corresponding to an input $v(t) \in \mathcal{V}$ such that $x(t)=x$ for some t, $t \geq 0$, then $\phi(x(t+h)) \leq \phi(x(t))$ for $h \geq 0$.

Definition 4. We say that $\phi(x)$ is a Lyapunov function in the strong sense for the system (13), iff it is continuous and positive definite and there exists an open neighborhood of the origin $\mathcal{Q}$ such that if $x \notin \mathcal{Q}$ and $x(\cdot)$ is a solution of (13) corresponding to an input $v(t) \in \mathcal{V}$ such that $x(t)=x$ for some t, $t \geq 0$, then $\phi(x(t+h)) < \phi(x(t))$ for $h \geq 0$.

If (13) has a weak Lyapunov function ϕ, then the set $\mathcal{B}_\phi(\xi) = \{x:\phi(x) \leq \xi\}$ is a positively invariant set whenever $\xi > 0$ is such that $\mathcal{Q} \subset \mathcal{B}_\phi(\xi)$. We remind that, from a geometric point of view, the boundary of the set $\mathcal{B}_\phi(\xi)$ is the surface of level ξ of $\phi(\,)$.

When the set $\mathcal{Q}$ needs to be explicitly mentioned, we say that ϕ is a Lyapunov function outside $\mathcal{Q}$. Let us consider now

$$\xi^* = \inf \{\xi: \mathcal{Q} \subset \mathcal{B}_\phi(\xi),\ \xi > 0\}.$$

We see that $\phi(\,)$ is a Lyapunov function outside $\text{int}\{\mathcal{B}_\phi(\xi^*)\}$. So, even if Definition 3 introduces a generic set, $\mathcal{Q}$ may be chosen as the open set $\text{int}\{\mathcal{B}_\phi(\xi)\}$ for a suitable ξ.

It known that if a system of the form (13) (see for instance [19][20]) has a Lyapunov function in the strong sense, then all its solutions are ultimately bounded in every set containing $\mathcal{B}_\phi(\xi^*)$ in its interior.

Assume now that a C-set $\mathcal{S} \subset \mathcal{R}^n$ is assigned and let $\psi(\lambda)$ be a strictly increasing real continuous function, defined for $\lambda \geq 0$, such that $\psi(0)=0$, $\psi(1)=1$. We can consider the function $\phi^\psi_{\mathcal{S}}(x): \mathcal{R}^n \to \mathcal{R}$ (see Fig. 2)

$$\phi^\psi_{\mathcal{S}}(x) = \begin{cases} 0 & \text{if } x=0 \\ \psi(\lambda) & \text{where } \lambda^{-1}x \in \partial\mathcal{S} \text{ otherwise.} \end{cases} \tag{14}$$

This function has the property that the surface of level $\psi(\lambda)$ is obtained by enlarging by a factor $\lambda>0$ the boundary of $\mathcal{P}$ so that $\lambda\mathcal{P}=\{x: \phi_{\mathcal{P}}^{\psi}(x)\leq\psi(\lambda)\}$ and in particular $\mathcal{P}=\{x: \phi_{\mathcal{P}}^{\psi}(x)\leq\psi(1)=1\}$.

It is immediately seen that $\phi_{\mathcal{P}}^{\psi}(x)$ is zero only for x=0, and it is positive definite on $\mathcal{R}^n$ so it may be a candidate Lyapunov function.

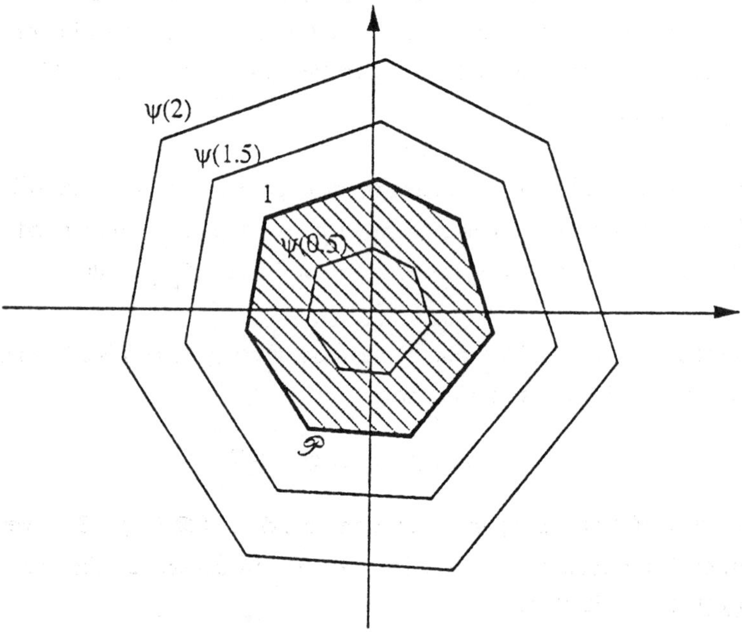

Fig. 2. Level surfaces of the function $\phi_{\mathcal{P}}^{\psi}$.

It should be noticed that the nonincreasing property of $\phi_{\mathcal{P}}^{\psi}(x)$ introduced in Definition 3 does not depend on the weighting function $\psi(\lambda)$ in the sense that if $\phi_{\mathcal{P}}^{\psi}(x)$ is a Lyapunov function, then for every other function $\psi_1(\lambda)$ with the properties above $\phi_{\mathcal{P}}^{\psi_1}(x)$ is a Lyapunov function, while this condition is strictly related to the shape of the set $\mathcal{P}$. For instance, the quadratic function $\omega(x)=x^TWx$, where W is a symmetric positive definite matrix, is obtained by choosing $\mathcal{P}$ as the ellipsoid $\{x^TWx\leq1\}$, and $\psi(\lambda)=\lambda^2$. If such a function is a

Lyapunov function for the system, then $[\omega(x)]^{1/2}$ (which is that derived for $\psi(\lambda)=\lambda$) is a Lyapunov function too.

Clearly the choice of $\psi(\lambda)$ is crucial if we want to have a "measure of the convergence speed" of the state x(t) by analyzing the decay of $\phi^{\psi}_{\mathcal{S}}(x(t))$.

For the choice $\psi(\lambda)\equiv\lambda$, we use the notation $\phi_{\mathcal{S}}(x) = \phi^{\lambda}_{\mathcal{S}}(x)$. In this case $\phi_{\mathcal{S}}(x)$ is convex, positively homogeneous and it known as the Minkowsky functional of the C-set $\mathcal{S}$ [18]. Such a functional introduces a distance in the state space defined as $\delta(x_1,x_2) = \phi_{\mathcal{S}}(x_1-x_2)$. If $\mathcal{S}$ is balanced (i.e. $x\in\mathcal{S} \Rightarrow -x\in\mathcal{S}$) then $\phi_{\mathcal{S}}(x)$ is a norm.

Proposition 2. The C-set $\mathcal{S}$ is PI for (13) if and only if there exists an open set containing the origin $\mathcal{Q}\subset\mathcal{S}$ such that $\phi^{\psi}_{\mathcal{S}}(x)$ is a Lyapunov function outside $\mathcal{Q}$ for (13).

Proof. If. Let $x(0)\in\partial\mathcal{S}$. We have that $x(0)\notin\mathcal{Q}$ since $\mathcal{Q}$ is open and $\phi^{\psi}_{\mathcal{S}}(x(0))=1$. If $\phi^{\psi}_{\mathcal{S}}(\cdot)$ is a Lyapunov function in the weak sense outside $\mathcal{Q}\subset\mathcal{S}$, then any solution x(t) of (13) corresponding to x(0) fulfills the property $\phi^{\psi}_{\mathcal{S}}(x(t))\leq 1$ for $t\geq 0$, and so $x(t)\in\mathcal{S}$, $t>0$. The same property holds for all initial states in $\mathcal{S}$. Indeed, if $x(0)\in\mathcal{S}$, then we may consider any line for x(0) and we may denote x_1 and x_2 the intersections of this line with $\partial\mathcal{S}$. For any admissible disturbance $v(\cdot)$, the solutions corresponding to $x_1(0)=x_1$ and $x_2(0)=x_2$ and $v(\cdot)$, fulfill the inclusion $x_1(t)\in\mathcal{S}$, $x_2(t)\in\mathcal{S}$. On the other hand, the solution x(t) related to x(0), lies on the segment for $x_1(t)$ and $x_2(t)$ and then, by convexity, $x(t)\in\mathcal{S}$.

Only if. Let $\mathcal{S}$ be a PI C-set and let $\mathcal{Q}=\mathrm{int}\{\mathcal{S}\}$. We prove that if $x(0)\notin\mathcal{Q}$ then $\phi^{\psi}_{\mathcal{S}}(x(t)) \leq \phi^{\psi}_{\mathcal{S}}(x(0))$, $t\geq 0$. In fact, $\mathcal{S}$ is an absorbing set (i.e. for any x $\exists$ $\eta>0$ such that $\eta x\in\mathcal{S}$) and so if $x(0)\notin \mathrm{int}\{\mathcal{S}\}$ then there exists ξ, $0<\xi\leq 1$ such that $\xi x(0)\in\partial\mathcal{S}$ By linearity, if x(t) is the solution corresponding to x(0) and $v(\cdot)$, $x^*(t)=\xi x(t)$ is the solution corresponding to $\xi x(0)$ and $\xi v(\,)$. Since $\mathcal{V}$ contains the origin, $\xi v(t)\in\mathcal{V}$ and this means that $x^*(t)\in\mathcal{S}$, $t\geq 0$. If there exists $t_1\geq 0$ such that $\phi^{\psi}_{\mathcal{S}}(x(t_1)) > \phi^{\psi}_{\mathcal{S}}(x(0))=\psi(\xi^{-1})$, it is equivalent to $x(t_1)\notin\xi^{-1}\mathcal{S}$ and so $\xi x(t_1) = x^*(t_1)\notin\mathcal{S}$: a contradiction. □

The proposition above may be rephrased by saying that the C-set $\mathcal{P}$ is PI if and only if $\phi^{\psi}_{\mathcal{P}}(x)$ is a Lyapunov function outside int$\{\mathcal{P}\}$. From Proposition 2, it comes also the following important fact: if a set $\mathcal{P}$ is PI for (13) then every set $\lambda\mathcal{P}$, $\lambda>1$, is PI. Clearly, if v=0, this property holds for all $\lambda\geq 0$.
If $\mathcal{P}$ has the additional property that for $x(0)\in\mathcal{P}$ then $x(t)\in$ int$\{\mathcal{P}\}$ t>0, we call this condition strong positive invariance according to [16]. This fact is necessary and sufficient for $\phi^{\psi}_{\mathcal{P}}(x)$ to be a Lyapunov function in the strong sense outside $\mathcal{Q}$ =int$\{\mathcal{P}\}$.
Proposition 2 establishes a connection between the positive invariance property of a C-set and the existence of a (in general not differentiable) Lyapunov function. We see that by assuring the existence of a PI C-set we guarantee the stability.
Although we have considered the particular class of positive definite functions of the form (14), the considerations above lead us to the conclusion that this choice is not restrictive if linear systems are considered. For any Lyapunov function $\phi(\ )$ (in the weak or in the strong sense) outside $\mathcal{Q}$ we can consider ξ^* as it has been defined above and we can generate a Lyapunov function outside $\mathcal{Q}$ of the class (14) starting from the PI C-set conv$\{\mathcal{B}_{\phi}(\xi^*)\}$.
The characteristics of a Lyapunov function are usually described in terms of differential properties. There is a connection between the differential properties of a Lyapunov function $\phi(x)$ and the geometric properties of the sets $\mathcal{B}_{\phi}(\xi)$, $\xi>\xi^*$.
Let us consider the continuous-time case first.
Let $\mathcal{S}$ be a C-set and for any $x\in\partial\mathcal{S}$ let $\mathcal{C}_{\mathcal{S}}(x)$ be the tangent cone to $\mathcal{S}$ in x (see [21] for a definition). We say that (13b) is subtangential (see [22]) to $\mathcal{S}$ iff

$$x' = Ax + Ev \in \mathcal{C}_{\mathcal{S}}(x) \text{ for all } v\in\mathcal{V} \text{ and } x\in\partial\mathcal{S}, \tag{15}$$

where $\mathcal{C}_{\mathcal{S}}(x)$ denotes the tangent cone to $\mathcal{S}$ in x (see Fig. 3). We recall that if $\mathcal{S}$ has a smooth boundary surface and it is represented as $\mathcal{S}=\{x: \phi(x)\leq\gamma\}$ where $\phi(x)$ is a differentiable function, then the tangent cone is equal to the tangent plane $\mathcal{C}_{\mathcal{S}}(x) = \{z: (\nabla\phi(x),z)\leq 0\}$, where $\nabla[\phi(x)]$ denotes the gradient of $\phi(x)$.

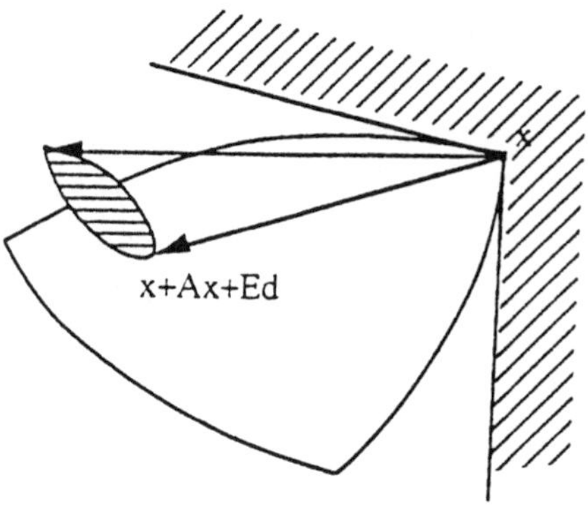

Fig. 3. Subtangentiality condition.

Condition (15) is equivalent to the positive invariance of $\mathcal{S}$ [23]. Such a condition is also equivalent to the following property concerning any trajectory x(t) of the system through x (say such that x(t)=x for some t)

$$D^+ \phi^{\psi}_{\mathcal{S}}(x) = \limsup_{\tau \to 0^+} \frac{\phi^{\psi}_{\mathcal{S}}(x(t+\tau)) - \phi^{\psi}_{\mathcal{S}}(x(t))}{\tau} \leq 0, \tag{16}$$

namely the condition for the Lyapunov derivative to be nonpositive which has to be verified for all $v(\cdot)$: $v(t) \in \mathcal{V}$, $t \geq 0$, and $x \in \partial\mathcal{S}$.

On the other hand, the condition of Definition 3 for ϕ is equivalent to property (16) say $D^+\phi(x) \leq 0$, to be fulfilled for each $v(\cdot)$ and each x outside $\mathcal{Q}$. So, from a geometric point of view, we can say that a function $\phi(\cdot)$ is a Lyapunov function outside $\text{int}\{\mathcal{B}_\phi(\xi^*)\}$ iff (13b) is subtangential to $\mathcal{B}_\phi(\xi)$ for $\xi \geq \xi^*$.

In the discrete-time case, condition (15) may be naturally replaced by the boundary contractivity condition

$$Ax + Ev \in \mathcal{S} \quad \text{for all } v \in \mathcal{V} \text{ and } x \in \partial\mathcal{S}, \tag{17}$$

while (16) is replaced by

$$\delta\phi^{\psi}_{\mathcal{S}}(t) = \phi^{\psi}_{\mathcal{S}}(t+1) - \phi^{\psi}_{\mathcal{S}}(x(t)) \leq 0. \tag{18}$$

say the condition for the Lyapunov difference to be nonpositive.

In order to take advantage of the above conditions, a convenient way to check them would be desirable. This is why many contributions in literature are devoted to particular classes of Lyapunov functions and positively invariant sets.
The most commonly considered Lyapunov functions are the quadratic ones. Such a choice is clearly motivated by the well known Lyapunov equations which is one of the most powerful methods to check if a quadratic function is a Lyapunov one.
Recently, [24] [25] [26] [27] [28] attention has been given to the class of polyhedral Lyapunov functions and the corresponding PI sets. With the term polyhedral function we mean a function given as in (14) where $\mathcal{S}$ is a polyhedral C-set and $\psi(\lambda)=\lambda$.
The main feature of this functions is that to check if a polyhedral set is PI (and so the corresponding function is a Lyapunov one) is a linear programming problem. The extension of these results to systems with disturbances was presented in [16]. In the following chapters we show the main properties of both quadratic and polyhedral functions, especially in view of their application to the constrained regulator problem for systems with disturbances.

V. BOUNDING ELLIPSOIDS

The idea of using ellipsoidal sets as bounding sets for the Disturbed Linear Constrained Regulator Problem was introduced in [5]. The ellipsoidal sets are attractive since they may be finitely represented by a set of n(n-1)/2 parameters.
Let us consider a linear time-invariant system of the form

$$\frac{d}{dt}\,x(t) = A\,x(t)$$

and let $\mathcal{S}=\{x^TWx\leq 1\}$ be an ellipsoid assigned by a definite positive symmetric matrix W. The positive invariance condition for such a set is obtained from condition (15). Noting that the tangent cone to x_0 on the surface of $\mathcal{S}$ is the half space $\{x\colon x_0Wx \leq 0\}$, we immediately derive the condition $x_0W\,A\,x_0 \leq 0$ which equivalent to

$$W A + A^T W = -Q \qquad (19)$$

for some positive semidefinite Q. Of course, equation (19) which is known as Lyapunov equation, is so famous that need not to be discussed. Here we wish only to remark that equation (19) contains all the informations concerning the subtangentiality conditions for the system.

The equation (19) is constructive in the sense that it may also be used to find a PI ellipsoid. Given a positive semidefinite Q, if there exists a positive definite solution W, then the corresponding ellipsoid is a PI set for the continuous-time system.

In the same way, for the discrete-time case

$$x(t+1) = A\,x(t)$$

we have to consider the discrete-time Lyapunov equation

$$A^T W A - W = -Q$$

that may be easily shown to be equivalent the boundary contractivity condition.

In both discrete and continuous-time cases the existence of PI ellipsoidal sets for the system is equivalent to stability. In the next chapter we shall see that the conditions for the existence of polyhedral PI sets are no much more restrictive.

Let now consider the perturbed case (13). A condition for the invariance of the ellipsoidal set for a linear (possibly time-variant) system is given in [5], where it is assumed that the disturbances are constrained in an ellipsoidal set. We consider here the time-invariant case.

Theorem 1. [5]. Let W and Q be positive definite symmetric matrices. The ellipsoidal set $\mathcal{S} = \{ x^T W^{-1} x \leq 1 \}$ is PI for the continuous-time system $x'(t) = A\,x(t) + E\,v(t)$ where $v(t)$ varies in the set $\mathcal{V} = \{v^T Q^{-1} v \leq 1\}$, if for some $\beta > 0$, the following equation holds:

$$A W + W A^T + \beta W + \frac{1}{\beta} E Q E^T = 0 \qquad (20)$$

We now prove Theorem 1 using the subtangentiality condition (15) that can be written as

$$\alpha(x,v) = x^T W^{-1} x' = x^T W^{-1}(A x + E v) \leq 0,$$

for all $x \in \partial\mathcal{S}$ and $v \in \mathcal{V}$, that is such that $\{ x^T W^{-1} x = 1 \}$ and $\{ v^T Q^{-1} v \leq 1 \}$.

Denoting by $z=W^{-1}x$, we have $z^T W z = 1$, and also

$$\alpha(x,v) = z^T (AW z + E v) = \frac{1}{2} z^T (AW+WA^T) z + z^T E v =$$
$$= \frac{1}{2}[-\beta z^T W z - \frac{1}{\beta} z^T EQE^T z] + z^T E v = \frac{1}{2}[-\beta - \frac{1}{\beta} z^T EQE^T z] + z^T E v.$$

The term in square brackets may be thought has a function of β and for $\beta>0$ it reaches a maximum in $\beta_o=\sqrt{z^T EQE^T z}$ where it is equal to $-2\beta_o$. So we have

$$\alpha(x,v) \leq -\sqrt{z^T EQE^T z} + z^T E v.$$

Denoting by $y = QE^T z \in \mathscr{R}^m$ and introducing in $\mathscr{R}^m$ the inner product $(v,w) = v^T Q^{-1} w$ and the induced norm $\| v \| = \sqrt{(v,v)}$ we have that $\| v \| \leq 1$. Simple computations yield

$$\alpha(x,v) \leq -\sqrt{(y,y)} + (y, v) \leq - \| y \| + \| y \| \| v \| \leq 0.$$

Condition (20) is only sufficient for $\mathscr{S}$ to be PI in the sense that, if the ellipsoidal set $\mathscr{S}$ is PI, W and Q do not necessarily fulfill equation (20) for some β as shown by the following counterexample.

Counterexample. Let us consider the following system:

$$A = \begin{bmatrix} -1 & 0 \\ 0 & -1 \end{bmatrix} \quad E = \begin{bmatrix} 1 \\ 1 \end{bmatrix} \quad W^{-1} = \begin{bmatrix} 1/2 & 0 \\ 0 & 1/2 \end{bmatrix} \quad Q^{-1} = [1]$$

It is immediately seen that

$$A W + W A^T + \beta W + \frac{1}{\beta} E Q E^T = \begin{bmatrix} 1 & 0 \\ 0 & 1 \end{bmatrix} (2\beta-4) + \frac{1}{\beta} \begin{bmatrix} 1 & 1 \\ 1 & 1 \end{bmatrix}$$

which is different from zero for any positive β. On the other hand, the set $\mathscr{S} = \{x^T W^{-1} x \leq 1\}$ is a circle of radius $\sqrt{2}$. The condition $x \in \partial\mathscr{S}$ is equivalent to the constraint $(x_1^2 + x_2^2)/2=1$, moreover we have $| v | \leq 1$, and thus

$$\alpha(x,v) = x^T W^{-1} x' = -\frac{1}{2}(x_1^2 + x_2^2) + \frac{1}{2}(x_1 + x_2) v \leq -1 + \frac{1}{2}|x_1 + x_2| \leq 0$$

and then the system is subtangetial to $\mathscr{S}$ which is therefore PI.

As shown in [7] Eq. (20) may be constructive in the sense that if it has a positive definite solution W, we derive the ellipsoid $\mathcal{P} = \{ x^T W^{-1} x \le 1 \}$ which is PI. A sufficient condition that assures the existence of a positive definite solution W is that A is asymptotically stable, and that the pair (A,E) is reachable.

Conversely, assume that W is positive definite, so is W^{-1}, then the system is asymptotically stable. To verify this assertion one must simply note that if we set v=0, the function $x^T W^{-1} x$ is a Lyapunov one in the strong sense since for $x \neq 0$ we have $\alpha (x,0) \le - \beta x^T W^{-1} x/2 < 0$.

The discrete-time version of Eq. (20) is given by the following theorem.

Theorem 2. [5]. The ellipsoidal set $\mathcal{P} = \{ x^T W^{-1} x \le 1 \}$ is PI for the system (13a) where v(t) varies in the ellipsoidal set $\mathcal{V} = \{ v^T Q^{-1} v \le 1 \}$ if for some β, $0<\beta<1$, the following equation holds:

$$- W + \frac{1}{1-\beta} AWA^T + \frac{1}{\beta} EQE^T = 0 . \tag{21}$$

We see that one important feature of the class of ellipsoidal sets is that the subtangentiality condition and the boundary contraction condition are assured by an algebraic linear condition. In the next chapter we consider polyhedral sets and we shall see that sufficient and necessary conditions for the positive invariance may be stated in terms of linear programming.

VI. BOUNDING POLYTOPES

Although the idea of using polyhedral sets as bounding sets for the system state was introduced many years ago [6], the ellipsoidal sets have been more frequently considered for many years. The main motivation is that the algorithms involving polyhedral sets if applyed to large dimension systems require a large amount of computing power and storage memory. It is natural that the techniques which use such kind of sets have been promoted by the increase of computer performance.

The reason way polyhedral sets are attractive is that any compact and convex set of $\mathcal{R}^n$ can be arbitrarily closely approximated by a polyhedral set. Moreover they are finitely represented in the sense that a finite number of parameters are necessary for their representation. Moreover we have that many operations such as intersection, convex hull and sum between polyhedral sets produce as results sets which are again of the polyhedral type. Such a property, conversely, does not hold for ellipsoids, and this is the reason way many results concerning the constrained control involving ellipsoidal sets turn out to be too restrictive for the problem. Such a shortcoming has been pointed out in [5] where to overcome the problem it is proposed to replace the set resulting from any of the operation mentioned above by an approximating ellipsoidal one. In this way, we keep the simplicity of the set representation but loose possible solutions.

Recently the problem of constrained control for systems without disturbances has been approached via polyhedral sets [24][25][26][27][28]. These results are essentially based on the fundamental property that the positive invariance of a set of the polyhedral type is equivalent to the existence of feasible solutions of a linear programming problem. An extension to perturbed systems of the form (13) has been proposed in [16] and [29]. We discuss here these results.

For a polytope $\mathcal{S}$ containing the origin in its interior we consider a representation in terms of delimiting planes of the form

$$\mathcal{S} = \{ \Phi_i x \leq \Theta_i, \Theta_i > 0, i=1,..,s \} \text{ or synthetically } \{ \Phi x \leq \Theta, \Theta > 0 \}. \quad (22)$$

We also consider its representation in terms of vertices

$$\mathcal{S} = \text{conv} \{ w_j, j=1,..,r \}. \quad (23)$$

Both representations (22) and (23) are useful to give equivalent conditions for the positive invariance of $\mathcal{S}$.

A. Invariance conditions for Polyhedral sets.

The conditions of positive invariance shown in the previous chapter for ellipsoidal sets are only sufficient. We show here that if we consider a polyhedral C-set, we can give necessary and sufficient conditions for its positive invariance.

Let us assume that $\mathcal{V}$ is a polytope expressed as $\mathcal{V} = \text{conv} \{v_k, k=1,..,l \}$. The following result concerns the discrete-time system (13a) [16].

Theorem 3. The polytope $\mathcal{S}$ is PI for system (13a) if and only if

$$A w_j + E v_k \in \mathcal{S}, \quad w_j \in \text{vert}\{\mathcal{S}\}, j=1,..,r, \ v_k \in \text{vert}\{\mathcal{V}\}, \ k=1,..,l. \quad (24)$$

Condition (24) may be easily checked if also the plane description (22) of $\mathcal{S}$ is available. In this case, we have a set of $s \cdot r \cdot l$ linear inequalities to be verified. Anyway, representation (22) is not strictly necessary because we can check such condition using the vertex description only. If the plane representation (22) holds, then we may consider the following alternative formulation. For each plane of $\mathcal{S}$ consider the term

$$\Delta_i = \max_{v \in \mathcal{V}} \Phi_i E v$$

The determination of Δ_i is a simple linear programming problem having the polytope $\mathcal{V}$ as feasible set. Let us define the polyhedron

$$\mathcal{S}^* = \{ \Phi_i x \le \Theta_i - \Delta_i = \Theta_i^*, \ i=1,..,s\} \quad (25)$$

that we may rewrite as $\mathcal{S}^* = \{\Phi x \le \Theta^* = \Theta - \Delta\}$.

It should be noticed that (25) is defined even if $\mathcal{V}$ is not assumed to be a polytope but is a C- set. In this more general case, the computation of the shift term Δ_i is a convex optimization problem. For instance, if $\mathcal{V}$ is an ellipsoid defined as $\mathcal{V} = \{v^T Q^{-1} v \le 1\}$ we have

$$\Delta_i = \sqrt{\Phi_i E Q E^T \Phi_i^T}.$$

The following result holds for a generic disturbance constraint C-set $\mathcal{V}$ [16].

Theorem 4. The polytope $\mathcal{S}$ is PI for system (13a) if and only if the following condition holds

$$A w_j \in \mathcal{S}^*, \quad \text{for all } w_j \in \text{vert}\{\mathcal{S}\}, \ j=1,..,r. \quad (26)$$

Using condition (26), we may decompose the problem in two stages: the computation of the shift terms Δ_i and the check of (26) for each vertex of $\mathcal{S}$. The number of inequalities introduced by condition (26) is $s \cdot r$ and so it is l times smaller than those introduced by condition (24).

In the continuous-time case, we have the following theorem [16].

Theorem 5. The polytope $\mathcal{S}$ is PI for system (13b) if and only if the following condition holds

$$Aw_j + E\,v_k \in \mathcal{C}_{\mathcal{S}}(w_j), \quad w_j \in \mathrm{vert}\{\mathcal{S}\},\ j=1,...,r, \quad v_k \in \mathrm{vert}\{\mathcal{V}\},\ k=1,...,l\,. \quad (27)$$

Since $\mathcal{S}$ is a polytope, the tangent cone $\mathcal{C}_{\mathcal{S}}(w_j)$ is defined by the inequalities $\Phi_i\, x \le 0$, obtained by considering all the planes of $\mathcal{S}$ that contain w_j, say such that $\Phi_i\, w_j = \Theta_i$.

Theorem 5 may be interpreted as an extreme point property: it says that if $\mathcal{S}$ is a polytope, the subtangentiality condition (15) is to be checked only on the vertices of $\mathcal{S}$ instead of all the points of its boundary .

We may reconsider the problem by introducing the discrete-time Euler approximating system

$$x(t+\tau) = [\, I + \tau A\,]\, x(t) + \tau E\, v(t),\ \tau > 0. \quad (28)$$

System (28) allows to transform the problem for continuous-time systems in an equivalent discrete-time problem. We have the following theorem [16].

Theorem 6. The polytope $\mathcal{S}$ is PI for system (13b) if and only if there exists $\tau>0$ such that it is PI for the discrete-time system (28).

However, the open question is how to choose τ. Applying condition (24), we easily see that if $\mathcal{S}$ is PI for (28) for some τ_0, then it is PI for any τ, $0<\tau<\tau_0$. In this sense the condition of invariance for the continuous-time problem is equivalent to the invariance condition for the discrete-time system (28) if τ is chosen sufficiently small.

A suitable value for τ_0 may be found in the following way. Condition (27) requires that the vectors (Aw_j+Ev_k) belong to the tangent cone $\mathcal{C}_{\mathcal{S}}(w_j)$ in w_j, and it is equivalent to the property that , for all $\tau>0$, the vectors $w_j + \tau\,(Aw_j + E\,v_k)$ do not intersect $\mathcal{C}_{\mathcal{S}}(w_j)$ that is the faces of $\mathcal{S}$ that contain w_j. If τ is such that the length of each vector $\tau(Aw_j+Ev_k)$ is less than the distance of w_j from all the planes that do not contain w_j, the corresponding boundaries cannot be violated and so we obtain that (27) is equivalent to the condition

$$w_j + \tau\,(Aw_j + E\,v_k) \in \mathcal{S}$$

which is exactly the condition (24) applied to (28). We recall that the distance of w_j from the plane $\Phi_i\,x = \Theta_i$ is given by $|\,\Phi_i\,w_j - \Theta_i\,| / \|\Phi_i\|$. So we can take

$$\tau_0 = \frac{\min_{ij}\ |\,\Phi_i\,w_j - \Theta_i\,| / \|\Phi_i\|}{\max_{jk}\ \|\,A\,w_j + E\,v_k\,\|}$$

where the minimum on the numerator is obtained by considering all the vertices of $\mathcal{S}$ and for each of them all the planes that do not contain it, while the maximum on the denominator is computed by exhaustively considering all the vertices of $\mathcal{S}$ and $\mathcal{V}$. So, as a corollary of Theorem 6, we can say that $\mathcal{S}$ is PI for (13b) if and only if it is PI for (28) for any τ, $0<\tau\le\tau_0$.

All the conditions above require the representation of $\mathcal{S}$ in terms of vertices. The condition given by the following theorem requires only the plane representation of $\mathcal{S}$. Such condition extends the results in [27] [24] where the case v=0 is considered.

Theorem 7. The polytope $\mathcal{S} = \{\,\Phi x \le \Theta\,\}$ is PI for the discrete-time system (13a) if and only if there exists a matrix H of dimensions sxs such that

i)	$H \ge 0$	(i.e. $h_{ij} \ge 0$ i,j =1,..,s)	
ii)	$H\,\Phi = \Phi\,A$		(29)
iii)	$H\,\Theta \le \Theta^*$		

where Θ^* is that defined in (25).

The proof of Theorem 7 is given in [29].

In the continuous-time case we have a similar result. Let us denote by $\mathcal{M}_s$ the class of the sxs matrices having nonnegative nondiagonal entries, that is if $H\in\mathcal{M}_s$ then $h_{ij}\ge 0$ if $i\ne j$.

Theorem 8. The polytope $\mathcal{S} = \{ \Phi x \leq \Theta \}$ is PI for the continuous-time system (13b) if and only if there exists a matrix H such that

$$\begin{array}{lll} \text{i)} & H \in \mathcal{M}_s & (\text{i.e. } h_{ij} \geq 0, \text{ for } i \neq j, \ i,j = 1,..,s) \\ \text{ii)} & H\Phi = \Phi A & \\ \text{iii)} & H\Theta \leq -\Delta & \end{array} \qquad (30)$$

where Δ is the vector having as components the shift terms Δ_i defined above.

This result for the particular case v=0 was presented in [30].
The conditions of Theorem 7 and Theorem 8 do not require the knowledge of the vertices of $\mathcal{S}$ but, on the other hand, they introduce the unknowns h_{ij} and they require to check if a polyhedral set in the space of the sxs matrices is nonempty, while the conditions of the Theorems 3-6 lead to a set of inequalities which have to be verified.

B. Invariant Polyhedral sets and system spectral properties.

In chapter IV we have investigated the connection between PI C-sets and Lyapunov functions. In this section, we analyze the spectral properties of systems having Lyapunov functions of the polyhedral type. We first consider the case v=0, then we generalize these results to systems of the form (13).
As it is known, a linear discrete-time system of the form $x(t+1) = A\,x(t)$ has an ellipsoidal PI set if and only if it is stable that is its eigenvalues are all contained in the closed unit circle. The same result holds in the continuous-time case, in the sense that a necessary and sufficient condition for the existence of a PI ellipsoidal set is that the system eigenvalues have nonpositive real part.
The existence of PI C-sets of polyhedral type requires conditions for the eigenvalues of A which are a little bit more stringent but in practice largely acceptable. As shown in [31] if the system $x(t+1)=Ax(t)$ has real poles only, then a necessary and sufficient condition for the existence of such kind of sets is the stability. For systems with real and complex poles, a sufficient condition is that each pole fulfills the following property [31]:

$$\text{If } \lambda \in \sigma(A) \text{ then } |\mathrm{Re}(\lambda)| + |\mathrm{Im}(\lambda)| \leq 1$$

An equivalent condition is given by the following theorem.

Theorem 9. The discrete-time system $x(t+1)=Ax(t)$ has a PI polyhedral C-set if and only if it is stable and each eigenvalue λ which is on the boundary of the unit circle has rational phase that is

$$\lambda = e^{j\theta}, \; \theta = 2\pi\, p/q, \text{ for some integers p and q.} \tag{31}$$

To give a proof, we first consider a particular case. Let us consider a matrix of the form:

$$A = \begin{bmatrix} \cos(\theta) & -\sin(\theta) \\ \sin(\theta) & \cos(\theta) \end{bmatrix} \tag{32}$$

It is to note that (32) has eigenvalues $\lambda=e^{\pm j\theta}$. The second order system $x(t+1)=Ax(t)$ has a PI polyhedral C-set if and only if $\theta/2\pi$ is a rational number. Indeed, for any initial state $x(0)$ the solution vector $x(k)$ fulfills the property $\|x(k)\| = \|x(0)\|$ and then $x(k)$, $k=1,2,..$ lies on a circle of radius $\|x(0)\|$. The angle between $x(k)$ and $x(k+h)$ is given by $h\theta$ for every k and h. If $\theta/2\pi=p/q$ for some integers p and q, then we have $x(q)=x(0)$. Consider now the convex hull of the points $x(k)$ $\mathcal{S}= \mathrm{conv}\{x(h), h=0,..,q-1\}$ which is a regular polygon. Applying condition (24) (with $v=0$) we immediately see that $\mathcal{S}$ is PI. Conversely if $\theta/2\pi$ is an irrational number, we have that $x(k)\neq x(h)$ if $k\neq h$ and it is then immediate that the only PI C-sets are circles of arbitrary radius.

Assume now that A is asymptotically stable, and consider an arbitrary polyhedral C-set of the form $\mathcal{S} = \{ \Phi x \leq \Theta \}$. Then there exists k_o such that

$$\mathcal{S} = \{ \Phi A^k x \leq \Theta, \; k=0, .., k_o \}.$$

is the maximal PI set contained in $\mathcal{S}$. We shall prove this claim in section D, where systems with disturbances will be considered.

Proof of Theorem 9. Sufficiency. Since the system is stable, we may consider a state transformation represented by an invertible matrix T, such that

$$T^{-1} A T = \mathrm{diag}\{A_1, A_2\} = \begin{bmatrix} A_1 & 0 \\ 0 & A_2 \end{bmatrix}, \tag{33}$$

in which the eigenvalues of A_1 are all in the interior of the unit circle while $A_2 = \mathrm{diag}\{A_2^{(i)}, i=1,..,r\}$, where $A_2^{(i)}$, is either a 1x1 matrix with coefficient equal to 1 or -1, otherwise it is a 2x2 matrix having the form (32). According to (33), we may consider the partition of $x = [x_1^T, x_2^T]^T$ $x_1 \in \mathscr{R}^{n_1}$, $x_2 \in \mathscr{R}^{n_2}$. We may find in $\mathscr{R}^{n_1}$ a convex polyhedral C-set $\mathscr{S}_1$ which is PI for $x_1(t+1)=A_1x_1(t)$. Let now consider $x_2(t+1)=A_2x_2(t)$. For any subsystem corresponding to the block $A_2^{(i)}$ we may consider the PI set $\mathscr{S}_2^{(i)}$ that is given by any interval symmetric with respect to the origin if $A_2^{(i)}$ is of order one and by a regular C-polygon as described above if $A_2^{(i)}$ is of the second order since its eigenvalues have rational phase. The polyhedral C-set given by $\mathscr{S}_2 = (\mathscr{S}_2^{(1)} \times \mathscr{S}_2^{(2)} \times .. \times \mathscr{S}_2^{(r)})$ is clearly PI for $x_2(t+1)=A_2x_2(t)$ and then $\mathscr{S} = (\mathscr{S}_1 \times \mathscr{S}_2)$ is PI for the system. The condition $0 \in \mathrm{int}\{\mathscr{S}\}$ is obvious. Since any PI polyhedron is mapped by any linear transformation into another polyhedron that is still PI in the new reference frame, the first part of the proof is completed. For necessity, we must first notice that the stability is required because, as it has been seen in chapter IV, it is necessary for the system to have PI C-sets, so we can assume that the system is represented as in (33). Let us suppose that a polytope $\mathscr{S}$ containing the origin in its interior exists and let us consider the intersection $\mathscr{S}_2^{(i)}$ of $\mathscr{S}$ with the eigenspace $\mathscr{E}_2^{(i)}$ relative to the block $A_2^{(i)}$. $\mathscr{S}_2^{(i)}$ is also a convex polytope with a nonempty relative interior. If we select the subsystem related to $A_2^{(i)}$, $\mathscr{S}_2^{(i)}$ is a PI (one or two dimensional) polyhedral C-set. So, if $A_2^{(i)}$ is of the second order, it is of the form (32) and its eigenvalues fulfill condition (31). Condition (31) is obvious if $A_2^{(i)}$ is of the first order. □

Let us consider now the continuous-time case. We first notice that, system stability is necessary for the existence of PI C-sets.

If the system $x'(t)=Ax(t)$ is asymptotically stable, then there exists τ such that the eigenvalues of $[I+\tau A]$ (which are $\{1+\tau\lambda_i, \lambda_i \in \sigma(A), i=1,...,n\}$) are in the open unit circle and so we can construct a PI polyhedral C-set for $x(t+\tau) = [I+\tau A]\, x(t)$ which is PI, from Theorem 5, for the continuous-time system.

If the system is stable, we may assume the matrix A to be decomposed as follows

$$\mathrm{diag}\{A_1, A_2\} = \begin{bmatrix} A_1 & 0 \\ 0 & A_2 \end{bmatrix}, \tag{34}$$

in which all the eigenvalues of A_1 are in the open left half plane while the matrix $A_2 = \mathrm{diag}\{A_2^{(i)},\ i=1,..,r\}$, where $A_2^{(i)}$, is either a 1x1 matrix with coefficient equal to 0, otherwise it is a 2x2 matrix of the form:

$$A_2^{(i)} = \begin{bmatrix} 0 & -\omega_i \\ \omega_i & 0 \end{bmatrix} . \tag{35}$$

The trajectories of second order system represented by $A_2^{(i)}$ of the form (35) are circles if $\omega_i \neq 0$ and, if this is the case, a polyhedral PI C-set cannot exist for such a subsystem and then for (34). Conversely, if $\omega_i=0$ for all i, A_2 is a zero matrix and every set is PI for the corresponding subsystem. Therefore we can choose a polyhedral C-set $\mathcal{S}_1$ for the first subsystem and any C-set $\mathcal{S}_2$ contained in the space related to A_2, to obtain $\mathcal{S}= \mathcal{S}_1 \times \mathcal{S}_2$ which a PI polyhedral C-set. Therefore we have the following result.

Theorem 10. The continuous-time system $x'(t)=Ax(t)$ has a PI polyhedral C-set if and only if it is stable, and each eigenvalue λ on the imaginary axis is equal to zero that is

$$\text{if } \mathrm{Re}(\lambda)=0 \text{ then } \lambda=0 . \tag{36}$$

Let now consider systems with disturbances. We say that the system (13a) or (13b) is stable under persistent disturbance (SUPD), if there exist two neighborhoods $\mathcal{N}$ and $\mathcal{M}$ such that $\mathcal{M} \subset \mathcal{N}$ and for all initial conditions $x(0) \in \mathcal{M}$ and all $v(\cdot)$ having values in its constraint set we have that $x(t) \in \mathcal{N}$ for $t>0$. Such a condition is equivalent to say that the system is stable and all marginally stable eigenvalues are not reachable eigenvalues of (A,E). Let us consider the following lemma.

Lemma 1. Assume that the system (13a) (or (13b)) is asymptotically stable. Then there exists a positively invariant polyhedral C-set.

Proof. To prove this result one may observe that, for any arbitrary polyhedral C-set $\mathcal{S} \subset \mathcal{R}^n$ and any positive parameter ξ the existence of a PI polyhedral C-set contained in $\xi\mathcal{S}$ is equivalent to the existence PI polyhedral C-set contained in $\mathcal{S}$ if the disturbance varies in $\xi^{-1}\mathcal{V}$. In section C it is shown that for a sufficiently large ξ this problem has a solution. The computation of $\mathcal{S}$ may be performed with the methods suggested in section D. □

The following theorem holds.

Theorem 11. There exist PI C-sets of the polyhedral type for system (13a) (respectively (13b)) if and only if the system is stable under persistent disturbances, and the marginally stable eigenvalues fulfill condition (31) (respectively (36)).

Proof. If the system has a PI polyhedral C-set then it must be SUPD. So, representing the system as in (34), we have that the intersection of the polyhedral C-set with the subspace related to A_2, whose vectors are unreachable state of (A,E), is a polytope with nonempty relative interior so the eigenvalues of A_2 fulfill condition (31) (or (36)). Conversely, we know how to construct a PI polyhedral C-set for the subsystem $\wp x_2(t)=A_2x_2(t)$, $x_2(t)\in\mathcal{R}^{n_2}$ (which is unaffected by the disturbance) if condition (31) (respectively (36)) is fulfilled. The existence of a polyhedral PI C-set for the asymptotically stable subsystem $\wp x_1(t) = A_1(t)\, x_1 + E_1 v(t)$, $x_1(t)\in\mathcal{R}^{n_1}$, is assured by Lemma 1. □

In view of the application to our problem, it is not so important if the system has a generic PI set of the polyhedral type but it is important to check if there exists a PI C-set $\mathcal{S}$ contained in the constraint set $\mathcal{X}$.

It should be noticed that in the case v=0, if a system has a PI C-set and $\mathcal{X}$ contains the origin in its interior, the existence of a PI C-set $\mathcal{S}$ is equivalent to the existence of a PI C-set $\mathcal{S}'$ contained in $\mathcal{X}$. $\mathcal{S}'$ may be obtained from $\mathcal{S}$ by contracting it of a suitable scale factor ξ, $0\leq\xi\leq 1$, where ξ is chosen sufficiently small in order to assure $\mathcal{S}'=\xi\mathcal{S}\subset\mathcal{X}$. The set $\mathcal{S}'$ is still a PI set as it can be immediately seen. For instance, if the system is asymptotically stable, the set $\mathcal{S}$ may be determined by solving the Lyapunov equation (22) (respectively (20)) with a positive definite Q to obtain an ellipsoidal PI set of the form $\{x^TWx\leq\xi^2\}$. Unfortunately, these properties fail in the general case $v(\cdot)\neq 0$.

In the next two sections we show how the maximum disturbance size assuring the existence of such a set may be computed and how the maximal PI set contained in $\mathcal{X}$ can be found.

C. Maximum disturbance size.

As it has been pointed out in the introduction, an important feature of a control system is the disturbance rejecting property. To introduce a measure of such a property we choose an output measure μ_e, a disturbance measure μ_v and we define as the disturbance rejection measure the number

$$\rho = \frac{\mu_v(v(\cdot))}{\mu_e(e(\cdot))}.$$

The term μ_e may take into account a penalization for the control action since the control u=KC x may be considered as a system output.

Clearly the number ρ is strictly related to the choice of the measures μ_e and μ_v.

In the classical control theory, the disturbance is chosen in a given class of functions which are thought to be sufficiently representative for the practical situations for instance steps, impulses, sine or cosine functions. The output measure is chosen correspondingly. For example for a unit step input we can choose the steady state output norm or the maximum overshoot.

In our approach we assume that the disturbance (and then the output) is unknown a-priory, and so we have to introduce a measure for a family of functions.

For instance, if a statistic distribution of the noise is available, we can choose such measures as the variances of the output and the input respectively.

In the H_∞ approach both disturbance and output are measured with the L_2 norm. Taking zero initial condition and assuming system stability, ρ is the inverse of the H_∞ norm of the disturbance-error transfer function [32].

The l_1 (respectively L_1 in the continuous-time case) control theory [11][12][13][14][15] considers the l_∞ norm for input and output. The central point of this theory is that the inverse of ρ which is the supremum of the l_∞ (L_∞) norm in the family of outputs corresponding to inputs in the l_∞ (L_∞) unit ball is equal to the l_1 (L_1) norm of the impulse response.

In view of Problem 1 it is natural to choose a measure which is induced by the disturbance and output constraint sets.

So, assuming that the constraint set $\mathscr{E}$ and $\mathscr{V}$ for the error and the disturbance are C-sets, a measure for v and e is derived by considering the Minkowsky functional of $\mathscr{E}$ and $\mathscr{V}$

$$\mu_v(v(\cdot)) = \sup_{t\geq 0} \phi_{\mathscr{V}}(v(t))$$
$$\mu_e(e(\cdot)) = \sup_{t\geq 0} \phi_{\mathscr{E}}(e(t))$$

The condition $e(t)\in\mathscr{E}$ is equivalent to $\phi_{\mathscr{E}}(e(t))\leq 1$, so $\mu_e=1$ and then $\rho=\mu_v$.
The index ρ is correctly defined if we fix a set of possible initial conditions. Assuming that such a set is given, if the condition $\rho\geq 1$ holds, then the requirements of the problem are fulfilled. In this case, we have that there exists a PI set containing the specified set of initial conditions which is contained in the bounding region for the states induced by $\mathscr{E}$ and $\mathscr{U}$ say

$$\{x:\ Nx\in\mathscr{E},\ K\,x\in\mathscr{U}\}\,. \tag{37}$$

The determination of the index ρ is equivalent to the evaluation of the maximum value of ε such that there exists a positively $\varepsilon\mathscr{V}$-invariant set contained in the state constraint set (37). The choice of linear constraints on input and output induces constraints (37) of the polyhedral type on the state so we consider the following problem.
Given a polyhedral set $\mathscr{P}$ containing the origin in its interior, find the maximum value of ε such that assuming $v(t)\in\varepsilon\mathscr{V}$ there exists a PI set $\mathscr{S}\subset\mathscr{P}$ containing the origin.
Let us denote by $\mathscr{R}(\varepsilon)$ the set of all states which are reachable from the origin with constrained input $v(t)\in\varepsilon\mathscr{V}$. It is immediate that $\mathscr{R}(\varepsilon)$ is a PI set. Moreover, if a PI set $\mathscr{S}$ containing the origin exists, then $\mathscr{R}(\varepsilon)\subset\mathscr{S}$ and in this sense $\mathscr{R}(\varepsilon)$ is the minimal PI set containing the origin. Then ρ is simply derived by considering the maximum value of ε such that $\mathscr{R}(\varepsilon)\subset\mathscr{P}$.
A technique for the evaluation of the reachability set is presented in [33]. However, the evaluation of $\mathscr{R}(\varepsilon)$ is not necessary for the evaluation of ρ.
Let now assume for $\mathscr{P}$ a representation of the form

$$\mathscr{P}=\{\ x:\ \Omega_i\, x\leq\Xi_i,\ \Xi_i>0,\ i=1,..,s\ \}\ \text{or synthetically}\ \{\ \Omega x\leq\Xi,\ \Xi>0\ \}. \tag{38}$$

Let us consider the discrete-time case first. To assure that for x(0)=0, the condition $x(t)\in \mathcal{P}$ is fulfilled for all t>0, we must consider the expression of the forced response which is given by

$$x(t) = \sum_{h=0}^{t-1} A^{t-h-1} E\, v(h). \tag{39}$$

The condition $x(t)\in \mathcal{P}$ which is equivalent to $\Omega_i\, x(t) \le \Xi_i$, i=1,..,s, in view of Eq. (39) may be rewritten as

$$\sum_{h=0}^{t-1} \Omega_i\, A^{t-h-1} E\, v(h) \le \Xi_i \qquad i=1,..,s,\ t=0,1,2, \ldots$$

We have that the values v(h), h=1,..,t-1, are arbitrary, so such a condition may be equivalently rewritten as

$$\sum_{k=0}^{t-1} \max_{v_k\in \varepsilon\mathcal{V}} \{\Omega_i\, A^k E\, v_k\} \le \Xi_i \qquad i=1,..,s,\ t=0,1,2, \ldots$$

From linearity we have

$$\varepsilon \sum_{k=0}^{t-1} \max_{v_k\in \mathcal{V}} \{\Omega_i\, A^k E\, v_k\} \le \Xi_i \qquad i=1,..,s,\ t=0,1,2, \ldots$$

The sum in the left member of the expression above is a nondecreasing function of t (note that $0\in\mathcal{V}$ so the terms of the sum are nonnegative). So we get [29]

$$\rho = \min_i \frac{\Xi_i}{\sum_{h=0}^{\infty} \max_{v_k\in \mathcal{V}} \{\Omega_i A^k E v_k\}} \tag{40}$$

However, expression (40) is defined only if the series on the denominator is a convergent one. If this is not the case then we may assume $\rho=0$.

There are many conditions that guarantee convergence. For instance, if we assume that $\mathcal{P}$ is a polyhedron of the form $-\Xi_i^- \le \Omega_i\, x \le \Xi_i^+$, $\Xi_i^- > 0$, $\Xi_i^+ > 0$, (such kind of polyhedra are naturally derived if lower and upper bound on the output variables are taken) an equivalent condition for the series to be convergent is that the system represented by the triplet (A,E,Ω) (where Ω is considered as an output matrix) is BIBO stable. It is easy to see that such a condition holds for generic bounds of the form (38) as long as $\mathcal{V}$ contains the origin in its interior.

Conversely, it may happen that all the terms of the sum are zero, and in this case we assume $\rho=\infty$ (perfect disturbance decoupling). Assuming that $\mathcal{V}$ contains the origin in its interior, this condition is equivalent to the condition that the reachable modes of (A,E) are unobservable modes of (A,Ω) (see [34] [35] [36]).

The importance of expression (40) lies in the fact that in this way the evaluation of the reachability set $\mathcal{R}(\varepsilon)$ is not required. Note that expression (40) requires the numerical evaluation of a series and such a problem may be easily solved by a digital computer. The computation of each of the terms of such a series is a convex programming problem. If $\mathcal{V}$ is a polytope it requires only the evaluation of the maximum value of $\Omega_i A^h E v_k$ with $v_h \in \mathrm{vert}\{\mathcal{V}\}$. If $\mathcal{V}$ is the ellipsoid $\{v^T Q^{-1} v \le 1\}$, such a maximum has the same expression of the shift term Δ_i defined in (25) in the ellipsoidal case say $(\Omega_i A^h E Q E^T (A^h)^T \Omega_i^T)^{1/2}$.

It must be remarked that expression (40) includes the particular case of symmetric constraints of the rectangular shape that have as Minkowsky functional the $\| \ \|_\infty$ norm. If an output of the form e=Nx is given, the constraint

$$\| e(t) \|_\infty \le 1,\ t \ge 0, \text{ or equivalently } -e_0 \le Nx(t) \le e_0\ t \ge 0,$$

where $e_0=[1,1,..,1]^T$ is equivalent to assume that $e(\cdot)$ is contained in the unit ball in l_∞. If the same constraint is chosen for $v(\cdot)$ say $v(t) \in \mathcal{V} = \{v: \|v\|_\infty \le 1\}$ $t \ge 0$, we can easily derive

$$\rho^{-1} = \max_i \max_{\|v_k\|_\infty \le 1,\ t>0} \left| \sum_{k=0}^{t-1} N_i A^k E v_k \right| = \| (NA^h E) \|_{l_1},$$

where the last term denotes the l_1 norm of the disturbance-output impulse response (see for instance [13]). This result is known as the central point of the l_1 control theory. Using expression (40) we consider a more general shape of (possibly non symmetrical) polyhedral constraints.

In the continuous-time case, we derive in the same way the following expression for ρ

$$\rho = \min_i \frac{\Xi_i}{\int_0^\infty \max_{v(\alpha) \in \mathcal{V}} \{\Omega_i \exp(A\alpha) E v(\alpha)\}\, d\alpha} \qquad (41)$$

The convergence conditions for the integral appearing in (41) are, of course the same as the discrete-time case. If convergence is assumed, the integral may be easily evaluated by approximating methods.
Again, as a particular case, according with the L_1 theory (see [11] [32] [14]) , if we assume for the disturbance $v(\cdot)$ and the output $e(\cdot)$ the L_∞ norm, we have that ρ as defined in (41) is the inverse of the L_1 norm of the impulse response.

D. The maximal invariant set.

In the previous section, we have found an expression of the maximal disturbance size that allows the permanence of the state in its admissible region assuming that $x(0)=0$. For a disturbance measure less then this value it is assured the existence of a PI and admissible set containing the origin which included in the state constraint set.
The simple determination of the index ρ does not give any information about such a PI set. The importance of such a set is that it is a set in which the state may be contained at each time with the condition that it will be kept within the boundaries in the future. In particular we are interested in the maximal PI set, say the set of all initial states for which the permanence of the state within its constraints (we include in this constraints also those induced by the control constraints) is guaranteed on an infinite horizon.
Such a set (if it is not empty) is convex if the constraints are so. This property is immediate since, as we have seen, if a set is PI also it convex hull is so.
We consider the discrete-time case first. Let us assume that the polyhedral set

$$\mathcal{P} = \{ \Omega x \leq \Xi, \Xi > 0 \}$$

containing the origin is assigned. For any fixed measure ε of the disturbance let us introduce the following sequence of sets:

$$\mathcal{P}_k = \{ \Omega A^k x \leq \Xi - \varepsilon \sum_{h=0}^{k-1} \Delta^{(h)} = \Xi^{(k)} \} \tag{42}$$

where it has been denoted by $\Delta^{(h)}$ the vector having components

$$\Delta_i^{(h)} = \max_{v \in \mathcal{V}} \{\Omega_i A^h E v\} \tag{43}$$

Let us introduce the sets

$$\mathcal{S}_j = \bigcap_{h=0}^{j} \mathcal{P}_k = \{ \ \Omega A^k x \leq \ \Xi^{(k)}, k=0,1,..\ j \ \} \tag{44}$$

We see from (44) that the sets $\mathcal{S}_j$ may be recursively derived as follows. Assuming for $\mathcal{S}_j$ the representation $\mathcal{S}_j = \{ \ \Phi^{(j)} \ x \leq \Theta^{(j)} \ \}$ we have that

$$\begin{aligned} \mathcal{S}_0 &= \mathcal{P} = \ \{\Phi^{(0)} \ x \leq \Theta^{(0)}\} = \{ \ \Omega x \leq \Xi, \ \Xi > 0 \ \}. \\ \mathcal{S}_{j+1} &= \{ \ \Phi^{(j)} \ Ax \leq \Theta^{(j)} - \varepsilon \ \Delta_\Theta{}^{(j)}\} \cap \{ \ \Omega x \leq \Xi \ \} \end{aligned} \tag{45}$$

where the i-th component of the shift term vector $\Delta_\Theta{}^{(j)}$ is defined as

$$[\Delta_\Theta{}^{(j)}]_i = \max_{v \in \mathcal{V}} \Phi_i{}^{(j)} E \ v.$$

With obvious meaning of the term, let us consider the set $\mathcal{S}_\infty$. We have the following theorem.

Theorem 12. The set $\mathcal{S}_\infty$ is the maximal PI set contained in $\mathcal{P}$.

Proof. Let $\mathcal{M}$ be the maximal PI set contained in $\mathcal{P}$. We have that the set $\mathcal{S}_j$ is the set of all initial states x(0) for which the solution of x(t+1)=Ax(t)+Ev(t) fulfills the property $x(t) \in \mathcal{P}$, t=0,1,..,j. For j=0 it is obvious. For j=1 we have that $x(1) \in \mathcal{P}$ iff $\Omega[Ax(0)+Ev] \leq \Xi$, for all $v \in \mathcal{V}$ that is $\Omega Ax(0) \leq \Xi^{(1)}$ so $x(0) \in \mathcal{S}_1$. By induction assume that $x(t) \in \mathcal{P}$, t=0,1,..,j is equivalent to $x(0) \in \mathcal{S}_j$, then the condition $x(t) \in \mathcal{P}$, t=0,1,..,j+1, is equivalent to $x(1) \in \mathcal{S}_j$, and $x(0) \in \mathcal{P}$. We have therefore that the constraints $\Phi^{(j)}[Ax(0)+Ev] \leq \Theta^{(j)}$ and $\Phi^{(0)}x(0) \leq \Theta^{(0)}$ hold for x(0) for all $v \in \mathcal{V}$ which are equivalent to those in (45) so we get $x(0) \in \mathcal{S}_{j+1}$. So the condition $x(t) \in \mathcal{P}$, $t \geq 0$, implies $x(0) \in \mathcal{S}_j$ for all $j \geq 0$, then $x(0) \in \mathcal{S}_\infty$ so $\mathcal{M} \subset \mathcal{S}_\infty$. Conversely assume that x(0) is not contained in $\mathcal{M}$, this means that there exist v(·) and $t \geq 0$ such that $x(t) \neq \mathcal{P}$, and then $x(0) \neq \mathcal{S}_t$, so $x(0) \neq \mathcal{S}_\infty$, then $\mathcal{S}_\infty \subset \mathcal{M}$. We have therefore $\mathcal{M} = \mathcal{S}_\infty$. □

In general, the maximal PI set $\mathcal{S}_\infty$ is given by the intersection of an countable number of sets. It is clear that a desirable property is that such a set is generated by a finite number of inequalities. From (45) we can see that such a condition holds if for a finite j we have that $\mathcal{S}_j = \mathcal{S}_{j+1}$ in this case we have $\mathcal{S}_j = \mathcal{S}_\infty$.

There are some conditions involving the stability of the system and the measure of the disturbance ε that assure this property. As pointed out in the previous section, we are mainly interested in PI sets contained in $\mathcal{P}$ and containing the origin, and the existence of such kind of sets is assured if $\varepsilon \leq \rho$. We have the following theorem [29].

Theorem 13. Assume that the system is asymptotically stable and that the set $\mathcal{P}$ is a C-set. Then assuming that $v(t) \in \varepsilon\mathcal{V}$, with $\varepsilon < \rho$, then $\mathcal{S}_\infty$ is finitely generated that is there exists j such that $\mathcal{S}_\infty = \mathcal{S}_j$.

Proof. For $\varepsilon < \rho$, the vector sequence $\Xi^{(k)}$ has nonnegative terms (i.e. $\Xi_j^{(k)} \geq 0$), it is nonincreasing and it converges to the vector $\Xi^{(\infty)}$ having strictly positive components. We have that $\Xi^{(\infty)} \leq \Xi^{(k)}$ and so the set $\{\Omega A^k x \leq \Xi^{(\infty)}\}$ is contained in the set $\mathcal{P}_k = \{\Omega A^k x \leq \Xi^{(k)}\}$. Since A is asymptotically stable, we have that there exists k such that the set $\{\Omega A^k x \leq \Xi^{(\infty)}\}$ contains $\mathcal{P}$, but then $\mathcal{P}_k$ also contains $\mathcal{P}$. From (44) we have immediately $\mathcal{P}_k = \mathcal{P}_{k+1}$ and then $\mathcal{S}_k = \mathcal{S}_{k+1}$. □

Even if the state constraint set induced by the output and input constraints is not in general bounded, we can ever reduce our problem to this case by introducing new constraints. The assumption that all the state variables are bounded is quite reasonable.

We can relax the assumption of a bounded $\mathcal{P}$ if we assume a representation of the form (which is of practical significance)

$$\mathcal{P} = \{ -e^- \leq Nx \leq e^+, \quad e^-, e^+ > 0 \}. \tag{46}$$

Theorem 14. Assume that the system is asymptotically stable and that the set $\mathcal{P}$ is represented as in (46). Then assuming that $v(t) \in \varepsilon\mathcal{V}$, with $\varepsilon < \rho$, the maximal PI set is finitely generated that is there exists j such that $\mathcal{S}_\infty = \mathcal{S}_j$.

Proof. Apply to the system a linear state transformation leading to the decomposition $x(t) = [x_1(t)^T \; x_2(t)^T]^T$, where $x_1(t)$ contains the observable state variables of (A,N) and $x_2(t)$ contains the unobservable ones. Correspondingly, the matrices A,E,N assume the form

$$\begin{bmatrix} A_1 & 0 \\ A^* & A_2 \end{bmatrix}, \qquad \begin{bmatrix} E_1 \\ E_2 \end{bmatrix}, \qquad \begin{bmatrix} N_1 & 0 \end{bmatrix},$$

With this state transformation the constraints (46) take the form

$$e_-^{(k)} \le N_1 A_1^k x_1 \le e_+^{(k)} \tag{47}$$

and so only the observable part of the state is constrained. Assuming that n_1 is the dimension of $x_1(t)$, the inequalities (47) $k=0,1,.., n_1-1$ generate a polytope $\mathcal{P}^{(1)}$ in the observable subspace of the system which contains the origin in its relative interior. Moreover, it is immediate to see that these constraints are exactly those we derive by considering the subsystem (A_1,E_1,N_1). The determination of the maximal PI set contained in $\mathcal{P}$ is then equivalent to the determination of the maximal PI set contained in $\mathcal{P}^{(1)}$ for the subsystem (A_1,E_1,N_1). By applying Theorem 13 the proof is complete. □

The results above assure the maximal PI set to be finitely generated if $\varepsilon<\rho$. In some cases, it may happen that such a property holds even for $\varepsilon=\rho$. Let us consider the following example.

Example 1. Let us consider the following second order system

$$A = \frac{1}{2}\begin{bmatrix} 1 & -1 \\ 1 & 1 \end{bmatrix} \quad E=\begin{bmatrix} 1 \\ 1 \end{bmatrix}$$

$$\mathcal{V} = \{ -1 \le v \le 1 \} \; \mathcal{P} = \{ -4 \le x_1 , x_2 \le 4 \}$$

We have that for $v \in \varepsilon\mathcal{V}$ the sets $\mathcal{S}_k$ are given by $-\Xi^{(k)} \le A^k x \le \Xi^{(k)}$ where

$$\Xi^{(k)} = \begin{bmatrix} 4 \\ 4 \end{bmatrix} - \varepsilon \{ \underbrace{\begin{bmatrix} 1 \\ 1 \end{bmatrix} + \begin{bmatrix} 0 \\ 1 \end{bmatrix} + \begin{bmatrix} 1/2 \\ 1/2 \end{bmatrix} + \begin{bmatrix} 1/2 \\ 0 \end{bmatrix} + \begin{bmatrix} 1/4 \\ 1/4 \end{bmatrix} + \ldots}_{k \text{ terms}} \}$$

With easy computations we see that

$$\Xi^{(k)} \to \Xi^{(\infty)} = \begin{bmatrix} 4 \\ 4 \end{bmatrix} - \varepsilon \begin{bmatrix} 8/3 \\ 10/3 \end{bmatrix}$$

The vector $\Xi^{(\infty)}$ must have nonnegative components so we get $\rho=\frac{6}{5}$ and we easily check that for $\varepsilon=\rho$ the condition $\mathcal{S}_3 = \mathcal{S}_4$ holds. The maximal PI set $\mathcal{S}_\infty$ for $\varepsilon=6/5$ is therefore the polyhedron $\mathcal{S}_3$ which is represented in Fig. 4.

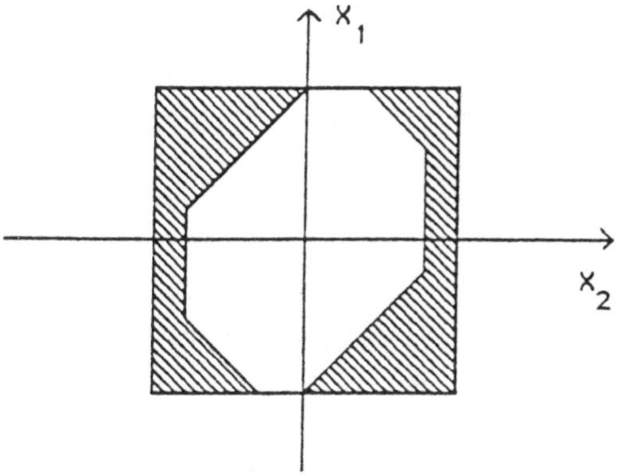

Fig. 4 The maximal PI set.

Let us consider now the continuous-time case. In this case the constraints corresponding to (42) are easily shown to be

$$\mathscr{P}_{(t)} = \{\ \Omega_i \exp(At)\, x \leq \Xi_i - \varepsilon \int_0^t \Delta_i(\alpha)\, d\alpha\ ,\ i=1,..,s\} \tag{48}$$

where $\Delta_i(\alpha) = \max_{v(\alpha)\in\mathscr{V}}\{\Omega_i \exp(A\alpha)Ev(\alpha)\}$. For any $t\geq 0$, the inequalities (48) form a polyhedral set. The maximal PI set is given by the intersection of all these sets $\mathscr{S}_\infty=\{x:\ x\in\mathscr{P}_{(t)},\ t\geq 0\}$. In general such a set is not polyhedral.

However, we can approximate the maximal PI set by sampling the parameter t with an appropriate interval and thus reducing the problem to that of handling a countable number of sets. If a polyhedron is derived with this method, it contains the maximal PI set and it approximates it as closely as the sampling interval is small. It must be remarked that a polyhedron derived in this way is not PI (in general it contains the maximal PI set).

As an alternative approach, we can approximate the maximal PI set by applying the method presented for the discrete-time case to the Euler approximating system (28). If A is asymptotically stable, the matrix $[I+\tau A]$ is so for a sufficiently small τ. Using this discretization we can find a polyhedral approximation of the maximal PI set. The main advantage of this method is that such an approximating set is a PI set and therefore it is an internal approximation of the maximal PI set.

Theorem 15. Let $\mathcal{P}$ be a polyhedral C-set and let the system $x'(t) = A\,x(t) + E\,v(t)$ with $v \in \varepsilon\mathcal{V}$, be asymptotically stable. For $\varepsilon<\rho$, let $\mathcal{S}_\infty(\varepsilon)$ be the corresponding maximal PI set for such a system and let $\mathcal{S}_\infty(\tau,\varepsilon)$ be the maximal PI set for the Euler approximating system (28) with sampling interval τ. The following properties hold

i) There exists $\tau>0$ such that the maximal PI set $\mathcal{S}_\infty(\tau,\varepsilon)$ is nonempty.

ii) The set $\mathcal{S}_\infty(\tau,\varepsilon)\subset\mathcal{S}_\infty(\varepsilon)$ converges to $\mathcal{S}_\infty(\varepsilon)$ as $\tau\to 0^+$, in the sense that for each $\lambda>0$ there exists τ^* such that $\mathcal{S}_\infty(\varepsilon) \subset (1+\lambda)\mathcal{S}_\infty(\tau,\varepsilon)$ for $\tau<\tau^*$.

The proof of the theorem is based on the following lemma. Consider the system

$$x'(t) = A\,x(t) + E\,v(t) + w(t) \tag{49}$$

where $w(t)$ is a fictitious unknown disturbance such that $\|w(t)\| < w_0$. If the C-set $\mathcal{S}$ is PI for system (49), then there exists τ^* such that for $0<\tau\le\tau^*$ $\mathcal{S}$ is PI for the Euler approximating system (28). The proof of the lemma which involves the compactness of $\mathcal{S}$ and the properties of the tangent cones is tedious and it is not presented.

Let $\mathcal{R}(\varepsilon)$ be the set of all reachable states from the origin with constrained input $v(t)\in\varepsilon\mathcal{V}$. We have that $\mathcal{R}(\rho) \subset \mathcal{P}$ and so for every $\varepsilon^*<\rho$, $\mathcal{R}(\varepsilon^*) = \varepsilon^*/\rho\,\mathcal{R}(\rho) \subset \mathrm{int}\{\mathcal{P}\}$, since $\mathcal{P}$ is a C-set. Since system (49) is asymptotically stable, then w_o may be chosen sufficiently small such that the set $\mathcal{R}^*(\varepsilon^*,w_o)$ of all states that are reached by (49) from the origin fulfills the property $\mathcal{R}^*(\varepsilon^*,w_o)\subset\mathcal{P}$. Clearly, the set $\mathcal{R}^*(\varepsilon^*,w_o)$ is invariant for (49), so we can take a sufficiently small τ in order to assure that $[I+\tau A]$ is asymptotically stable and that $\mathcal{R}^*(\varepsilon^*,w_o)$ is PI for the Euler discrete-time system. So, from Theorem 13, for every $\varepsilon\le\varepsilon^*$ $\mathcal{S}_\infty(\tau,\varepsilon)$ is nonempty and it is a polyhedral C-set for $\varepsilon<\varepsilon^*$. Since $\varepsilon<\varepsilon^*$ and $\varepsilon^*<\rho$ are arbitrary, (i) follows. To prove convergence, let $\lambda>0$ be arbitrary. Let, $\mathcal{S}_\infty^*(\varepsilon,w_o)$ be the maximal PI set contained in $\mathcal{P}$ for (49). We prove that we can choose w_o such that $\mathcal{S}_\infty(\varepsilon) \subset (1+\lambda)\mathcal{S}_\infty^*(\varepsilon,w_o)$. The set $(1+\lambda)\mathcal{S}_\infty^*(\varepsilon,w_o)$ is

$$\{x:\ \Omega_i \exp(At)x \le (1+\lambda)\,(\,\Xi_i - \varepsilon\int_0^t \Delta_i(\alpha)d\alpha\ -\ w_o\int_0^t \Delta_i^*(\alpha)d\alpha),\ i=1,\dots,s,\ t\ge 0\} \tag{50}$$

where $\Delta_i^*(\alpha) = \max_{\|w(\alpha)\|\le 1} \Omega_i \exp(A\alpha)\, w(\alpha) = \|\,\Omega_i \exp(A\alpha)\,\|$.

Let $x \in \mathscr{S}_\infty(\varepsilon)$. We have that

$$\Omega_i \exp(At)\, x \;\leq\; \Xi_i - \varepsilon \int_0^t \Delta_i(\alpha)\, d\alpha, \qquad i=1,..,s.$$

Since $\varepsilon<\rho$, the right term of the inequality above is positive, so we can take w_o such that

$$0 < w_o \leq \min_i\; \lambda\, [\, \Xi_i - \varepsilon \int_0^t \Delta_i(\alpha)\, d\alpha \;] \,/\, [\, (1+\lambda) \int_0^t \Delta_i^*(\alpha)\, d\alpha \,],$$

to see with easy computations that x fulfills (50), so $x \in (1+\lambda)\mathscr{S}_\infty^*(\varepsilon,w_o)$ and then $\mathscr{S}_\infty(\varepsilon) \subset (1+\lambda)\mathscr{S}_\infty^*(\varepsilon,w_o)$. On the other hand, $\mathscr{S}_\infty^*(\varepsilon,w_o)$ is PI for system (49) and so there exists τ^* such that $\mathscr{S}_\infty^*(\varepsilon,w_o)$ is PI for the Euler approximating system if $0<\tau\leq\tau^*$. Thus $\mathscr{S}_\infty^*(\varepsilon,w_o)\subset\mathscr{S}_\infty(\tau,\varepsilon)$ then $(1+\lambda)\mathscr{S}_\infty^*(\varepsilon,w_o)\subset(1+\lambda)\mathscr{S}_\infty(\tau,\varepsilon)$ and therefore $\mathscr{S}_\infty(\varepsilon) \subset (1+\lambda)\mathscr{S}_\infty(\tau,\varepsilon)$. □

It is well known that in general the discretization of a continuous-time linear system based on the exponential matrix leads to a better approximation of the solution than that one obtained via the Euler approximating system. From Theorem 15 we see that the Euler approximating system, is more useful for our problem of determining a PI set for the continuous-time system.

VII. CONTROL SYNTHESIS VIA NONSMOOTH OPTIMIZATION.

In the previous chapters it has been shown that the existence of a PI set is of fundamental importance for the problem of the constrained control for systems with disturbances since it represents an effective bounding sets for the system state. In particular, the polyhedral sets have been considered as candidate positively invariant sets and some properties of this kind of sets have been illustrated. Analysis results have been presented: we have seen how to compute a PI set contained in the state constraint set when the compensator is assigned.

In this chapter we present some examples in order to describe how a design technique based on these results may be derived. In [16] it has been shown that the assignment of a PI polyhedral compact and convex set is a linear programming problem. In other words, all feedback matrices that assure the positive invariance and admissibility of a given polytope are the solutions of a set of linear inequalities.

If we consider discrete-time systems, from Theorem 4 and (25) we see that all the feedback state matrices assuring the invariance of $\mathcal{S} = \{\Phi x \leq \Theta\}$ with disturbances constrained as $v(t) \in \varepsilon\mathcal{V}$ fulfill the constraints

$$\Phi_i \, [A+BK] \, w_k \leq \Theta_i - \varepsilon \, \Delta_i = \Theta_i^* \quad i=1,..,s, \;\; k=1,.., r, \tag{51}$$

where $w_k \in \text{vert}\,\{\mathcal{S}\}$. If $\mathcal{U}$ is a polyhedron of the form $\mathcal{U} = \{u\colon \Sigma u \leq \Gamma\}$ the admissibility conditions introduce the additional constraints for K

$$\Sigma \, K w_k \leq \Gamma, \qquad\qquad k = 1,.., r. \tag{52}$$

Expressions (51)-(52) represent a sets of linear inequalities whose solutions are the feedback matrices that solve the problem with the assigned $\mathcal{S}$. We may wish to maximize the value of ε. In this way, a linear programming problem is obtained. For the continuous-time case we may similarly apply the criteria of Theorems 5 and 6.

However, the solution set of (51)-(52) with the assigned set $\mathcal{S}$ may be empty. To overcome this shortcoming, we may consider another design method based on the maximization of the index ρ of (40) and (41) with respect to the parameters of the compensator.

It may be shown by simple examples that the function $\rho(K)$ is in general nonsmooth, so a nonsmooth optimization procedure is to be used. As it is known, the nonsmooth optimization problems are normally solved by iterative methods. The reader is referred to [37] to have a complete description. Such iterative methods have the shortcoming that they may converge to local maxima instead of the global one. To solve this problem, we can apply the procedure many times starting from different initial points in the parameter space. Reasonable initial values of the compensator parameters to start the iteration may be derived by considering other synthesis methods, for example the pole assignment, assuring in this way that the initial compensator stabilizes the plant. As observed above if an optimal positive ρ is found, the BIBO stability is assured by the corresponding compensator.

When the optimal feedback gains are evaluated, we may determine, as a second step, a positively invariant set for the problem as described above.

The synthesis method suggested here, has the advantage that it generates an optimal compensator whose order may be fixed a-priory. Moreover, we can take in account constraints for the compensator parameters which are due to the physical nature of the compensator. This feature is of practical importance in those cases in which the compensator has a fixed structure and its parameters must be chosen in a fixed range.

Example 2. Boiler-Turbine Model .
Let us consider the following continuous-time system representing a boiler-turbine system. This example was considered in [7] where the state feedback control problem was solved using the set theoretic control technique based on bounding ellipsoidal sets. The system matrices are given by

$$A=\begin{bmatrix} -0.0075 & 0.0075 & 0 \\ 0.1086 & -0.149 & 0 \\ 0 & 0.1415 & -0.1887 \end{bmatrix},\quad B=\begin{bmatrix} 0.0037 \\ 0 \\ 0 \end{bmatrix},\quad E=\begin{bmatrix} 0 \\ -0.0538 \\ 0.1187 \end{bmatrix}$$

The state variable are: x_1 the drum pressure, x_2 the throttle pressure, x_3 the reheat pressure. The control variable u is the firing rate, and the disturbance input is the control valve position. The constraints for the state the control and the disturbance are

$$\mathscr{X} = \{ x: |x_1| \le 0.1,\ |x_2| \le 0.05,\ |x_3| \le 0.1 \}$$

$$\mathscr{U} = \{ u: |u| \le 0.25 \} \quad \mathscr{V} = \{ v: |v| \le 0.1 \}$$

The solution derived by the set theoretic approach based on the Lagrange Multipliers method proposed in [7] is

$$K_{ST} = [\,-37.85 \quad -4.39 \quad 0.475\,].$$

The corresponding closed-loop eigenvalues are $\{-0.1566\pm j0.0242, -0.1719\}$. In [7] it is proved that this control assures that no boundary violation occurs for a disturbance having size ε=1.27. Such a result was compared with that obtained via the linear quadratic method. Assuming a cost index of the form

$$J=\int_0^{\infty} \{ x(t)^T P\, x(t) + u(t)^T Q\, u(t) \}\, dt$$

with $P = \mathrm{diag}\{ 1/0.1^2, 1/0.05^2, 1/0.1^2 \}$ and $Q = [\,1/0.25^2\,]$, the solution

$$K_{LQ} = [\,-3.811 \quad -0.4034 \quad 0.0134\,].$$

is found. The corresponding closed-loop eigenvalues are {-0.0160, -0.1537, 0.1888}. In [7] it is shown that if a step function of amplitude 1.27 is applied, the state boundaries are violated.

Using expression (41) we can compute the maximum value of the disturbance size which is $\rho \cong 0.86$ for the linear-quadratic solution K_{LQ} and $\rho \cong 1.46$ for K_{ST}.

By applying a direct search method procedure to solve the nonsmooth optimization problem we found the feedback matrix

$$K_{NS} = [\,-45.66 \quad -8.295 \quad -0.663\,].$$

The corresponding maximum disturbance size is $\rho \cong 1.49$. The corresponding eigenvalues are -0.19, $-0.15 \pm j0.052$.

Example 3. D-C Electric motor.

Let us consider the following continuous-time model representing the direct current electric motor represented in Fig. 5.

$$A = \begin{bmatrix} -0.07 & -0.86 \\ 0.06 & -0.0085 \end{bmatrix}, \quad B = \begin{bmatrix} 1 \\ 0 \end{bmatrix}, \quad E = \begin{bmatrix} 0 \\ 0.06 \end{bmatrix}$$

Its state variables are $i = x_1$ the armature current, $\omega = x_2$ the angular speed. The control variable is the armature voltage, and the disturbance is the load torque.

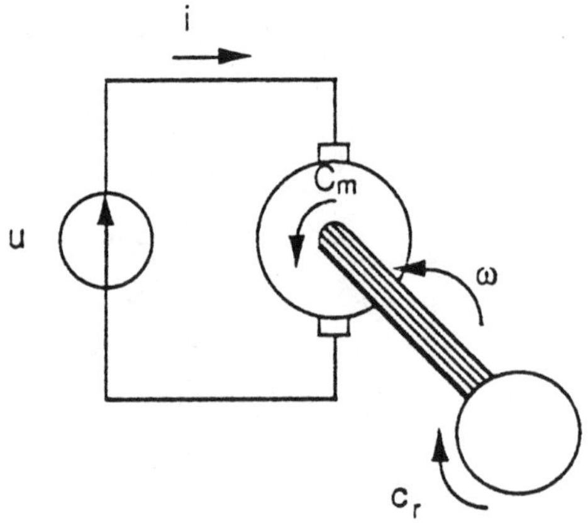

Fig. 5. Direct current electric motor.

We assume that the admissible fluctuations from the work point for the state and control variables are 100% of the nominal value for the armature current, voltage and the load torque disturbance and 20% of the nominal value for the speed. Since adimensional variable are used we have the following constraints

$$\mathscr{X} = \{ x: |x_1| \le 1,\ |x_2| \le 0.2 \},$$
$$\mathscr{U} = \{ u: |u| \le 1 \},\quad \mathscr{V} = \{ v: |v| \le 1 \}.$$

The system eigenvalues are $\{-0.039 \pm j\ 0.225\}$.

We tackle this problem via Euler approximating system. Let us fix $\tau=1$ second. We derive the following matrices

$$I+\tau A = \begin{bmatrix} 0.93 & -0.860 \\ 0.06 & 0.9915 \end{bmatrix},\quad \tau B = \begin{bmatrix} 1 \\ 0 \end{bmatrix},\quad \tau E = \begin{bmatrix} 0 \\ 0.06 \end{bmatrix}$$

We first consider a state feedback control. The index ρ for $K = [\ 0\ \ 0\ ]$ is $\rho=0.068$. Applying the optimization procedure it has been derived the feedback matrix

$$K_{NL} = [\ -1.41\ \ -6.56\ ].$$

The corresponding index is $\rho=1.028$. Let us set $\varepsilon=1$. The corresponding maximal PI and admissible set which is represented in figure Fig. 6 is given by

$$\{ -1 \le x_1 \le 1,\ -1 \le 0.48x_1 + 7.42\ x_2 \le 1,\ -1 \le 1.41x_1 + 6.56\ x_2 \le 1 \}.$$

From Theorem 6, we have that the derived set is positively invariant for the continuous-time system with the the evaluated feedback control K_{NL}, so K_{NL} solves the problem (for $\varepsilon=1$) for the continuous-time system. The corresponding continuous-time closed-loop system eigenvalues are $\{-1.054, -0.435\}$.

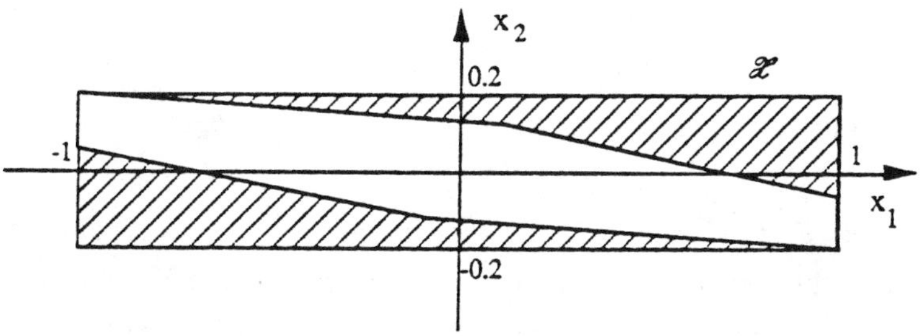

Fig. 6. Maximal positively invariant set for $\varepsilon=1$.

Let now assume that the full state measurement is not available, but only the angular speed ω is measured. This is equivalent to introduce an output matrix C = [0 1]. If a static output feedback is considered (i.e. a proportional feedback control) we find

$$K = [\ 0.718\ ]$$

that gives an index ρ=0.159. The evaluated closed-loop eigenvalues for the continuous-time system are {-0.0392$\pm$ j0.0867}.

Consider now a dynamic compensator of the first order. The optimization procedure derives the following compensator

$$F = [\ -0.3127\ ] \qquad G = [\ 3.585\ ]$$

$$H = [\ 2.341\ ] \qquad K = [\ -5.885\ ]$$

which assures a disturbance rejection measure of ρ=1.0281. The corresponding closed-loop eigenvalues are { -0.8009, -0.5222 , -0.0681 } .

The second order compensator derived by the procedure is

$$F = \begin{bmatrix} -0.0600 & -1.367 \\ 0.2082 & -0.0295 \end{bmatrix} \qquad G = \begin{bmatrix} 5.325 \\ 3.377 \end{bmatrix}$$

$$H = [\ -0.1758 \quad 1.579\ ] \qquad K = [\ -4.759\ ]$$

that assures a value of ρ=1.0283. The corresponding closed loop-system eigenvalues for the continuous-time system are {-0.8407$\pm$j0.5691, -0.4179,-0.0688}.

We see that in this case, the first order compensator provides in practice the same performance as the second order one.

VIII. CONCLUSIONS

One of the oldest and most important problems arising in control applications is to preserve the system from undesirable effects due to the presence of external disturbances. Moreover the designer has often to cope with the the presence of physical bounds on the system output and control variables.

In this contribution, we have examined the conditions that a compensator must fulfill in order to assure that the closed-loop system state and control are kept within their constraints in the presence of an unknown but bounded disturbance. The central point of the discussion is that the resulting closed-loop system must have a positively invariant and admissible set contained in the system state constraint set. Such a set represents an effective constraint set in the sense that for each state contained in it at at a given time, it is assured that the state and control constraints are fulfilled at all future times. Moreover, to each positively invariant set which is convex, compact and contains the origin in its interior, a Lyapunov function is associated, and this fact is important for the system stability.

In the literature, ellipsoidal sets and the corresponding quadratic Lyapunov functions have been most frequently considered as bounding sets for the system state. Here we have examined the properties of the polyhedral sets.

Constraints of the polyhedral type naturally arise from many practical problems in which upper and lower bounds on input and output variables are assigned. The most interesting feature of these sets is that we can check the positive invariance of a polyhedral set by solving a linear programming problem. The conditions involved are necessary and sufficient. Moreover, they have the advantage that they can arbitrarily closely approximate the maximal positively invariant set contained in the state constraint region.

If polyhedral constraint are assigned, we can compute the maximum disturbance size assuring that no boundary violations occur. Such a value is an important index of performance for the compensator. A design method may be derived by choosing the parameters of the compensator in order to maximize this index.

REFERENCES

1. H. S. Witsenhausen, "A minmax Control Problem for Sampled Linear Systems", IEEE Transactions on Automatic Control," Vol. AC-13, 5-21 (1968).

2. E. Davison "The Output Control of Linear Time-Invariant Multivariable Systems with Unmeasurable Arbitrary Disturbances," IEEE Transaction on Automatic Control, Vol. AC-17, 621-629 (1972).

3. E. Davison and I. J. Ferguson, " The Design of Controllers for the Multivariable Robust Servomechanism Problem Using Parameter Optimization Methods," IEEE Transaction on Automatic Control, Vol. AC-26, 93-110 (1981).
4. C. D. Johnson, "Theory of Disturbance-Accomodating Controllers," in "Control and Dynamic Systems," Vol. (C.T. Leondes, ed.), Academic Press, San Diego, 387-489 (1975).
5. J. D. Glover anf F. C. Schweppe, "Control of Linear Dynamic Systems with Set-Constrained Disturbances," IEEE Transaction on Automatic Control, Vol. AC-16, 411-423 (1971).
6. D.P. Bertsekas and I.B. Rhodes, "On the Minmax Reachability of Target Sets and Target Tubes," Automatica, Vol. 7, 233-247 (1971).
7. P.B. Usoro, F.C. Schweppe, L.A. Gould, D.N. Wormley, "A Lagrange Approach to Set Theoretic Control Synthesis", IEEE Transactions on Automatic Control, Vol. AC-27, 393-399 (1982).
8. P.B. Usoro, F.C. Schweppe, L.A. Gould, D.N. Wormley, "Ellipsoidal Set-Theoretic Control Synthesis," ASME Journal of Dynamic Systems, Measurement and Control, Vol. 104, 331-336 (1982).
9. A.G. Parlos, A.F. Henry, F.C. Schweppe, L.A. Gould, D.D. Lanning, "Nonlinear Multivariable Controller of Nuclear Power on the Unknown-but Bounded Disturbances," IEEE Transaction on Automatic Control, Vol. AC-33, 130-137 (1988).
10. B. A. Francis, J. W Helton, and G. Zames, "H^∞ Optimal Feedback Controllers for Linear Multivariable Systems," IEEE Transaction on Automatic Control, Vol. AC-29, 888-900 (1984).
11. M. Vidiasagar, "Further Results on the Optimal Rejection of Persistent Bounded Disturbances," IEEE Transactions on Automatic Control, Vol. AC-36, 642-652, (1991).
12. M. Vidiasagar, "Optimal Rejection of Persistent Bounded Disturbances," IEEE Transactions on Automatic Control, Vol. AC-31, 527-534 (1991).
13. M. Dahleh and J. B. Pearson, " l^1 - Optimal Controllers for MIMO Discrete-Time Systems," IEEE Transaction on Automatic Control, Vol. AC-32, 314-322 (1987).

14. M. Dahleh and J. B. Pearson, " L^1 - Optimal Controllers for MIMO Continuous-Time Systems," IEEE Transaction on Automatic Control, Vol. AC-32, 889-895 (1987).
15. J. B. Perason and B. Bahmieh, "On Minimizing Maximum Errors," IEEE Transaction on Automatic Control, Vol. AC-35, 598-601 (1990).
16. F. Blanchini, "Feedback Control for Linear Systems with State and Control Bounds in the Presence of Disturbance," IEEE Transaction on Automatic Control, Vol. 35, 1231-1235, (1990).
17. A. Feuer and M. Heymann, "Admissible Sets in Linear Feedback Systems with Bounded Controls," International Journal of Control, Vol. 23, 381-392, (1976).
18. D. Luenberger, "Optimization by Vector Space Methods," John Wiley and Sons, Inc., New York: Wiley (1969).
19. Leitmann, G., "Guaranteed Asymptotic Stability for Some Linear Systems with Bounded Uncertainties," Journal of Dynamic System, Measurement, and Control, Vol. 101, (1979).
20. B. R. Barmish and G. Leitmann, "On Ultimate Boundedness Control of Uncertain Systems in the Absence of Matching Assumptions," IEEE Transaction on Automatic Control, Vol. AC-27, 153-158, 1982.
21. J.B. Aubin, A. Cellina, "Differential Inclusions," Spinger-Verlag, Berlin (1984).
22. A. Feuer and M. Heymann, " Ω-Invariance in Control Systems with Bounded Controls," Journal of Mathematical Analysis and Applications, Vol. 53, 267-276 (1976).
23. Nagumo, M., "Uber die Lage der Integralkurven gewohnlicher Differential-gleichungen," Proceedings of the Physical-Mathematical Society, Japan, Vol. 24, 1942.
24. A. Benzaouia and C. Burgat, "Regulator Problem for Linear Discrete-Time Systems with non-symmetrical Constrained Control," International Journal of Control, vol. 48, 2441-2451 (1988).
25. A. Benzaouia and C. Burgat, "The Regulator Problem for a Class of Linear Systems with Constrained Control," System and Control Letters, Vol. 10, 357-363 (1988).

26. J. Hennet and J. Beziat, "A Class of Invariant Regulators for the Discrete-time Linear Constrained Regulation Problem," Automatica, Vol. 27, 549-554 (1991).
27. G. Bitsoris, "On the Positive Invariance of Polyhedral Sets for Discrete-Time Systems," System and Control Letters, vol. 11, 243-248 (1988).
28. M. Vassilaky, J. Hennet and G. and Bitsoris, "Feedback Control of Discrete-Time Systems Under State and Control Constraints," International Journal of Control, Vol. 47, 1727-1735 (1988).
29. F. Blanchini, "Constrained Control for Uncertain Linear Systems," Proceedings of the 29-th Conference on Decision and Control, Honolulu, Usa, Vol. 6, (1990).
30. M.Vassilaki and G.Bitsoris, "Constrained Regulation of Linear Continuous-time Dynamical Systems," System and Control Letters, vol. 47, 247-252 (1989).
31. G. Bitsoris, "Positively Invariant Polyhedral sets of Discrete-Time Linear Systems," International Journal of Control, Vol. 47, 1713-1726 (1988).
32. C. A. Desoer and M. Vidiasagar, "Feedback systems: Input-Output Properties," New York: Academic, 1975.
33. J. B. Lasserre, "A Complete Characterization of Reachable sets for Constrained Linear Time-Varying Systems," IEEE Transactions on Automatic Control, Vol. AC-32, 836-838, (1987)
34. G. Basile, G. Marro, A. Piazzi, "Revisiting the Regulator Problem in the Geometric Approach, Part 1, Disturbance Localization by Dynamic Compensation," Journal of Optimization Theory and Applications, Vol. 53, 9-21, (1987).
35. G. Basile, G. Marro, A. Piazzi, "Revisiting the Regulator Problem in the Geometric Approach, Part 2, Asymptotic Tracking and Regulation in the Presence of Disturbances," Journal of Optimization Theory and Applications, Vol. 53, 23-36, (1987).
36. J. C. Willems and C. Commault, "Disturbance Decoupling by Measurement Feedback with Stability or Pole Placement, SIAM Journal on Control and Optimization, Vol. 19, 490-540 (1981).
37. E. Gill, W. Philips, M. Murray and A. Wright, "Practical Optimization," Academic Press, New York, 1981.

H^∞ Super-Optimal Solutions

Da-Wei Gu
Ian Postlethwaite
Engineering Department
Leicester University
LE1 7RH, UK

Mi-Ching Tsai
Mechanical Engineering Department
National Cheng Kung University
Taiwan 70101, ROC

1 Introduction

A central problem in control system design is how to obtain desirable performance in the presence of uncertainties which are inevitable in any mathematical model of an engineering process. These uncertainties range from known modelling approximations to unknown parameters and disturbance signals. Therefore, the most important property of a feedback design is robustness which is the ability to maintain performance and stability in the face of perturbations representing uncertainty. Unlike linear quadratic design [2], the $\mathbf{H}^\infty$ approach [33], which meets frequency-domain performance specifications, is ideal for handling a prescribed set of uncertainties. The approach is also useful for singular value analysis and loop-shaping synthesis techniques [4] [8]. The $\mathbf{H}^\infty$-norm of a stable transfer function matrix $G(s)$, denoted $\|G\|_\infty$, is the maximum over all frequencies of the largest singular value of the frequency response $G(j\omega)$. Its popularity stems from two important results:

CONTROL AND DYNAMIC SYSTEMS, VOL. 51

- A sufficient condition for closed-loop stability to be robust against a set of plant perturbations is given by a bound on the $\mathbf{H}^\infty$-norm of a stable closed-loop transfer function.

- The $\mathbf{H}^\infty$-norm of a stable transfer function matrix represents a bound on the maximum energy gain from the input signals to the output.

A minimization criterion using the $\mathbf{H}^\infty$-norm represents a worst case design in the frequency-domain [14] and is especially suitable for robust control. In standard $\mathbf{H}^\infty$ optimization only the maximum singular value ($\mathbf{H}^\infty$-norm) of a cost function is minimized. For multivariable systems, there is in general a set of optimal solutions achieving the same cost. Naturally, in practice, one would like to choose the solution which gives in some sense the best design. But there is no easy way of defining and finding such a solution. A stronger minimization criterion can, however, be posed to get a *unique* solution [29]. In this criterion, it is proposed to minimize not just the maximum singular value ($\mathbf{H}^\infty$-norm) of a transfer function but all singular values in turn from the largest down to the smallest. This results in a unique solution to what is now called the $\mathbf{H}^\infty$ *super-optimization* problem in which all singular values are minimized. Although the criterion was initially motivated by a mathematician's search for uniqueness, it also has some engineering significance as argued in [16] [6]. In [32] Yue states that a standard $\mathbf{H}^\infty$ controller is sometimes unnecessarily complex; for instance, if the plant is diagonal then the $\mathbf{H}^\infty$ controller may not be diagonal whereas the super-optimal $\mathbf{H}^\infty$ controller would always be diagonal. In general, however, a super optimal $\mathbf{H}^\infty$ controller will be more complex (higher McMillan degree) than the corresponding standard $\mathbf{H}^\infty$ controller.

It is well known that $\mathbf{H}^\infty$ optimal control problems have equivalent model-matching problems the solutions of which are sometimes simpler to find. In this paper we will consider directly the model-matching formulation. Model-matching problems can be divided into three kinds depending on the structure of the problem: *1-block*, *2-block* and *4-block* formulations [3]. We will consider only 1 and 2-block problems in this paper. Model-matching problems in which all singular values are minimized are sometimes called *strengthened* model-matching problems to distinguish them from those in which only the largest singular value is minimized.

A conceptual algorithm, based on operator theory, was first given by Young in [29] for calculating the unique super-optimal solution in the 1-block case. Later Kwakernaak [16] proposed an implementable polynomial approach for finding the super-optimal solution to a mixed sensitivity problem (a 2-block problem). Limebeer *et al* [18] gave a state-space approach for solving the 1-block problem and this was extended to more general problems in [15] . In this article, we summarize much of our own work [23] [11] [12] [13] [20] on a state-space approach to $\mathbf{H}^{\infty}$ super-optimal solutions for both 1 and 2-block problems. In the 2-block case, we also discuss the convergence of our algorithm and present a convergence proof which has thus far been absent from the literature.

Following this introductory section, a state-space approach is described in Section 2 for finding $\mathbf{H}^{\infty}$ super-optimal solutions for 1-block model-matching problems. In Section 3, the 2-block model-matching problem is considered. As in standard $\mathbf{H}^{\infty}$ optimization the 2-block problem requires an iterative algorithm (γ-iteration) at each layer of optimization. The algorithm and its convergence are considered in detail in Section 4. Conclusions are given in Section 5.

Nomenclature

$j\Re$ ($j\Re_e$)	the imaginary axis (including infinity)
$\mathbf{L}^2/\mathbf{L}^{\infty}$	frequency-domain Lebesque spaces
$\mathbf{H}^2/\mathbf{H}^{\infty}$	Hardy spaces
$\mathcal{G}\mathbf{H}^{\infty}$	the group of units of $\mathbf{H}^{\infty}$
$\|G\|_2$	$\mathbf{L}^2/\mathbf{H}^2$ norm
$\|G\|_{\infty}$	$\mathbf{L}^{\infty}/\mathbf{H}^{\infty}$ norm
$\|\Gamma\|$	the norm of operator Γ
Γ^*	adjoint of operator Γ

A^*	complex conjugate transpose of matrix A
A^T	transpose of matrix A
$\lambda(A)$	eigenvalues of A
$s_i(A)$	the ith singular value of A ordered so that $s_1(A) \geq s_2(A) \geq \cdots$
$s_i^\infty(G)$	$:= \sup_\omega s_i[G(j\omega)]$
$\sigma_i(G)$	the ith Hankel singular value of G
$\mathcal{H}_G$	*Hankel* operator generated by $G \in \mathbf{L}^\infty$
$\mathcal{T}_G$	*Toeplitz* operator generated by $G \in \mathbf{L}^\infty$
$G^\sim(s)$	$:= G(-s)^T$
$\deg(G)$	the McMillan degree of proper real-rational G
Prefix $\Re$	real-rational
subscript $\perp$	orthogonal complement
$\left[\begin{array}{c\|c} A & B \\ \hline C & D \end{array}\right]$	the state-space realization of a real-rational transfer function matrix *i.e.* $C(sI - A)^{-1}B + D$

$\Re$ is used as a prefix to denote real-rational. The subscripts n and $p \times q$ (as in $\Re_n$ and $\Re_{p\times q}$) will denote the corresponding n-vectors and $p \times q$ matrices. Dimensions will appear explicitly as subscripts when necessary. Thus $\Re\mathbf{L}^\infty$ denotes the space of real-rational matrices (of appropriate dimensions) with elements analytic on $j\Re_e$. $\Re\mathbf{H}^\infty$ is the subspace of $\Re\mathbf{L}^\infty$ consisting of all real-rational asymptotically stable matrices; while $\Re\mathbf{H}_-^\infty$ consists of matrices analytic in the open left-half-complex-plane. The matrix $G \in \Re\mathbf{L}^\infty$ is called *all-pass* if $G^\sim G = I$. In addition, when G is stable, it is called *inner*. Note that G need not be square, but $p \geq q$. If G is a nonsquare inner matrix, then a matrix $G_\perp$ can be found such that $[\,G \quad G_\perp\,]$ is square and inner; $G_\perp$ is called a complementary inner factor (CIF). Symbols in **boldface** capital letters are either sets or spaces.

2 1-Block Super-Optimal Solution

In this section, we will consider the 1-block strengthened model-matching problem: for a given $F \in \Re\mathbf{H}^{\infty}{}_{p\times q}$ $(p \geq q)$, find a $Q \in \Re\mathbf{H}^{\infty}_{-}{}_{p\times q}$ such that the sequence

$$s_1^{\infty}(F-Q),\ s_2^{\infty}(F-Q),\ s_3^{\infty}(F-Q),\ \cdots,\ s_q^{\infty}(F-Q) \tag{1}$$

is minimised with respect to the lexicographic ordering. We will denote the optimal sequence by $(\bar{s}_1^{\infty}, \bar{s}_2^{\infty}, \cdots, \bar{s}_q^{\infty})$ and call them the s-numbers of F.

Clearly

$$\min_{Q\in\Re\mathbf{H}_{-}^{\infty}} \|F-Q\|_{\infty} =: \bar{s}_1^{\infty}. \tag{2}$$

Now let the set of optimal solutions to (2) be denoted by $\{\mathbf{Q_{opt}}\}$. That is, for any $Q_a \in \{\mathbf{Q_{opt}}\}$,

$$\|F-Q_a\|_{\infty} = \bar{s}_1^{\infty} \ . \tag{3}$$

Assume that the realization (A, B, C, D) of F is minimal and balanced and that its gramians are both equal to $\Sigma = \text{diag}[\sigma_i]$ where σ_i is the ith Hankel singular value of F. Based on this notation, we have $\bar{s}_1^{\infty} = \sigma_1$ from

$$\min_{Q\in\Re\mathbf{H}_{-}^{\infty}} \|F-Q\|_{\infty} = \|\mathcal{H}_F\| \ \ (=\sigma_1) \ . \tag{4}$$

We first present a numerical algorithm for solving the 1-block super-optimal problem which follows closely the steps of the algorithm given in [29] except that for each step we provide a reliable implementation using state-space models. When the matrix F has several Hankel singular values all reaching the value of the largest it is possible to determine simultaneously all the corresponding s-numbers of F, which are also of the same value. Based on this a so called *frame approach* can be used to reduce the computations required to find the super-optimal Q. The approach is useful when the s-numbers of F have multiplicity greater than one.

The section consists of several subsections. In §2.1 a state-space realization is given for a maximizing vector of the Hankel operator of F. It is also shown how the maximizing vector is related to any optimal solution Q_{opt} of problem (2). §2.2 describes how to construct a square inner matrix from a maximizing vector, and in §2.3 it is shown how square inner matrices can be used to diagonalize, layer by layer, the cost function $F - Q$ in (2). By this technique the strengthened model-matching problem is successively reduced to a similar problem of one less dimension at each step, resulting in an implementable algorithm for the super-optimal solution. The frame approach is described in §2.4. An illustrative example is given in §2.5.

2.1 Maximizing Vectors

Definition 1 *Let* $\mathbf{X}$ *and* $\mathbf{Y}$ *be Hilbert spaces and let* $\Gamma : \mathbf{X} \to \mathbf{Y}$ *be a bounded operator. A* **maximizing vector** *for* Γ *is a nonzero vector* $x \in \mathbf{X}$ *such that*

$$\|\Gamma x\|_2 = \|\Gamma\| \ \|x\|_2 \ .$$

$(x, \ \Gamma x)$ is a maximizing vector pair of Γ. If x is an input vector of Γ satisfying

$$(\Gamma^* \Gamma) x = s^2 x$$

then s is called a *singular value* of Γ and x is called the corresponding *singular vector*.

It is well known [31] that a maximizing vector $v(-s) \in \Re\mathbf{H}_{\perp}^{2}$ of the Hankel operator $\mathcal{H}_F$ satisfies

$$\|\mathcal{H}_F v(-s)\|_2 = \sigma_1 \ \|v(-s)\|_2 \ . \tag{5}$$

The next theorem characterizes the properties of the maximizing vector pair of a Hankel operator.

Theorem 1 ([30]) *Let* $v(-s) \in \Re\mathbf{H}_{\perp}^{2}$ *and* $w(s) \in \Re\mathbf{H}^{2}$ *be a maximizing vector pair of* $\mathcal{H}_F$. *Then any* $Q_a \in \{\mathbf{Q}_{\mathbf{opt}}\}$ *satisfies*

$$(F(s) - Q_a(s))\, v(-s) = \sigma_1 \ w(s) \ . \tag{6}$$

Further,

$$\|v(j\omega)\|_2 = \|w(j\omega)\|_2, \quad \forall \ \omega \in \Re \ . \tag{7}$$

■

It should be noted that for any $\omega \in \Re$, $v(j\omega)$ and $w(j\omega)$ are vectors in C_q and C_p respectively and $\|v(j\omega)\|_2$ and $\|w(j\omega)\|_2$ are Euclidean norms.

In engineering terms a system model in balanced form decouples the system internally in the sense that the system can be looked at functionally as the connection of two dual sub-systems : the input sub-system $I(sI-A)^{-1}B$ and the output sub-system $C(sI-A)^{-1}I$. From the orthogonal property of a balanced system and the descending order of the diagonal elements in the gramians, the internal signal which will be magnified maximally through the output sub-system has zeros on all channels except the first. On the other hand, to reach this particular internal signal, the external input signal should be such as to excite only the first channel of the input sub-system. Hence, one of the merits of the balanced realization is that it provides some insight into how to find a *maximizing* vector.

The following result from [22] gives a state-space realization for a maximizing vector pair of $\mathcal{H}_F$.

Lemma 1 *Let (A,B,C,D) be a balanced realization of F and define*

$$v(s) = \left[\begin{array}{c|c} A^T & \frac{1}{\sqrt{\sigma_1}}e_1 \\ \hline B^T & 0 \end{array}\right] \tag{8}$$

$$w(s) = \left[\begin{array}{c|c} A & \frac{1}{\sqrt{\sigma_1}}e_1 \\ \hline C & 0 \end{array}\right] \tag{9}$$

where e_1 denotes the n-dimensional vector with 1 *as the first entry and* 0*'s as all the other entries. Then $(v(-s),w(s))$ is a maximizing vector pair of $\mathcal{H}_F$.* ■

Remark 1 $v(s)$ and $w(s)$ defined in Lemma 1 are unit vectors in $\Re\mathbf{H}^2$, *i.e.*,

$$\|v\|_2 = \|w\|_2 = \frac{1}{\sigma_1}e_1^T\Sigma e_1 = 1 .$$

From Glover [9], a set of $\{\mathrm{Q_{opt}}\}$ can be found explicitly in terms of the realization matrices (A,B,C,D) . We list these formulae in Lemma 2, since they will be used later to construct an optimizer in our algorithm.

Lemma 2 ([9]) *Partition the minimal balanced realization* (A, B, C, D) *of* F *to conform with* $\Sigma = \begin{bmatrix} \sigma_1 I_k & 0 \\ 0 & \Sigma_2 \end{bmatrix}$ *where* k *is the multiplicity of* σ_1. *Define*

$$\begin{aligned}
\tilde{A} &= (\Sigma_2^2 - \sigma_1^2 I)^{-1}(\sigma_1^2 A_{22}^T + \Sigma_2 A_{22}\Sigma_2 - \sigma_1 C_2^T U B_2^T) \\
\tilde{B} &= (\Sigma_2^2 - \sigma_1^2 I)^{-1}(\Sigma_2 B_2 + \sigma_1 C_2^T U) \\
\tilde{C} &= C_2\Sigma_2 + \sigma_1 U B_2^T \\
\tilde{D} &= D - \sigma_1 U
\end{aligned}$$

where U *satisfies*

$$U^T U \leq I_q \tag{10}$$

$$C_1 = -U B_1^T \ . \tag{11}$$

Then $Q_{opt}(s) = (\tilde{A}, \tilde{B}, \tilde{C}, \tilde{D})$ *is an optimal solution.* ■

Note that the (1,1) and (2,2) blocks of Σ in the above lemma are reversed compared with Theorem 6.3 of [9]. In addition, U need not be unitary even when F is square.

Using the state-space models of Lemma 2, we can easily verify the result of Theorem 1, *i.e.*,

$$(F(s) - Q_{opt}(s))\, v(-s) = \mathcal{H}_F v(-s) = \sigma_1 w(s)$$

for any optimal solution Q_{opt} which is in terms of the choice of U satisfying (10) and (11).

Remark 2 Theorem 1 also implies that any $Q_a \in \{\mathbf{Q_{opt}}\}$ and any maximizing vector $v(-s)$ of $\mathcal{H}_F$ satisfy

$$[F(s)v(-s)]_- = Q_a(s)v(-s) \ ,$$

where $[F(s)v(-s)]_- + [F(s)v(-s)]_+ = F(s)v(-s)$, and $[F(s)v(-s)]_-$ is the unstable part of $F(s)v(-s)$. Also $v(-s)$ is a singular vector of the error system $E(s)$ defined by $F(s) - Q_a(s)$.

Theorem 1 has shown an important property of a maximizing vector pair. That is, $w^*w(j\omega) = v^*v(j\omega)$, $\forall\omega \in \Re$. The following corollary states that there always exists a maximizing vector pair such that $w^*w(j\omega) = v^*v(j\omega) > 0$, $\forall\omega \in \Re$. First we need a special form of minimal balanced realization of a stable transfer function matrix.

Lemma 3 ([**25**]) *For a given $G \in \Re\mathbf{H}^\infty$ with dimensions $p \times q$ and Hankel singular values $(\sigma_1, \cdots, \sigma_k, \sigma_{k+1}, \cdots, \sigma_n)$ such that $\sigma_1 = \cdots = \sigma_k > \sigma_{k+1}$ and $\sigma_n > 0$, there exists a minimal balanced realization*

$$\begin{aligned} A &= \begin{bmatrix} a_{11} & a_{12} & \dots & a_{1n} \\ a_{21} & a_{22} & \dots & a_{2n} \\ \vdots & \vdots & \ddots & \vdots \\ a_{n1} & a_{n2} & \dots & a_{nn} \end{bmatrix} \\ B &= [b_1^T, b_2^T, \dots, b_n^T]^T \; ; \quad b_i^T \in \Re_q \\ C &= [c_1, c_2, \dots, c_n] \; ; \qquad c_i \in \Re_p \end{aligned}$$

such that

$$a_{ii} = -\frac{1}{2\sigma_i}\|b_i^T\|_2^2 = -\frac{1}{2\sigma_i}\|c_i\|_2^2 < 0, \quad i = 1, 2, \dots, n \quad .$$

Moreover, if r is the rank of $[c_1, c_2, \dots, c_k]$, then

$$\begin{aligned} r &= rank([b_1^T, b_2^T, \dots, b_k^T]) \\ &= rank([c_1, c_2, \dots, c_r]) \\ &= rank([b_1^T, b_2^T, \dots, b_r^T]) \\ &\leq \min(p, q, k) \quad . \end{aligned}$$

■

Corollary 1 *Let $F = (A, B, C, D)$ be given as in the above Lemma 3 and the maximizing vector pair $(v(-s), w(s))$ be defined as in Lemma 1. Then*

$$\|v(j\omega)\|_2 = \|w(j\omega)\|_2 > 0, \; \forall \; \omega \in \Re \; . \tag{12}$$

Proof: See [25] for a proof. ■

When the largest Hankel singular value, σ_1, of F has multiplicity greater than one, we can see that the maximizing vector pair given by Lemma 1 does not guarantee the property shown in Corollary 1 if (A, B, C, D) is an arbitrary balanced realization of F. The reason is that in this case not all truncated subsystems are stable. However, the balanced system constructed by Lemma 3 ensures the truncated subsystems are stable and hence the property of Corollary 1. We now give an example to illustrate this fact.

Example 1 Consider $F(s) = \frac{s}{s^2+s+1}$ which has a balanced realization given by

$$F(s) = \left[\begin{array}{c|c} A & B \\ \hline C & 0 \end{array}\right] = \left[\begin{array}{cc|c} -1 & 1 & 1 \\ -1 & 0 & 0 \\ \hline 1 & 0 & 0 \end{array}\right].$$

Its Hankel singular values are $\sigma_1 = \sigma_2 = 0.5$. By Lemma 1, we obtain

$$v_1(-s) = -\frac{1}{\sqrt{\sigma_1}} B^T(sI + A^T)^{-1} e_1 = -\sqrt{2}\,\frac{s}{s^2 - s + 1} \quad (13)$$

$$w_1(s) = \frac{1}{\sqrt{\sigma_1}} C(sI - A)^{-1} e_1 = \sqrt{2}\,\frac{s}{s^2 + s + 1}. \quad (14)$$

It is clear that $v_1(0) = w_1(0) = 0$ since $a_{22} = 0$. And, we also have that

$$v_2(-s) = -\frac{1}{\sqrt{\sigma_1}} B^T(sI + A^T)^{-1} e_2 = \sqrt{2}\,\frac{-1}{s^2 - s + 1} \quad (15)$$

$$w_2(s) = \frac{1}{\sqrt{\sigma_1}} C(sI - A)^{-1} e_2 = \sqrt{2}\,\frac{1}{s^2 + s + 1} \quad (16)$$

are a maximizing vector pair of $\mathcal{H}_F$.

Here we have shown that the maximizing vector pair of $\mathcal{H}_F$ is not unique when the multiplicity of $\sigma_i(F)$ is not equal to one. By the state similarity transformation $\begin{bmatrix} \frac{1}{\sqrt{2}} & -\frac{1}{\sqrt{2}} \\ \frac{1}{\sqrt{2}} & \frac{1}{\sqrt{2}} \end{bmatrix}$, another realization, which satisfies the conditions of Lemma 3, is

$$F(s) = \left[\begin{array}{cc|c} -\frac{1}{2} & \frac{3}{2} & \frac{1}{\sqrt{2}} \\ -\frac{1}{2} & -\frac{1}{2} & -\frac{1}{\sqrt{2}} \\ \hline \frac{1}{\sqrt{2}} & -\frac{1}{\sqrt{2}} & 0 \end{array}\right].$$

We now see that $a_{11} = a_{22} = -0.5$ and the maximizing vector pair

$$\begin{aligned} v(-s) &= \frac{-s-1}{s^2-s+1} \\ w(s) &= \frac{s+1}{s^2+s+1} \end{aligned}$$

has the property of Corollary 1.

From now on we will assume that the balanced realization of F is given by Lemma 3, and hence the corresponding maximizing vector pair has the property of Corollary 1.

2.2 Inner Transformation Matrices

A major feature of super-optimal algorithms is the diagonalization of the cost function $(F - Q)$ in (1) by norm-preserving transformations. Vectors v and w defined in Lemma 1 play an important role in this diagonalization. Two square inner matrix functions derived from the inner parts of the inner-outer factorizations (IOF) of v and w are required. But since v and w are strictly proper, the usual IOF routines [5] are not applicable. In this subsection, we give a construction based on the work of Anderson [1].

From Theorem 1 and Lemma 1, we obtain

$$w^{\sim}\ w = v^{\sim}\ v = \phi + \phi^{\sim} \tag{17}$$

where

$$\phi(s) = \left[\begin{array}{c|c} A & e_1 \\ \hline e_1^T & 0 \end{array}\right] \left(= \left[\begin{array}{c|c} A^T & e_1 \\ \hline e_1^T & 0 \end{array}\right]\right) . \tag{18}$$

Corollary 1 implies that it is possible to find a stable and minimum phase $\alpha_o(s)$ such that

$$v(s) = v_i(s)\ \alpha_o(s) \tag{19}$$

$$w(s) = w_i(s)\ \alpha_o(s) \tag{20}$$

where $v_i(s)$ and $w_i(s)$ are inner. Because ϕ is strictly proper we can not apply directly the usual state-space methods developed (e.g. [5]) for the spectral factorization of $\phi + \phi^{\sim}$. However an improper $\chi(s)$ satisfying

$$\chi(s)\phi(s) = 1 \tag{21}$$

can be introduced where $\chi(s) = s + \chi_1(s)$ and $\chi_1(\infty) \neq 0$. The spectral factorization of $\chi_1 + \chi_1^{\sim}$ will help to solve the spectral factorization of $\phi + \phi^{\sim}$. Lemma 4 below gives such a χ_1. We first outline the construction of χ.

Let matrix $A \in \Re_{n\times n}$ be partitioned as

$$A = \begin{bmatrix} a_{11} & A_{12} \\ A_{21} & A_{22} \end{bmatrix} \tag{22}$$

where $a_{11} < 0$ and A_{22} is stable. And let ϕ defined in (18) be expressed as

$$\begin{aligned} \phi(s) &= \sum_{i=1}^{\infty} e_1^T A^{i-1} e_1 s^{-i} \\ &= \sum_{i=1}^{\infty} h_{-i} s^{-i} \end{aligned} \tag{23}$$

where

$$h_{-i} := e_1^T A^{i-1} e_1, \quad \forall\, i \tag{24}$$

are usually called the *Markov parameters* of ϕ.

Suppose that $\chi(s)$ is the function satisfying $\chi(s)\phi(s) = 1$ and let $\chi(s)$ be expressed as

$$\chi(s) = t_1 s + t_0 + \sum_{i=1}^{\infty} t_{-i}\, s^{-i}$$

where parameters t_i are to be determined. Then, from $\phi(s)\chi(s) = 1$, we obtain the identity

$$\begin{bmatrix} h_{-1} & h_{-2} & h_{-3} & h_{-4} & \cdots \\ 0 & h_{-1} & h_{-2} & h_{-3} & \cdots \\ 0 & 0 & h_{-1} & h_{-2} & \cdots \\ 0 & 0 & 0 & h_{-1} & \cdots \\ 0 & 0 & 0 & 0 & \cdots \end{bmatrix} \begin{bmatrix} t_1 & t_0 & t_{-1} & t_{-2} & \cdots \\ 0 & t_1 & t_0 & t_{-1} & \cdots \\ 0 & 0 & t_1 & t_0 & \cdots \\ 0 & 0 & 0 & t_1 & \cdots \\ 0 & 0 & 0 & 0 & \cdots \end{bmatrix} = I \tag{25}$$

which is a product of two infinite Toeplitz matrices. Comparing both sides of (25), we have

$$t_1 = \frac{1}{h_{-1}}$$

$$
\begin{aligned}
t_0 &= \frac{-1}{h_{-1}}(t_1 h_{-2}) \\
t_{-1} &= \frac{-1}{h_{-1}}(t_0 h_{-2} + t_1 h_{-3}) \\
t_{-2} &= \frac{-1}{h_{-1}}(t_1 h_{-2} + t_0 h_{-3} + t_1 h_{-4}) \\
&\vdots \\
t_{-n} &= \frac{-1}{h_{-1}}(t_{n-1} h_{-2} + \cdots + t_0 h_{-(n+1)} + t_1 h_{-(n+2)}) \\
&\vdots
\end{aligned}
$$

Then from (24),

$$
\begin{aligned}
h_{-1} &= e_1^T e_1 = 1 \\
h_{-2} &= e_1^T A e_1 = a_{11} \\
h_{-3} &= e_1^T A^2 e_1 = a_{11}^2 + A_{12}A_{21} \\
h_{-4} &= e_1^T A^3 e_1 = a_{11}^3 + a_{11}^2 A_{12}A_{21} + A_{12}A_{22}A_{21}
\end{aligned}
$$

we obtain

$$
\begin{aligned}
t_1 &= 1 \\
t_0 &= -a_{11} \\
t_{-1} &= -A_{12}A_{21} \\
t_{-2} &= -A_{12}A_{22}A_{21} \quad .
\end{aligned}
$$

By induction, we find that

$$\chi(s) = s + (-a_{11}) + \sum_{i=1}^{\infty} (-A_{12})A_{22}^{i-1}A_{21}\; s^{-i}$$

satisfies $\phi(s)\chi(s) = 1$. We summarize these results next.

Lemma 4 *Let $\chi_1(s)$ and $\chi(s)$ be defined as*

$$\chi_1(s) = \left[\begin{array}{c|c} A_{22} & A_{21} \\ \hline -A_{12} & -a_{11} \end{array}\right] \tag{26}$$

$$\chi(s) = s + \chi_1(s) \quad . \tag{27}$$

Then, $\chi(s)\phi(s) = 1$.

Proof: See [23] or [25] for a proof. ■

With χ and χ_1 of Lemma 4, we can see that

$$\chi + \chi^{\sim} = \chi_1 + \chi_1^{\sim} \ . \tag{28}$$

Then from Corollary 1, we obtain

$$\begin{aligned} \phi^{\sim}(\chi_1^{\sim} + \chi_1)\phi &= \phi^{\sim}(\chi\phi) + (\chi\phi)^{\sim}\phi \\ &= \phi^{\sim} + \phi \\ &= w^{\sim} w \\ &> 0, \quad \forall s = j\omega, \ \omega \in (-\infty, \infty) \ . \end{aligned}$$

Since $\chi_1(\infty) = -a_{11} > 0$, the spectral factorization of $\chi_1 + \chi_1^{\sim}$ can be found in [5]. Their realizations are given in the following lemma.

Lemma 5 *Let χ_1 be defined as in Lemma 4. Then*

$$\chi_1(j\omega) + \chi_1^*(j\omega) > 0, \quad \forall \ \omega \in \Re_e \ . \tag{29}$$

The spectral factors ψ_l and ψ_r such that

$$\chi_1 + \chi_1^{\sim} = \psi_r \ \psi_r^{\sim} = \psi_l^{\sim} \ \psi_l$$

are given as

$$\psi_l = \left[\begin{array}{c|c} A_{22} & A_{21} \\ \hline H_l & \sqrt{-2a_{11}} \end{array}\right] \tag{30}$$

$$\psi_r = \left[\begin{array}{c|c} A_{22} & H_r \\ \hline -A_{12} & \sqrt{-2a_{11}} \end{array}\right] \tag{31}$$

where

$$H_l = -\frac{1}{\sqrt{-2a_{11}}}(A_{12} + A_{21}^T P_l) \tag{32}$$

$$H_r = \frac{1}{\sqrt{-2a_{11}}}(A_{21} + P_r A_{12}^T) \tag{33}$$

$$P_l = Ric \begin{bmatrix} A_{22} + A_{21}(-2a_{11})^{-1}A_{12} & A_{21}(-2a_{11})^{-1}A_{21}^T \\ -A_{12}^T(-2a_{11})^{-1}A_{12} & -(A_{22} + A_{21}(-2a_{11})^{-1}A_{12})^T \end{bmatrix} \tag{34}$$

$$P_r = Ric \begin{bmatrix} (A_{22} + A_{21}(-2a_{11})^{-1}A_{12})^T & A_{12}^T(-2a_{11})^{-1}A_{12} \\ -A_{21}(-2a_{11})^{-1}A_{21}^T & -(A_{22} + A_{21}(-2a_{11})^{-1}A_{12}) \end{bmatrix} \tag{35}$$

Remark 3 It can be proved that $\chi_1(s)$ is a unit (scalar) in $\Re\mathbf{H}^\infty$ and hence $P_l \geq 0$ and $P_r \geq 0$. Also, we can show [25] that $a_{11} < 0$ and A_{22} is a stable matrix and hence $A_{sc} = A_{22} - A_{21}a_{11}^{-1}A_{12}$ is stable. Since the zeros of $\chi_1(s)$ are the eigenvalues of A_{sc}, $\chi_1(s)$ is a unit in $\Re\mathbf{H}^\infty$. In fact, P_r and P_l are the controllability and observability gramians of the pairs (A_{22}, H_r) and (H_l, A_{22}) respectively.

Since $\chi_1^\sim + \chi_1$ is scalar, we know, $\psi_r = \psi_l$. Let α_o be defined as

$$\alpha_o(s) = \psi_l(s)\ \phi(s)\ \ (= \phi(s)\psi_r(s)) \quad . \tag{36}$$

Then

$$\begin{aligned} \alpha_o^*\ \alpha_o &= (\psi_l\ \phi)^\sim(\psi_l\ \phi) \\ &= \phi^\sim(\chi_1^\sim + \chi_1)\phi \\ &= \phi^\sim(\chi^\sim + \chi)\phi \\ &= w^\sim\ w \\ &= v^\sim\ v \ . \end{aligned}$$

This shows that if ϕ is stable and minimum phase, then α_o defined by (36) is the outer part of w and v. Actually, ϕ in (18) can be rewritten as

$$\phi(s) = \frac{det(sI - A_{22})}{det(sI - A)} \quad .$$

Thus it is obvious that ϕ is stable and minimum phase since A_{22} and A are stable, and α_o defined by (36) is the outer part of both w and v.

For the inner parts, we define

$$v_i = v\ \frac{\chi}{\psi_r} \tag{37}$$

$$w_i = w\ \frac{\chi}{\psi_l} \ . \tag{38}$$

Since w and v are strictly proper, $w\chi$ and $v\chi$ make sense. The inner and outer parts of w and v are given in the next lemma.

Lemma 6 *Let v and w be given as in Lemma 1 and partition*

$$B = \begin{bmatrix} B_1 \\ B_2 \end{bmatrix}, \qquad C = \begin{bmatrix} C_1 & C_2 \end{bmatrix}$$

where B_1 is a row vector and C_1 is a column vector. Then the IOFs

$$\begin{aligned} w &= w_i \, \alpha_o \\ v &= v_i \, \alpha_o \end{aligned}$$

are given by

$$\alpha_o = \left[\begin{array}{cc|c} a_{11} & A_{12} & 1 \\ A_{21} & A_{22} & 0 \\ \hline \sqrt{-2a_{11}} & H_l & 0 \end{array}\right] \tag{39}$$

$$= \left[\begin{array}{cc|c} a_{11} & A_{12} & \sqrt{-2a_{11}} \\ A_{21} & A_{22} & H_r \\ \hline 1 & 0 & 0 \end{array}\right] \tag{40}$$

and

$$w_i = \left[\begin{array}{c|c} A_{22} - \frac{1}{\sqrt{-2a_{11}}} A_{21} H_l & \frac{1}{\sqrt{-2a_{11}}} A_{21} \\ \hline \frac{1}{\sqrt{\sigma_1}}(C_2 - \frac{1}{\sqrt{-2a_{11}}} C_1 H_l) & \frac{1}{\sqrt{-2a_{11}\sigma_1}} C_1 \end{array}\right] \tag{41}$$

$$v_i = \left[\begin{array}{c|c} (A_{22} + \frac{1}{\sqrt{-2a_{11}}} H_r A_{12})^T & \frac{1}{\sqrt{-2a_{11}}} A_{12}^T \\ \hline \frac{1}{\sqrt{\sigma_1}}(B_2 + \frac{1}{\sqrt{-2a_{11}}} H_r B_1)^T & \frac{1}{\sqrt{-2a_{11}\sigma_1}} B_1^T \end{array}\right] . \tag{42}$$

where H_l, H_r, P_l and P_r are defined as in Lemma 5.

Furthermore, the observability gramian, denoted by Y_o, of w_i is

$$Y_o = \frac{\Sigma_2}{\sigma_1} - P_l \tag{43}$$

and the observability gramian, denoted by X_o, of v_i is

$$X_o = \frac{\Sigma_2}{\sigma_1} - P_r \quad . \tag{44}$$

Proof: A proof can be found in [25]. ∎

Remark 4 We can show that the controllability and observability gramians for

$$\left(A, \begin{bmatrix} \sqrt{-2a_{11}} \\ H_r \end{bmatrix}\right) \quad \text{and} \quad ([\sqrt{-2a_{11}} \quad H_l],\ A)$$

given in Lemma 6, are

$$\begin{bmatrix} 1 & 0 \\ 0 & P_r \end{bmatrix} \quad \text{and} \quad \begin{bmatrix} 1 & 0 \\ 0 & P_l \end{bmatrix}$$

respectively. Then it is clear that $\|\alpha_o\|_2 = \|w\|_2 = \|v\|_2 = 1$.

Remark 5 We know that $\frac{\Sigma_2}{\sigma_1} \leq I$ and $P_r \geq 0,\ P_l \geq 0$. This implies that $0 \leq Y_o \leq I$ and $0 \leq X_o \leq I$. Equivalently, we obtain that $P_r \leq I$ and $P_l \leq I$. This gives a useful upper bound while we are finding the solutions of (34) and (35).

To get square, inner transformation matrices, we need to construct the complementary inner factors of v_i and w_i respectively. These can be found in the next lemma.

Lemma 7 *Let w_i and v_i in (41) and (42) be denoted by*

$$v_i = \left[\begin{array}{c|c} A_v & B_v \\ \hline C_v & D_v \end{array}\right]$$

$$w_i = \left[\begin{array}{c|c} A_w & B_w \\ \hline C_w & D_w \end{array}\right].$$

The complementary inner factor of w_i can be chosen as

$$W_\perp = \left[\begin{array}{c|c} A_w & B_{w\perp} \\ \hline C_w & D_{w\perp} \end{array}\right]$$

with $D_{w\perp}$ and $B_{w\perp}$ satisfying, respectively,

$$D_{w\perp} D_{w\perp}{}^T = I - D_w D_w^T$$

$$Y_o B_{w\perp} + C_w^T D_{w\perp} = 0$$

where Y_o is given as (43).

The complementary inner factor of v_i can be chosen as

$$V_\perp = \left[\begin{array}{c|c} A_v & B_{v\perp} \\ \hline C_v & D_{v\perp} \end{array}\right]$$

with $D_{v\perp}$ and $B_{v\perp}$ satisfying, respectively,

$$D_{v\perp} {D_{v\perp}}^T = I - D_v D_v^T$$

$$X_o B_{v\perp} + C_v^T D_{v\perp} = 0$$

where X_o is given as (44).

Thus, let

$$V_s = [v_i \;\; V_\perp]$$

$$W_s = [w_i \;\; W_\perp] \; .$$

Both V_s and W_s are square, inner matrices.

Remark 6 The square inner matrices W_s, V_s constructed in Lemma 7 are not always minimal since $0 \leq Y_o$, $0 \leq X_o$. However, based on a known solution Y_o (X_o), we can easily find one minimal realization of W_s (V_s); constructions are given in [25] (§4.5).

2.3 Diagonalization

The basic idea used in this subsection is that the singular values of the cost function $(F - Q)$ can be minimized one at a time using a succession of norm-preserving transformations. The transformations reduce the cost function, layer by layer, to a diagonal form (or pseudo-diagonal form if non-square), where the magnitudes of the diagonal elements are constant over all frequencies and equal to the optimum singular values. At each layer, a model-matching problem as in (1), but with dimension one less than at the previous layer, is solved using Lemma 2.

The norm-preserving transformations for the first layer of diagonalization are derived from the V_s and W_s constructed in Lemma 7. The result is summarized in Theorem 2 below.

Theorem 2 *Let V_s, W_s be given by Lemma 7. Then for any $Q_a \in \{\mathbf{Q_{opt}}\}$,*

$$W_s^T(-s)(F(s) - Q_a(s))V_s(-s) = \begin{bmatrix} \sigma_1 g_1(s) & 0 \\ 0 & G_2(s) \end{bmatrix} \tag{45}$$

where $g_1(s) \in \Re\mathbf{L}^{\infty}$ is all-pass and is independent of Q_a, and

$$\|G_2\|_{\infty} \leq \sigma_1 , \quad G_2(s) \in \Re\mathbf{L}^{\infty}{}_{(p-1)\times(q-1)} . \tag{46}$$

Proof: See [23] for a proof. ■

The following corollary is useful in obtaining the model-matching problem (or Nehari problem) for the second layer. The proof follows directly from Theorem 2 and the fact that $W_{\perp}^T(-s)$, $Q_1(s)$, $Q_2(s)$, $V_{\perp}(-s)$ all lie in $\Re\mathbf{H}_{-}^{\infty}$.

Corollary 2 *Let Q_1, $Q_2 \in \{\mathbf{Q_{opt}}\}$. Then*

$$W_s^T(-s)(Q_1(s) - Q_2(s))V_s(-s) \in \begin{bmatrix} 0 & 0 \\ 0 & \Re\mathbf{H}_{-}^{\infty}{}_{(p-1)\times(q-1)} \end{bmatrix} . \tag{47}$$

■

To obtain the next sub-problem, we choose an optimal solution from Lemma 2, and denote it as Q_{opt}. Then, using Theorem 2 and Corollary 2, for any other solution $Q_a \in \{\mathbf{Q_{opt}}\}$

$$\begin{aligned} \|F - Q_a\|_{\infty} &= \|(F - Q_{opt}) - (Q_a - Q_{opt})\|_{\infty} \\ &= \|W_s^T(-s)(F - Q_{opt})V_s(-s) \; - \; W_s^T(-s)(Q_a - Q_{opt})V_s(-s)\|_{\infty} \\ &= \| \begin{bmatrix} \sigma_1 g_1 & 0 \\ 0 & G_2 \end{bmatrix} - \begin{bmatrix} 0 & 0 \\ 0 & \tilde{Q}_2 \end{bmatrix} \|_{\infty} \end{aligned}$$

where

$$G_2 = W_{\perp}^T(-s)(F - Q_{opt})V_{\perp}(-s) \in \Re\mathbf{L}^{\infty}{}_{(p-1)\times(q-1)} \tag{48}$$

and

$$\tilde{Q}_2 = W_{\perp}^T(-s)(Q_a - Q_{opt})V_{\perp}(-s) \in \Re\mathbf{H}_{-}^{\infty}{}_{(p-1)\times(q-1)} . \tag{49}$$

Now define the set

$$\mathbf{\Xi} = \{Q_a - Q_{opt} : Q_a \in \{\mathbf{Q_{opt}}\} \} . \tag{50}$$

Then we know that $Q_{sup} - Q_{opt} \in \Xi$, where Q_{sup} is the super-optimal solution. By the definition of the super-optimal solution, we need to solve

$$\min_{\tilde{Q} \in \Xi} \|G_2 - W_\perp^T(-s)\tilde{Q}V_\perp(-s)\|_\infty \,. \tag{51}$$

The next step is to characterize the set Ξ. The characterization given below is based on the properties of a structural inner matrix, defined in [25] (§2.3).

Theorem 3 *Let right coprime factorizations of $W_\perp(-s)$ and $V_\perp(-s)$ be*

$$W_\perp(-s) = N_{W_\perp}(s)M_{W_\perp}^{-1}(s) \tag{52}$$

$$V_\perp(-s) = N_{V_\perp}(s)M_{V_\perp}^{-1}(s) \tag{53}$$

where $N_{W_\perp}$, $N_{V_\perp}$ are structural inner, and $M_{W_\perp}$, $M_{V_\perp}$ are (square) inner. (The formulae can be found in Lemma 2.8 in [25].) Then

$$\Xi \subseteq N_{W_\perp}(-s)\Re\mathbf{H}^\infty_{-\,(p-1)\times(q-1)}N_{V_\perp}^T(-s) \,. \tag{54}$$

Proof: A proof can be found in [23] or [25]. ∎

Next, let us consider the following optimization problem which (as we will see below) results in a solution to the original problem

$$\min_{\tilde{Q}} \|G_2 - W_\perp^T(-s)\tilde{Q}V_\perp(-s)\|_\infty, \quad \forall\, \tilde{Q} \in N_{W_\perp}(-s)\Re\mathbf{H}^\infty_{-\,(p-1)\times(q-1)}N_{V_\perp}^T(-s) \tag{55}$$

or, equivalently,

$$\min_{Q_2} \|G_2 - M_{W_\perp}(-s)Q_2M_{V_\perp}^T(-s)\|_\infty, \quad \forall\, Q_2 \in \Re\mathbf{H}^\infty_{-\,(p-1)\times(q-1)} \tag{56}$$

where G_2 is defined in (48). Let the optimal solution set of (56) be denoted by $\{\mathbf{Q_{opt,2}}\}$. The characterization of Ξ suggests that any element in the set

$$Q_{opt} + N_{W_\perp}(-s)\{\mathbf{Q_{opt,2}}\}N_{V_\perp}^T(-s) \tag{57}$$

satisfies the first two layers of the super-optimal problem defined in (1). This can be verified as follows.

For all $Q_2 \in \{\mathbf{Q_{opt,2}}\}$, we have

$$\begin{aligned}
&\|F-(Q_{opt}+N_{W_\perp}(-s)Q_2N^T_{V_\perp}(-s))\|_\infty \\
&= \|W_s^T(-s)(F-Q_{opt})V_s(-s) - W_s^T(-s)N_{W_\perp}(-s)Q_2N^T_{V_\perp}(-s)V_s(-s)\|_\infty \\
&= \|\begin{bmatrix} \sigma_1 g_1 & 0 \\ 0 & G_2 \end{bmatrix} - \begin{bmatrix} 0 & 0 \\ 0 & M_{W_\perp}(-s)Q_2M^T_{V_\perp}(-s) \end{bmatrix}\|_\infty \\
&= \|\begin{bmatrix} \sigma_1 g_1 & 0 \\ 0 & G_2 - M_{W_\perp}(-s)Q_2M^T_{V_\perp}(-s) \end{bmatrix}\|_\infty \ .
\end{aligned}$$

Since Q_2 is an optimal solution to (56),

$$\|G_2 - M_{W_\perp}(-s)Q_2M^T_{V_\perp}(-s)\|_\infty \ \le \ \|G_2\|_\infty \ \le \ \sigma_1 \tag{58}$$

and hence

$$\|F-(Q_{opt}+N_{W_\perp}(-s)Q_2N^T_{V_\perp}(-s))\|_\infty = \sigma_1 \ .$$

This shows that the solution $Q_{opt}+N_{W_\perp}(-s)Q_2N^T_{V_\perp}(-s)$ satisfies optimality in the *first* layer of (1).

Furthermore, following the same procedure as with $F-Q_{opt}$ for $F-Q_{sup}$, it is easy to see that the second singular value of $F-Q_{sup}$ is equivalent to

$$\|G_2 - W_\perp^T(-s)(Q_{sup}-Q_{opt})V_\perp(-s)\|_\infty \ . \tag{59}$$

But $Q_{sup}-Q_{opt} \in \mathbf{\Xi}$, and by Theorem 3

$$Q_{sup}-Q_{opt} \ \in \ N_{W_\perp}(-s)\Re\mathbf{H}^\infty_{-\,(p-1)\times(q-1)}N^T_{V_\perp}(-s) \ . \tag{60}$$

This implies

$$\begin{aligned}
&\|G_2 - W_\perp^T(-s)(Q_{sup}-Q_{opt})V_\perp(-s)\|_\infty \\
&\quad = \min_{\tilde{Q}\in\Xi} \|G_2 - W_\perp^T(-s)\tilde{Q}V_\perp(-s)\|_\infty \\
&\quad \ge \min_{Q_2} \|G_2 - M_{W_\perp}(-s)Q_2M^T_{V_\perp}(-s)\|_\infty
\end{aligned}$$

where Q_2 is over $\Re\mathbf{H}^\infty_{-\,(p-1)\times(q-1)}$. However Q_{sup} is the super-optimal solution. So

$$\|G_2 - W_\perp^T(-s)(Q_{sup}-Q_{opt})V_\perp(-s)\|_\infty = \|G_2 - M_{W_\perp}(-s)Q_2M^T_{V_\perp}(-s)\|_\infty \ . \tag{61}$$

This shows that $Q_{opt} + N_{W_\perp}(-s)Q_2N^T_{V_\perp}(-s)$ also satisfies optimality in the *second* layer of (1). Consequently, the super-optimal solution is in the set (57). If $\min(p,q) = 2$, then (57) gives the super-optimal solution.

We will now show that sub-problem (56) can be transformed to a problem exactly like (2), a solution of which is immediate from Lemma 2. Since $M_{W_\perp} \in \Re\mathbf{H}^\infty{}_{(p-1)\times(p-1)}$ and $M_{V_\perp} \in \Re\mathbf{H}^\infty{}_{(q-1)\times(q-1)}$ are inner, problem (56) is equivalent to

$\min_{Q_2}\|M^T_{W_\perp}(s)G_2(s)M_{V_\perp}(s) - Q_2(s)\|_\infty$

$$= \min_{Q_2}\|N^T_{W_\perp}(s)(F(s)-Q_{opt}(s))N_{V_\perp}(s)-Q_2(s)\|_\infty \quad (62)$$

where Q_2 varies over the whole set $\Re\mathbf{H}^\infty_-{}_{(p-1)\times(q-1)}$.

Obviously this sub-problem is the same as (2) (only with dimensions reduced by one) if the variable Q_2 absorbs the anti-stable part of the first entry. This demonstrates, therefore, that the whole procedure above can be performed repeatedly to achieve the super-optimal solution.

An Algorithm

We now present an algorithm which is a summary of the above development. Assume $F(s) \in \Re\mathbf{H}^\infty{}_{p\times q}$ and $rank(F) = q > 1$.

Step 1 Construct the minimal balanced realization of $F(s)$ and hence its maximizing vector pair $(v_1(-s), w_1(s))$ as in Lemma 1.

Step 2 Choose an optimal $Q_{opt}(s)$ using the state-space formulae in Lemma 2.

Step 3 Assign $U_V(s) = I_q$, $U_W(s) = I_p$, $Q_{sup}(s) = Q_{opt}(s)$ and $E_1(s) = F(s) - Q_{opt}(s)$.

Step 4 For $j = 1$ to q

1. Perform IOFs of $v_j(s)$ and $w_j(s)$ using Lemma 6.
2. Compute the structural inner $N_{V_\perp}(s)$ and $N_{W_\perp}(s)$ (see Theorem 3).

3. Decompose $\tilde{F}_j(s) = N^T_{W_\perp}(s)E_j(s)N_{V_\perp}(s) = [\tilde{F}_j(s)]_+ + [\tilde{F}_j(s)]_-$ and let $F_{j+1}(s) = [\tilde{F}_j(s)]_+$.
4. Set $U_V(s) = U_V(s)N_{V_\perp}(s)$ and $U_W(s) = U_W(s)N_{W_\perp}(s)$.
5. Construct the minimal balanced realization for the sub-problem $F_{j+1}(s)$ and determine its maximizing vector $(v_{j+1}(-s), w_{j+1}(s))$ as Step 1.
6. Find $Q_{opt,j+1}(s)$ corresponding to $F_{j+1}(s)$ as in step (2).
7. $E_{j+1}(s) = F_{j+1}(s) - Q_{opt,j+1}(s)$.
8. $Q_{sup}(s) = Q_{sup}(s) + U_W(-s)(Q_{opt,j+1}(s) + [\tilde{F}_j(s)]_-)U_V^T(-s)$.

2.4 A Frame Approach

The solution procedure for the 1-block strengthened model-matching problem described in the previous subsections is numerically implementable, and gets one s-number per iteration. When the matrix F has several Hankel singular values all reaching the value of the largest it is possible to determine immediately the corresponding s-numbers of F, which are also of the same value. Based on this a *frame approach* is proposed to reduce the number of computations required to find the super-optimal Q. The approach is based on interpreting s-numbers as largest gains between appropriately defined spaces and making special use of the properties of a balanced realization. The approach is useful in 2-block super-optimal problems which we will discuss in the next section because there the solutions are solved via a sequence of 1-block problems for which an increasing number of the largest s-numbers reach the value one.

An alternative interpretation or definition for the s-numbers of F is described below for the first two layers.

We know that $\bar{s}_1^\infty = \|\mathcal{H}_F\|$, but what about $\bar{s}_2^\infty$? The minimum norm of (62) is equivalent to the norm of a Hankel operator induced by

$$M^T_{W_\perp} G_2 M_{V_\perp} \;=\; N^T_{W_\perp}(F - Q_{opt})N_{V_\perp} \;=:\; \tilde{F} \;.$$

From the definition

$$\begin{aligned} \mathcal{H}_{\tilde{F}} &: \Re\mathbf{H}^2_{\perp q-1} \mapsto \Re\mathbf{H}^2{}_{p-1} \\ \mathcal{H}_{\tilde{F}}(x) &= \mathcal{P}_{\Re\mathbf{H}^2{}_{p-1}}(\tilde{F}x)\ , \quad \forall\, x \in \Re\mathbf{H}^2_{\perp q-1} \end{aligned}$$

thus $\bar{s}_2^{\infty} = \|\mathcal{H}_{\tilde{F}}\|$, and there exists a maximizing vector $\tilde{v}_2 \in \Re\mathbf{H}^2_{\perp p-1}$ of $\mathcal{H}_{\tilde{F}}$ corresponding to the singular value $\bar{s}_2^{\infty}$. In the following we will see that there exists a vector $v_2 \in \Re\mathbf{H}^2_{\perp p}$ which is pointwise orthogonal to the maximizing vector of $\mathcal{H}_F$ such that

$$\|\mathcal{H}_F v_2\|_2 = \|\mathcal{H}_{\tilde{F}}\tilde{v}_2\|_2 = \bar{s}_2^{\infty}\ .$$

For simplicity, a function $G(-s)$ will be denoted as G^- in the following.

Let vector N_{w_i} be the CIF of the structural inner $N_{W_\perp}$. Then we know that N_{w_i} is structural inner (Corollary 2.32 in [25]). Furthermore, from the properties of a structural inner function (Corollary 2.27 in [25]), we get

$$\Re\mathbf{H}^2_{\perp q-1} = N_{V_\perp}^{\sim}\Re\mathbf{H}^2_{\perp q} \tag{63}$$

$$\Re\mathbf{H}^2{}_{p-1} = N_{W_\perp}^{T}\Re\mathbf{H}^2{}_{p}\ . \tag{64}$$

Here we also need the following property for a square inner matrix:

Lemma 8 *Let $N(s)$ be square inner. Then*

$$\mathcal{P}_{(N\mathbf{H}^2)}x = \mathcal{M}_N\mathcal{P}_{\mathbf{H}^2}\mathcal{M}_{N^*}x, \quad \forall\, x \in \mathbf{L}^2$$

where $\mathcal{P}_{N\mathbf{H}^2}$ is the orthogonal projection from $\mathbf{L}^2$ to the subspace $N\mathbf{H}^2$.

Proof: Follows from results in [28]. ∎

The following deduction will use (63), (64), Theorem 2, Lemma 8 and the fact that $\Re\mathbf{H}^2{}_p \cap W_\perp\Re\mathbf{L}^2{}_{p-1}$ is the orthogonal complement of the kernel of the projection operator $\mathcal{P}_{\Re\mathbf{H}^2{}_p}(I - w_i w_i^{\sim})$ in $\Re\mathbf{L}^{\infty}{}_p$. Note that for a given function $x \in \Re\mathbf{L}^2{}_p$ there exist $y_1 \in \Re\mathbf{L}^2$ and $y_2 \in \Re\mathbf{L}^2{}_{p-1}$ such that $x = w_i y_1 + W_\perp y_2$.

We can now deduce that:

$$
\begin{array}{lll}
 & \max_x \| \mathcal{P}_{\Re\mathbf{H^2}_{p-1}} \tilde{F} x \|_2 & x \in \Re\mathbf{H}^2_{\perp q-1} \\
= & \max_x \| \mathcal{P}_{\Re\mathbf{H^2}_{p-1}} N^T_{W_\perp}(F - Q_{opt}) N_{V_\perp} x \|_2 & x \in \Re\mathbf{H}^2_{\perp q-1} \\
= & \max_x \| \mathcal{P}_{\Re\mathbf{H^2}_{p-1}} N^T_{W_\perp}(F - Q_{opt}) N_{V_\perp} N_{V_\perp}^{\sim} x \|_2 & x \in \Re\mathbf{H}^2_{\perp q} \\
= & \max_x \| \mathcal{P}_{(N^T_{W_\perp} \Re\mathbf{H^2}_p)} N^T_{W_\perp}(F - Q_{opt}) N_{V_\perp} N_{V_\perp}^{\sim} x \|_2 & x \in \Re\mathbf{H}^2_{\perp q} \\
= & \max_x \| \mathcal{P}_{(N^T_{W_\perp} \Re\mathbf{H^2}_p)} N^T_{W_\perp}(F - Q_{opt})(I - v_i^- v_i^T) x \|_2 & x \in \Re\mathbf{H}^2_{\perp q} \\
= & \max_x \| \mathcal{P}_{(N^T_{W_\perp} \Re\mathbf{H^2}_p)} N^T_{W_\perp}(F - Q_{opt}) x \|_2 & x \in \Re\mathbf{H}^2_{\perp q} \\
= & \max_x \| \mathcal{P}_{(N^T_{W_\perp} \Re\mathbf{H^2}_p)} N^T_{W_\perp} F x \|_2 & x \in \Re\mathbf{H}^2_{\perp q} \cap V_\perp^- \Re\mathbf{L^2}_{q-1} \\
= & \max_x \left\| \mathcal{P}_{\left(\begin{bmatrix} N^T_{w_i} \\ N^T_{W_\perp} \end{bmatrix} \Re\mathbf{H^2}_p \right)} \begin{bmatrix} 0 \\ N^T_{W_\perp} \end{bmatrix} F x \right\|_2 & x \in \Re\mathbf{H}^2_{\perp q} \cap V_\perp^- \Re\mathbf{L^2}_{q-1} \\
= & \max_x \left\| \begin{bmatrix} N^T_{w_i} \\ N^T_{W_\perp} \end{bmatrix} \mathcal{P}_{\Re\mathbf{H^2}_p} N^-_{W_\perp} N^T_{W_\perp} F x \right\|_2 & x \in \Re\mathbf{H}^2_{\perp q} \cap V_\perp^- \Re\mathbf{L^2}_{q-1} \\
= & \max_x \| \mathcal{P}_{\Re\mathbf{H^2}_p} (I - w_i w_i^{\sim}) F x \|_2 & x \in \Re\mathbf{H}^2_{\perp q} \cap V_\perp^- \Re\mathbf{L^2}_{q-1} \\
= & \max_x \| \mathcal{P}_{(\Re\mathbf{H^2}_p \cap W_\perp \Re\mathbf{L^2}_{p-1})} F x \|_2 & x \in \Re\mathbf{H}^2_{\perp q} \cap V_\perp^- \Re\mathbf{L^2}_{q-1}
\end{array}
$$

The space $\Re\mathbf{H}^2_{\perp q} \cap V_\perp(-s)\Re\mathbf{L^2}_{q-1}$ is the subspace of $\Re\mathbf{H}^2_{\perp q}$ which is *pointwise orthogonal* to $v(-s)$, $\forall$ s in $j\Re$. Similarly, $\Re\mathbf{H^2}_p \cap W_\perp(s)\Re\mathbf{L^2}_{p-1}$ is the subspace of $\Re\mathbf{H^2}_p$ which is *pointwise orthogonal* to $w(s)$, $\forall$ s in $j\Re$. Hence, the second s-number is the largest *gain* of the vectors in the pointwise orthogonal complement of $v(-s)$ in $\Re\mathbf{H}^2_{\perp q}$ through the multiplication of F and the projection onto the pointwise orthogonal complement of $w(s)$ in $\Re\mathbf{H^2}_p$. A similar situation is also true for the other s-numbers [29]. Therefore, if we can find r maximizing vectors pointwise orthogonal to each other and all reaching the value $\bar{s}_1^\infty$, then we have

$$\bar{s}_1^\infty = \cdots = \bar{s}_r^\infty .$$

Based on this idea, we can propose a frame approach which speeds up the calculation of the super-optimal solution.

In the following, we consider the situation which arises when the value of the largest s-number is shared by $r > 1$ s-numbers. When this happens a *frame* of r maximizing vectors can be used to solve r layers of optimality in one iteration. We will assume that $F(s) = (A, B, C, D)$ is a minimal balanced realization constructed by Lemma 3.

2.4.1 Maximizing Vector Frame

We begin with the maximizing vectors. Define

$$v_j(s) = \left[\begin{array}{c|c} A^T & \frac{1}{\sqrt{\sigma_1}} e_j \\ \hline B^T & 0 \end{array}\right] \quad , \quad j = 1, 2, \cdots, k$$

and

$$w_j(s) = \left[\begin{array}{c|c} A & \frac{1}{\sqrt{\sigma_1}} e_j \\ \hline C & 0 \end{array}\right] \quad , \quad j = 1, 2, \cdots, k$$

where e_j denotes the n-dimensional vector with 1 as the jth entry and 0's as all the other entries, and k is the multiplicity of σ_1. Then we have the following theorem.

Lemma 9 *All the maximizing vector pairs of $\mathcal{H}_F$ corresponding to the largest Hankel value σ_1 are generated by $\{v_j(-s)\}_{j=1}^k$ and $\{w_j(s)\}_{j=1}^k$ respectively. That is, for any unit vector pair $x(-s) \in \Re\mathbf{H}^2_{\perp q}$ and $y(s) \in \Re\mathbf{H}^2{}_p$ satisfying*

$$\sigma_1 y(s) = \mathcal{H}_F x(-s)$$

there exist $\alpha_j, \beta_j \in \Re$ with $\sum_{j=1}^k \alpha_j^2 = 1$ and $\sum_{j=1}^k \beta_j^2 = 1$ such that

$$x(s) = \sum_{j=1}^{k} \alpha_j \; v_j(s)$$

and

$$y(s) = \sum_{j=1}^{k} \beta_j \; w_j(s) \quad .$$

Proof: See [13] or [25] for a proof. ■

Consider the $F(s)$ shown in Example 1 again. All maximizing vectors $(x(-s), y(s))$ can be generated by

$$\{(x, y) := (\alpha v_1 + \beta v_2, \; \alpha w_1 + \beta w_2); \quad \alpha^2 + \beta^2 = 1\}$$

where v_1, w_1, v_2 and w_2 are given by (13)-(16).

For the cases

a) $\alpha = \beta = \frac{1}{\sqrt{2}}$

$$x(s) = \frac{s-1}{s^2+s+1}, \qquad y(s) = \frac{s+1}{s^2+s+1}$$

b) $\alpha = \frac{\sqrt{3}}{2}, \beta = \frac{-1}{2}$

$$x(s) = \frac{1}{\sqrt{2}}\frac{\sqrt{3}s+1}{s^2+s+1}, \qquad y(s) = \frac{1}{\sqrt{2}}\frac{\sqrt{3}s-1}{s^2+s+1}$$

each is a maximizing vector pair of $\mathcal{H}_F$.

The following theorem generalizes the results of Lemma 1 and Corollary 1.

Theorem 4 *The maximizing (vector) frame $(V(-s), W(s))$ of $\mathcal{H}_F$ corresponding to s-number $\bar{s}_1^\infty (= \sigma_1)$ is given by*

$$V(s) = \left[\begin{array}{c|c} A^T & \frac{1}{\sqrt{\sigma_1}}E_r \\ \hline B^T & 0 \end{array}\right] \tag{65}$$

and

$$W(s) = \left[\begin{array}{c|c} A & \frac{1}{\sqrt{\sigma_1}}E_r \\ \hline C & 0 \end{array}\right] \tag{66}$$

where $E_r = \begin{bmatrix} I_r \\ 0 \end{bmatrix}$. *Moreover,*

$$(V^*V)(j\omega) > 0 \ \ \textit{and} \ \ (W^*W)(j\omega) > 0 \quad \forall\, \omega \in \Re \tag{67}$$

Proof: See [13] or [25] for a proof. ■

This theorem shows that for any frequency ω the column vectors of $V(j\omega)$ and $W(j\omega)$ are linearly independent in the complex Euclidean spaces $\mathcal{C}_\mathbf{q}$ and $\mathcal{C}_\mathbf{p}$ respectively.

2.4.2 Formulae for the Frame Approach

Analogous to the results given in §2.2, we now extend the IOF to a group of maximizing vectors.

By direct calculation, we have

$$\begin{aligned} W^{\sim} W &= \Phi_w^{\sim} + \Phi_w \\ V^{\sim} V &= \Phi_v^{\sim} + \Phi_v \end{aligned}$$

where

$$\begin{aligned} \Phi_v &= \left[\begin{array}{c|c} A^T & E_r \\ \hline E_r^T & 0 \end{array}\right] \\ \Phi_w &= \left[\begin{array}{c|c} A & E_r \\ \hline E_r^T & 0 \end{array}\right] . \end{aligned}$$

Let Z_w and Z_v be defined by

$$\begin{aligned} Z_v &= s + Z_{v1} \\ Z_w &= s + Z_{w1} \end{aligned}$$

where

$$\begin{aligned} Z_{v1} &= \left[\begin{array}{c|c} A_{22}^T & A_{12}^T \\ \hline -A_{21}^T & -A_{11}^T \end{array}\right] \\ Z_{w1} &= \left[\begin{array}{c|c} A_{22} & A_{21} \\ \hline -A_{12} & -A_{11} \end{array}\right] \end{aligned}$$

and $A = \begin{bmatrix} A_{11} & A_{12} \\ A_{21} & A_{22} \end{bmatrix}$ is partitioned with A_{11} of dimension: $r \times r$, where r is defined as in Lemma 3. Then, we can easily prove that

$$\begin{aligned} Z_v \Phi_v &= I \\ \Phi_w Z_w &= I \end{aligned}$$

and

$$Z_{v1}(-j\omega) + Z_{v1}^T(j\omega) = Z_{w1}^T(-j\omega) + Z_{w1}(j\omega) > 0, \quad \forall \omega \in \Re_e \ .$$

The next lemma provides corresponding formulae for the IOF. These results are the matrix case of Lemmas 5 and 6.

Lemma 10 *Let the IOFs of V and W be denoted by*

$$\begin{aligned} V &= V_i V_o \\ W &= W_i W_o \ . \end{aligned}$$

Then

$$V_i = \left[\begin{array}{c|c} (A_{22} + H_r R_D^{-\frac{1}{2}} A_{12})^T & A_{12}^T R_D^{-\frac{1}{2}} \\ \hline \frac{1}{\sqrt{\sigma_1}}(B_2 + H_r R_D^{-\frac{1}{2}} B_1)^T & \frac{1}{\sqrt{\sigma_1}} B_1^T R_D^{-\frac{1}{2}} \end{array}\right]$$

$$V_o = \left[\begin{array}{c|c} A^T & E_r \\ \hline [R_D^{\frac{1}{2}} \ \ H_r^T] & 0 \end{array}\right]$$

where

$$\begin{aligned} R_D &= -(A_{11}^T + A_{11}) > 0 \\ H_r &= (A_{21} + P_r A_{12}^T)\ R_D^{-\frac{1}{2}} \end{aligned}$$

$$P_r = Ric \begin{bmatrix} (A_{22} + A_{21} R_D^{-1} A_{12})^T & A_{12}^T R_D^{-1} A_{12} \\ -A_{21} R_D^{-1} A_{21}^T & -(A_{22} + A_{21} R_D^{-1} A_{12}) \end{bmatrix}$$

and

$$W_i = \left[\begin{array}{c|c} A_{22} - A_{21} R_D^{-\frac{1}{2}} H_l & A_{21} R_D^{-\frac{1}{2}} \\ \hline \frac{1}{\sqrt{\sigma_1}}(C_2 - C_1 R_D^{-\frac{1}{2}} H_l) & \frac{1}{\sqrt{\sigma_1}} C_1 R_D^{-\frac{1}{2}} \end{array}\right]$$

$$W_o = \left[\begin{array}{c|c} A & E_r \\ \hline [R_D^{\frac{1}{2}} \ \ H_l] & 0 \end{array}\right]$$

where

$$H_l = -R_D^{-\frac{1}{2}}(A_{12} + A_{21}^T P_l)$$

$$P_l = Ric \begin{bmatrix} A_{22} + A_{21} R_D^{-1} A_{12} & A_{21} R_D^{-1} A_{21}^T \\ -A_{12}^T R_D^{-1} A_{12} & -(A_{22} + A_{21} R_D^{-1} A_{12})^T \end{bmatrix} .$$

To get square, inner matrices $V_s = [V_i \ \ V_\perp]$ and $W_s = [W_i \ \ W_\perp]$, the complementary inner factors, $V_\perp$ and $W_\perp$, can be found as in [5].

2.5 An Example

Consider the example in [30] transformed into the s-domain as

$$F(s) = \begin{bmatrix} \frac{-(-s+1)(-(2+\sqrt{3})s+(\sqrt{3}-2))}{(s+1)(s+1)} & \frac{-s+1}{s+1} \\ \frac{-(2+\sqrt{3})s+(\sqrt{3}-2)}{s+1} & 1 \end{bmatrix} .$$

In step 1, the minimal balanced realization is given as

$$F(s) = \left[\begin{array}{c|c} A & B \\ \hline C & D \end{array}\right] = \left[\begin{array}{cc|cc} -1.6928 & -0.5422 & -3.3737 & -0.5886 \\ 0.8853 & -0.3072 & 0.5506 & 0.7521 \\ \hline -3.2160 & 0.1425 & -3.7321 & -1.0000 \\ -1.1771 & -0.9212 & -3.7321 & 1.0000 \end{array}\right] .$$

The Hankel singular values are

$$\Sigma = \begin{bmatrix} 3.4641 & 0 \\ 0 & 1.4142 \end{bmatrix} .$$

Then we obtain $\bar{s}_1^\infty = 3.4641$.

In step 2, we obtain

$$P_l = 0.0992 \ , \qquad P_r = 0.2646$$

and

$$H_r = 0.4032 \ , \qquad H_l = 0.2469 \ .$$

In step 3 , two square inner function are constructed as

$$W_s(s) = \left[\begin{array}{c|cc} A_w & B_w & B_{w\perp} \\ \hline C_w & D_w & D_{w\perp} \end{array}\right] = \left[\begin{array}{c|cc} -0.4260 & 0.4812 & 1.5892 \\ \hline 0.3084 & -0.9391 & -0.3437 \\ -0.4101 & -0.3437 & 0.9391 \end{array}\right]$$

and

$$V_s(s) = \left[\begin{array}{c|cc} A_v & B_v & B_{v\perp} \\ \hline C_v & D_v & D_{v\perp} \end{array}\right] = \left[\begin{array}{c|cc} -0.4260 & -0.2947 & -2.4175 \\ \hline -0.1014 & -0.9851 & -0.1719 \\ 0.3348 & -0.1719 & 0.9851 \end{array}\right] .$$

In step 4, we have

$$N_{V_\perp}(s) = \left[\begin{array}{c|c} -0.4260 & -2.4175 \\ \hline 0.0408 & -0.1719 \\ 0.0124 & 0.9851 \end{array}\right]$$

and

$$N_{W_\perp}(s) = \left[\begin{array}{c|c} -0.4260 & 1.5892 \\ \hline -0.1242 & -0.3437 \\ -0.0934 & 0.9391 \end{array}\right]$$

The balanced realization of the stable part of $N_{W_\perp}^T(s)(F(s) - Q_{opt}(s))N_{V_\perp}(s)$ is solved again by step 1 and the maximal Hankel singular value in this layer, *i.e.* $\bar{s}_2^\infty$, is 0.7889. Since $q = r = 2$, the super-optimal solution is

$$Q_{sup}(s) = \left[\begin{array}{c|cc} 0.4260 & 0.3179 & 0.0963 \\ \hline 1.0365 & -0.5740 & -0.1738 \\ 0.7796 & -2.4318 & 0.4748 \end{array}\right]$$

or

$$Q_{sup}(s) = \begin{bmatrix} \frac{-0.5740s+0.5740}{s-0.4260} & \frac{-0.1738s+0.1738}{s-0.4260} \\ \frac{-2.4318s+1.2837}{s-0.4260} & \frac{0.4748s-0.1272}{s-0.4260} \end{bmatrix} .$$

3 2-Block Super-Optimal Solutions

In this section, we consider the 2-block strengthened model-matching problem. The 2-block case arises in many practical problems, for instance, in optimal disturbance attenuation with control weighting and in any control problem with more controlled outputs than control inputs. A state-space algorithm for 2-block super-optimal solutions is developed in this section and is based on results of the previous section for the less practical 1-block problem. A feature of the procedure is an optimality criterion which is a natural generalization of that used in the γ-iteration scheme of Doyle [5] for the standard $\mathbf{H}^\infty$ optimization problem.

Without loss of generality we consider the 2-block super-optimal problem in the following form:

$$\inf_{Q(s)\in\Re\mathbf{H}^\infty} \left\| \begin{matrix} R_1(s) - Q(s) \\ R_2(s) \end{matrix} \right\|_\infty \tag{68}$$

where $R_1 \in \Re\mathbf{L}^\infty{}_{p_1\times m}$ and $R_2 \in \Re\mathbf{L}^\infty{}_{p_2\times m}$. where $p_1 \geq m$. We will restrict R_1 to be anti-stable and R_2 to be stable since we can always find its stable factor $\hat{R}_2$ such that $\hat{R}_2^\sim \hat{R}_2 = R_2^\sim R_2$. Clearly, if $R_2 = 0$, this problem will be 1-block.

Let $R = \begin{bmatrix} R_1 \\ R_2 \end{bmatrix}$ and define an operator Γ_R, mapping $\mathbf{H^2}$ to $\mathbf{H^2_\perp} \oplus \mathbf{H^2}$, as

$$\Gamma_R = \begin{bmatrix} \check{\mathcal{H}}_{R_1} \\ \mathcal{T}_{R_2} \end{bmatrix} \tag{69}$$

where $\check{\mathcal{H}}_{R_1}$ is the *dual Hankel operator* generated by R_1, defined as

$$\begin{aligned} \check{\mathcal{H}}_{R_1} \quad &: \quad \mathbf{H^2} \longrightarrow \mathbf{H^2_\perp} \\ f \quad &\longrightarrow \quad \check{\mathcal{H}}_{R_1} f = (\mathcal{P}_{\mathbf{H^2_\perp}} \mathcal{M}_{R_1}) f \quad , \end{aligned}$$

The operator $\mathcal{M}$ is the multiplicative (Laurent) operator; ; and $\mathcal{T}_{R_2}$ is the Toepliz operator generated by R_2. Γ_R is, in general, infinite dimensional [5] [27] and the operator norm of Γ_R is given by

$$\|\Gamma_R\| = \lambda_{max}^{\frac{1}{2}}[\check{\mathcal{H}}_{R_1}^* \check{\mathcal{H}}_{R_1} + \mathcal{T}_{R_2^* R_2}]$$

where $\lambda_{max}(\bullet)$ denotes the largest eigenvalue. It is well known [5] that

$$\inf_{Q \in \Re\mathbf{H}^\infty} \left\| \begin{matrix} R_1 - Q \\ R_2 \end{matrix} \right\|_\infty = \|\Gamma_R\| \quad =: \gamma_{opt,1}.$$

As shown in [5] and [27], there does not always exist a closed-form solution to $\gamma_{opt,1}$; an exception arises when $R_2 = cI$, where c is a scalar constant [10]. However, the problem can be reduced to a succession of 1-block problems whose explicit solutions approach as close as we like a solution to (68) using an iterative technique called γ-iteration. In other words, we can try iteratively different values of γ until the accuracy of γ to $\gamma_{opt,1}$ is satisfactory and hence obtain a (sub)optimal Q.

In the MIMO case, there will be a set of optimal Q's attaining the infimum of (68) and the standard $\mathbf{H}^\infty$ optimization approach will accept any one of them as its solution. The optimal Q's always make the largest singular value of the cost function, $\begin{bmatrix} R_1 - Q \\ R_2 \end{bmatrix}(j\omega)$, a constant over all frequencies [14] [7], and among the optimal solutions, there is a set of optimizers which make all the singular values of the cost equal over all frequencies. These are called *equalizing* (or *all-pass*) solutions. But, for practical design, the equalizing solution seems not to be the best and may even be considered the worst [16]. Hence a stronger minimization criterion is needed. The problem then arises of how to define the criterion mathematically and how to choose one solution among all solutions to meet the criterion. Such a problem definition for the 2-block problem is defined below, and its solutions will be given in the following subsection.

Definition 2 *The 2-block strengthened model-matching problem is defined as follows: find a Q such that the sequence*

$$s_1^\infty \begin{bmatrix} R_1 - Q \\ R_2 \end{bmatrix}, \; s_2^\infty \begin{bmatrix} R_1 - Q \\ R_2 \end{bmatrix}, \; \ldots, \; s_m^\infty \begin{bmatrix} R_1 - Q \\ R_2 \end{bmatrix} \tag{70}$$

is minimised with respect to the lexicographic ordering where

$$s_i^\infty \begin{bmatrix} R_1 - Q \\ R_2 \end{bmatrix} := \sup_\omega \; s_i \begin{bmatrix} R_1 - Q \\ R_2 \end{bmatrix}(j\omega) \; . \tag{71}$$

The Q attaining this minimum is called the super-optimal solution.

Again we will denote the optimal sequence by $(\bar{s}_1^\infty, \bar{s}_2^\infty, \cdots, \bar{s}_m^\infty)$ and call them the *s-numbers* of R. We will also use $Q_{sup,k}$ to denote the optimal solutions for the first k layers, e.g. $Q_{sup,2}$ denotes the optimal solutions for the first two layers. For the kth super-optimal solution $Q_{sup,k}$ and each $i \leq k$, we will have $s_i[(R - Q_{sup,k})(j\omega)] = \bar{s}_i^\infty \; \forall \, \omega$. The s-numbers can be interpreted as the largest energy gains from appropriately defined input spaces to the output [21]. Intuitively then, if disturbance rejection is an objective reflected in the cost function it is better (more robust) to minimize all the singular values rather than just the largest one.

A philosophy proposed in [11] for the 2-block super-optimal is to work out successively the optimal singular values $\gamma_{opt,i}$ (i.e. the ith s-number $\bar{s}_i^\infty$) by an iterative method and then synthesize the super-optimal solution. There are 4 main steps for the solution of the 2-block case:

1. Find an input maximizing singular vector of the corresponding operator.
2. Find a norm-preserving transformation to operate on $\begin{bmatrix} R_1 - Q \\ R_2 \end{bmatrix}$ to isolate the i-1 largest s-numbers without changing the singular value structure.
3. Solve for the ith s-number using iterative procedures.
4. After finding all s-numbers, construct the super-optimal solution by solving a 1-block super-optimization problem.

We will now consider each one of these steps, in turn, in the following four subsections.

3.1 Maximizing Singular Vectors

The following lemma links the maximizing singular vector of $\mathcal{H}_{(R_1\Phi_1^{-1})}$ to that of Γ_R (the first layer).

Lemma 11 *Let* $\Phi_1^\sim \Phi_1 = \gamma_{opt,1}^2 I - R_2^\sim R_2$ *and let* (ξ, w) *be a unit maximizing vector pair of* $\check{\mathcal{H}}_{(R_1 \Phi_1^{-1})}$. *Then there exists a* $Q_{opt} \in \Re\mathbf{H}^\infty$ *such that*

$$(R_1 - Q_{opt})\Phi_1^{-1}\xi = w \ .$$

In addition, $\Phi_1^{-1}\xi$ *is a maximizing singular vector of* Γ_R.

Proof: A proof can be found in [26]. ■

3.2 Diagonalizing Transformations

From Lemma 11, we obtain that $\Phi_1^{-1}\xi$ is a maximizing singular vector of Γ_R and hence

$$(\Phi_1^{-1}\xi)^\sim (R_1 - Q_{opt})^\sim (R_1 - Q_{opt})(\Phi_1^{-1}\xi) \ = \ (\Phi_1^{-1}\xi)^\sim \Phi_1^\sim \Phi_1 \ (\Phi_1^{-1}\xi). \qquad (72)$$

where Q_{opt} (i.e. $Q_{sup,1}$) is an optimal solution for the first layer. It should be noted that $\|\xi\|_2 = 1$ and $\|w\|_2 = 1$, but generally $\|\Phi_1^{-1}\xi\|_2 \neq 1$.

Let

$$\xi = \xi_i \ \xi_o$$

be an inner-outer factorization of ξ, and define

$$u = \Phi_1^{-1} \ \xi_i \ . \qquad (73)$$

Factorize u into

$$u = u_i \ u_o$$

where u_i is inner (vector), and $u_o \in \mathcal{G}\mathbf{H}^\infty$ is outer (scalar). Then, from (72),

$$u_i^\sim (R_1 - Q_{opt})^\sim (R_1 - Q_{opt}) u_i \ = \ u_i^\sim \Phi_1^\sim \Phi_1 u_i \qquad (74)$$

and

$$u_i^\sim \left((R_1 - Q_{opt})^\sim (R_1 - Q_{opt}) \ + \ R_2^\sim R_2\right) u_i \ = \ \gamma_{opt,1}^2 u_i^\sim u_i = \gamma_{opt,1}^2 \ . \qquad (75)$$

A transformation matrix for the first layer is hence found as

$$U_1 = [u_i \ \ U_\perp] \qquad (76)$$

where $U_\perp$ is a complementary inner factor (CIF) of u_i such that U_1 is square and inner, and

$$s_1^\infty\left(\begin{bmatrix} R_1 - Q_{opt} \\ R_2 \end{bmatrix} U_1\right) = \gamma_{opt,1} .$$

Now, consider the matrix

$$U_1^\sim((R_1 - Q_{opt})^\sim(R_1 - Q_{opt}) + R_2^\sim R_2)U_1. \tag{77}$$

The $(1,1)$-block of matrix (77) has reached the maximum eigenvalue already, by (75). Therefore, the off-diagonal elements in the first row and the first column of matrix (77) are zeros and the largest eigenvalue of the $(2,2)$-block is less than or equal to $\gamma_{opt,1}^2$. This shows that there exists a square, inner and real-rational matrix $U_1(s)$ such that, for any Q_{opt} which reaches the first layer of optimality,

$$U_1^\sim(R_1 - Q_{opt})^\sim(R_1 - Q_{opt})U_1 + U_1^\sim R_2^\sim R_2 U_1 = \begin{bmatrix} \gamma_{opt,1}^2 & \\ & G_{Q_{opt}} \end{bmatrix}$$

where $G_{Q_{opt}}^\sim = G_{Q_{opt}}$ and $\|G_{Q_{opt}}\|_\infty \le \gamma_{opt,1}^2$.

The next theorem gives the diagonalizing property up to the first $k-1$ layers by a norm-preserving inner matrix U_{k-1} when Q reaches the first $k-1$ layers of optimality.

Theorem 5 **([12])** *For any $Q \in \{\mathbf{Q}_{\mathbf{sup,k-1}}\}$, i.e., any Q which reaches the first $k-1$ layers of optimality, we have*

$$U_{k-1}^\sim[(R_1 - Q)^\sim(R_1 - Q) + R_2^\sim R_2]U_{k-1} = \begin{bmatrix} \gamma_{opt,1}^2 & & & \\ & \ddots & & \\ & & \gamma_{opt,k-1}^2 & \\ & & & G_{Q,k} \end{bmatrix}$$

where $\|G_{Q,k}\|_\infty \le \gamma_{opt,k-1}^2$ and the inner matrix U_{k-1} is constructed from the recursive formulae below.

Recursive Formulae for calculating the required inner transformation matrices.

Consider $k \geq 2$ and assume that $\gamma_{opt,1}$, Φ_1 and $U_1 = [u_i \;\; U_\perp] = \tilde{U}_1$ have been found. Then let $\xi_1, = \xi_i$ and $\eta_1 = \begin{bmatrix} b_1^{-1} \\ 0 \\ \vdots \\ 0 \end{bmatrix}$ where $b_1 = u_o^{-1}$.

1.

$$\Phi_k^\sim \Phi_k = \begin{bmatrix} \gamma_{opt,1}^2 & & & \\ & \ddots & & \\ & & \gamma_{opt,k-1}^2 & \\ & & & \gamma_{opt,k}^2 I \end{bmatrix} - U_{k-1}^\sim R_2^\sim R_2 U_{k-1} \tag{78}$$

where $\Phi_k \in \mathcal{G}\mathbf{H}^\infty$ is a spectral factor (i.e. $\Phi_k, \Phi_k^{-1} \in \Re\mathbf{H}^\infty$).

2. Find an all-pass vector ξ_k such that

$$[\,\Phi_k \eta_1 \quad \cdots \quad \Phi_k \eta_{k-1} \quad \xi_k\,] \tag{79}$$

is all-pass; and $(R_1 - Q)U_{k-1}\Phi_k^{-1}\xi_k$ reaches optimality 1, *i.e.*

$$\xi_k^\sim (\Phi_k^{-1})^\sim U_{k-1}^\sim (R_1 - Q)^\sim (R_1 - Q) U_{k-1} \Phi_k^{-1} \xi_k = \xi_k^\sim \xi_k \; = \; 1 \tag{80}$$

for any Q reaching the first k layers of optimality.

3. Find a CIF $U_{k,\perp}$ of $[\,0 \quad I_{m-k+1}\,]\,\Phi_k^{-1}\xi_k b_k$, *i.e.*,

$$[\,[\,0 \quad I_{m-k+1}\,]\,\Phi_k^{-1}\xi_k b_k \quad U_{k,\perp}\,] \in \Re\mathbf{H}^\infty{}_{(m-k+1)\times(m-k+1)}$$

is an inner matrix, where b_k is the denominator of the right coprime factorization of $[\,0 \quad I_{m-k+1}\,]\,\Phi_k^{-1}\xi_k$ with inner numerator.

4.

$$\tilde{U}_k = \begin{bmatrix} I_{k-1} & 0 & 0 \\ 0 & [\,0 \quad I_{m-k+1}\,]\,\Phi_k^{-1}\xi_k b_k & U_{k,\perp} \end{bmatrix} \tag{81}$$

$$U_k = U_{k-1}\tilde{U}_k \tag{82}$$

$$\eta_k = \begin{bmatrix} [\,I_{k-1} \quad 0\,]\,\Phi_k^{-1}\xi_k \\ b_k^{-1} \\ 0 \\ \vdots \\ 0 \end{bmatrix} \tag{83}$$

Remark 7 From the 1-block super-optimal solution theory [23], such ξ_k, in step 2, can be found as $\xi_k = N_k^{\sim} v_k$ where N_k is the inner part of $U_{k-1}\Phi_k^{-1}$ and v_k is a maximizing vector of $\check{\mathcal{H}}_{(R_1 U_{k-1}\Phi_k^{-1} N_k^{\sim})}$. More properties about the formulae can be found in [12].

3.3 Spectral Factorizations

The spectral factor Φ_k of (78) for $k \geq 2$ has a recursive form given by

$$\begin{aligned}
\Phi_k^{\sim}\Phi_k &= \begin{bmatrix} \gamma_{opt,1}^2 & & & \\ & \ddots & & \\ & & \gamma_{opt,k-1}^2 & \\ & & & \gamma_k^2 I \end{bmatrix} - \tilde{U}_{k-1}^{\sim} U_{k-2}^{\sim} R_2^{\sim} R_2 U_{k-2} \tilde{U}_{k-1} \\
&= \begin{bmatrix} \gamma_{opt,1}^2 & & & \\ & \ddots & & \\ & & \gamma_{opt,k-1}^2 & \\ & & & \gamma_k^2 I \end{bmatrix} \\
&\quad - \tilde{U}_{k-1}^{\sim} (\begin{bmatrix} \gamma_{opt,1}^2 & & \\ & \ddots & \\ & & \gamma_{opt,k-1}^2 I \end{bmatrix} - \Phi_{k-1}^{\sim}\Phi_{k-1}) \tilde{U}_{k-1} \\
&= \tilde{U}_{k-1}^{\sim}\Phi_{k-1}^{\sim}\Phi_{k-1}\tilde{U}_{k-1} - \begin{bmatrix} 0_{(k-1)\times(k-1)} & \\ & (\gamma_{opt,k-1}^2 - \gamma_k^2) I \end{bmatrix}.
\end{aligned} \tag{84}$$

Based on this property, we are able to construct Φ_k using Φ_{k-1}, a spectral factor of one less dimension than Φ_k. In following we consider $k = 2$ and give the constructions of Φ_2 and the corresponding Ψ.

Factorize ξ_i into

$$\xi_i(-s) = N_{\xi_i}(s) M_{\xi_i}^{-1}(s) \tag{85}$$

where $M_{\xi_i}(s)$ is a scalar inner function and $N_{\xi_i}(s)$ is vector and *structural inner* [24] (i.e. $N_{\xi_i}(s)$ has no transmission zeros). Let

$$U_1 = [u_i \;\; U_{\perp} M_{\xi_i}] \tag{86}$$

and suppose that $\forall\, s$ in $j\Re_e$

$$\Phi_2^{\sim}\,\Phi_2 = \begin{bmatrix} \gamma_{opt,1}^2 & 0 \\ 0 & \gamma_2^2 I \end{bmatrix} - U_1^{\sim} R_2^{\sim} R_2 U_1 > 0, \quad \forall\, \gamma_2 \in [\bar{s}_2^{\infty},\ \bar{s}_1^{\infty}] \tag{87}$$

Note that since M_{ξ_i} is *scalar* and *inner*, it does not affect the solution of the spectral factorizations of (87), and is only introduced for theoretical convenience.

We know from (84) that there exists a spectral factor Φ_2 of the form

$$\Phi_2 = \begin{bmatrix} \phi_{11} & \phi_{12} \\ 0 & \phi_{22} \end{bmatrix} . \tag{88}$$

and hence restrict the solution of (87) to this particular form. By direct calculations, we will obtain the following three equalities:

$$\phi_{11}^{\sim}\phi_{11} = (u_o^{-1})^{\sim} u_o^{-1} \tag{89}$$

$$\phi_{12} = M_{\xi_i}\xi_i^{\sim}\Phi_1 U_{\perp} \tag{90}$$

$$\begin{aligned} \phi_{22}^{\sim}\phi_{22} &= (\gamma_2^2 - \gamma_{opt,1}^2)I + U_{\perp}^{\sim}\Phi_1^{\sim}(I - \xi_i^{\sim}\xi_i)\Phi_1 U_{\perp} \\ &= (\xi_{\perp}^{\sim}\Phi_1 U_{\perp})^{\sim}(\xi_{\perp}^{\sim}\Phi_1 U_{\perp}) - (\gamma_{opt,1}^2 - \gamma_2^2)I . \end{aligned} \tag{91}$$

where $\xi_{\perp}$ is the complementary inner part of ξ. Since $\phi_{11} = u_o^{-1} \in \mathcal{G}\mathbf{H}^{\infty}$ and $\phi_{12} = M_{\xi_i}\xi_i^{\sim}\Phi_1 U_{\perp}$ is stable where $M_{\xi_i}\xi_i^{\sim} = N_{\xi_i}^T$, to complete the solution of (87), we only need compute the spectral factor ϕ_{22} of (91) which is one-dimensional less. The assumption (87) implies $\phi_{22}^{\sim}\phi_{22} > 0$, $\forall\ s$ in $j\Re_e$, so ϕ_{22} which is clearly a function of γ_2 can be solved [5].

As mentioned in Remark 7, we need to solve the following spectral factorization:

$$\begin{aligned} \Psi^{\sim}\Psi &= U_1\Phi_2^{\sim}\Phi_2 U_1^{\sim} \\ &= [u_i \ \ U_{\perp}] \begin{bmatrix} \gamma_{opt,1}^2 & 0 \\ 0 & \gamma_2^2 I \end{bmatrix} [u_i \ \ U_{\perp}]^{\sim} - R_2^{\sim}R_2 . \end{aligned} \tag{92}$$

i.e., to find $\Psi \in \mathcal{G}\mathbf{H}^{\infty}$ and inner N_2 such that $U_1\Phi_2^{-1} = \Psi^{-1}N_2$. Let Φ_2 have the form of (88). Then

$$\begin{aligned} U_1\Phi_2^{-1} &= [u_i \ \ U_{\perp}M_{\xi_i}] \begin{bmatrix} u_o & -u_o\xi_i^{\sim}\Phi_1 U_{\perp}M_{\xi_i}\phi_{22}^{-1} \\ 0 & \phi_{22}^{-1} \end{bmatrix} \\ &= [\,\Phi_1^{-1}\xi_i \quad (I - \Phi_1^{-1}\xi_i\xi_i^{\sim}\Phi_1)U_{\perp}M_{\xi_i}\phi_{22}^{-1}\,] \\ &= [\,\Phi_1^{-1}\xi_i \quad \Phi_1^{-1}(I - \xi_i\xi_i^{\sim})\Phi_1 U_{\perp}M_{\xi_i}\phi_{22}^{-1}\,] \\ &= \Phi_1^{-1}\ [\,\xi_i \quad \xi_{\perp}\,] \begin{bmatrix} 1 & 0 \\ 0 & \xi_{\perp}^{\sim}\Phi_1 U_{\perp}M_{\xi_i}\phi_{22}^{-1} \end{bmatrix} . \end{aligned} \tag{93}$$

Factorize

$$\Psi_2^{-1} N_2 = [\ \xi_i \quad \xi_\perp\ ] \begin{bmatrix} 1 & 0 \\ 0 & \xi_\perp^\sim \Phi_1 U_\perp M_{\xi_i} \phi_{22}^{-1} \end{bmatrix} . \tag{94}$$

where $\Psi_2 \in \mathcal{G}\mathbf{H}^\infty$ and N_2 is a square inner matrix (i.e. a left coprime factorization with inner numerator). From (91) and (94), we have

$$\Psi_2^\sim \Psi_2 = [\ \xi_i \quad \xi_\perp\ ] \begin{bmatrix} 1 & 0 \\ 0 & (\xi_\perp^\sim \Phi_1 U_\perp)^{-1\sim} \phi_{22}^\sim \phi_{22} (\xi_\perp^\sim \Phi_1 U_\perp)^{-1} \end{bmatrix} [\ \xi_i \quad \xi_\perp\ ]^\sim$$

$$= [\ \xi_i \quad \xi_\perp\ ] \begin{bmatrix} 1 & 0 \\ 0 & I - (\gamma_{opt,1}^2 - \gamma_2^2)(\xi_\perp^\sim \Phi_1 U_\perp)^{-1\sim} (\xi_\perp^\sim \Phi_1 U_\perp)^{-1} \end{bmatrix} [\ \xi_i \quad \xi_\perp\ ]^\sim$$

$$= I - (\gamma_{opt,1}^2 - \gamma_2^2)\ \xi_\perp (\xi_\perp^\sim \Phi_1 U_\perp)^{-1\sim} (\xi_\perp^\sim \Phi_1 U_\perp)^{-1} \xi_\perp^\sim . \tag{95}$$

Therefore we obtain

$$U_1 \Phi_2^{-1} = \Phi_1^{-1} \Psi_2^{-1} N_2 = \Psi^{-1} N_2 \tag{96}$$

where

$$\Psi = \Psi_2 \Phi_1 . \tag{97}$$

This shows that by solving (95), we will obtain Ψ_2 and hence Ψ. In addition, from (94),

$$N_2 = \Psi_2 [\ \xi_i \quad \xi_\perp\ ] \begin{bmatrix} 1 & 0 \\ 0 & \xi_\perp^\sim \Phi_1 U_\perp M_{\xi_i} \phi_{22}^{-1} \end{bmatrix} =: [\ x_i \quad X_\perp\ ] . \tag{98}$$

It can be seen from (95) that $\|\Psi_2\|_\infty = 1$, $\|\Psi_2 \xi\|_2 = 1$, and $\Psi_2 \xi_i$ and each column of $\Psi_2 \xi_\perp$ are pointwise orthogonal since $\xi_i^\sim \Psi_2^\sim \Psi_2 \xi_\perp = 0$. An upper bound on the McMillan degree of Ψ_2 is given in the following lemma. This will help to prevent dimension expansion in the solution process.

Lemma 12 *Let*

$$\Psi_2^\sim \Psi_2 = I - (\gamma_{opt,1}^2 - \gamma_2^2)\ \xi_\perp (\xi_\perp^\sim \Phi_1 U_\perp)^{-1\sim} (\xi_\perp^\sim \Phi_1 U_\perp)^{-1} \xi_\perp^\sim > 0,\ \forall\ s \text{ in } j\Re_e .$$

Then there exists a spectral factor $\Psi_2 \in \mathcal{G}\mathbf{H}^\infty$ *with* $deg(\Psi_2) \le deg(\xi_i) \le deg(R_1) - 1$.

Proof: See [25] for a proof. ■

The degree bound in Lemma 12 is useful in the γ_2 iteration, because to solve Ψ_2 we need to form the product $(\xi_\perp^\sim \Phi_1 U_\perp)^{-1}\xi_\perp^\sim$ whose state-space realization will not necessarily be minimal. The bound gives the desired order of Ψ_2. It has been shown in [25] that the CIF $\xi_\perp$ is not minimal if ξ_i is not structural inner. Therefore $deg(\xi_\perp) < deg(\xi_i)$ and $deg(\Psi_2) < deg(\xi_i)$. This suggests that we can construct a CIF $\xi_\perp$ from the structural inner part of ξ_i as shown in (85) such that Ψ_2 has a lower degree (before finding a reduced (minimal) order model).

3.4 γ_k-iteration

A generalized iterative scheme, namely γ_k-iteration, can be used to solve the 2-block super-optimal problem. The basic idea is that having minimized the first $k-1$ layers the kth layer can be solved by iterations on an equivalent 1-block problem.

To do this, we need a criterion to judge whether or not we have reached the optimal singular values. Such a criterion is given by the following theorem which lies at the heart of the γ_k-iteration scheme.

Theorem 6 *Assume that $\gamma_k \leq \gamma_{opt,k-1}$ satisfies*

$$\begin{bmatrix} \gamma_{opt,1}^2 & & & \\ & \ddots & & \\ & & \gamma_{opt,k-1}^2 & \\ & & & \gamma_k^2 I \end{bmatrix} - U_{k-1}^\sim R_2^\sim R_2 U_{k-1} > 0, \ \forall \ s \text{ in } j\Re_e \tag{99}$$

where U_{k-1} is as in Theorem 5. Then there exists a $Q \in \{\mathbf{Q_{sup,k-1}}\}$ such that

$$\gamma_{Q,k} \leq \gamma_k$$

if and only if

$$\inf_{Q\in\Re\mathbf{H}^\infty} s_j^\infty((R_1 - Q)U_{k-1}\Phi_k^{-1}) \leq 1, \ \forall \ j = 1, 2, \cdots, k \tag{100}$$

where $\gamma_{Q,k} := s_k^\infty\left(\begin{bmatrix} R_1 - Q \\ R_2 \end{bmatrix}\right)$, *and* Φ_k *is the spectral factor of the left hand side of (99).*

Proof: (*Necessity*) Since $Q \in \{\mathbf{Q_{sup,k-1}}\}$, we obtain using Theorem 5

$$\gamma_{Q,k} \leq \gamma_k$$

$$\Longleftrightarrow \quad U_{k-1}^\sim[(R_1 - Q)^\sim(R_1 - Q) + R_2^\sim R_2]U_{k-1} \leq \begin{bmatrix} \gamma_{opt,1}^2 & & & \\ & \ddots & & \\ & & \gamma_{opt,k-1}^2 & \\ & & & \gamma_k^2 I \end{bmatrix}$$

$$\Longleftrightarrow \quad U_{k-1}^\sim(R_1 - Q)^\sim(R_1 - Q)U_{k-1} \leq \begin{bmatrix} \gamma_{opt,1}^2 & & & \\ & \ddots & & \\ & & \gamma_{opt,k-1}^2 & \\ & & & \gamma_k^2 I \end{bmatrix} - U_{k-1}^\sim R_2^\sim R_2 U_{k-1}$$

$$\Longleftrightarrow \quad \Phi_k^{-\sim}U_{k-1}^\sim(R_1 - Q)^\sim(R_1 - Q)U_{k-1}\Phi_k^{-1} \leq I$$

$$\Longleftrightarrow \quad s_j^\infty((R_1 - Q)U_{k-1}\Phi_k^{-1}) \leq 1, \ \forall\, j = 1, 2, \cdots, k \ .$$

Thus, $\gamma_{Q,k} \leq \gamma_k$ implies

$$\inf_{\hat{Q} \in RH} s_j^\infty((R_1 - \hat{Q})U_{k-1}\Phi_k^{-1}) \leq 1, \ \forall\, j = 1, 2, \cdots, k \ .$$

(*Sufficiency*) If (100) holds, there exists a $\tilde{Q} \in \Re\mathbf{H}^\infty$ such that

$$s_j^\infty((R_1 - \tilde{Q})U_{k-1}\Phi_k^{-1}) \leq 1, \ \forall\, j = 1, 2, \cdots, k \ . \tag{101}$$

For such a $\tilde{Q}$ we will now show that $\tilde{Q} \in \{\mathbf{Q_{sup,k-1}}\}$.

From the constructions of U_{k-1} in the recursive formulae section 6.2, we have

$$U_{k-1} = U_{k-2}\tilde{U}_{k-1} = \cdots = U_1\tilde{U}_2\cdots\tilde{U}_{k-1}$$

where the $\tilde{U}_j's$, $j = 2, \cdots, k-1$ are given in (81). Alternatively, we can write

$$U_{k-1} = U_1\hat{U}_2 = U_2\hat{U}_3 = \cdots = U_{k-2}\hat{U}_{k-1}$$

where

$$\hat{U}_j = \tilde{U}_j \cdots \tilde{U}_{k-1}, \quad j = 2, \cdots, k-1$$

and

$$\hat{U}_j = \begin{bmatrix} I_{j-1} & 0 \\ 0 & \Delta_j \end{bmatrix} .$$

Note that Δ_j is an $(m-j+1) \times (m-j+1)$ inner matrix. Therefore, from (101), we have

$$\Phi_k^{-\sim} U_{k-1}^{\sim} (R_1 - \tilde{Q})^{\sim} (R_1 - \tilde{Q}) U_{k-1} \Phi_k^{-1} \leq I$$

$$\iff \quad U_{k-1}^{\sim} (R_1 - \tilde{Q})^{\sim} (R_1 - \tilde{Q}) U_{k-1} \leq \Phi_k^{\sim} \Phi_k$$

$$\iff \quad U_{k-1}^{\sim} (R_1 - \tilde{Q})^{\sim} (R_1 - \tilde{Q}) U_{k-1} \leq \begin{bmatrix} \gamma_{opt,1}^2 & & & \\ & \ddots & & \\ & & \gamma_{opt,k-1}^2 & \\ & & & \gamma_k^2 I \end{bmatrix} - U_{k-1}^{\sim} R_2^{\sim} R_2 U_{k-1}$$

$$\iff \quad U_{k-1}^{\sim} [(R_1 - \tilde{Q})^{\sim} (R_1 - \tilde{Q}) + R_2^{\sim} R_2] U_{k-1} \leq \begin{bmatrix} \gamma_{opt,1}^2 & & & \\ & \ddots & & \\ & & \gamma_{opt,k-1}^2 & \\ & & & \gamma_k^2 I \end{bmatrix} . \tag{102}$$

Now since U_{k-1} is inner,

$$s_1^{\infty} \begin{pmatrix} R_1 - \tilde{Q} \\ R_2 \end{pmatrix} \leq \gamma_{opt,1} .$$

However, $\gamma_{opt,1}$ is the optimal singular value for the first layer. Thus,

$$s_1^\infty \begin{pmatrix} R_1 - \tilde{Q} \\ R_2 \end{pmatrix} = \gamma_{opt,1} \; .$$

That is, $\tilde{Q} \in \{\mathbf{Q_{sup,1}}\}$.

Consequently, the left-hand side of (102) is

$$\hat{U}_2^\sim U_1^\sim ((R_1 - \tilde{Q})^\sim (R_1 - \tilde{Q}) + R_2^\sim R_2) U_1 \hat{U}_2$$

$$= \begin{bmatrix} 1 & \\ & \Delta_2^\sim \end{bmatrix} \begin{bmatrix} \gamma_{opt,1}^2 & \\ & G_{\tilde{Q},1} \end{bmatrix} \begin{bmatrix} 1 & \\ & \Delta_2 \end{bmatrix}$$

$$= \begin{bmatrix} \gamma_{opt,1}^2 & \\ & \Delta_2^\sim G_{\tilde{Q},1} \Delta_2 \end{bmatrix}$$

which from (102) implies that the largest eigenvalue of $\Delta_2^\sim G_{\tilde{Q},1} \Delta_2$ is less than or equal to $\gamma_{opt,2}^2$, i.e.,

$$s_2^\infty \begin{pmatrix} R_1 - \tilde{Q} \\ R_2 \end{pmatrix} \leq \gamma_{opt,2} \; .$$

Again since $\gamma_{opt,2}$ is the optimal value of the second layer we have $\tilde{Q} \in \{\mathbf{Q_{sup,2}}\}$, and the left-hand side of (102) is now

$$\begin{bmatrix} 1 & & \\ & 1 & \\ & & \Delta_3^\sim \end{bmatrix} \begin{bmatrix} \gamma_{opt,1}^2 & & \\ & \gamma_{opt,2}^2 & \\ & & G_{\tilde{Q},2} \end{bmatrix} \begin{bmatrix} 1 & & \\ & 1 & \\ & & \Delta_3 \end{bmatrix} .$$

Following this procedure, we conclude that $\tilde{Q} \in \{\mathbf{Q_{sup,k-1}}\}$. Moreover, using the 'if and only if' deduction at the beginning of this proof (the necessity part), we also obtain

$$\gamma_{\tilde{Q},k} \leq \gamma_k \; .$$

■

From Theorem 6, we are able to find the optimal[1] γ_k by computing the numbers

$$t_j := \inf_{Q\in\Re\mathbf{H}^\infty} s_j^\infty((R_1 - Q)U_{k-1}\Phi_k^{-1}), \quad j = 1, 2, \cdots, k \,. \tag{103}$$

We next demonstrate how we can calculate the optimal γ_k by iteratively calculating the t_j numbers defined in (103). We do this by equivalently iterating on the Hankel singular values of an appropriately defined Hankel operator. The approach is based on the following two remarks.

Remark 8 For $R_1 \in \Re\mathbf{L}^\infty$ and $W \in \mathcal{G}\mathbf{H}^\infty$, we have

$$\inf_{Q\in\Re\mathbf{H}^\infty} \|(R_1 - Q)W\|_\infty = \|\check{\mathcal{H}}_{(R_1 W)}\| \,.$$

This follows directly from $Q_{opt} = ([R_1 W]_+ + \hat{Q}_{opt})W^{-1} \in \Re\mathbf{H}^\infty$ where $\hat{Q}_{opt} \in \Re\mathbf{H}^\infty$ is the zero-order optimal Hankel approximation [9] of $R_1 W$, and $[\bullet]_+$ denotes the stable part.

Remark 9 For $W \in \Re\mathbf{H}^\infty{}_{m\times m}$ and $WW^\sim > 0$, $\forall$ s in $j\Re_e$, we have

$$\inf_{Q\in\Re\mathbf{H}^\infty} s_i^\infty((R_1 - Q)W) \le \sigma_i([R_1 W_o]_-), \forall\, i = 2, 3, \cdots, m$$

where $W_o \in \mathcal{G}\mathbf{H}^\infty$ is the outer part of W, and $[\bullet]_-$ denotes the anti-stable part. If the maximizing vectors of the Hankel operator induced by $[R_1 W_o]_-$ form a *frame* with the number r [13], then

$$\inf_{Q\in\Re\mathbf{H}^\infty} s_i^\infty((R_1 - Q)W) = \sigma_i([R_1 W_o]_-), \forall\, i = 1, 2, \cdots, r$$

By performing a left coprime factorization with inner numerator of $U_{k-1}\Phi_k^{-1}$, we will obtain

$$\Psi^{-1} N_k = U_{k-1}\Phi_k^{-1}$$

where N_k is a square inner matrix and $\Psi \in \mathcal{G}\mathbf{H}^\infty$. Then the condition

$$\|\check{\mathcal{H}}_{(R_1\Psi^{-1})}\| = \inf_{Q\in\Re\mathbf{H}^\infty} \|(R_1 - Q)\Psi^{-1}\|_\infty > 1$$

[1]In practice, we will only be able to get as close as desired to the optimal γ_k because of finite precision in the numerical procedure.

implies that for any Q

$$\|(R_1 - Q)U_{k-1}\Phi_k^{-1}\|_\infty = \|(R_1 - Q)\Psi^{-1}\|_\infty > 1$$

which implies that $\gamma_k < \gamma_{opt,k}$ from Theorem 6.

For $\gamma_k \geq \gamma_{opt,k}$, the proof of Theorem 6 indicates that the first $k-1$ s-numbers in (103) are equal to one and hence the corresponding Hankel operator $\check{\mathcal{H}}_{(R_1\Psi^{-1})}$ has at least $k-1$ Hankel singular values equal to one. The following lemma considers the case $k = 2$ for notational simplicity and will be used later in the convergence proof.

In the γ_2-iteration, we need to verify that the largest Hankel singular value of $\check{\mathcal{H}}_{(R_1\Phi_1^{-1}\Psi_2^{-1})}$ is equal to one with multiplicity 2. Since $R_1\Phi_1^{-1}$ is known at this stage, the iterative scheme only depends on the solution of Ψ_2 which is a function of γ_2. Recall that $\|\Psi_2\|_\infty = 1$, $\|\Psi_2\xi\|_2 = 1$, and $\Psi_2\xi_i$ and each column of $\Psi_2\xi_\perp$ are pointwise orthogonal since $\xi_i^\sim\Psi_2^\sim\Psi_2\xi_\perp = 0$.

Lemma 13 *Let $[u_i \;\; U_\perp]$ be defined as in Lemma 11. Then for any $\gamma_2 \leq \gamma_{opt,1}$ satisfying*

$$\Psi^\sim\Psi = [u_i \;\; U_\perp]\begin{bmatrix}\gamma_{opt,1}^2 & 0 \\ 0 & \gamma_2^2 I\end{bmatrix}[u_i \;\; U_\perp]^\sim - R_2^\sim R_2 > 0\ , \ \forall\ s \text{ in } j\Re_e$$

$\check{\mathcal{H}}_{(R_1\Psi^{-1})}$ and $\check{\mathcal{H}}_{(R_1\Phi_1^{-1})}$ have at least one Hankel singular value the same.

Proof: A proof can be found in [26]. ■

In practice, the conclusion of Lemma 13 is useful because of the numerical problems associated with achieving Hankel singular values equal to one. As mentioned in the footnote on page 45 finite precision in the computation will make it difficult to reach the exact value of one for the largest Hankel singular value(s). However, by setting a precision tolerance for $\gamma_{opt,1}$ and computing the Ψ in a particular way such as $\Psi = \Psi_2\Phi_1$, we can retain the approximation to $\gamma_{opt,1}$ as one of the Hankel singular values of $\check{\mathcal{H}}_{(R_1\Psi^{-1})}$ and calculate an approximation to $\gamma_{opt,2}$ with the same precision as for $\gamma_{opt,1}$.

Applying N_2 as a norm-preserving transformation, the condition (100) for $k = 2$ is equivalent to checking the first two largest Hankel singular values of the operator $\check{\mathcal{H}}_{(R_1\Psi^{-1})}$. When $\gamma_2 = \gamma_{opt,2}$, we will obtain from Lemma 13 that $\check{\mathcal{H}}_{(R_1\Psi^{-1})}$ has two pointwise orthogonal maximizing vectors with respect to the Hankel singular value $\sigma_1 (= \sigma_2 = 1)$. Assume we have found the optimal γ_2. Then a maximizing frame, denoted by $[v_1 \;\; v_2]$, of $\check{\mathcal{H}}_{(R_1\Psi^{-1})}$ can be found. Note that the maximizing frame is unique up to right multiplication by a constant unitary matrix. Therefore we may construct a maximizing frame of the form

$$[v_1 \;\; v_2] = [x_i \;\; X_\perp] \begin{bmatrix} \xi_o & 0 \\ 0 & \tilde{\xi} \end{bmatrix}$$

where $\tilde{\xi} \in \Re\mathbf{H^2}_{m-1}$. With this construction, we then have

$$\begin{aligned} \Psi^{-1} v_1 &= U_1 \Phi_2^{-1} N_2^{\sim} v_1 \\ &= U_1 \Phi_2^{-1} \begin{bmatrix} \xi_o \\ 0 \end{bmatrix} \\ &= \xi \end{aligned}$$

and

$$\begin{aligned} [\,0 \quad I\,]\, \Phi_2^{-1} N_2^{\sim} v_2 &= [\,0 \quad I\,]\, \Phi_2^{-1} \begin{bmatrix} 0 \\ \tilde{\xi} \end{bmatrix} \\ &= \phi_{22}^{-1} \tilde{\xi}\,. \end{aligned}$$

Thus $\phi_{22}^{-1}\tilde{\xi}$ will be used to construct $\tilde{U}_2$. The computation procedure for $\tilde{U}_2$ is the same as U_1 obtained from $\Phi_1^{-1}\xi$. Using the $\tilde{U}_2$, we can further compute the (3,3)-entry of Φ_3 and a new Ψ from $U_2\Phi_3^{-1}$ by applying the same techniques shown earlier in this section. This shows that the results above can be extended to the case $k > 2$.

4 The 2-Block Algorithm and Its Convergence

4.1 An Algorithm

By using Theorem 6 as a super-optimality criterion and Theorem 5 for the spectral factorization, we can successively calculate the optimal singular values $\gamma_{opt,j}$ and also obtain the super-optimal solution. We now present the details of the algorithm; the conditions under which the algorithm converges are given in Theorem 8 in section 4.2.

In the following algorithm, we will assume $\gamma_{opt,j} > s_j^\infty(R_2) =: \gamma_{l,j}$, $j = 1, \cdots, m$ where $\gamma_{l,j}$ can be used as a lower bound for $\gamma_{opt,j}$. An upper bound for $\gamma_{opt,1}$ is $\gamma_{u,1} := \left\| \begin{bmatrix} \|\check{\mathcal{H}}_{R_1}\| \\ \|R_2\|_\infty \end{bmatrix} \right\|_2$ [3] and $\gamma_{u,j} := \gamma_{opt,j-1}$ will be an upper bound for $\gamma_{opt,j}$, $j > 1$.

The 2-Block Super-Optimal Algorithm:

Data: R_1, R_2, $U_0 = I$, $k = 1$, $flag = 0$, a selected tolerance $\epsilon > 0$ and $\gamma_{u,1}$, $\gamma_{l,j}$, $j = 1, \cdots, m$.

1. **If** $\gamma_{u,k} - \gamma_{l,k} < \epsilon$ and $flag = 0$, set $\gamma_{opt,k} = \gamma_{u,k}$ and calculate[2] $Q_{sup,k} = ([R_1\Psi^{-1}]_+ + \hat{Q}_{opt})\Psi$ where $\hat{Q}_{opt} \in \Re\mathbf{H}^\infty$ is the zero-order Hankel approximation of $[R_1\Psi^{-1}]_-$.
   Print $\bar{s}_j^\infty = \gamma_{opt,j}$, $j = 1, \cdots, k$, and $Q_{sup,k}$.
   Stop. (In this case, $Q_{sup,k}$ solves the first k layers of optimization.)

   Else, set $\gamma_k = (\gamma_{l,k} + \gamma_{u,k})/2$.

2. **If** the inequality in (104) below does not hold, set $t_1 = 1 + \epsilon$ and go to Step 4.

---

[2]If $k = 1$ (i.e., $\Psi = \Phi_1$), this algorithm will obtain a solution $Q_{sup,1}$ for which the first layer of optimality is attained subject to the tolerance ϵ.

Else, factorize, using formulae in section 3.3,

$$\Phi_k^{\sim}\Phi_k = \begin{bmatrix} \gamma_{opt,1}^2 & & & \\ & \ddots & & \\ & & \gamma_{opt,k-1}^2 & \\ & & & \gamma_k^2 I \end{bmatrix} - U_{k-1}^{\sim} R_2^{\sim} R_2 U_{k-1} > 0,$$

$$\forall s = j\Re_e \tag{104}$$

where the first k-1 elements on the diagonal in the constant matrix vanish when $k = 1$; and set $flag = 1$.

3. Find the left coprime factorization with inner numerator of $U_{k-1}\Phi_k^{-1}$, *i.e.*,

$$\Psi^{-1} N_k = U_{k-1}\Phi_k^{-1}$$

where N_k and Ψ are inner and outer respectively; then solve the 1-block super-optimal for $(R_1 - Q)\Psi^{-1}$, i.e., determine

$$t_j = \inf_{Q \in \Re \mathbf{H}^\infty} s_j^\infty[(R_1 - Q)\Psi^{-1}] \ , \ j = 1, \cdots, k \tag{105}$$

4. **If** $t_1 > 1$ set $\gamma_{l,k} = \gamma_k$ and go to Step 1.
   Else set $\gamma_{u,k} = \gamma_k$.
   If $\gamma_{u,k} - \gamma_{l,k} > \epsilon$, go to Step 1.

5. Set $\gamma_{opt,k} = \gamma_k$.

   If $m > k$, construct a square, inner matrix U_k such that, for any $Q_{sup,k}$,

$$U_k^{\sim}\left(\begin{bmatrix} R_1 - Q_{sup,k} \\ R_2 \end{bmatrix}\right)^{\sim}\left(\begin{bmatrix} R_1 - Q_{sup,k} \\ R_2 \end{bmatrix}\right) U_k$$

$$= \begin{bmatrix} \gamma_{opt,1}^2 & & & \\ & \ddots & & \\ & & \gamma_{opt,k}^2 & \\ & & & G_k \end{bmatrix} \tag{106}$$

for some $G_k \in \Re\mathbf{L}^\infty$ with $\|G_k\|_\infty \le \gamma_{opt,k}^2$. This can be achieved by using formulae (81) and (82).

Else, calculate the super-optimal solution

$$Q_{sup,m} = ([R_1\Psi^{-1}]_+ + \hat{Q}_{opt})\Psi \tag{107}$$

where $\hat{Q}_{opt}$ is the zero-order Hankel approximation of $[R_1\Psi^{-1}]_-$.
Print $\bar{s}_j^{\infty} = \gamma_{opt,j}$ $j = 1, \cdots, m$ and $Q_{sup,m}$.
Stop.

6. Set $k = k+1$, $flag = 0$, and $\gamma_{u,k} = \gamma_{opt,k-1}$.
Go to Step 1.

The computations in the algorithm can all be carried out using state-space models which are essential for numerical reliability. Also we can see that the solution procedure for the 2-block super-optimal problem actually consists of a series of 1-block problems, and the s-numbers of these 1-block problems will, in increasing number, approach the value of 1 subject to tolerance ϵ. Therefore the frame approach proposed in [13] and described in §2.4 will be useful in speeding up the computations.

4.2 Convergence

In the γ_2-iteration, when $\gamma_{opt,2} < \gamma_2 \le \gamma_{opt,1}$, we have by Theorem 6 and Lemma 13 that

$$\inf_{Q\in\Re\mathbf{H}^\infty} s_2^{\infty}((R_1 - Q)U_1\Phi_2^{-1}) = \inf_{Q\in\Re\mathbf{H}^\infty} s_2^{\infty}((R_1 - Q)\Psi^{-1}) \ (\le 1)$$

is a function of γ_2. And when $\gamma_2 < \gamma_{opt,2}$,

$$\inf_{Q\in\Re\mathbf{H}^\infty} s_1^{\infty}((R_1 - Q)U_1\Phi_2^{-1}) = \inf_{Q\in\Re\mathbf{H}^\infty} s_1^{\infty}((R_1 - Q)\Psi^{-1}) \ (> 1)$$

is a function of γ_2. Therefore, to find the optimal γ_2, we can compute the norm of $\check{\mathcal{H}}_{(R_1\Psi^{-1})}(I - \Psi_2\xi_i\xi_i^{\sim}\Psi_2^{\sim})$ by adjusting γ_2 and computing Ψ_2 from (95) until the corresponding operator norm reaches one. It is clear that Ψ_2 is a function of γ_2 and so is $\|\check{\mathcal{H}}_{(R_1\Psi^{-1})}(I - \Psi_2\xi_i\xi_i^{\sim}\Psi_2^{\sim})\|$. Thus, we define the function

$$f(\gamma_2) := \|\check{\mathcal{H}}_{(R_1\Psi^{-1})}(I - \Psi_2\xi_i\xi_i^{\sim}\Psi_2^{\sim})\| \,. \tag{108}$$

Calculating $f(\gamma_2)$ is equivalent to evaluating the largest Hankel singular value of $\check{\mathcal{H}}_{(R_1\Psi^{-1})}$ in the subspace which is pointwise orthogonal to $\Psi_2\xi$.

Let γ_2^2 be denoted by λ. Then $f^2(\gamma_2)$ is the largest eigenvalue of the following standard eigenvalue problem

$$(\check{\mathcal{H}}^*_{(R_1\Psi^{-1})}\ \check{\mathcal{H}}_{(R_1\Psi^{-1})})v_\lambda = \rho_\lambda v_\lambda \tag{109}$$

with the restriction that v_λ is pointwise orthogonal to $\Psi_2\xi$, $\forall\ s$ in $j\Re$. This means that we need to find the largest eigenvalue value in (109) with the corresponding eigenvector $v_\lambda \in \Re\mathbf{H^2}_m \bigcap X_\perp \Re\mathbf{L^2}_{m-1}$.

Since Ψ and Ψ^{-1} are in $\Re\mathbf{H}^\infty$, $\check{\mathcal{H}}_{(R_1\Psi^{-1})}$ is equal to $\check{\mathcal{H}}_{R_1}\mathcal{T}_{\Psi^{-1}}$ for such v_λ. We then obtain the following equivalent eigenvalue problems : $\forall\ \rho_\lambda \neq 0$,

$$
\begin{aligned}
& (\check{\mathcal{H}}^*_{(R_1\Psi^{-1})}\check{\mathcal{H}}_{(R_1\Psi^{-1})})v_\lambda = \rho_\lambda v_\lambda \\
\Longleftrightarrow\quad & \mathcal{T}^*_{\Psi^{-1}}\check{\mathcal{H}}^*_{R_1}\check{\mathcal{H}}_{R_1}\mathcal{T}_{\Psi^{-1}}v_\lambda = \rho_\lambda v_\lambda \\
\Longleftrightarrow\quad & \check{\mathcal{H}}^*_{R_1}\check{\mathcal{H}}_{R_1}u_\lambda = \rho_\lambda(\mathcal{T}^*_{\Psi^{-1}})^{-1}(\mathcal{T}_{\Psi^{-1}})^{-1}u_\lambda \\
\Longleftrightarrow\quad & \check{\mathcal{H}}^*_{R_1}\check{\mathcal{H}}_{R_1}u_\lambda = \rho_\lambda\mathcal{T}^*_\Psi\mathcal{T}_\Psi u_\lambda \\
\Longleftrightarrow\quad & \check{\mathcal{H}}^*_{R_1}\check{\mathcal{H}}_{R_1}u_\lambda = \rho_\lambda\mathcal{T}_{(\Psi^*\Psi)}u_\lambda \\
\Longleftrightarrow\quad & \mathcal{T}^{-1}_{(\Psi^*\Psi)}\check{\mathcal{H}}^*_{R_1}\check{\mathcal{H}}_{R_1}u_\lambda = \rho_\lambda u_\lambda \\
\Longleftrightarrow\quad & \check{\mathcal{H}}_{R_1}\mathcal{T}^{-1}_{(\Psi^*\Psi)}\check{\mathcal{H}}^*_{R_1}y_\lambda = \rho_\lambda y_\lambda \\
\Longleftrightarrow\quad & \check{\mathcal{H}}^*_{R_1}\left([u_i\ \ U_\perp]\begin{bmatrix}\gamma^2_{opt,1} & 0\\ 0 & \lambda I\end{bmatrix}\begin{bmatrix}u_i^*\\ U_\perp^*\end{bmatrix} - \mathcal{T}_{(R_2^*R_2)}\right)^{-1}\check{\mathcal{H}}_{R_1}y_\lambda = \rho_\lambda y_\lambda
\end{aligned}
$$

where with slight abuse of notation $[u_i\ \ U_\perp]$ is here considered to be a multiplicative (Laurent) operator. Note that $v_\lambda \in \Re\mathbf{H^2}_m \bigcap X_\perp \Re\mathbf{L^2}_{m-1}$ is equivalent to $U_\perp x_\lambda$ for some $x_\lambda \in \Re\mathbf{H^2}_{m-1}$.

Based on the above equivalent eigenvalue problems, the next theorem shows that $f(\gamma_2)$ is a strictly monotonically decreasing function.

Theorem 7 *Consider the case $k = 2$. For γ_2 satisfying*

$$[u_i\ \ U_\perp]\begin{bmatrix}\gamma^2_{opt,1} & 0\\ 0 & \gamma_2^2 I\end{bmatrix}[u_i\ \ U_\perp]^\sim - R_2^\sim R_2 > 0\ ,\ \forall\ s \text{ in } j\Re_e$$

$f(\gamma_2)$ is a strictly monotonically decreasing function with respect to γ_2.

Proof: First we define the operator

$$\Gamma_2 := U_\perp^* (\check{\mathcal{H}}_{R_1}^* \check{\mathcal{H}}_{R_1} + \mathcal{T}_{R_2^* R_2}) U_\perp \ .$$

Note that the norm of Γ_2, the largest eigenvalue of Γ_2, is equal to $\gamma_{opt,2}^2$.

The next problem is to find an eigenvector $x \in \Re\mathbf{H^2}_{m-1}$ of Γ_2 with eigenvalue $\lambda \neq 0$, i.e,

$$\lambda x = U_\perp^* (\check{\mathcal{H}}_{R_1}^* \check{\mathcal{H}}_{R_1} + \mathcal{T}_{R_2^* R_2}) U_\perp x \tag{110}$$

or

$$U_\perp^* (\check{\mathcal{H}}_{R_1}^* \check{\mathcal{H}}_{R_1}) U_\perp x = (\lambda I - U_\perp^* \mathcal{T}_{R_2^* R_2} U_\perp) x \ .$$

This implies

$$x = (\lambda I - U_\perp^* \mathcal{T}_{R_2^* R_2} U_\perp)^{-1} U_\perp^* (\check{\mathcal{H}}_{R_1}^* \check{\mathcal{H}}_{R_1}) U_\perp x$$

and

$$\check{\mathcal{H}}_{R_1} U_\perp x = \check{\mathcal{H}}_{R_1} U_\perp (\lambda I - U_\perp^* \mathcal{T}_{R_2^* R_2} U_\perp)^{-1} U_\perp^* (\check{\mathcal{H}}_{R_1}^* \check{\mathcal{H}}_{R_1}) U_\perp x \ . \tag{111}$$

Now let

$$\Gamma_\lambda := \lambda I - U_\perp^* \mathcal{T}_{R_2^* R_2} U_\perp$$

then from (111) we obtain $y = \check{\mathcal{H}}_{R_1} U_\perp x$. Clearly, the operator $\check{\mathcal{H}}_{R_1} U_\perp \Gamma_\lambda^{-1} U_\perp^* \check{\mathcal{H}}_{R_1}^*$ is compact, self adjoint, and positive semidefinite. Therefore consider the following eigenvalue problem: $\forall \ \rho_\lambda \neq 0$,

$$\rho_\lambda y_\lambda = \check{\mathcal{H}}_{R_1} U_\perp \Gamma_\lambda^{-1} U_\perp^* \check{\mathcal{H}}_{R_1}^* y_\lambda \ . \tag{112}$$

A necessary condition for λ to be an eigenvalue of Γ_2 is that $\check{\mathcal{H}}_{R_1} U_\perp \Gamma_\lambda^{-1} U_\perp^* \check{\mathcal{H}}_{R_1}^*$ has an eigenvalue $\rho_\lambda = 1$. Differentiating (112) with respect to λ and multiplying from the left by y_λ^* yields

$$\dot{\rho}_\lambda = - \frac{y_\lambda^* \check{\mathcal{H}}_{R_1} U_\perp \Gamma_\lambda^{-2} U_\perp^* \check{\mathcal{H}}_{R_1}^* y_\lambda}{y_\lambda^* y_\lambda} \ .$$

Obviously, $\dot{\rho}_\lambda \leq 0$.

To prove that $\dot{\rho}_\lambda < 0$, assume that $\dot{\rho}_\lambda = 0$. This implies that $U_\perp^* \check{\mathcal{H}}_{R_1}^* y_\lambda = 0$. Thus, $\rho_\lambda y_\lambda = 0$. We further have $\rho_\lambda = 0$. This is a contradiction. Therefore, ρ_λ is a *strictly monotonically decreasing function* of λ and the mapping from λ to a particular eigenvalue ρ_λ of $\check{\mathcal{H}}_{R_1} U_\perp \Gamma_\lambda^{-1} U_\perp^* \check{\mathcal{H}}_{R_1}^*$ is one to one.

Since $U_\perp x_\lambda$ for $x_\lambda \in \Re\mathbf{H^2}_{m-1}$ is equivalent to some $v_\lambda \in \Re\mathbf{H^2}_m \cap X_\perp \Re\mathbf{L^2}_{m-1}$, we now consider an equivalent problem as follows.

From the choice of norm-preserving transformation $[u_i \;\; U_\perp]$ and (110), we know

$$\lambda \begin{bmatrix} 0 \\ x \end{bmatrix} = [u_i \;\; U_\perp]^* (\check{\mathcal{H}}^*_{R_1} \check{\mathcal{H}}_{R_1} + \mathcal{T}_{R_2^* R_2}) [u_i \;\; U_\perp] \begin{bmatrix} 0 \\ x \end{bmatrix} .$$

That is,

$$\begin{bmatrix} \gamma^2_{opt,1} & 0 \\ 0 & \lambda I \end{bmatrix} \begin{bmatrix} 0 \\ x \end{bmatrix} = [u_i \;\; U_\perp]^* (\check{\mathcal{H}}^*_{R_1} \check{\mathcal{H}}_{R_1} + \mathcal{T}_{R_2^* R_2}) [u_i \;\; U_\perp] \begin{bmatrix} 0 \\ x \end{bmatrix}$$

and

$$[u_i \; U_\perp]^* \check{\mathcal{H}}^*_{R_1} \check{\mathcal{H}}_{R_1} [u_i \; U_\perp] \begin{bmatrix} 0 \\ x \end{bmatrix} = \left(\begin{bmatrix} \gamma^2_{opt,1} & 0 \\ 0 & \lambda I \end{bmatrix} - [u_i \;\; U_\perp]^* \mathcal{T}_{R_2^* R_2} [u_i \;\; U_\perp] \right) \begin{bmatrix} 0 \\ x \end{bmatrix} .$$

Hence

$$\begin{bmatrix} 0 \\ x \end{bmatrix} = \left(\begin{bmatrix} \gamma^2_{opt,1} & 0 \\ 0 & \lambda I \end{bmatrix} - [u_i \;\; U_\perp]^* \mathcal{T}_{R_2^* R_2} [u_i \;\; U_\perp] \right)^{-1} [u_i \; U_\perp]^* \check{\mathcal{H}}^*_{R_1} \check{\mathcal{H}}_{R_1} [u_i \; U_\perp] \begin{bmatrix} 0 \\ x \end{bmatrix}$$

and

$$\check{\mathcal{H}}_{R_1} [u_i \;\; U_\perp] \begin{bmatrix} 0 \\ x \end{bmatrix}$$
$$= \check{\mathcal{H}}_{R_1} \left([u_i \;\; U_\perp] \begin{bmatrix} \gamma^2_{opt,1} & 0 \\ 0 & \lambda I \end{bmatrix} [u_i \;\; U_\perp]^* - \mathcal{T}_{R_2^* R_2} \right)^{-1} \check{\mathcal{H}}^*_{R_1} \check{\mathcal{H}}_{R_1} [u_i \;\; U_\perp] \begin{bmatrix} 0 \\ x \end{bmatrix} \tag{113}$$

from which

$$y = \check{\mathcal{H}}_{R_1} \left([u_i \;\; U_\perp] \begin{bmatrix} \gamma^2_{opt,1} & 0 \\ 0 & \lambda I \end{bmatrix} [u_i \;\; U_\perp]^* - \mathcal{T}_{R_2^* R_2} \right)^{-1} \check{\mathcal{H}}^*_{R_1} y .$$

Therefore the eigenvalue problem considered is in the form of

$$\rho_\lambda y_\lambda = \check{\mathcal{H}}_{R_1} \left([u_i \;\; U_\perp] \begin{bmatrix} \gamma^2_{opt,1} & 0 \\ 0 & \lambda I \end{bmatrix} [u_i \;\; U_\perp]^* - \mathcal{T}_{R_2^* R_2} \right)^{-1} \check{\mathcal{H}}^*_{R_1} y_\lambda$$

and

$$\rho_\lambda y_\lambda = \check{\mathcal{H}}_{R_1} \mathcal{T}^{-1}_{(\Psi^* \Psi)} \check{\mathcal{H}}^*_{R_1} y_\lambda .$$

Finally since $X_\perp$ is a complementary inner factor of $\Psi_2 \xi_i$, $v_\lambda \in \Re\mathbf{H^2}_m \cap X_\perp \Re\mathbf{L^2}_{m-1}$ is equivalent to $U_\perp x_\lambda$ for some $x_\lambda \in \Re\mathbf{H^2}_{m-1}$.

Therefore we conclude by the equivalent eigenvalue problems mentioned earlier that the largest Hankel singular value of $\check{\mathcal{H}}_{(R_1\Psi^{-1})}$ corresponding to the input maximizing vector in the subspace which is pointwise orthogonal to $\Psi_2\xi$ is strictly monotonically decreasing with respect to γ_2. ■

In γ_2-iteration, the optimal γ_2 $(\gamma_{opt,2})$ can be found by computing Ψ_2 of (95) and evaluating whether the largest Hankel singular value of $\check{\mathcal{H}}_{(R_1\Phi_1^{-1}\Psi_2^{-1})}$ equals '1' with multiplicity two.

The next theorem summarizes the convergence property of the 2-block super-optimal algorithm.

Theorem 8 *Let*

$$\gamma_{opt,j} := \inf_{Q\in\Re\mathbf{H}^\infty} s_j^\infty\left(\begin{bmatrix} R_1 - Q \\ R_2 \end{bmatrix}\right), \quad j = 1, \cdots, m$$

where $R_1 \in \Re\mathbf{L}^\infty{}_{p_1\times m}$, $R_2 \in \Re\mathbf{L}^\infty{}_{p_2\times m}$ *and* $p_1 \geq m$. *For* $k \leq m$ *let* U_{k-1} *be constructed from formulae* (81) – (82) *such that, for any* Q_{opt} *at which the infimum over* $Q \in \Re\mathbf{H}^\infty$ *of*

$$\left(s_1^\infty\begin{bmatrix} R_1 - Q \\ R_2 \end{bmatrix}, s_2^\infty\begin{bmatrix} R_1 - Q \\ R_2 \end{bmatrix}, \cdots, s_{k-1}^\infty\begin{bmatrix} R_1 - Q \\ R_2 \end{bmatrix}\right)$$

is attained,

$$U_{k-1}^\sim\left(\begin{bmatrix} R_1 - Q_{opt} \\ R_2 \end{bmatrix}\right)^\sim\left(\begin{bmatrix} R_1 - Q_{opt} \\ R_2 \end{bmatrix}\right)U_{k-1} = \begin{bmatrix} \gamma_{opt,1}^2 & & & \\ & \ddots & & \\ & & \gamma_{opt,k-1}^2 & \\ & & & G_k \end{bmatrix}$$

for some $G_k \in \Re\mathbf{L}^\infty$ *with* $\|G_k\|_\infty \leq \gamma_{opt,k-1}^2$. *If there exist a lower bound* $\gamma_{l,k}$ $(< \gamma_{opt,k})$ *and an upper bound* $\gamma_{u,k}$ $(> \gamma_{opt,k})$ *for* $\gamma_{opt,k}$ *such that the* U_{k-1} *and* $\gamma_k \in [\gamma_{l,k}, \ \gamma_{u,k}]$ *satisfy*

$$\begin{bmatrix} \gamma_{opt,1}^2 & & & \\ & \ddots & & \\ & & \gamma_{opt,k-1}^2 & \\ & & & \gamma_k^2 I \end{bmatrix} - U_{k-1}^\sim R_2^\sim R_2 U_{k-1} > 0, \ \forall\ s \text{ in } j\Re_e \tag{114}$$

then the selection of γ_k *in Step 1 of the 2-block algorithm converges to* $\gamma_{opt,k}$.

Proof:

(i) **case** $k = 1$. The algorithm reduces to the standard 2-block γ-iteration and hence its convergence can be found for example in [3].

(ii) **case** $k = 2$. This follows from theorems 6 and 7, and the property of the particular construction of Ψ in Lemma 13 as we will now show.

Define, for a given γ_2,

$$G(\gamma_2) := R_1 \Psi^{-1}(I - \Psi_2 \xi_i \xi_i^{\sim} \Psi_2^{\sim})$$

where Ψ and Ψ_2 are given by (97) and (95), and ξ_i is the inner part of the maximizing vector of $\check{\mathcal{H}}_{(R_1 \Phi_1^{-1})}$ with respect to $\bar{\sigma}_1 := \|\check{\mathcal{H}}_{(R_1 \Phi_1^{-1})}\|$ ($= 1$, subject to tolerance ϵ, since $\gamma_{opt,1}$ is assumed to have been found). Then, from Lemma 13, we know that $\check{\mathcal{H}}_{(R_1 \Psi^{-1})}$ and $\check{\mathcal{H}}_{(R_1 \Phi_1^{-1})}$ have at least one Hankel singular value the same, i.e., $\bar{\sigma}_1$. Thus we obtain from Theorem 6 that

1) when $t_1 = \|\check{\mathcal{H}}_{G(\gamma_2)}\| > 1$, we have $\gamma_2 < \gamma_{opt,2}$ in which case $\Sigma_2 = \begin{bmatrix} t_1 & 0 \\ 0 & t_2 \end{bmatrix}$ where $t_2 = \bar{\sigma}_1$ (by Lemma 13) and

2) when $\Sigma_2 = \begin{bmatrix} t_1 & 0 \\ 0 & t_2 \end{bmatrix} \leq I$, we have $\gamma_2 \geq \gamma_{opt,2}$ in which case $t_1 = \bar{\sigma}_1$ and $t_2 = \|\check{\mathcal{H}}_{G(\gamma_2)}\| \leq 1$. (When $t_1 = t_2$, there exists a particular construction so that the maximizing vector corresponds to $t_1 = \bar{\sigma}_1$.)

Note that the condition $(t_1 + t_2)/2 > \bar{\sigma}_1$ ($\leq \bar{\sigma}_1$) is equivalent to $f(\gamma_2) > \bar{\sigma}_1$ ($\leq \bar{\sigma}_1$) where $f(\gamma_2) = \|\check{\mathcal{H}}_{G(\gamma_2)}\|$ as defined in (108), and if $\check{\mathcal{H}}_{(R_1 \Psi^{-1})}$ has two pointwise orthogonal maximizing vectors with respect to the Hankel singular value $\bar{\sigma}_1$, then $\|\check{\mathcal{H}}_{G(\gamma_2)}\| = \bar{\sigma}_1$.

We will conclude that the selection of γ_2 in Step 1 of the 2-block super-optimization algorithm converges to $\gamma_{opt,2}$ by showing that $\|\check{\mathcal{H}}_{G(\gamma_2)}\| = \bar{\sigma}_1$ if and only if $\gamma_2 = \gamma_{opt,2}$.

(Necessity): Assume that $\|\check{\mathcal{H}}_{G(\gamma_2)}\| = \bar{\sigma}_1$. This is situation 2) arising from Theorem 6 as discussed above which implies $\gamma_2 \geq \gamma_{opt,2}$ where $\gamma_{opt,2}$ is the minimal value of

$$\inf_{\hat{Q}\in\Re\mathbf{H}^\infty} s_2^\infty\left(\begin{bmatrix} R_1 - \hat{Q} \\ R_2 \end{bmatrix}\right) \quad \text{subject to} \quad s_1^\infty\left(\begin{bmatrix} R_1 - \hat{Q} \\ R_2 \end{bmatrix}\right) = \gamma_{opt,1}.$$

From Theorem 7, $f(\gamma_2)$ is a strictly monotonically decreasing function with respect to γ_2. Therefore,

$$\gamma_2 > \gamma_{opt,2} \quad \Longrightarrow \quad \|\check{\mathcal{H}}_{G(\gamma_{opt,2})}\| > \|\check{\mathcal{H}}_{G(\gamma_2)}\| \ .$$

This contradicts the definition of $\gamma_{opt,2}$. Hence $\gamma_2 = \gamma_{opt,2}$.

(Sufficiency): If $\gamma_2 = \gamma_{opt,2}$ but $\|\check{\mathcal{H}}_{G(\gamma_2)}\| < 1$, by continuity and monotonicity, $\exists\ \delta > 0$ such that

$$\|\check{\mathcal{H}}_{G(\bar{\gamma}_2)}\| \leq 1$$

where $\bar{\gamma}_2 = \gamma_2 - \delta$. This is impossible since $\gamma_{opt,2}$is the minimal norm. Hence $\|\check{\mathcal{H}}_{G(\gamma_2)}\| = 1$.

(iii) **case $k > 2$.** As for $k = 2$, but the constructions and notation become very complex. The main point is that Ψ $(= U_{k-1}\Phi_k^{-1}N_k^{\sim})$ can be constructed so that the operator $\check{\mathcal{H}}_{(R_1\Psi^{-1})}$ has at least $k-1$ Hankel singular values equal to 'one' .

■

Notice that if assumption (114) is true for $k = 1, 2, \cdots, m$, then the algorithm will produce a super-optimal solution for the 2-block problem. If however, the assumption is only valid up to some $k < m$, then the best solution we can achieve with this algorithm is Q_{opt} which reaches the first k layers of optimality. And, of course, the execution of the algorithm will stop at Step 1.

The condition in Theorem 8, regarding (114), is equivalent to

$$\begin{bmatrix} \gamma_{opt,1}^2 & & & \\ & \ddots & & \\ & & \gamma_{opt,k-1}^2 & \\ & & & \gamma_{opt,k}^2 I \end{bmatrix} - U_{k-1}^{\sim} R_2^{\sim} R_2 U_{k-1} > 0, \ \forall\ s \text{ in } j\Re_e \ .$$

Therefore we can conclude that theoretically the 2-block super-optimal problem has a unique super-optimal solution if

$$\Psi^{\sim}\Psi = U_m \begin{bmatrix} \gamma_{opt,1}^2 & & \\ & \ddots & \\ & & \gamma_{opt,m}^2 \end{bmatrix} U_m^{\sim} - R_2^{\sim} R_2 > 0, \quad \forall s \text{ in } j\Re_e \, .$$

The unique solution of a 2-block problem is given by the solution of the 1-block problem:

$$\inf_{Q \in \Re \mathbf{H}^\infty} s_i^\infty (R_1 \Psi^{-1} - Q) \, .$$

Note that $\check{\mathcal{H}}_{(R_1\Psi^{-1})}$ has m pointwise orthogonal maximizing vectors with respect to $\sigma_1 = 1$, and Ψ satisfies

$$I = ((R_1 - Q_{sup,m})\Psi^{-1})^{\sim}(R_1 - Q_{sup,m})\Psi^{-1} \quad .$$

Thus, when $R_2 = 0$ (the 1-block problem), we obtain

$$\Psi^{\sim}\Psi = U_m \begin{bmatrix} \gamma_{opt,1}^2 & & \\ & \ddots & \\ & & \gamma_{opt,m}^2 \end{bmatrix} U_m^{\sim}$$

where each column of U_m is the *direction* of the corresponding input maximizing vector.

Compared with standard $\mathbf{H}^\infty$ optimization, the algorithm not only finds a super-optimal solution, but also gives the corresponding maximizing vectors. This additional information may prove useful in multivariable robust controller design [17].

4.3 An Example

Let R_1 and R_2 be given by

$$R_1 = \left[\begin{array}{ccccc|cc} -0.1847 & 2.4347 & 4.6499 & -77.2055 & 52.3860 & 5.8600 & 9.8141 \\ -1.3259 & -0.2545 & 1.3299 & 7.9005 & 121.5585 & 9.3099 & -8.4518 \\ 0.3755 & 0.0594 & -0.8023 & 4.8947 & -26.8657 & -2.0831 & 0.7777 \\ 0.1402 & -0.1705 & -0.3555 & 5.6489 & 0.3384 & -0.1183 & -1.1213 \\ -0.0604 & 0.0218 & 0.4365 & -1.1862 & 5.3158 & 0.5237 & -0.1675 \\ \hline -0.0177 & -0.2099 & -3.5047 & 0.4207 & -13.4435 & 0 & 0 \\ 0.5191 & -0.2313 & -1.2999 & 7.1717 & -7.4449 & 0 & 0 \end{array}\right]$$

and

$$R_2 = \left[\begin{array}{ccccc|cc} -0.2500 & 0.1981 & -0.0229 & 4.3233 & -8.4484 & -5.8538 & 2.3952 \\ 0.9500 & -0.0040 & 0.0841 & 4.1215 & -9.0595 & -5.1656 & 1.8864 \\ 0.0307 & 0.1954 & -0.3218 & -3.0720 & -10.2447 & -3.1463 & -5.1514 \\ 0.0000 & -0.7242 & 0.7505 & -3.4873 & -0.0523 & 0.8005 & -1.5483 \\ 0.0000 & 0.6171 & 0.8808 & -0.3513 & -5.6601 & -1.4339 & -0.7192 \\ \hline 0.0000 & 0.0000 & -0.0000 & -0.3764 & 1.5890 & 1 & 0 \\ 0.0000 & -0.0000 & -0.0000 & 0.7427 & 0.8052 & 0 & 1 \end{array}\right]$$

Note that this example was derived from the one given in [16]. (See [25] for details.)

We summarize the procedure of the super-optimal solution for this example.

Step 1: Factorize $\Phi_1^{\sim}\Phi_1 = \gamma^2 I - R_2^{\sim}R_2$ and solve the 1-block super-optimal problem $(R_1 - Q)\Phi_1^{-1}$ until $\sigma_1\left([R_1\Phi_1^{-1}]_-\right) = 1$ by γ_1-iteration.

Step 2: Construct an input maximizing vector (or frame) ξ_1 of $\check{\mathcal{H}}_{(R_1\Phi_1^{-1})}$.

Step 3: While $rank(\xi_1) < 2$

1. Perform inner-outer factorization: $\xi_1 = \xi_i\ \xi_o$ and construct $\xi_\perp$.
2. Do inner-outer factorization: $\Phi_1^{-1}\xi_i = u_i\ u_o$ and construct $U_\perp$.

3. Factorize $\Psi_2^{\sim}\Psi_2 = I - (\gamma_{opt,1}^2 - \gamma_2^2)\ \xi_\perp(\xi_\perp^{\sim}\Phi_1 U_\perp)^{-1\sim}(\xi_\perp^{\sim}\Phi_1 U_\perp)^{-1}\xi_\perp^{\sim}$ and solve the 1-block problem $(R_1 - Q)\Phi_1^{-1}\Psi_2^{-1}$ until $\sigma_i\left([R_1\Phi_1^{-1}\Psi_2^{-1}]_-\right) = 1, i = 1, 2$ by γ_2-iteration.

Step 4: Compute $Q_{sup,2}$ where $\bar{s}_1^{\infty} = \gamma_{opt,1}$, $\bar{s}_2^{\infty} = \gamma_{opt,2}$, then stop.

By γ_1-iteration, Step 1, we obtain $\gamma_{opt,1} = 1.8068$ and hence the corresponding Φ_1 where the Hankel singular values of $[R_1\Phi_1^{-1}]_-$ are given by

$$[1.0000,\ 0.5494,\ 0.2598,\ 0.1146,\ 0.0396]\ .$$

The inner matrix $[\xi_i,\ \ \xi_\perp]$ in Step 3 is constructed as

$$[\xi_i\ \ \xi_\perp] = \left[\begin{array}{cccc|cc} -1.3463 & -0.0156 & 2.7261 & 0.0656 & 0.1693 & 11.2045 \\ -0.0971 & -1.6972 & 0.8286 & -1.8150 & -0.9510 & 133.6896 \\ -2.7474 & -0.1055 & -4.3380 & 0.3246 & 0.2499 & 24.2168 \\ 0.0975 & 1.7135 & -0.0192 & -2.2037 & 0.4018 & -74.4210 \\ \hline 0.5766 & -0.0649 & -0.3172 & -0.0366 & -0.9278 & 0.3731 \\ 1.0503 & -0.1170 & -0.8822 & -0.1011 & 0.3731 & 0.9278 \end{array}\right]$$

and Ψ_2 has degree 4 as predicted in Lemma 12.

By γ_2-iteration, we obtain $\gamma_{opt,2} = 1.1505$ where the Hankel singular values of $[R_1\Phi_1^{-1}\Psi_2^{-1}]_-$ are

$$[1.0000,\ 1.0000,\ 0.2619,\ 0.2390,\ 0.0419]\ .$$

A super-optimal solution $Q_{sup,2}$ can then be found using (107).

5 Conclusions

In this chapter, we have presented a state-space approach for finding $\mathbf{H}^\infty$ super-optimal solutions *i.e.* solutions to strengthened 1 and 2-block model-matching problems. Super-optimal control, in which all singular values of a given cost function are minimized, is attractive from a theoretical viewpoint because the solution is unique. In practice, the computations required are demanding and the resulting controllers are generally complex. For these reasons, it is hard to imagine practising control engineers using these tools in the near future. Nevertheless, if progress is to be made in robust control and the benefits of advanced control fully realized it is important that implementable algorithms are provided for any new theoretical developments. The contribution of this chapter is in providing state-space algorithms for solving super-optimal $\mathbf{H}^\infty$ control problems.

Acknowledgment

The authors would like to thank Professor N. J. Young for helpful comments on the proof of convergence of the algorithm in 2-block case, and the UK Science and Engineering Research Council for financial support.

References

[1] B.D.O. Anderson, "An algebraic solution to the spectral factorization problem," *IEEE Trans. Auto. Control*, vol. AC-12, pp. 410-414, 1967.

[2] B.D.O. Anderson and J.B. Moore, *Linear Optimal Control*, Prentice-Hall, Englewood Cliffs, New Jersey, 1971.

[3] C.C. Chu, J.C. Doyle and E.B. Lee, "The general distance problem in H^∞-optimal control theory," *Int. J. Control*, vol. 44, pp. 565-596, 1986.

[4] J.C. Doyle and G. Stein, "Multivariable feedback design: concepts for a classical/modern synthesis," *IEEE Trans. Auto. Control*, vol. 26, pp. 4-16, 1981.

[5] J.C. Doyle, "Lecture Notes in Advances in Multivariable Control," *ONR/Honeywell Workshop*, Minneapolis, 1984.

[6] Y.K. Foo and I. Postlethwaite, "All solutions, all-pass form solutions, and the 'best' solutions to an H^∞ optimization problem in robust control," *Systems and Control Letters*, vol. 7, pp. 261-268, 1985.

[7] B.A. Francis, *A Course in H^∞ Control Theory*, Lecture notes in control and information sciences, vol. 88, Springer Verlag, 1987.

[8] J.S. Freudenberg, D.P. Looze and J.B. Cruz, "Robustness analysis using singular value sensitivities," *Int. J. Control*, vol. 35, pp. 95-116, 1982.

[9] K. Glover, "All optimal Hankel-norm approximations of linear multivariable systems and their L^∞-error bounds," *Int. J. Control*, vol. 39, pp. 1115-1193, 1984.

[10] K. Glover and D. McFarlane, "Robust Stabilization of normalized coprime factor plant descriptions with H_∞-bounded uncertainty," *IEEE Trans. Auto. Control*, vol. 34, pp. 821-830, 1989.

[11] D.-W. Gu, M.C. Tsai and I. Postlethwaite, "An algorithm for super-optimal H^∞ design : the 2-block case," *Automatica*, vol. 25, pp. 215-221, 1989.

[12] D.-W. Gu, M.C. Tsai and I. Postlethwaite, "Improved formulated for the 2-block H^∞ super-optimal solution," *Automatica*, vol. 26, pp. 437-440, 1990.

[13] D.-W. Gu, M.C. Tsai and I. Postlethwaite, "A frame approach to the H^∞ super-optimal solution," *Proc. IEEE CDC, Tampa, Florida,* 1989.

[14] J.W. Helton, "Worst case analysis in the frequency domain: an H^∞ approach to control," *IEEE Trans. Auto. Control*, vol. 30, pp. 1154-1170, 1985.

[15] I.M. Jaimoukha, Ph.D. Thesis, Imperial College, London, 1990.

[16] H. Kwakernaak, "A polynomial approach to minimax frequency domain optimization of multivariable systems," *Int. J. Control*, vol. 44, pp. 117-156, 1986.

[17] N.A. Lehtomaki, D.A. Castanon, B.C. Levy, G. Stein, N.R. Sandell Jr. and M. Athans, "Robustness and modelling error characterization," *IEEE Trans. Auto. Control*, vol. 29, pp. 212-220, 1984.

[18] D.J.N. Limebeer, G.D. Halikias and K. Glover, "A state-space algorithm for the computation of superoptimal matrix interpolating functions," *Proc. MTNS Conference on Linear Circuits, Systems and Signal Processing: Theory and Applications*, North-Holland Amsterdam, 1988; *Int. J. Control*, vol. 50, pp. 2431-2466, 1989.

[19] B.C. Moore, "Principal component analysis in linear systems: controllability, observability, and model reduction," *IEEE Trans. Auto. Control*, vol. AC-26, pp. 17-32, 1981.

[20] I. Postlethwaite, M.C. Tsai and D.-W. Gu, "A state-space approach to discrete-time super-optimal H^∞ control problems," *Int. J. Control*, vol. 49, pp. 247-269, 1989.

[21] I. Postlethwaite, M.C. Tsai and D.-W. Gu, "Super-optimal H^∞ design," International Symposium on the Mathematical Theory of Networks and Systems, North-Holland, Amsterdam, 1989.

[22] L.M. Silverman and M. Bettayeb, "Optimal approximation of linear systems," *Proc. of the American Control Conference*, San Francisco, 1980.

[23] M.C. Tsai, D.-W. Gu, and I. Postlethwaite "A state-space approach to super-optimal H^∞ control problems," *IEEE Trans. Auto. Control*, vol. 33, pp. 833-843, 1988.

[24] M.C. Tsai, D.-W. Gu, I. Postlethwaite and B.D.O. Anderson, "A note on inner functions and a pseudo singular value decomposition in super-optimal H^∞ control*Int. J. Control*,," vol. 51, pp. 1119-1131, 1990.

[25] M.C. Tsai, "Super-optimal control system design for multivariable plants," D. Phil. Thesis, Oxford 1989.

[26] M.C. Tsai, D.-W. Gu and I. Postlethwaite, "On 2-Block Super-Optimal Solutions,," *Technical Report 90-5*, Engineering Department, Leicester University.

[27] M.S. Verma and E.A. Jonckheere, "L^∞ compensator with mixed-sensitivity as a broadband matching problem," *Systems and Control Letters*, vol. 4, pp. 125-130, 1984.

[28] F.B. Yeh, "Numerical solution of matrix interpolation problems," *PH. D. Thesis, Glasgow University*, 1983.

[29] N.J. Young, "The Nevanlinna-Pick problem for matrix-valued functions," *J. Operator Theory*, vol. 15-2, pp. 239-265, 1986.

[30] N.J. Young, "An algorithm for the super-optimal sensitivity minimising controller," *Proc. Workshop on New Perspectives in Industrial Control System Design using H^∞ Methods*, Oxford, 1986.

[31] N.J. Young, *An Introduction to Hilbert Space*, Cambridge University Press, 1988.

[32] A. Yue, "H^∞-design and the improvement of helicopter handling qualities," *D.Phil Thesis*, University of Oxford, 1989.

[33] G. Zames, "Feedback and optimal sensitivity: model reference transformations, multiplicative seminorms, and approximate inverses," *IEEE Trans. Auto. Control*, vol. 26, pp. 301-320, 1981.

Closed-Loop Transfer Recovery with Observer-Based Controllers

Part 1: Analysis

Ben M. Chen

Ali Saberi

School of Electrical Engineering and Computer Science
Washington State University
Pullman, Washington 99164-2752

Uy-Loi Ly

Department of Aeronautics and Astronautics, FS-10
University of Washington
Seattle, Washington 98195

I. INTRODUCTION

In feedback design many performance and robust stability objectives can be stated in the form of requirements placed on the maximum singular values of particular *closed-loop* transfer functions. The underlying idea of "loop shaping" is that the magnitude (or maximum singular value) of the *closed-loop* transfer function can be directly inferred from the singular value of a corresponding *open-loop* transfer function. The

CONTROL AND DYNAMIC SYSTEMS, VOL. 51

prominent design procedure under the terminology LQG/LTR [15] is one such design methodology in multivariable systems that is based on the concept of loop shaping. This design procedure is divided into two steps. The first step involves the design of a stabilizing state-feedback law that yields a loop transfer function satisfying the design specifications. The loop properties are usually described in relation to an *open-loop* system (e.g. for a loop transfer function broken at either the control or measurement paths). Such an open-loop transfer function defines the target loop shape. The second step is to match this target loop shape using an output-feedback design following a procedure called loop transfer recovery (LTR). This step involves the design of an output-feedback control law (typically an observer-based compensator) such that the resulting open-loop transfer function would have either exactly or approximately the same target loop shape as the one achieved under state feedback. In other words, the idea of LTR is to design a compensator to recover a specific open-loop transfer function.

In this paper we examine the idea of loop recovery from a different perspective. Namely, we develop the concept of recovery based on the closed-loop transfer function directly, as opposed to the open-loop transfer function found in the case of a traditional LTR design. The problem can be stated as follows. Suppose that one is able to synthesize a state-feedback law that yields satisfactory closed-loop performance. And let's define the closed-loop transfer function between the external input to the controlled output under state-feedback law to be the target closed-loop transfer function. Clearly from this definition, the closed-loop target transfer function is completely defined by the selection of a full-state feedback gain matrix. Now we would like to design an output-feedback control law with a closed-loop transfer function that matches either exactly or approximately the target closed-loop transfer function. In this respect, we are dealing with the problem of closed-loop transfer recovery (CLTR) instead of open-loop transfer recovery (LTR).

It should be pointed out that the procedure of CLTR can further be used as an effective tool in the design of multivariable control systems. For example, one can employ the procedure of CLTR in the synthesis of H_∞-norm-based control-laws. Namely, we start with a target closed-loop transfer function achieved in H_∞-optimization under state feedback. Then we proceed to the design of a compensator (with either a full-order, reduced-order, Luenberger or generalized observer-based structure) that recovers the desired target closed-loop transfer function.

Our study of the mechanism in CLTR is applicable to a general class of systems and aims at three important theoretical issues:

(a) characterization of the recovery error and the available freedom in the design of output-feedback control-laws for a given system and for an arbitrarily specified target closed-loop transfer function,

(b) development of necessary and/or sufficient conditions for a target closed-loop transfer function to be either exactly or asymptotically recoverable in a given system, and

(c) development of necessary and/or sufficient conditions on a given system such that it has at least one recoverable (either exactly or asymptotically) target closed-loop transfer function. These are some of the theoretical issues pertaining to the analysis of CLTR. Of course, one also needs to examine issues in CLTR that are related to systematic design algorithms for the recovery process.

This paper concerns mainly with the analysis of the CLTR mechanism. A sequel to this paper will address in details the design issues. The objective at hand is however to analyze methodically the mechanism of CLTR using an observer-based controller in its most general setting (i.e covering the cases of full-order, reduced-order and generalized observers). However, in order to limit the length of this paper, results are provided only for the full-order and reduced-order observer-based controllers. The basic methodology and tools used here are akin to those in [8], [3] and [4]. We would like to point out that the formulation of the controller structure can in many ways impact the recovery process. Identifying the appropriate controller structure for the recovery task remains a research topic for future investigation.

The paper is organized as follows. In section II, we define precisely the problem of closed-loop transfer recovery. Recognizing the importance of finite- and infinite-zero structure in the LTR problem, we recall in section III a special coordinate basis (s.c.b) of [12] and [10] that clearly displays the zero structure of a given system. Section IV deals with all the fundamental analyses of CLTR. In particular, subsection IV.A deals with the analysis of CLTR via full-order observer-based control-laws while in subsection IV.B we perform the same analysis for the case of reduced-order observer-based control-laws. In section V, we illustrate our results on a lateral autopilot design for a commercial transport airplane. Finally, we draw the conclusion of our work in section VI.

Throughout the paper, A' denotes the transpose of A, A^H denotes the conjugate transpose of A, I denotes an identity matrix while I_k denotes an identity matrix of dimension $k \times k$. $\lambda(A)$ and $\mathrm{Re}[\lambda(A)]$ denote respectively the set of eigenvalues and the set of real parts of eigenvalues of A. Similarly, $\sigma_{max}[A]$ and $\sigma_{min}[A]$ denote the maximum and minimum singular values of A respectively. $\mathrm{Ker}[V]$ and $\mathrm{Im}[V]$ denote respectively the kernel and the image of V. The open left-half and the closed right-half of the s-plane are denoted respectively by $\mathcal{C}^-$ and $\mathcal{C}^+$. Also, $\mathcal{R}_p$ denotes the subring of all proper rational functions of s while the set of matrices of dimension $\ell \times q$ whose elements belong to $\mathcal{R}_p$ is denoted by $\mathcal{M}^{\ell \times q}(\mathcal{R}_p)$.

II. PROBLEM STATEMENT

Let us consider a linear time-invariant system Σ,

$$\Sigma : \begin{cases} \dot{x} = A\,x + B_1\,w + B_2\,u, \\ z = C_1 x + D_{11} w + D_{12} u, \\ y = C_2 x + D_{21} w + D_{22} u, \end{cases} \tag{1}$$

where $x \in \Re^n$ is the state, $u \in \Re^m$ is the control input, $w \in \Re^k$ is the external signal or disturbance, $z \in \Re^\ell$ is the controlled output and $y \in \Re^p$ is the measurement output. For convenience, we also define Σ_{yw} to be the matrix quadruple (A, B_1, C_2, D_{21}) and Σ_{zu} for the matrix quadruple (A, B_2, C_1, D_{12}). Let us assume that the pair $(A,\ B_2)$ is stabilizable and the pair $(A,\ C_2)$ detectable. Without loss of generality, we also assume that $[C_1,\ D_{11},\ D_{12}]$, $[C_2,\ D_{21},\ D_{22}]$, $[B_1',\ D_{11}',\ D_{21}']'$ and $[B_2',\ D_{12}',\ D_{22}']'$ are of maximal ranks. Let F be a full-state feedback gain matrix such that under the state-feedback control

$$u = -Fx \tag{2}$$

(a) the closed-loop system is asymptotically stable, i.e. the eigenvalues of $A - B_2F$ lie in the left-half s-plane,

(b) the closed-loop transfer function from the disturbance w to the controlled output z, denoted by $T_{zw}(s)$, meets the given frequency dependent design specifications.

We also refer to $T_{zw}(s)$ as the target closed-loop transfer function given by

$$T_{zw}(s) = (C_1 - D_{12}F)(\Phi^{-1} + B_2F)^{-1}B_1 + D_{11} \tag{3}$$

where $\Phi = (sI_n - A)^{-1}$. Design of the appropriate full-state feedback gain matrix F can be done, for example, via H_2-, H_∞-theory or eigenstructure assignment. For design implementation, the next step in the design procedure is to recover the target closed-loop transfer function using only a measurement feedback controller. This is the problem of closed-loop transfer recovery (CLTR) and the focus of this paper.

The problem can be clearly stated using the configuration shown in figure 1 where $P(s)$ represents the transfer function matrix of the system Σ and $\mathbf{C}(s)$ the transfer function of an output-feedback controller. For a given $P(s)$ and a target closed-loop transfer function $T_{zw}(s)$ in (3), the problem is to find an internally stabilizing controller $\mathbf{C}(s)$ such that the recovery error defined as

$$E(s) := T_{zw}^o(s) - T_{zw}(s) \tag{4}$$

is either exactly or approximately equal to zero in the frequency region of interest. Here, $T_{zw}^o(s)$ represents the transfer function from w to z for the closed-loop system shown in figure 1. As we shall see, achieving

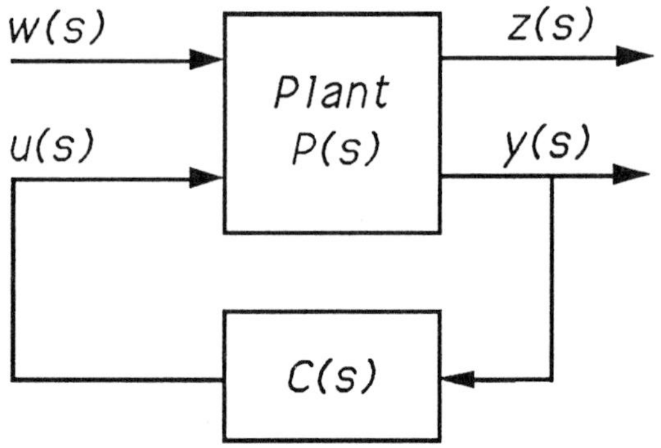

Figure 1: Plant with an Output-Feedback Controller.

exact closed-loop transfer recovery (ECLTR) is in general not possible. Hence, it is more appropriate to examine situation where approximate recovery can be achieved. An approximate CLTR is tied to the notion that recovery can be achieved to any degree of accuracy. In this process, one normally parameterizes the controller $\mathbf{C}(s)$ as a function of a positive scalar parameter σ thereby generating a family of controllers $\mathbf{C}(s,\sigma)$. We say asymptotic CLTR (ACLTR) is achieved if

$$T_{zw}^{o}(s,\sigma) \to T_{zw}(s)$$

as $\sigma \to \infty$ pointwise in s, or equivalently

$$E(s,\sigma) \to 0$$

as $\sigma \to \infty$ pointwise in s. From the point of view of design, once the conditions of ACLTR have been verified, a controller $\mathbf{C}(s,\sigma)$ with a particular value of σ can be found that will produce the desired level of recovery. Before we proceed to the analysis of CLTR, we need to provide precise meanings to the terminologies ECLTR and ACLTR. The following are definitions that characterize precisely the notions of ECLTR and ACLTR.

Definition 1. *The set of admissible target closed-loop transfer functions $\mathbf{T}(\Sigma)$ for the plant Σ is defined by*

$$\mathbf{T}(\Sigma) = \{T_{zw}(s) \in \mathcal{M}^{l\times k}(\mathcal{R}_p) \,|\, T_{zw}(s) \text{ is as defined in (3) and } \lambda(A - B_2F) \in \mathcal{C}^-\}.$$

Definition 2. *$T_{zw}(s) \in \mathbf{T}(\Sigma)$ is said to be exactly recoverable (ECLTR) if there exists a $\mathbf{C}(s) \in \mathcal{M}^{m\times p}(\mathcal{R}_p)$ such that*

(i) the closed-loop system comprising of $\mathbf{C}(s)$ and Σ as in (1) is asymptotically stable,

(ii) $T^o_{zw}(s) = T_{zw}(s)$.

Definition 3. *$T_{zw}(s) \in \mathbf{T}(\Sigma)$ is said to be asymptotically recoverable (ACLTR) if there exists a parameterized family of controllers $\mathbf{C}(s,\sigma) \in \mathcal{M}^{m\times p}(\mathcal{R}_p)$ where σ is a scalar parameter with positive values such that*

(i) the closed-loop system comprising of $\mathbf{C}(s,\sigma)$ and Σ as in (1) is asymptotically stable for all $\sigma > \sigma^*$ where $0 \leq \sigma^* < \infty$,

(ii) $T^o_{zw}(s,\sigma) \to T_{zw}(s)$ pointwise in s as $\sigma \to \infty$. Moreover, in the limits as $\sigma \to \infty$ the finite eigenvalues of the closed-loop system should remain in $\mathcal{C}^-$.[1]

Definition 4. *$T_{zw}(s)$ belonging to $\mathbf{T}(\Sigma)$ is said to be recoverable if $T_{zw}(s)$ is either exactly or asymptotically recoverable.*

Definition 5.

1. *The set of exactly recoverable target closed-loop transfer functions for the system Σ is denoted by $\mathbf{T}_{\mathrm{ER}}(\Sigma)$.*

2. *The set of recoverable (either exactly or asymptotically) target closed-loop transfer functions for the system Σ is denoted by $\mathbf{T}_{\mathrm{R}}(\Sigma)$.*

3. *The set of target closed-loop transfer functions which are only asymptotically recoverable but not exactly recoverable for the system Σ is denoted by $\mathbf{T}_{\mathrm{AR}}(\Sigma)$.*

Obviously, $\mathbf{T}_{\mathrm{R}}(\Sigma) = \mathbf{T}_{\mathrm{ER}}(\Sigma) \cup \mathbf{T}_{\mathrm{AR}}(\Sigma)$.

Remark 1. *The control-law $\mathbf{C}(s)$ in the above definitions is not restricted to any particular structure. However, in this paper we study*

[1] Here we have strengthened the notion of closed-loop stability by excluding those cases where, in the limits as $\sigma \to \infty$, some finite eigenvalues of the closed-loop system would be on the $j\omega$ axis. This avoids the problem of having an almost unstable behavior of the closed-loop system for large σ.

the closed-loop transfer recovery for two specific structures of $\mathbf{C}(s)$*; namely, full-order and reduced-order observer-based controllers. Furthermore, we label*

$$\{\mathbf{T}_{\mathrm{R}}(\Sigma),\ \mathbf{T}_{\mathrm{ER}}(\Sigma),\ \mathbf{T}_{\mathrm{AR}}(\Sigma)\}$$

with superscript f *(for full-order) and* r *(for reduced-order) as*

$$\{\mathbf{T}^f_{\mathrm{R}}(\Sigma),\ \mathbf{T}^f_{\mathrm{ER}}(\Sigma),\ \mathbf{T}^f_{\mathrm{AR}}(\Sigma)\}$$

and

$$\{\mathbf{T}^r_{\mathrm{R}}(\Sigma),\ \mathbf{T}^r_{\mathrm{ER}}(\Sigma),\ \mathbf{T}^r_{\mathrm{AR}}(\Sigma)\}$$

to signify results related to these particular controller structures.

The analysis of CLTR mechanism carried out here examines three fundamental issues. The first issue concerns with what can and what cannot be achieved for a given system and for an arbitrarily specified target closed-loop transfer function. For a given system, the second issue is to establish necessary and/or sufficient conditions on the target closed-loop transfer function so that it can be either exactly or asymptotically recovered. In another word, we characterize completely the set $\mathbf{T}_{\mathrm{R}}(\Sigma)$ of recoverable target closed-loop transfer functions. The third issue is to establish necessary and/or sufficient conditions on a given system such that it has at least one recoverable (either exactly or asymptotically) target closed-loop transfer function. That is, what are the conditions on a given system Σ so that the set of recoverable target closed-loop transfer $\mathbf{T}_{\mathrm{R}}(\Sigma)$ is nonempty?

III. PRELIMINARIES

We recall in this section a special coordinate basis (s.c.b) of a linear time-invariant system [12], [10]. Such a s.c.b has a distinct feature of explicitly displaying the finite and infinite zero structure of a given system and will play a very important role in both the analysis and the design of closed-loop transfer recovery. Consider the system characterized by

$$\tilde{\Sigma} : \begin{cases} \dot{\tilde{x}} = A\tilde{x} + B\tilde{u}, \\ \tilde{y} = C\tilde{x} + D\tilde{u}, \end{cases} \tag{5}$$

where $\tilde{x} \in \Re^n$, $\tilde{u} \in \Re^m$ and $\tilde{y} \in \Re^p$. Without loss of generality, we assume that matrices $[C,\ D]$ and $[B',\ D']'$ are of maximal rank. It is simple to verify that there exist non-singular transformations U and V such that

$$UDV = \begin{bmatrix} I_{m_0} & 0 \\ 0 & 0 \end{bmatrix}, \tag{6}$$

where m_0 is the rank of matrix D. Hence, hereafter and without loss of generality, it is assumed that matrix D has the form given on the right-hand side of (6). One can now rewrite the system of (5) as

$$\begin{cases} \dot{\tilde{x}} = A\tilde{x} + \begin{bmatrix} B_0 & B_1 \end{bmatrix} \begin{bmatrix} \tilde{u}_0 \\ \tilde{u}_1 \end{bmatrix}, \\ \begin{bmatrix} \tilde{y}_0 \\ \tilde{y}_1 \end{bmatrix} = \begin{bmatrix} C_0 \\ C_1 \end{bmatrix} \tilde{x} + \begin{bmatrix} I_{m_0} & 0 \\ 0 & 0 \end{bmatrix} \begin{bmatrix} \tilde{u}_0 \\ \tilde{u}_1 \end{bmatrix}, \end{cases} \tag{7}$$

where the matrices B_0, B_1, C_0 and C_1 have appropriate dimensions. We have the following theorem.

Theorem 1 (s.c.b). *Consider the system $\tilde{\Sigma}$ characterized by $(A,\ B,\ C,\ D)$. There exist nonsingular transformations Γ_1, Γ_2 and Γ_3, an integer $m_f \le m - m_0$ and integer indexes q_i, $i = 1$ to m_f, such that*

$$\tilde{x} = \Gamma_1 x \ , \quad \tilde{y} = \Gamma_2 y \ , \quad \tilde{u} = \Gamma_3 u$$

$$x = [x_a',\ x_b',\ x_c',\ x_f']' \ , \quad x_a = [(x_a^-)', (x_a^+)']'$$

$$x_f = [x_{f_1}',\ x_{f_2}',\ \cdots,\ x_{f_{m_f}}']'$$

$$y = [y_0',\ y_f',\ y_b']' \ , \quad y_f = [y_1,\ y_2,\ \cdots,\ y_{m_f}]'$$

$$u = [u_0',\ u_f',\ u_c']' \ , \quad u_f = [u_1,\ u_2,\ \cdots,\ u_{m_f}]',$$

we have the following system of equations:

$$\dot{x}_a^- = A_{aa}^- x_a^- + B_{a0}^- y_0 + L_{af}^- y_f + L_{ab}^- y_b \tag{8}$$

$$\dot{x}_a^+ = A_{aa}^+ x_a^+ + B_{a0}^+ y_0 + L_{af}^+ y_f + L_{ab}^+ y_b \tag{9}$$

$$\dot{x}_b = A_{bb}x_b + B_{b0}y_0 + L_{bf}y_f \quad , \quad y_b = C_b x_b \tag{10}$$

$$\dot{x}_c = A_{cc}x_c + B_{c0}y_0 + L_{cb}y_b + L_{cf}y_f + B_c[E_{ca}^- x_a^- + E_{ca}^+ x_a^+] + B_c u_c \tag{11}$$

$$y_0 = C_{0a}^- x_a^- + C_{0a}^+ x_a^+ + C_{0b}x_b + C_{0c}x_c + C_{0f}x_f + u_0 \tag{12}$$

and for each $i = 1$ to m_f,

$$\dot{x}_{f_i} = A_{f_i}x_{f_i} + L_{i0}y_0 + L_{if}y_f + B_{f_i}\,[u_i + E_{ia}x_a + E_{ib}x_b + E_{ic}x_c + \sum_{j=1}^{m_f} E_{ij}x_{f_j}] \tag{13}$$

$$y_i = C_{f_i}x_{f_i} \quad , \quad y_f = C_f x_f . \tag{14}$$

Here the states x_a^-, x_a^+, x_b, x_c and x_f are respectively of dimension n_a^-, n_a^+, n_b, n_c and

$$n_f = \sum_{i=1}^{m_f} q_i$$

while x_i is of dimension q_i for each $i = 1$ to m_f. The control vectors u_0, u_f and u_c are respectively of dimension m_0, m_f and $m_c = m - m_0 - m_f$ while the output vectors y_0, y_f and y_b are respectively of dimension $p_0 = m_0$, $p_f = m_f$ and $p_b = p - p_0 - p_f$. The matrices A_{f_i}, B_{f_i} and C_{f_i} have the following form:

$$A_{f_i} = \begin{bmatrix} 0 & I_{q_i - 1} \\ 0 & 0 \end{bmatrix}, \quad B_{f_i} = \begin{bmatrix} 0 \\ 1 \end{bmatrix}, \quad C_{f_i} = [1,\ 0,\ \cdots,\ 0]. \tag{15}$$

(Obviously for the case when $q_i = 1$, $A_{f_i} = 0$, $B_{f_i} = 1$ and $C_{f_i} = 1$.) Furthermore, we have $\lambda(A_{aa}^-) \in \mathbb{C}^-$, $\lambda(A_{aa}^+) \in \mathbb{C}^+$, the pair (A_{cc}, B_c) is controllable and the pair (A_{bb}, C_b) is observable. Also, assuming that x_i are arranged such that $q_i \leq q_{i+1}$, the matrix L_{if} has the particular form,

$$L_{if} = [L_{i1}, L_{i2}, \cdots, L_{i\,i-1}, 0, 0, \cdots, 0].$$

Also, the last row of each L_{if} is identically zero.

Proof : This follows from theorem 2.1 of [12] and [10]. ∎

We can rewrite the s.c.b given by theorem 1 in a more compact form,

$$\Gamma_1^{-1}(A - B_0C_0)\Gamma_1 = \begin{bmatrix} A_{aa}^- & 0 & L_{ab}^-C_b & 0 & L_{af}^-C_f \\ 0 & A_{aa}^+ & L_{ab}^+C_b & 0 & L_{af}^+C_f \\ 0 & 0 & A_{bb} & 0 & L_{bf}C_f \\ B_cE_{ca}^- & B_cE_{ca}^+ & L_{cb}C_b & A_{cc} & L_{cf}C_f \\ B_fE_a^- & B_fE_a^+ & B_fE_b & B_fE_c & A_f \end{bmatrix},$$

$$\Gamma_1^{-1}\left[B_0 \quad B_1 \right]\Gamma_3 = \begin{bmatrix} B_{a0}^- & 0 & 0 \\ B_{a0}^+ & 0 & 0 \\ B_{b0} & 0 & 0 \\ B_{c0} & 0 & B_c \\ B_{f0} & B_f & 0 \end{bmatrix},$$

$$\Gamma_2^{-1}\begin{bmatrix} C_0 \\ \\ C_1 \end{bmatrix}\Gamma_1 = \begin{bmatrix} C_{0a}^- & C_{0a}^+ & C_{0b} & C_{0c} & C_{0f} \\ 0 & 0 & 0 & 0 & C_f \\ 0 & 0 & C_b & 0 & 0 \end{bmatrix}$$

and

$$\Gamma_2^{-1}D\Gamma_3 = \begin{bmatrix} I_{m_0} & 0 & 0 \\ 0 & 0 & 0 \\ 0 & 0 & 0 \end{bmatrix}.$$

In what follows, we state some important properties of the s.c.b which are pertinent to our present work.

Property 1. *The given system $\tilde{\Sigma}$ is right-invertible if and only if x_b and hence y_b are nonexistent ($n_b = 0$, $p_b = 0$), left-invertible if and only if x_c and hence u_c are nonexistent ($n_c = 0$, $m_c = 0$), invertible if and only if both x_b and x_c are nonexistent. Moreover, $\tilde{\Sigma}$ is degenerate if and only if it is neither left- nor right-invertible.*

Property 2. *We note that (A_{bb}, C_b) and (A_{f_i}, C_{f_i}) form observable pairs. Unobservability could arise only in the variables x_a and x_c. In fact, the system $\tilde{\Sigma}$ is observable (detectable) if and only if (A_{obs}, C_{obs}) is an observable (detectable) pair, where*

$$A_{obs} = \begin{bmatrix} A_{aa} & 0 \\ B_cE_{ca} & A_{cc} \end{bmatrix}, \quad C_{obs} = \begin{bmatrix} C_{a0} & C_{c0} \\ E_a & E_c \end{bmatrix},$$

$$A_{aa} = \begin{bmatrix} A_{aa}^- & 0 \\ 0 & A_{aa}^+ \end{bmatrix},$$

$$C_{a0} = [C_{a0}^-, C_{a0}^+], \quad E_a = [E_a^-, E_a^+], \quad E_{ca} = [E_{ca}^-, E_{ca}^+].$$

Similarly, (A_{cc}, B_c) and (A_{f_i}, B_{f_i}) form controllable pairs. Uncontrollability could arise only in the variables x_a and x_b. In fact, $\tilde{\Sigma}$ is controllable (stabilizable) if and only if (A_{con}, B_{con}) is a controllable (stabilizable) pair, where

$$A_{con} = \begin{bmatrix} A_{aa} & L_{ab}C_b \\ 0 & A_{bb} \end{bmatrix}, \quad B_{con} = \begin{bmatrix} B_{a0} & L_{af} \\ B_{b0} & L_{bf} \end{bmatrix},$$

$$B_{a0} = \begin{bmatrix} B_{a0}^- \\ B_{a0}^+ \end{bmatrix}, \quad L_{ab} = \begin{bmatrix} L_{ab}^- \\ L_{ab}^+ \end{bmatrix}, \quad L_{af} = \begin{bmatrix} L_{af}^- \\ L_{af}^+ \end{bmatrix}.$$

Property 3. *Invariant zeros of $\tilde{\Sigma}$ are the eigenvalues of A_{aa}. Moreover, the stable and the unstable invariant zeros of $\tilde{\Sigma}$ are the eigenvalues of A_{aa}^- and A_{aa}^+, respectively.*

There are interconnections between the s.c.b and various invariant and almost invariant geometric subspaces. To show these interconnections, we define

- $\mathcal{V}^g(A, B, C, D)$ — the maximal subspace of $\Re^n$ which is $(A + BF)$–invariant and contained in $\mathrm{Ker}(C + DF)$ such that the eigenvalues of $(A + BF)|\mathcal{V}^g$ are contained in $\mathcal{C}_g \subseteq \mathcal{C}$ for some F.

- $\mathcal{S}^g(A, B, C, D)$ — the minimal $(A + KC)$–invariant subspace of $\Re^n$ containing in $\mathrm{Im}(B + KD)$ such that the eigenvalues of the map which is induced by $(A + KC)$ on the factor space $\Re^n/\mathcal{S}^g$ are contained in $\mathcal{C}_g \subseteq \mathcal{C}$ for some K.

For the cases that $\mathcal{C}_g = \mathcal{C}$, $\mathcal{C}_g = \mathcal{C}^-$ and $\mathcal{C}_g = \mathcal{C}^+$, we replace the index g in $\mathcal{V}^g$ and $\mathcal{S}^g$ by '$*$', '$-$' and '$+$', respectively. Various components of the state vector of s.c.b have the following geometrical interpretations.

Property 4.

1. *$x_a^- \oplus x_a^+ \oplus x_c$ spans $\mathcal{V}^*(A, B, C, D)$.*
2. *$x_a^- \oplus x_c$ spans $\mathcal{V}^-(A, B, C, D)$.*
3. *$x_a^+ \oplus x_c$ spans $\mathcal{V}^+(A, B, C, D)$.*

4. $x_c \oplus x_f$ *spans* $\mathcal{S}^*(A, B, C, D)$.
5. $x_a^- \oplus x_c \oplus x_f$ *spans* $\mathcal{S}^+(A, B, C, D)$.
6. $x_a^+ \oplus x_c \oplus x_f$ *spans* $\mathcal{S}^-(A, B, C, D)$.

IV. GENERAL ANALYSIS

In this section, we deal with the general analysis of closed-loop transfer recovery in the cases of full- and reduced-order observer-based controllers. We will study three fundamental issues in closed-loop transfer recovery. The first issue is concerned with what can and what cannot be achieved for a given system and for an arbitrarily target closed-loop transfer function. The second issue deals with the development of necessary and/or sufficient conditions a target closed-loop transfer function must satisfy in order for it to be recovered either exactly or asymptotically. And the third issue is concerned with the development of necessary and/or sufficient conditions on a given system such that it has at least one (either exactly or asymptotically) recoverable target closed-loop transfer function.

To begin, let us consider Luenberger observer-based controllers which include as special cases the full-order and reduced-order observer-based controllers. Lemma 1 in this section gives an explicit expression for the recovery error function, between the target closed-loop transfer function and the one realized by a Luenberger observer-based controller. Lemma 2 expresses CLTR in terms of a transfer function matrix denoted here as the recovery matrix. This formulation plays a central role in the development of our results. Without loss of generality, we assume that $D_{22} = 0$ with justification given in Appendix A. Let us now consider the following Luenberger observer-based controller,

$$\begin{cases} \dot{v} = Lv + G_1 y + G_2 u, \\ \hat{x} = Pv + Jy, \\ u = -F\hat{x} \end{cases} \tag{16}$$

where $v \in \Re^r$ with r being the order of the controller and $\hat{x} \in \Re^n$. It is well known that, in the disturbance-free case (i.e. $w = 0$) the variable

$\hat{x}$ is an asymptotic estimate of the state x provided that the matrix L is a stability matrix and there exists a matrix $Q \in \Re^{r \times n}$ satisfying the following conditions:

$$QA - LQ = G_1 C_2, \qquad G_2 = QB_2, \qquad JC_2 + PQ = I_n. \tag{17}$$

Let $T^{\ell}_{zw}(s)$ denote the closed-loop transfer function from w to z with a general Luenberger observer-based controller. Then we have the following lemmas.

Lemma 1. *With the observer defined in (16) and (17), the recovery error function between the target closed-loop transfer $T_{zw}(s)$ and the one realized by Luenberger observer-based controller $T^{\ell}_{zw}(s)$ is given by*

$$E_\ell(s) = T^{\ell}_{zw}(s) - T_{zw}(s) = T_{zu}(s) \cdot M_\ell(s) \tag{18}$$

where

$$T_{zu}(s) = (C_1 - D_{12}F)(\Phi^{-1} + B_2 F)^{-1} B_2 + D_{12}, \tag{19}$$

is the closed-loop transfer function from u to z under state feedback and

$$M_\ell(s) = F[P(sI - L)^{-1}(QB_1 - G_1 D_{21}) - J D_{21}]. \tag{20}$$

Proof : The result follows directly after some simple algebra. ∎

Lemma 2. *Given that the system Σ_{zu} is left-invertible. Then*

1. *Exact recovery takes place (i.e. $E_\ell(j\omega) = 0 \ \forall \omega \in \Re$) if and only if $M_\ell(j\omega) = 0 \ \forall \omega \in \Re$.*

2. *Asymptotic recovery is achievable (i.e. $||E_\ell(j\omega)||$ can be made arbitrarily small for all $\omega \in \Re$) if and only if $||M_\ell(j\omega)||$ can be made arbitrarily small for all $\omega \in \Re$.*

Proof : It is obvious. ∎

Note that the conditions given in lemma 2 are not necessary for $E_\ell(s)$ to be zero or small if the system Σ_{zu} is not left-invertible. Nonetheless, they remain as sufficient conditions for the recovery error to be zero or small. To see this, let us examine the following example.

Example 1 : Consider a system characterized by

$$\dot{x} = \begin{bmatrix} -1 & 1 \\ 1 & -1 \end{bmatrix} x + \begin{bmatrix} 0 & 1 \\ 1 & 0 \end{bmatrix} u + \begin{bmatrix} 0 \\ 1 \end{bmatrix} w,$$

$$y = [\,0 \quad 1\,]\, x,$$

$$z = [\,0 \quad 1\,]\, x.$$

Let the target closed-loop transfer function be specified by

$$F = \begin{bmatrix} 1 & 0 \\ 0 & 1 \end{bmatrix},$$

and let

$$L = \begin{bmatrix} -1 & 0 \\ 1 & -1 \end{bmatrix}, \quad G_1 = \begin{bmatrix} 1 \\ 0 \end{bmatrix}, \quad G_2 = B_2, \quad P = Q = I_2 \text{ and } J = 0.$$

Then it is simple to verify that

$$T_{zu}(s) = [\,\tfrac{1}{s+1} \quad 0\,], \quad M_\ell(s) = \begin{bmatrix} 0 \\ \frac{1}{s+1} \end{bmatrix}.$$

and $E_\ell(s) = T_{zu}(s)M_\ell(s) = 0$. Hence exact recovery can take place even though $M_\ell(s) \neq 0$.

As seen from the above example when the system Σ_{zu} is not left-invertible, one may be able to find an observer-based controller such that $M_\ell(s)$ is nonzero and the recovery error $E_\ell(s)$ is equal to zero. However, this situation is fairly limited and may not generally be valid for other non-left-invertible systems. Thus, a general analysis in closed-loop transfer recovery entails a detailed study of the matrix $M_\ell(s)$ which depends on both the state-feedback matrix F and the observer parameters. Since the state-feedback gain matrix F is considered given, the only degree-of-freedom left for closed-loop transfer recovery is in the selection of the observer parameters. First of all, the observer parameters must be selected such that closed-loop stability of the observer-based control system is guaranteed. The remaining degrees-of-freedom in choosing the observer parameters are then used to achieve CLTR. That is, one uses the available freedom in the observer design to make the norm of $M_\ell(j\omega)$ either zero or small over the range of frequencies of interest. Due to the significance of the matrix $M_\ell(s)$ in CLTR, we refer $M_\ell(s)$ as the recovery matrix.

In the next subsections, we focus our attention to full-order and reduced-order observer-based controllers and perform a complete analysis of CLTR for each of these cases.

A. Full-Order Observer-Based Controller

In this subsection, we consider the problem of closed-loop transfer recovery using a full-order observer-based controller design. State-space description of a full-order observer-based controller is given by

$$\begin{cases} \dot{\hat{x}} = (A - KC_2)\hat{x} + B_2 u + Ky, \\ u = -F\hat{x}, \end{cases} \tag{21}$$

where the full-order observer gain matrix K is chosen such that $A - KC_2$ is asymptotically stable. The transfer function of this full-order observer-based controller is

$$-u(s) = \mathbf{C}(s)y(s),$$

where

$$\mathbf{C}(s) = F(sI_n - A + B_2F + KC_2)^{-1}K.$$

Note that the full-order observer-based controller is a special case of the Luenberger observer-based controller in (16) and (17) with

$$\begin{cases} L = A - KC_2, & G_1 = K, & G_2 = B_2, \\ P = I_n, & J = 0, & Q = I_n. \end{cases} \tag{22}$$

From lemma 1 it follows that the recovery error and the recovery matrix, denoted here by $E_f(s)$ and $M_f(s)$ respectively, are given by

$$E_f(s) = T_{zu}(s)M_f(s), \tag{23}$$

where $T_{zu}(s)$ is as defined in (19) and

$$M_f(s) = F(\Phi^{-1} + KC_2)^{-1}(B_1 - KD_{21}). \tag{24}$$

(The subscript ℓ in (18) and (20) is replaced by f to signify the specific case of a full-order observer-based controller.)

1. Analysis For Arbitrary Target Closed-Loop Transfer Functions

Study of equations (23) and (24) will provide a clear insight into the basic mechanism of CLTR. In fact, these equations indicate that examination of the recovery error $E_f(s)$ can be done generally in terms of the study of $M_f(s)$. Lemma 1 and the expression for $M_f(s)$ given in (24) will therefore form the basis of our investigation. Since the state-feedback gain matrix F is considered given, the only degree-of-freedom for closed-loop transfer recovery is in the selection of the observer gain K. First of all, in order to guarantee the overall closed-loop stability, K must be selected such that the observer-dynamic matrix

$$A_o = A - KC_2 \tag{25}$$

is asymptotically stable (i.e. $\lambda(A_o) \in \mathcal{C}^-$). The remaining degrees-of-freedom in choosing K can then be used for the purpose of achieving CLTR. Now in view of (24) and lemma 2, exact closed-loop transfer recovery (ECLTR) is possible for an *arbitrarily* given F if $\tilde{M}_f(j\omega) \equiv 0$ where $M_f(s) = F\tilde{M}_f(s)$ and

$$\tilde{M}_f(s) = (sI_n - A_o)^{-1}(B_1 - KD_{21}). \tag{26}$$

Since the matrix $(j\omega I_n - A_o)^{-1}$ is nonsingular, $\tilde{M}(j\omega) \equiv 0$ clearly implies that $B_1 - KD_{21} \equiv 0$. The class of systems in which $B_1 - KD_{21}$ can be rendered exactly zero is rather restrictive. Hence, one would most likely resort to an approach based on asymptotic closed-loop transfer recovery (ACLTR), i.e. to render $\tilde{M}_f(j\omega)$ approximately zero in some sense. As mentioned in section 2, to analyze whether ACLTR is possible we need to parameterize the controller with a tuning parameter σ. In the case of a full-order observer-based controller, this parameterization can be simply introduced in the observer gain matrix $K(\sigma)$. The resulting family of controllers parameterized by $K(\sigma)$ is

$$\mathbf{C}(s,\sigma) = F[sI_n - A + B_2F + K(\sigma)C_2]^{-1}K(\sigma). \tag{27}$$

In this formulation $M_f(s)$ and $\tilde{M}_f(s)$ are now functions of σ, denoted respectively by $M_f(s,\sigma)$ and $\tilde{M}_f(s,\sigma)$. In our analysis and for the sake of clarity, we assume that A_o is nondefective. This allows us to expand $\tilde{M}_f(s,\sigma)$ and hence $M_f(s,\sigma)$ in a dyadic form,

$$\tilde{M}_f(s,\sigma) = \sum_{i=1}^{n} \frac{\tilde{R}_i}{s-\lambda_i} \tag{28}$$

where the residue matrix $\tilde{R}_i$ is given by

$$\tilde{R}_i = W_i V_i^H [B_1 - K(\sigma) D_{21}]. \tag{29}$$

Here W_i and V_i are respectively the right- and left-eigenvectors associated with the eigenvalue λ_i of A_o and

$$WV^H = V^H W = I_n.$$

The matrices W and V can be partitioned as follows,

$$W = [W_1,\, W_2,\, \cdots,\, W_n] \quad \text{and} \quad V = [V_1,\, V_2,\, \cdots,\, V_n]. \tag{30}$$

In general, all λ_i, V_i and W_i are functions of σ. However for economy of notation, we will omit the dependence on σ explicitly unless it is needed for clarity.

In what follows, we examine conditions under which the i-th term in the dyadic expansion of $\tilde{M}_f(s,\sigma)$ in (28) can be made zero or small. There are basically two ways to achieve this:

1. The first possibility is to assign the closed-loop eigenvalue λ_i to any finite value in C^- while simultaneously rendering the residue $\tilde{R}_i$ zero either exactly or asymptotically, i.e.

$$\tilde{R}_i = W_i(\sigma) V_i^H(\sigma)[B_1 - K(\sigma) D_{21}] = 0$$

or

$$\tilde{R}_i(\sigma) = W_i(\sigma) V_i^H(\sigma)[B_1 - K(\sigma) D_{21}] \to 0$$

as $\sigma \to \infty$. This procedure involves a finite eigenstructure assignment of A_o.

2. The other possibility is to make

$$\frac{\tilde{R}_i}{s-\lambda_i} \to 0$$

pointwise in s as $\sigma \to \infty$ by placing the eigenvalue $\lambda_i(\sigma)$ asymptotically at infinity and making sure that the residue $\tilde{R}_i(\sigma)$ is uniformly bounded as $\sigma \to \infty$. It is important to recognize that placing λ_i asymptotically at infinity alone will not give the desired result unless the residue $\tilde{R}_i$ is also bounded. This amounts to assigning $W_i(\sigma)$ and $V_i(\sigma)$ such that

$$\tilde{R}_i(\sigma) = W_i(\sigma)V_i^H(\sigma)[B_1 - K(\sigma)D_{21}]$$

remains bounded while $|\lambda_i| \to \infty$ as $\sigma \to \infty$. Thus, this procedure deals with an infinite eigenstructure assignment of A_o.

Two fundamental questions immediately arise in the application of an eigenstructure assignment technique to this problem of closed-loop transfer recovery:

(1) How many left-eigenvectors of A_o can be assigned to the null space of $[B_1 - K(\sigma)D_{21}]'$? and,

(2) How many eigenvalues of A_o can be placed at asymptotically infinite locations in $\mathcal{C}^-$ and at the same time the residues associated with these eigenvalues are also finite?

The answers to these two questions are given in the following two lemmas.

In the analysis that follows, we shall apply the s.c.b transformation developed in section 3 to the system Σ_{yw}. Let n_a^- and n_a^+ be respectively the number of stable and unstable invariant zeros of Σ_{yw} and n_f the number of infinite zeros of Σ_{yw}. Moreover, we let

$$n_c := \dim\{\mathcal{V}^*(A, B_1, C_2, D_{21}) \cap \mathcal{S}^*(A, B_1, C_2, D_{21})\}$$

and

$$n_b := n - n_a^- - n_a^+ - n_c - n_f .$$

These integers (i.e. n_a^-, n_a^+, n_b, n_c and n_f) can be readily obtained from the s.c.b of Σ_{yw}.

Lemma 3. *For any given $K(\sigma)$ such that A_o is stable, let λ_i be an eigenvalue of A_o and V_i its corresponding left-eigenvector. Then the* *maximum number of $\lambda_i \in \mathcal{C}^-$ which satisfy the condition*

$$\tilde{R}_i = W_i V_i^H [B_1 - K(\sigma) D_{21}] = 0$$

is $(n_a^- + n_b)$. A total of n_a^- of these eigenvalues λ_i coincide with the stable invariant zeros of the system Σ_{yw} and the remaining n_b of these eigenvalues can be assigned arbitrarily to any location in $\mathcal{C}^-$. The eigenvectors V_i that correspond to these $(n_a^- + n_b)$ eigenvalues span the subspace

$$\Re^n / \mathcal{S}^-(A, B_1, C_2, D_{21}).$$

Moreover, n_a^- of these eigenvectors corresponding to the eigenvalues at the stable invariant zeros are simply the left-state zero directions and span the subspace

$$\mathcal{V}^*(A, B_1, C_2, D_{21}) / \mathcal{V}^+(A, B_1, C_2, D_{21}).$$

Proof : See [3].

Lemma 4. *For any given $K(\sigma)$ such that A_o is stable, let λ_i be an eigenvalue of A_o, V_i and W_i its corresponding left- and right-eigenvectors respectively. The maximum number of eigenvalues of A_o that can be assigned arbitrarily to asymptotically infinite locations in $\mathcal{C}^-$ while at the same time the residue matrix*

$$\tilde{R}_i = W_i V_i^H [B_1 - K(\sigma) D_{21}]$$

remain bounded as $|\lambda_i| \rightarrow \infty$ is $(n_b + n_f)$. Furthermore, the left-eigenvectors V_i of these asymptotically infinite eigenvalues span the subspace

$$\Re^n / \mathcal{V}^*(A, B_1, C_2, D_{21}).$$

Proof : See [3]. ∎

Remark 2. *Consider the case when Σ_{yw} is right-invertible and D_{21} is of maximal rank. Clearly this covers the special case where Σ_{yw} is a non-strictly proper single-input and single-output system. For this case, we have $n_b + n_f = 0$ and hence there is no eigenvalue λ_i of A_o that can be assigned to an infinite location and at the same time the corresponding residue matrix $\tilde{R}_i$ is bounded.*

As implied by lemma 3, there are n_b eigenvalues where one can assign arbitrarily to any locations in $\mathcal{C}^-$ and still maintain $\tilde{R}_i \equiv 0$. These n_b eigenvalues are therefore among the $(n_b + n_f)$ eigenvalues indicated in lemma 4. That is, there is a set of n_b eigenvalues which can be placed arbitrarily at either asymptotically finite locations in $\mathcal{C}^-$ according to lemma 3 or at asymptotically infinite locations in $\mathcal{C}^-$ according to lemma 4. For practical design considerations, such as limited controller bandwidth and sensor noise reduction, one often keeps these n_b eigenvalues at stable and reasonably finite locations.

Combining the results of lemmas 3 and 4, one can deduce all the conditions under which various terms of $\tilde{M}_f(s,\sigma)$ can be made zero either exactly or asymptotically. There are altogether $(n_a^- + n_b + n_f)$ eigenvalues which can be assigned either at finite or at asymptotically infinite locations so that the corresponding terms of $\tilde{M}_f(s,\sigma)$ in its dyadic expansion (28) are either exactly or asymptotically zero. Thus, a question arises as to under what conditions $(n_a^- + n_b + n_f)$ is equal to n the system dimension. It is easy to see that $n_a^- + n_b + n_f = n$ if and only if Σ_{yw} is left-invertible ($n_c = 0$) and of minimum phase[2] ($n_a^+ = 0$). If Σ_{yw} is not left-invertible and/or of nonminimum phase, then there are exactly $n_e \equiv n_a^+ + n_c$ terms of $\tilde{M}_f(s,\sigma)$ which cannot in general be rendered zero. To fully understand the behavior of $\tilde{M}_f(s,\sigma)$, we need to partition it into four parts,

$$\tilde{M}_f(s,\sigma) = \tilde{M}_-(s,\sigma) + \tilde{M}_b(s,\sigma) + \tilde{M}_\infty(s,\sigma) + \tilde{M}_e(s,\sigma), \tag{31}$$

where

$$\tilde{M}_-(s,\sigma) = \sum_{i=1}^{n_a^-} \frac{\tilde{R}_i}{s-\lambda_i},$$

$$\tilde{M}_b(s,\sigma) = \sum_{i=n_a^-+1}^{n_a^-+n_b} \frac{\tilde{R}_i}{s-\lambda_i},$$

$$\tilde{M}_\infty(s,\sigma) = \sum_{i=n_a^-+n_b+1}^{n_a^-+n_b+n_f} \frac{\tilde{R}_i}{s-\lambda_i},$$

---

[2] A system is said to be of nonminimum phase if at least one of its invariant zeros is in the closed right-half plane, otherwise it is said to be of minimum phase.

and

$$\tilde{M}_e(s,\sigma) = \sum_{i=n_a^-+n_b+n_f+1}^{n} \frac{\tilde{R}_i}{s-\lambda_i}.$$

Let $\Lambda_-(\sigma)$, $\Lambda_b(\sigma)$, $\Lambda_\infty(\sigma)$ and $\Lambda_e(\sigma)$ be the sets of eigenvalues of A_o associated with the parts $\tilde{M}_-(s,\sigma)$, $\tilde{M}_b(s,\sigma)$, $\tilde{M}_\infty(s,\sigma)$ and $\tilde{M}_e(s,\sigma)$ respectively . Corresponding to each of these partitions of eigenvalues, we define the associated right- and left-eigenvectors of A_o in the sets $W_-(\sigma)$, $W_b(\sigma)$, $W_\infty(\sigma)$, $W_e(\sigma)$ and $V_-(\sigma)$, $V_b(\sigma)$, $V_\infty(\sigma)$, $V_e(\sigma)$ respectively. Also as a convenient notation, we will use an overbar on a variable to denote its limit (whenever it exists) as $\sigma \to \infty$. For example, $\overline{\tilde{M}_e}(s)$ and $\overline{W}_e$ denote respectively the limits of $\tilde{M}_e(s,\sigma)$ and $W_e(\sigma)$ as $\sigma \to \infty$. Various parts of $\tilde{M}_f(s,\sigma)$ have now the following interpretation:

1. $\tilde{M}_-(s,\sigma)$ contains n_a^- terms with eigenvalues in the set $\Lambda_-(\sigma)$. In accordance with lemma 3, there exists a gain $K(\sigma)$ such that $\tilde{M}_-(s,\sigma)$ is identically zero by placing elements of $\Lambda_-(\sigma)$ at the stable invariant zeros of Σ_{yw} and the set of left-eigenvectors $V_-(\sigma)$ to be the left-state zero directions. In fact, $K(\sigma)$ can be selected such that $\Lambda_-(\sigma)$ and $V_-(\sigma)$ approach asymptotically the set of system stable invariant zeros and their left-state zero directions as $\sigma \to \infty$. In this case, we have $\tilde{M}_-(s,\sigma) \to 0$ as $\sigma \to \infty$.

2. $\tilde{M}_b(s,\sigma)$ contains n_b terms with eigenvalues in the set $\Lambda_b(\sigma)$. In accordance with lemmas 3 and 4, there exists a gain $K(\sigma)$ such that $\tilde{M}_b(s,\sigma)$ is identically zero by assigning elements of $\Lambda_b(\sigma)$ arbitrarily at either finite or infinite locations in $\mathcal{C}^-$ asymptotically as $\sigma \to \infty$. As discussed earlier, in order to limit the controller bandwidth, we will assume hereafter that these eigenvalues are assigned to asymptotically finite locations. Also, $K(\sigma)$ can be designed so that $\tilde{M}_b(s,\sigma) \to 0$ as $\sigma \to \infty$.

3. $\tilde{M}_\infty(s,\sigma)$ contains n_f terms with eigenvalues in the set $\Lambda_\infty(\sigma)$. In accordance with lemma 4, there exists a gain $K(\sigma)$ such that $\tilde{M}_\infty(s,\sigma) \to 0$ as $\sigma \to \infty$ by assigning elements of $\Lambda_\infty(\sigma)$ arbitrarily to asymptotically infinite locations in $\mathcal{C}^-$.

4. $\tilde{M}_e(s,\sigma)$ contains the remaining $n_e \equiv n_a^+ + n_c$ terms with eigenvalues in the set $\Lambda_e(\sigma)$. This term does not exist (i.e. $n_e = 0$) when the system Σ_{yw} is left-invertible and of minimum phase. In view of lemmas 3 and 4, $\tilde{M}_e(s,\sigma)$ cannot in general be rendered zero either asymptotically or otherwise by any assignment of $\Lambda_e(\sigma)$ and the associated sets of right- and left-eigenvectors $W_e(\sigma)$ and $V_e(\sigma)$. However, as to be discussed later, selection of the full-order observer design gain $K(\sigma)$ can be done so that the error term $\tilde{M}_e(s,\sigma)$ has a particular frequency-shaped properties or some desired H_2- and H_∞-norms. Note that the eigenvalues of A_o in Λ_e can be assigned to any locations in $\mathcal{C}^-$, except of course for those corresponding to the stable but unobservable eigenvalues of A, since (A, C_2) is assumed to be a detectable pair. These arbitrary locations can either be asymptotically finite or infinite.

As the above discussion indicates, lemmas 3 and 4 form the heart of the underlying mechanism of CLTR. They enable us to decompose $\tilde{M}_f(s,\sigma)$ and hence $M_f(s,\sigma)$ into four distinct parts exhibiting clearly conditions under which closed-loop recovery is or is not possible. Although the results presented so far do not directly provide methods for obtaining the gain $K(\sigma)$, they do however give the essential closed-loop eigenstructure and hence guidelines in the assignment of the eigenvalues and eigenvectors of A_o. These guidelines can in turn be used to formulate a systematic method for designing the full-order observer gain $K(\sigma)$. In a sequel paper [2], we will discuss in details three possible methods for designing $K(\sigma)$. They are:

(1) A method based on the minimization of the H_2-norm of $M_f(s,\sigma)$,

(2) A method based on the minimization of the H_∞-norm of $M_f(s,\sigma)$ and,

(3) Asymptotic time-scale and eigenstructure assignment (ATEA) method.

The latter method is an extension of the one given in [11] and [9] which allows designers a great flexibility in the shaping of $\tilde{M}_f(s,\sigma)$. Putting these aside, we come back to the problem of characterizing achievable closed-loop transfer recovery. To do this, we simply assume that the observer gain $K(\sigma)$ was given and has been chosen from the set $\mathcal{K}^*(A, B_1, C_2, D_{21}, \sigma)$ defined below.

Definition 6. *$\mathcal{K}^*(A, B_1, C_2, D_{21}, \sigma)$ is a set of gains $K(\sigma) \in \Re^{n \times p}$ such that*

(1) *$A_o(\sigma) = A - K(\sigma)C_2$ is stable for all $\sigma > \sigma^*$ where $0 \leq \sigma^* < \infty$,*

(2) *The finite eigenvalues of $A_o(\sigma)$ remain in $\mathcal{C}^-$ as $\sigma \to \infty$,*

(3′) *If $n_f = 0$ then $\tilde{M}_-(s,\sigma)$ and $\tilde{M}_b(s,\sigma)$ are identically zero for all σ,*

(3″) *If $n_f \neq 0$ then, as $\sigma \to \infty$, $\tilde{M}_-(s,\sigma)$ and $\tilde{M}_b(s,\sigma)$ are either identically zero or asymptotically zero. Moreover, the eigenvalues in the set $\Lambda_-(\sigma)$ and $\Lambda_b(\sigma)$ tend to finite locations in $\mathcal{C}^-$ and,*

(4) *$\tilde{M}_\infty(s,\sigma) \to 0$ as $\sigma \to \infty$.*

It is obvious that $\mathcal{K}^*(A, B_1, C_2, D_{21}, \sigma)$ is a nonempty set.

Remark 3. *In the case where the system Σ_{yw} does not have any infinite zeros (i.e. $n_f = 0$), every element $K(\sigma)$ of $\mathcal{K}^*(A, B_1, C_2, D_{21}, \sigma)$ is independent of σ and furthermore $\|K(\sigma)\| \leq \alpha < \infty$ for all σ. On the other hand, if the system Σ_{yw} has at least one infinite zero (i.e. $n_f \neq 0$), then $K(\sigma)$ of $\mathcal{K}^*(A, B_1, C_2, D_{21}, \sigma)$ must be a function of σ and $\|K(\sigma)\| \to \infty$ as $\sigma \to \infty$.*

Theorem 2 given below characterizes the asymptotic behavior of the achieved loop transfer function for $K(\sigma) \in \mathcal{K}^*(A, B_1, C_2, D_{21}, \sigma)$. We have the following theorem.

Theorem 2. *Consider the closed-loop system Σ^c comprising of the system Σ and a full-order observer-based controller. Let (A, B_2) be stabilizable and (A, C_2) be detectable. Then, given any F such that $A-BF$ is asymptotically stable and for a gain $K(\sigma) \in \mathcal{K}^*(A, B_1, C_2, D_{21}, \sigma)$, the closed-loop system Σ^c is asymptotically stable. Moreover, as $\sigma \to \infty$,*

$$E_f(s,\sigma) \to T_{zu}(s) F \overline{\tilde{M}}_e(s). \tag{32}$$

Proof : Expression (32) follows from lemmas 3 and 4, as well as the properties of $\mathcal{K}^*(A, B_1, C_2, D_{21}, \sigma)$. ∎

In view of theorem 2, $\overline{\tilde{M}_e(s)}$ can be termed as the limit of the recovery matrix. We have the following corollaries of theorem 2.

Corollary 1. *Let the system* Σ_{yw} *be left-invertible and of minimum phase. Then* $\mathbf{T}^f_{\mathrm{R}}(\Sigma) = \mathbf{T}(\Sigma)$. *Moreover, for any gain*

$$K(\sigma) \in \mathcal{K}^*(A, B_1, C_2, D_{21}, \sigma),$$

the corresponding full-order observer-based controller achieves closed-loop transfer recovery for any given $T_{zw}(s) \in \mathbf{T}(\Sigma)$.

Proof : For a left-invertible and minimum-phase system Σ_{yw}, we have $n_a^+ = 0$ and $n_c = 0$. Thus $n_a^+ + n_c = 0$ and $\tilde{M}_e(s,\sigma)$ is nonexistent. Hence, the results are obvious. ∎

Remark 4. *Results of corollary 1 are exactly those of Fujita* et al *[6].*

Corollary 2. *Let the system* Σ_{yw} *be left-invertible and of minimum phase. And let* D_{21} *be of maximal column rank. Then*

$$\mathbf{T}^f_{\mathrm{ER}}(\Sigma) = \mathbf{T}^f_{\mathrm{R}}(\Sigma) = \mathbf{T}(\Sigma) \ \ \text{and} \ \ \mathbf{T}^f_{\mathrm{AR}}(\Sigma) = \emptyset.$$

Moreover, the observer gain

$$K(\sigma) \in \mathcal{K}^*(A, B_1, C_2, D_{21}, \sigma)$$

is independent of σ *and the corresponding full-order observer-based controller achieves exact closed-loop transfer recovery (ECLTR) for any given* $T_{zw}(s) \in \mathbf{T}(\Sigma)$.

Proof : When the system Σ_{yw} is left-invertible and of minimum phase and D_{21} is of maximal column rank, then we have $n_a^+ = 0$, $n_c = 0$ and $n_f = 0$. Note that a left-invertible system Σ_{yw} has no infinite zeros if and only if D_{21} is of maximal rank. Thus, both $\tilde{M}_e(s,\sigma)$ and $\tilde{M}_\infty(s,\sigma)$ are nonexistent. Hence the results of corollary 2 are obvious. ∎

Remark 5. *If the system* Σ_{yw} *is a non-strictly proper and of minimum-phase single-input single-output system, then* $\mathbf{T}^f_{\mathrm{ER}}(\Sigma) = \mathbf{T}^f_{\mathrm{R}}(\Sigma)$ *and* $\mathbf{T}^f_{\mathrm{AR}}(\Sigma) = \emptyset$.

Remark 6. *Whenever ECLTR is feasible, the corresponding full-order observer-gain matrix* $K(\sigma) \in \mathcal{K}^*(A, B_1, C_2, D_{21}, \sigma)$ *is finite and constant for all* σ *and hence* $\mathbf{C}(s,\sigma) = \mathbf{C}(s)$.

2. Analysis For Recoverable Target Closed-Loop Transfer Functions

In the previous subsection, the analysis of closed-loop transfer recovery does not take into account any knowledge of the state-feedback gain matrix F. It is essentially a study of the matrix $\tilde{M}_f(s)$ or $\tilde{M}_f(s,\sigma)$ as to when it can or cannot be rendered zero using a full-order observer-based controller. This section complements the analysis of the previous subsection by taking directly into account the knowledge of F. Obviously then, the analysis of this section is a study of $M_f(s) = F\tilde{M}_f(s)$ or $M_f(s,\sigma) = F\tilde{M}_f(s,\sigma)$. Two basic issues have been addressed:

(1) What class of target closed-loops can be recovered exactly (or asymptotically) for a given system Σ? Or equivalently, what are the necessary and sufficient conditions a target closed-loop transfer function $T_{zw}(s)$ must satisfy so that it can exactly (or asymptotically) be recoverable for the given system? and,

(2) What are the necessary and sufficient conditions on the system Σ so that it has at least one recoverable target loop?

Answers to these questions would enable designers to identify the appropriate number and type of control inputs and measurement outputs in the plant model needed in the CLTR task. To answer the questions, we introduce an auxiliary system Σ_E where now the condition for the set of exactly recoverable target loops $\mathbf{T}^f_{\mathrm{ER}}(\Sigma)$ to be nonempty is equivalent to the auxiliary system Σ_E being stabilizable by a static output-feedback control. Similarly, another auxiliary system Σ_A can be introduced for the ACLTR case. Here, it will be shown that the set of recoverable target loops $\mathbf{T}^f_{\mathrm{AR}}(\Sigma)$ is nonempty if and only if Σ_A is stabilizable by a static output-feedback control.

In what follows, we derive conditions for ECLTR and ACLTR in terms of geometric properties. We also give the necessary and sufficient conditions for the sets $\mathbf{T}^f_{\mathrm{ER}}(\Sigma)$ and $\mathbf{T}^f_{\mathrm{AR}}(\Sigma)$ to be non-empty. We have the following theorems.

Theorem 3. *Let the system Σ_{zu} be left-invertible and stabilizable and Σ_{yw} be detectable. Then, an admissible target closed-loop transfer function $T_{zw}(s)$ of Σ (i.e. $T_{zw}(s) \in \mathbf{T}(\Sigma)$) is exactly recoverable by a* *full-order observer-based controller if and only if*

$$\mathcal{S}^-(A, B_1, C_2, D_{21}) \subseteq Ker(F).$$

That is,

$$\mathbf{T}^f_{\mathrm{ER}}(\Sigma) = \{\, T_{zw}(s) \in \mathbf{T}(\Sigma) \mid \mathcal{S}^-(A, B_1, C_2, D_{21}) \subseteq Ker(F) \,\}.$$

Proof : See Appendix B. ∎

The following theorem characterizes the non-emptiness of $\mathbf{T}^f_{\mathrm{ER}}(\Sigma)$.

Theorem 4. *Let the system Σ_{zu} be left-invertible and stabilizable and Σ_{yw} be detectable. Let $\overline{C}_E$ be any full-rank matrix of dimension $(n_a^- + n_b) \times n$ such that*

$$Ker(\overline{C}_E) = \mathcal{S}^-(A, B_1, C_2, D_{21}).$$

Then, the system Σ has at least one exactly recoverable target closed-loop transfer function (i.e. $\mathbf{T}^f_{\mathrm{ER}}(\Sigma)$) is nonempty if and only if the auxiliary system Σ_E characterized by the matrix triple $(A, B_2, \overline{C}_E)$ is stabilizable by a static output-feedback controller.

Proof : See Appendix C. ∎

Theorems 5 and 6 below state the results for the case of ACLTR.

Theorem 5. *Let the system Σ_{zu} be left-invertible and stabilizable and Σ_{yw} be detectable. Then, an admissible target closed-loop transfer function $T_{zw}(s)$ of Σ (i.e. $T_{zw}(s) \in \mathbf{T}(\Sigma)$) is recoverable (either exactly or asymptotically) by a full-order observer-based controller if and only if*

$$\mathcal{V}^+(A, B_1, C_2, D_{21}) \subseteq Ker(F).$$

That is,

$$\mathbf{T}^f_{\mathrm{R}}(\Sigma) = \{\, T_{zw}(s) \in \mathbf{T}(\Sigma) \mid \mathcal{V}^+(A, B_1, C_2, D_{21}) \subseteq Ker(F) \,\}.$$

Proof : See Appendix D. ∎

As in theorem 4, theorem 6 below characterizes the non-emptiness of $\mathbf{T}^f_{\mathrm{R}}(\Sigma)$.

Theorem 6. *Let the system Σ_{zu} be left-invertible and stabilizable and Σ_{yw} be detectable. Let $\overline{C}_{A}$ be any full-rank matrix of dimension $(n_a^- + n_b + n_f) \times n$ such that*

$$Ker(\overline{C}_{A}) = \mathcal{V}^{+}(A, B_1, C_2, D_{21}).$$

Then, the system Σ has at least one asymptotically recoverable target closed-loop transfer function (i.e. $\mathbf{T}_{\mathrm{R}}^{f}(\Sigma)$ is nonempty) if and only if the auxiliary system Σ_{A} characterized by the matrix triple $(A, B_2, \overline{C}_{A})$ is stabilizable by a static output-feedback controller.

Proof : It follows along the same lines as in theorem 4. ∎

B. Reduced-Order Observer-Based Controller

In this section, let us consider the problem of closed-loop transfer recovery using reduced-order observer-based controllers. Without loss of generality and for simplicity of presentation, we assume that the matrices C_2 and D_{21} are already in the form

$$C_2 = \begin{bmatrix} 0 & C_{2,02} \\ I_{p-m_0} & 0 \end{bmatrix} \quad \text{and} \quad D_{21} = \begin{bmatrix} D_{21,0} \\ 0 \end{bmatrix}, \tag{33}$$

where m_0 is the rank of D_{21}. The system Σ can be rewritten as,

$$\begin{cases} \begin{bmatrix} \dot{x}_1 \\ \dot{x}_2 \end{bmatrix} = \begin{bmatrix} A_{11} & A_{12} \\ A_{21} & A_{22} \end{bmatrix} \begin{bmatrix} x_1 \\ x_2 \end{bmatrix} + \begin{bmatrix} B_{1,1} \\ B_{1,2} \end{bmatrix} w + \begin{bmatrix} B_{2,1} \\ B_{2,2} \end{bmatrix} u, \\ \begin{bmatrix} y_0 \\ y_1 \end{bmatrix} = \begin{bmatrix} 0 & C_{2,02} \\ I_{p-m_0} & 0 \end{bmatrix} \begin{bmatrix} x_1 \\ x_2 \end{bmatrix} + \begin{bmatrix} D_{21,0} \\ 0 \end{bmatrix} w, \\ \quad z = C_1 \; x + \; D_{11} \; w + \; D_{12} \; u, \end{cases} \tag{34}$$

where $[x_1', \; x_2']' = x$ and $[y_0', \; y_1']' = y$. We note that $y_1 \equiv x_1$. Thus, one needs to estimate only the state x_2 in the reduced-order observer design. The procedure follows closely the development given in [7] and [1]. We first rewrite the state equation for x_1 in terms of the measured output y_1 and state x_2 as follows,

$$\dot{y}_1 = A_{11}y_1 + A_{12}x_2 + B_{1,1}w + B_{2,1}u, \tag{35}$$

where y_1 and u are known. Observation of x_2 is made via y_0 and

$$\tilde{y}_1 = A_{12}x_2 + B_{1,1}w = \dot{y}_1 - A_{11}y_1 - B_{2,1}u. \tag{36}$$

A reduced-order system for the estimation of the remaining state x_2 is given by

$$\begin{cases} \dot{x}_2 = A_{22}\ x_2 + \ B_{1,2}\ \ w + [A_{21},\ B_{2,2}] \begin{bmatrix} y_1 \\ u \end{bmatrix}, \\ \begin{bmatrix} y_0 \\ \tilde{y}_1 \end{bmatrix} = \begin{bmatrix} C_{2,02} \\ A_{12} \end{bmatrix} x_2 + \begin{bmatrix} D_{21,0} \\ B_{1,1} \end{bmatrix} w. \end{cases} \tag{37}$$

Based on equation (37), we can construct a reduced-order observer for the state x_2 as follows,

$$\dot{\hat{x}}_2 = A_{22}\hat{x}_2 + [A_{21},\ B_{2,2}] \begin{bmatrix} y_1 \\ u \end{bmatrix} + K_r \left(\begin{bmatrix} y_0 \\ \dot{y}_1 - A_{11}y_1 - B_{2,1}u \end{bmatrix} - \begin{bmatrix} C_{2,02} \\ A_{12} \end{bmatrix} \hat{x}_2 \right), \tag{38}$$

where K_r is the observer gain matrix for the reduced-order system. It is chosen such that

$$A_{or} = A_{22} - K_r \begin{bmatrix} C_{2,02} \\ A_{12} \end{bmatrix}$$

is asymptotically stable. In order to remove the dependency on $\dot{y}_1$, let us partition $K_r = [K_{r0},\ K_{r1}]$ to be compatible with the dimensions of the outputs $[y_0',\ \tilde{y}_1']'$ and at the same time define a new variable $v := \hat{x}_2 - K_{r1}\tilde{y}_1$. We obtain the following reduced-order observer-based controller,

$$\begin{cases} \dot{v} = A_{or}v + (B_{2,2} - K_{r1}B_{2,1})u \\ \qquad + [K_{r0},\ A_{21} - K_{r1}A_{11} + A_{or}K_{r1}] \begin{bmatrix} y_0 \\ y_1 \end{bmatrix}, \\ \hat{x} = \begin{bmatrix} 0 \\ I_{n-p+m_0} \end{bmatrix} v + \begin{bmatrix} 0 & I_{p-m_0} \\ 0 & K_{r1} \end{bmatrix} y, \\ u = -F\hat{x}, \end{cases} \tag{39}$$

We further note that the reduced-order observer-based controller given above is a special case of Luenberger observer-based controller of (16) with the following parameters,

$$\begin{cases} L = A_{or}, \quad G_1 = [K_{r0},\ A_{21} - K_{r1}A_{11} + A_{or}K_{r1}], \\ P = \begin{bmatrix} 0 \\ I_{n-p+m_0} \end{bmatrix}, \quad J = \begin{bmatrix} 0 & I_{p-m_0} \\ 0 & K_{r1} \end{bmatrix}, \\ G_2 = B_{2,2} - K_{r1}B_{2,1}, \quad Q = [-K_{r1},\ I_{n-p+m_0}]. \end{cases} \tag{40}$$

Now, let us partition F as

$$F = [F_1, \ F_2]$$

in conformity with $[x_1', \ x_2']'$. It follows from lemma 1 that the recovery error and the recovery matrix, denoted here by $E_r(s)$ and $M_r(s)$ respectively, are given by

$$E_r(s) = T_{zu}(s) \cdot M_r(s), \tag{41}$$

where $T_{zu}(s)$ is as defined in (19) and

$$M_r(s) = F_2(sI - A_r + K_r C_r)^{-1}(B_r - K_r D_r), \tag{42}$$

with

$$A_r = A_{22}, \quad B_r = B_{1,2}, \quad C_r = \begin{bmatrix} C_{2,02} \\ A_{12} \end{bmatrix}, \quad D_r = \begin{bmatrix} D_{21,0} \\ B_{1,1} \end{bmatrix}.$$

Remark 7. *The expression for $M_r(s)$ is identical to $M_f(s)$ of the full-order observer-based controller in (24), where F_2, (A_r, B_r, C_r, D_r) and K_r now take the place of F, (A, B_1, C_2, D_{21}) and K.*

We have the following important lemma regarding the properties of Σ_r characterized by (A_r, B_r, C_r, D_r).

Lemma 5.

1. *Σ_r is of (non-) minimum phase if and only if (A, B_1, C_2, D_{21}) is of (non-) minimum phase.*
2. *Σ_r is detectable if and only if Σ_{yw} is detectable.*
3. *Invariant zeros of Σ_r are the same as those of Σ_{yw}.*
4. *Orders of infinite zeros of Σ_r are reduced by one from those of Σ_{yw}.*
5. *Σ_r is left-invertible if and only if Σ_{yw} is left-invertible.*
6. $\begin{pmatrix} 0 \\ I \end{pmatrix} \mathcal{V}^+(A_r, B_r, C_r, D_r) = \mathcal{V}^+(A, B_1, C_2, D_{21})$.
7. $\begin{pmatrix} 0 \\ I \end{pmatrix} \mathcal{S}^-(A_r, B_r, C_r, D_r) = \mathcal{S}^-(A, B_1, C_2, D_{21}) \cap \mho$, *where* $\mho := \{x \mid C_2 x \in Im D_{21}\}$.

Proof See [5]. ∎

1. Analysis For Arbitrary Target Closed-Loop Transfer Functions

We note that with lemma 5 the analysis and design of full-order and reduced-order observer-based controllers for CLTR have been placed into the same framework. Now, let $\mathcal{K}^*(A_r, B_r, C_r, D_r, \sigma)$ be defined in a similar way as in definition 6. We have the following results which are analogous to the case of a full-order observer-based controller.

Theorem 7. *Consider the closed-loop system Σ^c comprising of the given system Σ and a reduced-order observer-based controller. Let (A, B_2) be stabilizable and (A, C_2) be detectable. Then, for any F such that $A - B_2F$ is asymptotically stable and for any gain*

$$K(\sigma) \in \mathcal{K}^*(A_r, B_r, C_r, D_r, \sigma),$$

the closed-loop system Σ^c is asymptotically stable. Moreover, as $\sigma \to \infty$,

$$E_r(s, \sigma) \to T_{zu}(s) F_2 \overline{\widetilde{M}}_{re}(s), \tag{43}$$

where $\overline{\widetilde{M}}_{re}(s)$ is for the reduced-order system Σ_r and can be derived following the procedure given in Section IV.A.1.

Proof : The proof follows along the same lines as in theorem 2 and the properties of Σ_r in lemma 5. ∎

In view of theorem 7, $\overline{\widetilde{M}}_{re}(s)$ can also be termed as the limit of the recovery matrix for the case of a reduced-order observer-based controller. We have the following corollaries of theorem 7.

Corollary 3. *Let Σ_{yw} be left-invertible and of minimum phase. Then $\mathbf{T}^r_{\mathrm{R}}(\Sigma) = \mathbf{T}(\Sigma)$. Furthermore, for any gain $K(\sigma) \in \mathcal{K}^*(A_r, B_r, C_r, D_r, \sigma)$, the corresponding reduced-order observer-based controller achieves closed loop transfer recovery for any given $T_{zw}(s) \in \mathbf{T}(\Sigma)$.*

Proof : The proof follows along the same lines as in corollary 1 and the properties of Σ_r in lemma 5. ∎

Corollary 4. *Let Σ_{yw} be left-invertible and of minimum phase with no infinite zero of order higher than one which implies that D_r is of* *maximal column rank. Then*

$$\mathbf{T}^r_{\mathrm{R}}(\Sigma) = \mathbf{T}^r_{\mathrm{ER}}(\Sigma) = \mathbf{T}(\Sigma)$$

and $\mathbf{T}^r_{\mathrm{AR}}(\Sigma) = \emptyset$. *Moreover, any gain*

$$K(\sigma) \in \mathcal{K}^*(A_r, B_r, C_r, D_r, \sigma)$$

is independent of σ *and the corresponding reduced-order observer-based controller achieves exact closed-loop transfer recovery (ECLTR) for any given* $T_{zw}(s) \in \mathbf{T}(\Sigma)$.

Proof : The proof follows along the same lines as in corollary 2 and the properties of Σ_r in lemma 5. ∎

2. Analysis For Recoverable Target Closed-Loop Transfer Functions

In what follows, we state in terms of geometric properties conditions under which ECLTR and ACLTR can be achieved using a reduced-order observer-based controller. As in the case of full-order observer-based controller, necessary and sufficient conditions are given that characterize the non-empty sets $\mathbf{T}^r_{\mathrm{ER}}(\Sigma)$ and $\mathbf{T}^r_{\mathrm{R}}(\Sigma)$. We have the following theorems.

Theorem 8. *Let the system* Σ_{zu} *be left-invertible and stabilizable and* Σ_{yw} *be detectable. Then an admissible target closed-loop transfer function* $T_{zu}(s)$ *of* Σ *(i.e.* $T_{zw}(s) \in \mathbf{T}(\Sigma)$*) is exactly recoverable by a reduced-order observer-based controller if and only if*

$$\mathcal{S}^-(A, B_1, C_2, D_{21}) \cap \mho \subseteq Ker(F).$$

That is,

$$\mathbf{T}^r_{\mathrm{ER}}(\Sigma) = \{\, T_{zw}(s) \in \mathbf{T}(\Sigma) \mid \mathcal{S}^-(A, B_1, C_2, D_{21}) \cap \mho \subseteq Ker(F)\}.$$

Proof : In view of lemma 5, we note that

$$\mathcal{S}^-(A, B_1, C_2, D_{21}) \cap \mho \subseteq Ker(F)$$

is equivalent to

$$\mathcal{S}^-(A_r, B_r, C_r, D_r) \subseteq Ker(F_2).$$

Hence the proof follows along the same lines as in theorem 3. ∎

Remark 8. *It is simple to observe from theorems 3 and 8 as well lemma 5 that* $\mathbf{T}^f_{\text{ER}}(\Sigma) \subseteq \mathbf{T}^r_{\text{ER}}(\Sigma)$. *That is, if a target closed-loop transfer function is exactly recoverable by a full-order observer-based controller, then it is also exactly recoverable by a reduced-order observer-based controller. But the reverse is not true in general.*

The following theorem characterizes the non-emptiness of $\mathbf{T}^r_{\text{ER}}(\Sigma)$ for reduced-order observer-based controllers.

Theorem 9. *Let the system* Σ_{zu} *be left-invertible and stabilizable and* Σ_{yw} *be detectable. Let* $\overline{C}_{re}$ *be any maximal rank matrix such that*

$$Ker(\overline{C}_{re}) = \mathcal{S}^-(A, B_1, C_2, D_{21}) \cap \mho.$$

Then the given system Σ *has at least one exactly recoverable target closed-loop transfer function using a reduced-order observer-based controller (i.e.* $\mathbf{T}^r_{\text{ER}}(\Sigma)$ *is nonempty) if and only if the auxiliary system* Σ_{re} *characterized by the matrix triple* $(A, B_2, \overline{C}_{re})$ *is stabilizable by a static output-feedback controller.*

Proof : It follows along the same lines as in theorem 4. ∎

Theorem 10 given below deals with ACLTR for reduced-order observer based controller.

Theorem 10. *Let the system* Σ_{zu} *be left-invertible and stabilizable and* Σ_{yw} *be detectable. Then an admissible target closed-loop transfer function* $T_{zw}(s)$ *of* Σ *(i.e.* $T_{zw}(s) \in \mathbf{T}(\Sigma)$*) is recoverable (either exactly or asymptotically) by a reduced-order observer-based controller if and only if*

$$\mathcal{V}^+(A, B_1, C_2, D_{21}) \subseteq Ker(F).$$

That is,

$$\mathbf{T}^r_{\text{R}}(\Sigma) = \{\, T_{zw}(s) \in \mathbf{T}(\Sigma) \mid \mathcal{V}^+(A, B_1, C_2, D_{21}) \subseteq Ker(F) \,\}.$$

Proof : In view of lemma 5, we note that

$$\mathcal{V}^{+}(A, B_1, C_2, D_{21}) \subseteq Ker(F)$$

is equivalent to

$$\mathcal{V}^{+}(A_r, B_r, C_r, D_r) \subseteq Ker(F_2).$$

Hence the proof follows along the same lines as in theorem 5. ∎

Remark 9. *It is trivial to see from theorems 5 and 10 that* $\mathbf{T}^f_{\mathrm{R}}(\Sigma) = \mathbf{T}^r_{\mathrm{R}}(\Sigma)$. *That is, if a target closed-loop transfer function is recoverable by a full-order observer-based controller, then it is also recoverable by a reduced-order observer-based controller. And the reverse is also true. Hence, it is obvious that the nonemptiness of* $\mathbf{T}^r_{\mathrm{AR}}(\Sigma)$ *is characterized by the same condition as in theorem 6.*

V. NUMERICAL EXAMPLE

The above analysis of CLTR is applied to the development of a localizer capture and track-hold design of a commercial transport. This numerical example is not intended to provide a complete illustration of all the analysis results discussed in the previous sections. The main reason for using this design problem is that it provides a realistic design situation where asymptotic and exact closed-loop transfer recovery using full-order and reduced-order observer-based controllers are applicable. For completeness, we provide a brief overview of the design procedure used in the synthesis of the chosen state-feedback gain F which defines the target closed-loop transfer function $T_{zw}(s)$ for closed-loop transfer recovery. Detailed description and design requirements for such a system have been extensively covered in literature (see for example the Special Issues in IEEE Control System Magazine [14]). It should be emphasized here that the analysis provided in previous sections are applicable to arbitrary state-feedback laws, regardless of the procedures from which these state-feedback laws are derived (i.e. H_2-, H_∞-norm based design methods, eigenstructure assignment or others).

Design model used in this example consists of the basic 4th-order lateral aircraft dynamics augmented with appropriate kinematic equations for the heading ψ and lateral track distance y_{track} along with a state for the integral of lateral track error $\int(y_{track} - y_c)dt$. State matrices describing the synthesis model without actuation dynamics in the notations of (1) are given below for a typical landing approach condition,

$$A = \begin{bmatrix} -0.2093 & 0.00077518 & 0.1003 & -0.991 & 0 & 0 & 0 \\ -7.492 & -3.44 & -0.0052035 & 0.9783 & 0 & 0 & 0 \\ 0 & 1 & 0 & -0.0012877 & 0 & 0 & 0 \\ 2.031 & -0.1696 & -0.0050925 & -0.2089 & 0 & 0 & 0 \\ 0 & 0 & 0 & 1 & 0 & 0 & 0 \\ 0 & 0 & 0 & 0 & 5.597 & 0 & 0 \\ 0 & 0 & 0 & 0 & 0 & 1 & 0 \end{bmatrix} \tag{44}$$

$$B_1 = \begin{bmatrix} 0.026222 & -0.0033036 \\ 9.2044 & 0.069525 \\ 0 & 0 \\ 0.10099 & 0.086836 \\ 0 & 0 \\ 0 & 0 \\ 0 & -1 \end{bmatrix},$$

$$B_2 = \begin{bmatrix} 0.0013217 & 0.063301 \\ 2.011 & 2.012 \\ 0 & 0 \\ 0.1304 & -1.393 \\ 0 & 0 \\ 0 & 0 \\ 0 & 0 \end{bmatrix},$$

$$C_1 = \begin{bmatrix} -19.856 & 0.07354 & 9.5153 & -94.015 & 0 & 0 & 0 \\ 0 & 0 & 14.142 & 0 & 0 & 0 & 0 \\ 0 & 0 & 0 & 20 & 0 & 0 & 0 \\ 0 & 0 & 0 & 0 & 0 & 1 & 0 \\ 0 & 0 & 0 & 0 & 0 & 0 & 1 \\ 0 & 0 & 0 & 0 & 0 & 0 & 0 \\ 0 & 0 & 0 & 0 & 0 & 0 & 0 \end{bmatrix},$$

$$D_{11} = \begin{bmatrix} 0 & 0 \\ -14.142 & 0 \\ 0 & 0 \\ 0 & -1 \\ 0 & 0 \\ 0 & 0 \\ 0 & 0 \end{bmatrix}, \quad D_{12} = \begin{bmatrix} 0.12539 & 6.0053 \\ 0 & 0 \\ 0 & 0 \\ 0 & 0 \\ 0 & 0 \\ 10 & 0 \\ 0 & 31.623 \end{bmatrix},$$

$$C_2 = \begin{bmatrix} 0 & 1 & 0 & 0 & 0 & 0 & 0 \\ 0 & 0 & 1 & 0 & 0 & 0 & 0 \\ 0 & 0 & 0 & 1 & 0 & 0 & 0 \\ 0 & 0 & 0 & 0 & 1 & 0 & 0 \\ 0 & 0 & 0 & 0 & 0 & 1 & 0 \\ 0 & 0 & 0 & 0 & 0 & 0 & 1 \end{bmatrix},$$

$$D_{21} = \begin{bmatrix} 0 & 0 \\ 0 & 0 \\ 0 & 0 \\ 0 & 0 \\ 0 & 0 \\ 0 & 0 \end{bmatrix}, \quad D_{22} = \begin{bmatrix} 0 & 0 \\ 0 & 0 \\ 0 & 0 \\ 0 & 0 \\ 0 & 0 \\ 0 & 0 \end{bmatrix}.$$

The state variables are

$$x = [\beta, p, \phi, r, \psi, y_{track}, \textstyle\int(y_{track} - y_c)dt]',$$

where β is the sideslip angle in degrees, p is the roll rate in degrees per second, ϕ is the roll angle in degrees, r is the yaw rate in degrees per second, ψ is the heading angle in degrees, y_{track} is the lateral track distance in feet and y_c is the command lateral track distance in feet. The control inputs

$$u = [\delta_{ac}, \delta_{rc}]'$$

consist of the aileron δ_{ac} and rudder δ_{rc} deflections in degrees. The disturbance inputs

$$w = [\phi_c, y_c]'$$

contain the bank angle command ϕ_c in degrees and the lateral track command y_c in feet. The measurement output variables are

$$y = [p, \phi, r, \psi, y_{track}, \textstyle\int(y_{track} - y_c)dt]'.$$

The controlled output variables z shown above are made up of weighted plant outputs z_p and control variables u in the following form

$$z = [(Q^{1/2}z_p)', (R^{1/2}u)']'.$$

The performance variables z_p include sideslip acceleration $\dot{\beta}$, bank angle deviation $(\phi - \phi_c)$, yaw rate r, lateral track deviation $(y_{track} - y_c)$ and integral track error $\int(y_{track} - y_c)dt$. The control variables u are included in the controlled output vector z to ensure that the resulting state-feedback design does not have excessive control gain and bandwidth. These control variables are scaled by a diagonal weighting matrix R. Note that in the design trade-offs, loop shapings are tuned on the performance variables z_p using a diagonal weighting matrix Q. Final selection of the diagonal weighting matrices Q and R is made after numerous design iterations that involve at each time closed-loop stability analysis, frequency responses of the transfer function $y_{track}(s)/y_c(s)$, time simulation to a lateral track command. It is observed in the design iterations that increasing penalty on the sideslip acceleration $\dot{\beta}$ will improve the aircraft turn-coordination, but at the expense of slower responses to bank angle and lateral track command inputs. In order to achieve good tracking performance and turn-coordination, responses of the controlled output vector z to the command inputs w must be kept as small as possible. State matrices for the desired controlled output vector $z(t)$ are given in equation (44).

Control-law synthesis is performed at one particular landing approach condition. A state-feedback law that yields satisfactory stability and closed-loop responses to a lateral track command y_c is obtained from the following H_∞-norm bound solution, i.e $\|T_{zw}(s)\|_{H_\infty} < 45$. An acceptable state-feedback gain matrix F is given below,

$$F = \begin{bmatrix} 0.10939 & 0.92375 & 4.2514 & 2.3791 & 3.9476 & 0.088639 & 0.0048466 \\ -1.1237 & 0.081467 & 0.32548 & -2.9659 & -1.6125 & -0.05404 & -0.0035325 \end{bmatrix} \tag{45}$$

Analysis of closed-loop transfer recovery for the above localizer capture and track-hold design proceeds as follows. First of all, we examine whether conditions for exact and asymptotic closed-loop recovery can be achieved with the given set of measurements and using an output-feedback observer-based control-law. It is simple to verify that the system Σ_{zu} is left-invertible. Thus according to Lemma 2, conditions for exact and asymptotic closed-loop recovery are governed completely by the existence of solutions that make the recovery error zero or arbitrarily small. Next we observe that, with the given measurement output $y(t)$ and disturbances $w(t)$, the system Σ_{yw} is left-invertible and has no invariant zeros. Thus from Corollary 1, a full-order observer-based controller can be used to achieve closed-loop transfer recovery for *any* closed-loop $T_{zw}(s)$ under state feedback and obviously the full-state feedback design defined in equation (45). Furthermore, from Corollary 3, the transfer function $T_{zw}(s)$ is also recoverable using a reduced-order observer-based controller. It can be determined from the s.c.b transformation that the system Σ_{yw} has two infinite-zeros of order 1. This result indicates that recovery using a full-order observer-based controller can only be achieved asymptotically. An acceptable observer-gain design K is given below and it is obtained using the ATEA design method [2].

$$K = \begin{bmatrix} 28.529 & 0.099657 & -3.6288 & 0 & -0.016212 & 34.789 \\ 9998.8 & 0.00010319 & 108.9 & 0 & 0 & 9.263 \\ 1 & 0.5 & -0.0012877 & 0 & 0 & 0 \\ 108.72 & -0.0094053 & 75.315 & 0 & 0 & -854.23 \\ 0 & 0 & 1 & 0.5 & 0 & 0 \\ 0 & 0 & 0 & 5.597 & 0.5 & 0 \\ 9.8159 & -0.00080954 & -854.25 & 0 & 0 & 9926.5 \end{bmatrix} \quad (46)$$

Development of different design methods for CLTR will be covered in a sequel paper [2]. It should be noted that in theory the observer gain matrix must be large in order to recover asymptotically the closed-loop performance (i.e. the case of ACLTR). In this particular design example, we notice that reducing the recovery error at low frequency does indeed involve a high observer gain design synthesized using either the ATEA or ARE-based methods. Singular value plots of the closed-loop transfer function $T_{zw}(j\omega)$ are shown in figure 2. The observer gain of equation (46) seems to provide reasonably small recovery error at low frequency and at the same time does not lead to excessive control responses to lateral track commands. Performance evaluation based on transient responses is depicted in figure 3 corresponding to the time simulation of system responses to a lateral track command

$$y_c = 1000(1 - e^{-0.065t})(\text{feet}).$$

This figure shows time responses of the full-state feedback design. Results corresponding to the above full-order observer design are the same within the resolution of the graph as those shown in figure 3.

Now we proceed to the problem of closed-loop transfer recovery using a reduced-order observer-based controller. It turns out that, for the above localizer capture and track-hold problem, one can actually achieved *exact* closed-loop recovery using a reduced-order observer-based controller. This result comes directly from Lemma 5 and the fact that the system Σ_{yw} is left-invertible, has no invariant zeros and has only infinite zeros of order 1. The ensuing reduced system Σ_r as defined in section IV.B has no infinite zeros. Hence, in this case, ECLTR

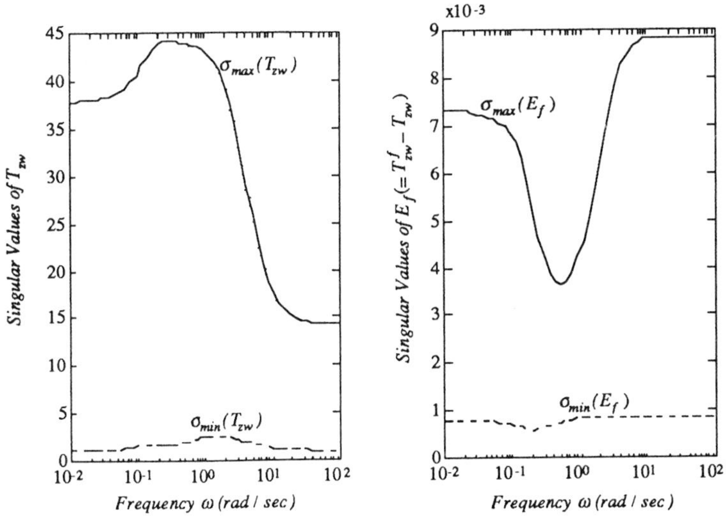

Figure 2: Singular Value Plots of $T_{zw}(j\omega)$ and $E_f(j\omega)$.

is possible. Again using the ATEA design method, we have obtained a reduced-order observer-based controller that yields exactly the same closed-loop transfer function $T_{zw}(s)$. State matrices of this controller are given below.

$$\begin{cases} \dot{v} = A_{cmp}v + B_{cmp}y, \\ -u = C_{cmp}v + D_{cmp}y, \end{cases} \tag{47}$$

where

$$A_{cmp} = -0.12265$$

$$B_{cmp} = [0.0095756 \quad 0.10031 \quad -0.81378 \quad 0.1095 \quad 0 \quad -0.0002034]$$

$$C_{cmp} = \begin{bmatrix} 0.10939 \\ -1.1237 \end{bmatrix}$$

$$D_{cmp} = \begin{bmatrix} 0.92406 & 4.2514 & 2.3791 & 3.9476 & 0.088639 & 0.0052271 \\ 0.078262 & 0.32548 & -2.9656 & -1.6125 & -0.05404 & -0.0074414 \end{bmatrix}.$$

Performance of this reduced-order controller is identical to that of the full-state feedback case (see Figures 2 and 3). The design is an output-feedback controller of first-order and having a controller pole at $s = -0.12$ *rad/sec.* Hence, the design concept of CLTR has enabled us to synthesize a low-order implementable output-feedback design for a

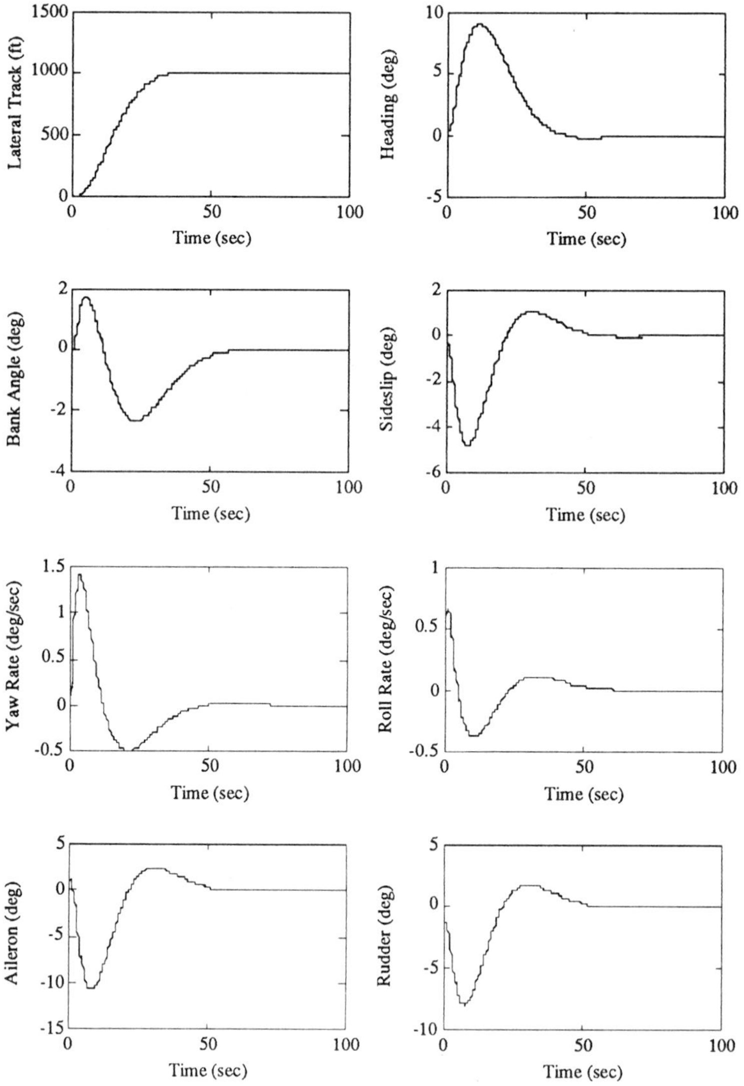

Figure 3: Aircraft Responses to a Lateral Track Command of 1000ft.

typical localizer capture and lateral track hold system starting from a satisfactory state-feedback control law. It should be pointed out that if actuator dynamics have been included into the design model, then exact closed-loop transfer recovery is no longer possible, even with a reduced-order observer-based controller since the infinite zeros of Σ_{yw} are no longer of order 1. However, the system is still asymptotically recoverable .

VI. CONCLUSIONS

In this paper, we deal with issues concerning the analysis of closed-loop transfer recovery using full-order and reduced-order observer-based controllers. There are several fundamental results given here. Based on the structural properties of the given system, we decompose the recovery matrix in the recovery error between the target closed-loop transfer function and that achieved by observer-based controllers, into three distinct parts for any arbitrarily specified admissible target closed-loop transfer function. The first part of recovery matrix can be rendered exactly zero by an appropriate finite eigenstructure assignment of observer dynamic matrix, while the second part can be rendered arbitrarily close to zero by an appropriate infinite eigenstructure assignment. The third part in general cannot be rendered zero, either exactly or asymptotically, by any means although there exists a multitude of ways to shape it.

The above analysis is general and applies to any arbitrarily specified target closed-loop transfer function. Results of the analysis enable designers to identify limitations of the given system in recovering the target closed-loop transfer function as a consequence of its structural properties, namely finite and infinite zero structures and invertibility. The next issue of our analysis concentrates on characterizing the required necessary and/or sufficient conditions on the target closed-loop transfer functions so that they are either exactly or asymptotically recoverable by means of observer-based controllers for the given system. Conditions developed here for a target closed-loop transfer function to be recoverable turn out to be constraints in its finite and infinite zero structures inherent of the system under consideration. The last issue covered in our analysis is to find the necessary and/or sufficient conditions on the given system such that it has at least one recoverable target closed-loop transfer function.

In a sequel, we will present design issues concerning the closed-loop transfer recovery.

Acknowledgement

The work of B. M. Chen and A. Saberi is supported in part by Boeing Commercial Airplane Group and in part by NASA Langley Research Center under grant contract NAG-1-1210. The work of U. Ly is supported in part by NASA Langley Research Center under grant contract NAG-1-1210.

A. APPENDIX A — JUSTIFICATION OF $D_{22} = 0$

The justification of the assumption $D_{22} = 0$ is as follows: Let us define a new auxiliary measurement output y_{new} as

$$y_{new} = y - D_{22}u = C_2x + D_{21}w.$$

Then we will show that a compensator

$$u(s) = K(s)y_{new}(s)$$

is equivalent to the compensator

$$u(s) = K(s)[I + D_{22}K(s)]^{-1}y(s)$$

under assumption that the closed-loop system is well-posed, i.e. the inverse of $I + D_{22}K(s)$ exists for almost all $s \in \mathcal{C}$. Let us consider the following relation,

$$\begin{aligned} u(s) &= K(s)y_{new}(s) \\ &= K(s)[y(s) - D_{22}u(s)] \\ &= K(s)y(s) - K(s)D_{22}u(s) \end{aligned}$$

This implies that

$$[I + K(s)D_{22}]u(s) = K(s)y(s).$$

Hence,

$$u(s) = [I + K(s)D_{22}]^{-1}K(s)y(s) = K(s)[I + D_{22}K(s)]^{-1}y(s). \quad (48)$$

Thus, whenever D_{22} is nonzero, one can define a new set of measurement output, namely y_{new}, and design a controller $K(s)$. Then the controller in (48) will yield the same performance when it is applied to the original system. ∎

B. APPENDIX B — PROOF OF THEOREM 3

Under the assumption that Σ_{zu} is left-invertible, it follows from Lemma 2 that an admissible target loop $T_{zw}(s)$ is exactly recoverable by a full-order observer-based controller (i.e. $E_f(s) = 0$) if and only if there exists an observer gain K such that $A - KC_2$ is asymptotically stable and the corresponding $M_f(s) = 0$. Thus, it is sufficient to show that $M_f(s) = 0$ if and only if

$$\mathcal{S}^-(A, B_1, C_2, D_{21}) \subseteq Ker(F).$$

To show this, let us consider an auxiliary system characterized by

$$\Sigma_{au} : \begin{cases} \dot{x} = A'x + C_2'u + F'w, \\ z = B_1'x + D_{21}'u. \end{cases} \quad (49)$$

Also, with a state-feedback law

$$u = -K'x,$$

the closed-loop transfer function from w to z, denoted here by $T_{zw}^{au}(s)$, is simply

$$T_{zw}^{au}(s) = M_f'(s).$$

Hence, the problem of finding an observer gain matrix such that $A - KC_2$ is asymptotically stable and $M_f(s) = 0$ is equivalent to the well-known disturbance decoupling problem. Then it follows from Stoorvogel [16] that the disturbance decoupling problem with internal stability is solvable to Σ_{au} in (49) if and only if

$$\mathcal{S}^-(A, B_1, C_2, D_{21}) \subseteq Ker(F).$$

This completes the proof of theorem 3. ∎

C. APPENDIX C — PROOF OF THEOREM 4

Without loss of generality, we assume that (A, B_1, C_2, D_{21}) is in the form of s.c.b as in theorem 1. Now in view of theorem 3, an exactly recoverable target closed-loop transfer function $T_{zw}(s)$ must satisfy the condition $\mathcal{S}^-(A, B_1, C_2, D_{21}) \subseteq Ker(F)$. This implies that $T_{zw}(s)$ is recoverable if and only if F is the form,

$$F = \begin{bmatrix} F_{a1}^- & 0 & F_{b1} & 0 & 0 \\ F_{a2}^- & 0 & F_{b2} & 0 & 0 \end{bmatrix}. \tag{50}$$

Thus condition that the given system has at least one exactly recoverable target closed-loop is simply equivalent to the existence of some appropriate matrix F_{a1}^-, F_{a2}^-, F_{b1} and F_{b2} such that $A - B_2F$ is asymptotically stable. Next in view of the properties of s.c.b, we note that $\overline{C}_E$ as defined in theorem 4 is of the form,

$$\overline{C}_E = \Gamma \begin{bmatrix} I_{n_a^-} & 0 & 0 & 0 & 0 \\ 0 & 0 & I_{n_b} & 0 & 0 \end{bmatrix},$$

where Γ is any non-singular matrix of dimension $(n_a^- + n_b) \times (n_a^- + n_b)$. It is now trivial to verify that existence of a matrix F of the form in (50) such that $A - B_2F$ is asymptotically stable, is equivalent to existence of a matrix G such that $A - B_2G\overline{C}_E$ is asymptotically stable. This is simply due to the fact $G\overline{C}_E$ has the same structure as F in (50). This completes the proof of theorem 4. ∎

D. APPENDIX D — PROOF OF THEOREM 5

Under the assumption that Σ_{zu} is left-invertible, it follows from Lemma 2 that an admissible target loop $T_{zw}(s)$ is asymptotically recoverable by a full-order observer-based controller if and only if there exists an observer gain $K(\sigma)$ such that $A - K(\sigma)C_2$ is asymptotically stable and the corresponding $M_f(s,\sigma) \to 0$ pointwise in s as $\sigma \to \infty$. Following the proof of theorem 3 in Appendix B, it is simple to see that such a problem is equivalent to the well-known almost disturbance decoupling problem with internal stability (ADDPS) and it is shown in Scherer [13] that ADDPS is solvable to Σ_{au} in (49) if and only if

$$\mathcal{V}^+(A, B_1, C_2, D_{21}) \subseteq Ker(F),$$

and we adhere to the notion of closed-loop stability by excluding those cases where, in the limits as $\sigma \to \infty$, the finite eigenvalues of the closed-loop system are on the $j\omega$ axis. This completes the proof of theorem 5. ∎

References

[1] B. M. Chen, A. Saberi., P. Bingulac and P. Sannuti, "Loop Transfer Recovery for Non-Strictly Proper Plants," *Control—Theory and Advanced Technology*, Vol. 6, No. 4, pp. 573-594 (1990).

[2] B. M. Chen, A. Saberi and U. Ly, "Closed-Loop Transfer Recovery With Observer-Based Controllers—Part 2: Design," To appear in *Control and Dynamic Systems: Advances in Theory and Applications*, Academic Press, Inc. (1991).

[3] B. M. Chen, A. Saberi and P. Sannuti, "Loop Transfer Recovery for General Nonminimum Phase Non-Strictly Proper Systems, Part 1 - Analysis," Submitted for publication (1991).

[4] B. M. Chen, A. Saberi and P. Sannuti, "Loop Transfer Recovery for General Nonminimum Phase Non-Strictly Proper Systems, Part 2 - Design," Submitted for publication (1991).

[5] B. M. Chen, A. Saberi and P. Sannuti, "Loop Transfer Recovery for General Nonminimum Phase Non-Strictly Proper Systems, Part 3 - Reduced-Order Observer Design," Submitted for publication (1991).

[6] M. Fujita, K. Uchida and F. Matsumura, "Asymptotic H_∞ Disturbance Attenuation Based on Perfect Observation," *Proceedings of American Control Conference*, pp. 3092-3097, San Diego, California (1990).

[7] H. Kwakernaak and R. Sivan, *Linear Optimal Control Systems*, John Wiley (1972).

[8] A. Saberi, B. M. Chen and P. Sannuti, "Theory of LTR for Nonminimum Phase Systems, Recoverable Target Loops, Recovery in a Subspace—Part 1: Analysis," *International Journal of Control*, Vol. 53, No. 5, pp. 1067-1115 (1991).

[9] A. Saberi, B. M. Chen and P. Sannuti, "Theory of LTR for Nonminimum Phase Systems, Recoverable Target Loops, Recovery in a Subspace—Part 2: Design," *International Journal of Control*, Vol. 53, No. 5, pp. 1117-1160 (1991).

[10] A. Saberi and P. Sannuti, "Squaring Down of Non-Strictly Proper Systems," *International Journal of Control*, 51, 3, pp. 621-629 (1990).

[11] A. Saberi and P. Sannuti, "Observer Design for Loop Transfer Recovery and for Uncertain Dynamical Systems," *IEEE Transactions on Automatic Control*, Vol. 35, No. 8, pp. 878-897 (1990).

[12] P. Sannuti and A. Saberi, "A Special Coordinate Basis of Multivariable Linear Systems - Finite and Infinite Zero Structure, Squaring Down and Decoupling," *International Journal of Control*, Vol.45, No. 5, pp. 1655-1704 (1987).

[13] C. Scherer, "H_∞-Optimization Without Assumptions on Finite or Infinite Zeros," Preprint (1989).

[14] Special Issues on Aerospace Control Systems, *IEEE Control Systems Magazine*, Vol. 10, No. 4, June (1990).

[15] G. Stein and M. Athans, "The LQG / LTR Procedure for Multivariable Feedback Control Design," *IEEE Transactions on Automatic Control*, AC-32, pp.105-114 (1987).

[16] A. A. Stoorvogel, "The Singular Linear Quadratic Gaussian Control Problem," Preprint, November (1990).

Closed-Loop Transfer Recovery with Observer-Based Controllers

Part 2: Design

Ben M. Chen

Ali Saberi

School of Electrical Engineering and Computer Science
Washington State University
Pullman, Washington 99164-2752

Uy-Loi Ly

Department of Aeronautics and Astronautics, FS-10
University of Washington
Seattle, Washington 98195

I. INTRODUCTION & PROBLEM STATEMENT

The problem of closed-loop transfer recovery (CLTR) has been discussed in an early sequel paper [1]. The basic problem addressed there is the analysis of closed-loop transfer recovery using full-order and reduced-order observer-based controllers. The design objective is to recover using output feedback (if possible) the closed-loop transfer function achieved under a full-state feedback design for a given set of

CONTROL AND DYNAMIC SYSTEMS, VOL. 51

disturbance input and controlled output variables. To be specific, let us consider a plant Σ defined by

$$\Sigma : \begin{cases} \dot{x} = A\,x + B_1\,w + B_2\,u, \\ z = C_1 x + D_{11} w + D_{12} u, \\ y = C_2 x + D_{21} w + D_{22} u, \end{cases} \tag{1}$$

where $x \in \Re^n$ is the state, $u \in \Re^m$ is the control input, $w \in \Re^k$ is the disturbance, $z \in \Re^\ell$ is the controlled output and $y \in \Re^p$ is the measurement output. For convenience, we also define the subsystem Σ_{yw} to represent the matrix quadruple (A, B_1, C_2, D_{21}) and the subsystem Σ_{zu} for the matrix quadruple (A, B_2, C_1, D_{12}). We assume that the pair $(A,\ B_2)$ is stabilizable and the pair $(A,\ C_2)$ detectable. Without loss of generality, the following matrices $[C_1,\ D_{11},\ D_{12}]$, $[C_2,\ D_{21},\ D_{22}]$, $[B_1',\ D_{11}',\ D_{21}']'$ and $[B_2',\ D_{12}',\ D_{22}']'$ are assumed of maximal ranks. As shown in [1], one can also assume without loss of generality that $D_{22} = 0$ as well. Let F be a full-state feedback gain matrix such that under the state-feedback control

$$u = -Fx \tag{2}$$

(a) the closed-loop system is asymptotically stable, i.e. the eigenvalues of $A - B_2F$ lie in the left-half s-plane,

(b) the closed-loop transfer function from the disturbance w to the controlled output z, denoted by $T_{zw}(s)$, meets the desired frequency dependent design specifications.

We also refer to $T_{zw}(s)$ as the target closed-loop transfer function given by

$$T_{zw}(s) = (C_1 - D_{12}F)(\Phi^{-1} + B_2F)^{-1}B_1 + D_{11} \tag{3}$$

where $\Phi = (sI - A)^{-1}$. The problem of closed-loop transfer recovery (CLTR) is then to find an internally stabilizing output-feedback controller $\mathbf{C}(s)$ such that the recovery error defined as

$$E(s) := T_{zw}^o(s) - T_{zw}(s), \tag{4}$$

is either exactly or approximately equal to zero in the frequency region of interest. Here, $T_{zw}^o(s)$ represents the transfer function from w

to z of the closed-loop system. As discussed in Part 1 [1], achieving exact closed-loop transfer recovery (ECLTR) is in general not possible. Hence, it is more appropriate to examine situation where approximate recovery can be achieved. An approximate CLTR is tied to the notion that recovery can be achieved to any degree of accuracy. In this process, one normally parameterizes the controller $\mathbf{C}(s)$ as a function of a positive scalar parameter σ thereby generating a family of controllers $\mathbf{C}(s, \sigma)$. We say asymptotic CLTR (ACLTR) is achieved if

$$T^o_{zw}(s, \sigma) \rightarrow T_{zw}(s)$$

as $\sigma \rightarrow \infty$ pointwise in s, or equivalently

$$E(s, \sigma) \rightarrow 0$$

as $\sigma \rightarrow \infty$ pointwise in s. From the point of view of design, once the conditions of ACLTR have been verified, one should be able to find a controller $\mathbf{C}(s, \sigma)$ with a particular value of σ that produces the desired level of recovery.

In Part 1 [1], we consider the CLTR problem using a full-order observer-based controller of the form,

$$\begin{cases} \dot{\hat{x}} = (A - KC_2)\hat{x} + B_2 u + Ky, \\ u = -F\hat{x}, \end{cases} \tag{5}$$

and a reduced-order observer-based controller of the form,

$$\begin{cases} \dot{v} = A_{or} v + (B_{2,2} - K_{r1} B_{2,1})u \\ \qquad\qquad +[K_{r0},\ A_{21} - K_{r1}A_{11} + A_{or}K_{r1}] \begin{bmatrix} y_0 \\ y_1 \end{bmatrix}, \\ \hat{x} = \begin{bmatrix} 0 \\ I_{n-p+m_0} \end{bmatrix} v + \begin{bmatrix} 0 & I_{p-m_0} \\ 0 & K_{r1} \end{bmatrix} y, \\ u = -F\hat{x}, \end{cases} \tag{6}$$

where K and $K_r = [K_{r0}, K_{r1}]$ are respectively the full-order and reduced-order observer gain matrices. The submatrices in (6) are defined in equations (33) to (38) of Part 1. While it is recognized that in most cases neither ECLTR nor ACLTR can be achieved using either a full-order or a reduced-order observer-based controller, the analysis of CLTR conducted in Part 1 provides however a detailed study of three fundamental issues related to the problem of CLTR. The first issue is concerned with what can and cannot be achieved for a given system and for an arbitrary target closed-loop transfer function. The second issue is to develop necessary and/or sufficient conditions for a target closed-loop transfer function to be recoverable either exactly or approximately. The third issue deals with the necessary and/or sufficient conditions on a given system such that it has at least one recoverable target closed-loop transfer function. Results of the analysis have identified some fundamental limitations of the given system as a consequence of its structural properties; namely, its finite and infinite-zero structure and invertibility. This enables designers to appreciate at the outset different design limitations incurred in the synthesis of output-feedback controllers; for example, how to select a meaningful set of measurements for the closed-loop transfer recovery design. Once we have chosen an appropriate set of measurement outputs, we can then proceed to the actual design of full-order or reduced-order observer-based controllers that will achieve as close as possible the desired target closed-loop transfer function. In this paper, we focus on three different design methods for closed-loop transfer recovery.

The paper is organized as follows. Section II reviews the necessary design constraints and the available design freedom. Section III develops the general ATEA method. Also, in section III, a simplified ATEA procedure is given for the design of exact closed-loop transfer recovery whenever it is feasible. Section IV examines design methods based on optimization. Here, two design methods have been considered; one that minimizes the H_2-norm of a recovery matrix while the other minimizes the respective H_∞-norm. Section V discusses the relative merits of ATEA and optimization-based designs. A numerical example based on a benchmark problem [16] is given in section VI, illustrating the practical usage of the analysis results in [1], and comparing different observer designs synthesized using methods discussed in sections III and IV. Conclusions are drawn in section VII.

As in [1], we adopt the following notations. A' denotes the transpose of A, A^H denotes the complex conjugate transpose of A, I denotes an identity matrix while I_k denotes the identity matrix of dimension $k \times k$. $\lambda(A)$ denotes the set of eigenvalues of A. Similarly, $\sigma_{max}[A]$ and $\sigma_{min}[A]$ denote respectively the maximum and minimum singular values of A. Ker$[V]$ and Im$[V]$ denote respectively the kernel and the image of V. The open-left, closed right-half s-planes and the $j\omega$ axis are denoted by $\mathcal{C}^-$, $\mathcal{C}^+$ and $\mathcal{C}^o$ respectively. Also, $\mathbf{T}^f_{\mathrm{ER}}(\Sigma)$ denotes the set of exactly recoverable target closed-loop transfer functions for any given system Σ using a full-order observer-based controller, $\mathbf{T}^f_{\mathrm{R}}(\Sigma)$ denotes the set of either exactly or asymptotically recoverable target closed-loop transfer functions, while $\mathbf{T}^f_{\mathrm{AR}}(\Sigma)$ denotes the set of target closed-loop transfer functions which are asymptotically recoverable but not exactly recoverable for the given system Σ using full-order observer-based controllers. Precise definitions of $\mathbf{T}^f_{\mathrm{ER}}(\Sigma)$, $\mathbf{T}^f_{\mathrm{R}}(\Sigma)$ and $\mathbf{T}^f_{\mathrm{AR}}(\Sigma)$ are given in [1].

II. DESIGN CONSTRAINTS AND FREEDOM

As shown in Part 1 [1], problem formulation of CLTR for the case of reduced-order observer-based controllers can be placed into the same framework as the one for full-order observer-based controllers. Thus, for simplicity of presentation, we will focus our development of CLTR design techniques only to the case of full-order observer-based controllers. In Part 1 [1], we have analyzed systematically when and under what conditions closed-loop transfer recovery (CLTR) is possible. According to the way we partitioned the recovery matrix [1], it is clear that any systematic design scheme for CLTR would involve, beside the requirement of closed-loop stability, placing some additional design constraints upon the observer gain matrix K. The goal is to make the recovery matrix as small as possible through a particular set of design constraints. Some constraints are related to the assignment of the finite, as well as the asymptotically infinite, eigenstructures of the observer dynamic matrix $A_o = A - K(\sigma)C_2$. After satisfying all the above constraints, some design freedom may still be left in the observer gain matrix K to assign other parts of the eigenstructure in the matrix A_o.

To see this, we recall that the recovery error between the target closed-loop transfer function $T_{zw}(s)$ and the one realized by a full-order observer-based controller of (5) is given by

$$E_f(s,\sigma) = T_{zu}(s)M_f(s,\sigma) \tag{7}$$

where $T_{zu}(s)$ is the closed-loop transfer function from u to z under state feedback as defined in (19) of [1] and

$$M_f(s,\sigma) = F(\Phi^{-1} + K(\sigma)C_2)^{-1}(B_1 - K(\sigma)D_{21}). \tag{8}$$

The matrix $M_f(s,\sigma)$ is called the *recovery matrix*. It plays a dominant role in the analysis of recovery. In fact, if the system Σ_{zu} is left-invertible, then according to lemma 2 of [1] the recovery error $E_f(j\omega,\sigma)$ is zero if and only if the recovery matrix $M_f(j\omega,\sigma)$ is zero. Thus, the study of $M_f(s,\sigma)$ is tantamount to the study of CLTR. Assuming for simplicity that A_o is nondefective, one can express $M_f(s,\sigma)$ in the following matrix partial fraction expansion,

$$M_f(s,\sigma) = \sum_{i=1}^{n} \frac{R_i(\sigma)}{s - \lambda_i(\sigma)} \tag{9}$$

where $R_i(\sigma)$ is the residue matrix given by

$$R_i(\sigma) = FW_i(\sigma)V_i^H(\sigma)[B_1 - K(\sigma)D_{21}] \tag{10}$$

with $W_i(\sigma)$ and $V_i(\sigma)$ being respectively the right- and left-eigenvectors associated with the eigenvalue $\lambda_i(\sigma)$ of A_o. These eigenvectors are scaled such that

$$W(\sigma)V^H(\sigma) = V^H(\sigma)W(\sigma) = I_n$$

where

$$W(\sigma) = [W_1(\sigma),\ W_2(\sigma),\ \cdots,\ W_n(\sigma)]$$

and

$$V(\sigma) = [V_1(\sigma),\ V_2(\sigma),\ \cdots,\ V_n(\sigma)]. \tag{11}$$

As evident from (9) and (10), one can render the i-th term of $M_f(s,\sigma)$ zero in one of two ways:

(1) Render the residue $R_i(\sigma)$ zero while λ_i is finite; or

(2) Place λ_i asymptotically to infinity while keeping the residue $R_i(\sigma)$ uniformly bounded.

The first approach is the problem of finite eigenstructure assignment while the latter concerns eigenstructure assignment for the asymptotically infinite eigenvalues of A_o. From the structural properties of the system Σ_{zw}, there may not exist enough design freedom to assign the needed eigenstructure in A_o to achieve ECLTR or ACLTR. Analysis of Part 1 reveals several guidelines as to when, how and to what extent such an assignment can be done. To review these guidelines, $M_f(s,\sigma)$ is partitioned into four parts [1],

$$M_f(s,\sigma) = M_-(s,\sigma) + M_b(s,\sigma) + M_\infty(s,\sigma) + M_e(s,\sigma), \tag{12}$$

where

$$M_-(s,\sigma) = \sum_{i=1}^{n_a^-} \frac{R_i^-(\sigma)}{s - \lambda_i^-(\sigma)}, \quad M_b(s,\sigma) = \sum_{i=1}^{n_b} \frac{R_i^b(\sigma)}{s - \lambda_i^b(\sigma)}$$

and

$$M_\infty(s,\sigma) = \sum_{i=1}^{n_f} \frac{R_i^\infty(\sigma)}{s - \lambda_i^\infty(\sigma)}, \quad M_e(s,\sigma) = \sum_{i=1}^{n_e} \frac{R_i^e(\sigma)}{s - \lambda_i^e(\sigma)}.$$

According to the above partitions of $M_f(s,\sigma)$, we define the corresponding subsets of eigenvalues, right- and left-eigenvectors of $A_o(\sigma)$,

$$\Lambda_-(\sigma) := \{\lambda_i^- \mid i = 1, \cdots, n_a^-\}, \quad \Lambda_b(\sigma) := \{\lambda_i^b \mid i = 1, \cdots, n_b\},$$

$$\Lambda_\infty(\sigma) := \{\lambda_i^\infty \mid i = 1, \cdots, n_f\}, \quad \Lambda_e(\sigma) := \{\lambda_i^e \mid i = 1, \cdots, n_e\},$$

$$V_-(\sigma) := \{V_i^- \mid i = 1, \cdots, n_a^-\}, \quad V_b(\sigma) := \{V_i^b \mid i = 1, \cdots, n_b\},$$

$$V_\infty(\sigma) := \{V_i^\infty \mid i = 1, \cdots, n_f\}, \quad V_e(\sigma) := \{V_i^e \mid i = 1, \cdots, n_e\},$$

$$W_-(\sigma) := \{W_i^- \mid i = 1, \cdots, n_a^-\}, \quad W_b(\sigma) := \{W_i^b \mid i = 1, \cdots, n_b\},$$

$$W_\infty(\sigma) := \{W_i^\infty \mid i = 1, \cdots, n_f\}, \quad W_e(\sigma) := \{W_i^e \mid i = 1, \cdots, n_e\}.$$

Hereafter, we will use an over bar on a certain variable to denote its limit as $\sigma \to \infty$ whenever it exists. For example, $\overline{M}_e(s)$ and $\overline{W}_e$ denote respectively the limits of $M_e(s,\sigma)$ and $W_e(\sigma)$ as $\sigma \to \infty$.

From [1], we showed that irrespective of the target closed-loop transfer function $T_{zw}(s)$, both $M_-(s,\sigma)$ and $M_b(s,\sigma)$ can be rendered zero either exactly or asymptotically as $\sigma \to \infty$ by an appropriate finite eigenstructure assignment of A_o. Also, $M_\infty(s,\sigma)$ can be rendered asymptotically zero as $\sigma \to \infty$ by an appropriate asymptotically infinite eigenstructure assignment of A_o. On the other hand, in general, $M_e(s,\sigma)$ can never be rendered zero although there exists abundant amount of freedom to shape $M_e(s,\sigma)$ within the given constraints. However, when specialized to a particular class of target closed-loop transfer functions, namely $T_{zw}(s) \in \mathbf{T}^f_{\mathrm{ER}}(\Sigma)$, $M_e(s,\sigma)$ can be rendered zero exactly. Similarly, $M_e(s,\sigma)$ can be rendered zero asymptotically as $\sigma \to \infty$ if $T_{zw}(s) \in \mathbf{T}^f_{\mathrm{R}}(\Sigma)$.

Next, we describe in details the design constraints and the available design freedom on the observer gain matrix K in the eigenstructure assignment of A_o for closed-loop transfer recovery. The discussion covers each of the partitions of $M_f(s,\sigma)$ given in equation (12).

(a) $M_-(s,\sigma)$ partition: For an arbitrary target closed-loop transfer function $T_{zw}(s)$, the term $M_-(s,\sigma)$ can be made identically zero (irrespective of the value of σ). To accomplish this, the set of n_a^- eigenvalues in $\Lambda_-(\sigma)$ and the corresponding set of left-eigenvectors $V_-(\sigma)$ of A_o must be selected to coincide respectively with the set of stable invariant zeros and their corresponding left-state zero directions of Σ_{yw}. It is also possible to render $M_-(s,\sigma)$ zero asymptotically as $\sigma \to \infty$. This is done by parameterizing $\Lambda_-(\sigma)$ and the corresponding set of left-eigenvectors $V_-(\sigma)$ of A_o so that in the limit $\overline{\Lambda}_-$ and $\overline{V}_-$ coincide respectively with the set of stable invariant zeros and their corresponding left-state zero directions of Σ_{yw}.

(b) $M_b(s,\sigma)$ partition : For an arbitrary target closed-loop transfer function $T_{zw}(s)$, the term $M_b(s,\sigma)$ can be rendered identically zero (irrespective of the value of σ). To accomplish this, the set of n_b eigenvalues in $\Lambda_b(\sigma)$ can be assigned arbitrarily to any asymptotically finite or infinite locations in $\mathcal{C}^-$, while the corresponding set of left eigenvectors $V_b(\sigma)$ of A_o is constrained to be in the null space of the matrix $[B_1 - K(\sigma)D_{21}]'$. Likewise, $M_b(s,\sigma)$ can be rendered asymptotically zero as $\sigma \longrightarrow \infty$. This can be done by selecting an arbitrary set of $\Lambda_b(\sigma)$ from either asymptotically finite or infinite locations in $\mathcal{C}^-$, while the corresponding set of left eigenvectors $V_b(\sigma)$ of A_o is chosen such that in the limit $\overline{V}_b$ is in the null space of the matrix $[B_1 - K(\sigma)D_{21}]'$. Note that, in practice, one should keep elements of $\overline{\Lambda}_b$ at finite locations in order to reduce the need for a high-bandwidth controller.

(c) $M_\infty(s,\sigma)$ partition : For an arbitrary target closed-loop transfer function $T_{zw}(s)$, the term $M_\infty(s,\sigma)$ can be rendered asymptotically zero as $\sigma \longrightarrow \infty$. The set of n_f eigenvalues in $\Lambda_\infty(\sigma)$ can be assigned to any asymptotically infinite locations in $\mathcal{C}^-$. That is, there exists a complete freedom in the way the eigenvalue $\lambda_i^\infty(\sigma) \in \Lambda_\infty(\sigma)$ tends to infinity as $\sigma \longrightarrow \infty$, i.e. the asymptotic direction and the rate at which each $\lambda_i^\infty(\sigma)$ goes to infinity can be chosen freely by the designer. However, for every $\lambda_i^\infty(\sigma) \in \Lambda_\infty(\sigma)$, the corresponding right and left eigenvectors $W_i^\infty(\sigma)$ and $V_i^\infty(\sigma)$ must be such that $W_i^\infty(\sigma)[V_i^\infty(\sigma)]^H[B_1 - K(\sigma)D_{21}]$ is uniformly bounded as $\sigma \longrightarrow \infty$. This constraint ensures that the residue $R_i^\infty(\sigma)$ remains uniformly bounded as $\sigma \longrightarrow \infty$ and thereby forces $\overline{M}_\infty(s)$ to be zero.

(d) $M_e(s,\sigma)$ partition: The term $M_e(s,\sigma)$ does not exist when $n_a^+ + n_c = 0$, i.e. when the system Σ_{yw} is of minimum phase and left-invertible. For the case where $n_a^+ + n_c \neq 0$, we examine *first* the problem of an arbitrary target closed-loop transfer function $T_{zw}(s)$. Here, the term $M_e(s,\sigma)$ can never be rendered zero although there exists abundant amount of freedom to assign the associated eigenvalues and eigenvectors. To be explicit, the set of $n_a^+ + n_c$ eigenvalues in $\Lambda_e(\sigma)$ can be assigned to any (either asymptotically finite or infinite) locations in $\mathcal{C}^-$, with the provision that any unobservable (and stable since the pair (A, C_2) is assumed to be detectable) eigenvalues of Σ_{yw} be included in the set $\Lambda_e(\sigma)$. Also, there exists a complete freedom consistent with the results of [7] in assigning the right- and left-eigenvector sets $W_e(\sigma)$ and $V_e(\sigma)$ and hence $\overline{W}_e$ and $\overline{V}_e$. But in general $\overline{\Lambda}_e$, $\overline{W}_e$ and $\overline{V}_e$ cannot be assigned such that $\overline{M}_e(s)$ is zero. However, there exists a multitude of ways to assign $\overline{\Lambda}_e$ and $\overline{W}_e$ (and hence $\overline{V}_e$). One possibility is to shape $\overline{M}_e(s)$ to have certain desired directional properties, or to try to make it as small as possible using some optimization techniques.

The above discussion on $M_e(s,\sigma)$ assumes that the target closed-loop transfer function $T_{zw}(s)$ is arbitrary. Apparently, one may be able to acquire additional design freedom by taking into account some structural properties of the full-state feedback gain matrix F. For example, as stated in theorem 5 of Part 1 [1], under the assumption that the system Σ_{zu} is left-invertible, then any admissible target closed-loop transfer function is recoverable (i.e. an element of $\mathbf{T}^f_{\mathrm{R}}(\Sigma)$) if and only if $\mathcal{V}^+(A, B_1, C_2, D_{21}) \subseteq Ker(F)$. Since $\mathcal{V}^+(A, B_1, C_2, D_{21})$ is the span of $\tilde{x}_a^+ \oplus \tilde{x}_c$, a given $T_{zw}(s)$ is recoverable if and only if the state-feedback gain matrix F is of the form,

$$\Gamma_3^{-1} F \Gamma_1 = \begin{bmatrix} F_{a1}^- & 0 & F_{b1} & 0 & F_{f1} \\ F_{a2}^- & 0 & F_{b2} & 0 & F_{f2} \end{bmatrix} \tag{13}$$

where Γ_3 and Γ_1 are nonsingular transformation matrices defined in theorem 1 of [1] for the system Σ_{yw}. Thus, whenever $T_{zw}(s)$ is an element of $\mathbf{T}^f_{\mathrm{R}}(\Sigma)$, it can be easily shown from the special structure of F in (13) that the term $\overline{M}_e(s)$ is identically zero, irrespective of the way we pick the set of $n_a^+ + n_c$ eigenvalues in $\Lambda_e(\sigma)$ and the associated right- and left-eigenvector sets $W_e(\sigma)$ and $V_e(\sigma)$. Similarly, as stated in theorem 3 of Part 1 [1] and again under the assumption that Σ_{zu} is left-invertible, then any admissible target closed-loop transfer function is exactly recoverable (i.e. an element of $\mathbf{T}^f_{\mathrm{ER}}(\Sigma)$) if and only if $\mathcal{S}^-(A,\ B_1,\ C_2,\ D_{21}) \subseteq \mathrm{Ker}(F)$. Since $\mathcal{S}^-(A,\ B_1,\ C_2,\ D_{21})$ is the span of $\tilde{x}_a^+ \oplus \tilde{x}_c \oplus \tilde{x}_f$, the closed-loop transfer function $T_{zw}(s)$ is exactly recoverable if and only if F is of the form,

$$\Gamma_3^{-1} F \Gamma_1 = \begin{bmatrix} F_{a1}^- & 0 & F_{b1} & 0 & 0 \\ F_{a2}^- & 0 & F_{b2} & 0 & 0 \end{bmatrix}. \tag{14}$$

Now, whenever $T_{zw}(s)$ is an element of $\mathbf{T}^f_{\mathrm{ER}}(\Sigma)$, then from the special structure of F given in (14), it can be shown that both $\overline{M}_e(s)$ and $\overline{M}_\infty(s)$ are zero, irrespective of the way we select the set of eigenvalues in $\Lambda_e(\sigma)$ and in $\Lambda_\infty(\sigma)$, and the associated set of eigenvectors $W_e(\sigma)$, $V_e(\sigma)$, $W_\infty(\sigma)$ and $V_\infty(\sigma)$. In fact, in this case, all eigenvalues of A_o can be assigned to finite locations in $\mathcal{C}^-$. Moreover, since we do not have to address the eigen assignment problem associated with asymptotically infinite eigenvalues, we no longer need to parameterize the observer gain K in terms of a tuning parameter σ.

III. DESIGN VIA 'ATEA'

The previous section summarizes the available design freedom as well as constraints associated with the problem of assigning the eigenstructure of observer dynamic matrix for closed-loop transfer recovery. We develop here a design procedure which follows the concept of asymptotic time-scale and eigenstructure assignment (ATEA) proposed originally in [12]. This concept has been successfully used to design full-order observers in the problem of loop transfer recovery for left-invertible and minimum-phase plants in [13], and for general strictly proper systems in [9]. In what follows, we will present a step-by-step ATEA design algorithm for general non-strictly proper systems. At first in subsection III.A, we give a design procedure for an arbitrary target closed-loop transfer function, i.e. without taking into account any specific characteristics of F. This is the most general design procedure. When a given $T_{zw}(s)$ is asymptotically recoverable, the state-feedback gain matrix F has the structure given in (13); this translates into additional freedom for selecting eigenvalues and eigenvectors of A_o. The procedure described in subsection III.A will yield a design that recovers $T_{zw}(s)$ only in the asymptotic sense. However, when exact recovery is possible, F has the structure given in (14) and in this case one can solve the ECLTR design problem with simply a finite eigenstructure assignment to A_o. For this case, the general ATEA design procedure of subsection III.A is greatly simplified and a design solution is presented in subsection III.B.

A. ACLTR Design Via ATEA

The ATEA design method is decentralized in nature. The original system is decomposed into several subsystems that can be addressed separately in the design for closed-loop transfer recovery. Basic underlying idea behind this method starts by expressing the system into the special coordinate basis (s.c.b) of Σ_{yw} (see theorem 1 of Part 1 [1] and also [10] and [11]). The finite eigenstructure of A_o is assigned by working with subsystems which represent the finite-zero structure of Σ_{yw} (cf. equations (8) and (9) of Part 1). Similarly, the asymptotically infinite eigenstructure of A_o is assigned by working with subsystems which represent the infinite-zero structure of Σ_{yw} (see equation (13) of Part 1 for each $i = 1$ to m_f).

There are two issues in formulating the observer-dynamic matrix $A_o = A - K(\sigma)C_2$ through the selection of $K(\sigma)$. The first one is related to eigenvalue assignment while the second one deals with eigenvector assignment. Let us first consider the problem of eigenvalue assignment. As discussed in section II, some eigenvalues of A_o are constrained while others can be freely assigned to any asymptotically finite or infinite locations in $\mathcal{C}^-$. To be specific,

(1) $\Lambda_-(\sigma)$ must coincide either exactly or in the asymptotic sense to the set of stable invariant zeros of Σ_{yw},

(2) $\Lambda_b(\sigma)$ and $\Lambda_e(\sigma)$ can be assigned freely to any asymptotically finite or infinite locations in $\mathcal{C}^-$ and,

(3) $\Lambda_\infty(\sigma)$ can only be assigned to asymptotically infinite locations in $\mathcal{C}^-$.

In order to conserve the controller bandwidth, both $\Lambda_b(\sigma)$ and $\Lambda_e(\sigma)$ should in practice be assigned to asymptotically finite locations. Let us next examine carefully the freedom available in assigning $\Lambda_\infty(\sigma)$. Clearly from the discussion in section II, a complete freedom is available in choosing each $\lambda_i^\infty(\sigma) \in \Lambda_\infty(\sigma)$ $(i = 1, \cdots, n_f)$. That is, both the asymptotic direction and the rate at which $\lambda_i^\infty(\sigma)$ goes to infinity can be set arbitrarily. In other words, the freedom available in assigning every asymptotically infinite eigenvalue $\lambda_i^\infty(\sigma)$ manifests itself in two ways:

(1) First, we choose the asymptotic directions along which these eigenvalues tend to infinity and,

(2) Secondly, we select the rates at which they tend to infinity.

To quantify both these choices, let $\Lambda_\infty(\sigma)$ for large values of σ be subdivided into r sets where $r \leq n_f$,

$$\frac{\Lambda_1}{\mu_1}, \frac{\Lambda_2}{\mu_2}, \cdots, \frac{\Lambda_r}{\mu_r}. \tag{15}$$

Here, $\Lambda_\ell(\ell = 1, \cdots, r)$ is a set of n_ℓ numbers in $\mathcal{C}^-$ and Λ_ℓ is closed under complex conjugation. Also $\sum_{\ell=1}^{r} n_\ell = n_f$. Apparently, the elements of Λ_ℓ define the asymptotic directions of the infinitely fast eigenvalues while the small parameters μ_ℓ $(\ell = 1, \cdots, r)$ which are a function of σ define the rates at which these eigenvalues tend to infinity.

In summary, regarding the eigenvalue assignment, we have the freedom to specify

(i) the asymptotic limits $\overline{\Lambda}_b$ and $\overline{\Lambda}_e$ of $\Lambda_b(\sigma)$ and $\Lambda_e(\sigma)$ and,

(ii) Λ_ℓ and μ_ℓ $(\ell = 1, \cdots, r)$.

Note that $\overline{\Lambda}_b$ and $\overline{\Lambda}_e$ together with $\overline{\Lambda}_-$ define all the asymptotically finite eigenvalues of A_o, while Λ_ℓ and μ_ℓ $(\ell = 1, \cdots, r)$ define the remaining asymptotically infinite eigenvalues.

Let us look now at the constraints and design freedom available in assigning the eigenvectors of A_o. The set of left-eigenvectors $\overline{V}_-$ is constrained to coincide with the corresponding set of left state-zero directions of the plant. Moreover, $Im\overline{V}_-$ coincides the subspace

$$\mathcal{V}^*(A, B_1, C_2, D_{21})/\mathcal{V}^+(A, B_1, C_2, D_{21}).$$

On the other hand, the set of eigenvectors $\overline{V}_b$ is constrained to be in the null space of $[B_1 - K(\sigma)D_{21}]'$. In view of the particular structure of s.c.b, it can be seen that every element $\overline{V}_i^b$ of $\overline{V}_b$ is constrained to be of the form

$$[0,\ 0,\ (V_i^b)^H,\ 0,\ 0]^H.$$

In other words, the set $\overline{V}_b$ can be represented in a matrix notation as

$$[0,\ 0,\ (V_b^b)^H,\ 0,\ 0]^H$$

where V_b^b is a $n_b \times n_b$ matrix. Thus the selection of $\overline{V}_b$ to be in the null space of $[B_1 - K(\sigma)D_{21}]'$ is equivalent to an arbitrary selection of V_b^b consistent with the freedom available for eigenvector assignment [7]. Again in view of the properties of s.c.b, we note that the columns of $\overline{V}_b$ span the subspace

$$\Re^n / \{\mathcal{S}^+(A, B_1, C_2, D_{21}) \cup \mathcal{S}^-(A, B_1, C_2, D_{21})\}.$$

There is also freedom available in specifying $\overline{W}_e$. It is shown in [4] that $Im\overline{W}_e$ coincides with the subspace $\mathcal{V}^+(A, B_1, C_2, D_{21})$. Again due to the special structure of s.c.b, $\overline{W}_e$ has the special matrix form

$$[(W_e^+)^H,\ 0,\ 0,\ (W_e^c)^H,\ 0]^H$$

where $W_{ee} \equiv [(W_e^+)^H,\ (W_e^c)^H]^H$ is a $n_e \times n_e$ matrix. Thus, an appropriate selection of $\overline{W}_e$ is equivalent to an arbitrary selection of W_{ee} consistent with the freedom available for eigenvector assignment [7].

Now, an assignment of both asymptotically finite and infinite eigenvalues and the corresponding eigenvectors to A_o can be viewed as a problem in asymptotic time-scale and eigenstructure assignment (ATEA). Further discussion on time-scale structure of a system can be found in [9]. In order to have a well-defined separation of time scales, we will assume throughout the paper that

$$\mu_\ell / \mu_{\ell+1} \to 0 \text{ as } \mu_{\ell+1} \to 0. \tag{16}$$

We emphasize that the freedom available in the asymptotically infinite eigenstructure of A_o is captured in the selection of an appropriate fast time-scale structure. The asymptotic directions of asymptotically infinite eigenvalues are specified in the sets Λ_ℓ ($\ell = 1, \cdots, r$) and $r \leq n_f$. Different values of time scales are defined by the small positive parameters μ_ℓ ($\ell = 1, \cdots, r$), which are function of a tuning parameter σ so that (16) holds as $\sigma \to \infty$. Note that there is another constraint on the infinite eigenstructure; namely for every asymptotically infinite eigenvalue $\lambda_i^\infty(\sigma)$, the corresponding right- and left-eigenvectors $W_i^\infty(\sigma)$ and $V_i^\infty(\sigma)$ of A_o must be such that

$$W_i^\infty(\sigma)[V_i^\infty(\sigma)]^H[B_1 - K(\sigma)D_{21}]$$

remains uniformly bounded as $\sigma \to \infty$. This constraint is automatically satisfied using the ATEA design procedure described in this section.

In what follows, we give a step-by-step design algorithm for ATEA. In view of the above discussion, input parameters to the algorithm are $\overline{\Lambda}_b$, V_b^b, $\overline{\Lambda}_e$, W_{ee}, Λ_ℓ and μ_ℓ ($\ell = 1, \cdots, r$), as well as the integer r. Among these, the primary ones are:

(1) $\overline{\Lambda}_e$ and W_{ee} which shape the error term $\overline{M}_e(s)$ and,

(2) Λ_ℓ and μ_ℓ ($\ell = 1, \cdots, r$), which define the time-scale structure of the observer and thus have a strong impact on the size of the resulting controller gain.

The remaining input parameters, namely $\overline{\Lambda}_b$ and V_b^b, are considered as secondary inputs.

The ATEA design algorithm can be divided into three steps. Steps 1 and 2 work with design at individual subsystem levels for the assignment of the asymptotically finite and infinite eigenstructures respectively. In step 3, designs at each subsystem level in steps 1 and 2 are then put together to form a complete design for closed-loop transfer recovery.

Step 1 : This step deals with the assignment of asymptotically finite eigenstructure (i.e., slow time-scale structure) and makes use of subsystems (8) to (11) of Part 1. $\lambda(A_{aa}^-)$ are the stable invariant zeros of Σ_{yw} and they form eigenvalues of A_o in the set $\overline{\Lambda}_-$, while the corresponding left-eigenvectors of A_o coincide with the left state-zero directions of Σ_{yw}. To place the set of eigenvalues $\overline{\Lambda}_b$ and left-eigenvectors $\overline{V}_b$, we choose a gain K_b such that $\lambda(A_{bb}^c)$ coincides with $\overline{\Lambda}_b$ while V_b^b coincides with the set of left-eigenvectors of A_{bb}^c where

$$A_{bb}^c = A_{bb} - K_bC_b. \tag{17}$$

Note that existence of such a K_b is guaranteed by the property 2 of section III in Part 1 [1] and as long as the eigenvector set $\overline{V}_b$ is consistent with the freedom available in the eigenvector assignment [7]. Next, in order to place the set of eigenvalues $\overline{\Lambda}_e$ and right-eigenvectors W_{ee}, let us first form the matrices A_{ee} and C_e as follows,

$$A_{ee} = \begin{bmatrix} A_{aa}^{+} & 0 \\ B_c E_{ca}^{+} & A_{cc} \end{bmatrix}, \quad C_e = \begin{bmatrix} C_{e0} \\ C_{e1} \end{bmatrix} = \begin{bmatrix} C_{0a}^{+} & C_{0c} \\ E_a^{+} & E_c \end{bmatrix}, \tag{18}$$

where

$$E_a^{+} = [(E_{1a}^{+})', (E_{2a}^{+})', \cdots, (E_{m_f a}^{+})']',$$

$$E_{ia} = [E_{ia}^{+}, E_{ia}^{-}], \quad E_c = [E_{1c}', E_{2c}', \cdots, E_{m_f c}']'.$$

Now, a gain $K_e = [K_{e0}, K_{e1}]$ can be chosen such that the set of eigenvalues and right-eigenvectors of A_{ee}^c coincide with $\overline{\Lambda}_e$ and W_{ee} respectively where

$$A_{ee}^c = A_{ee} - K_e C_e = A_{ee} - K_{e0} C_{e0} - K_{e1} C_{e1}. \tag{19}$$

Again, note that existence of such a K_e is guaranteed by the property 2 of section III of [1] and as long as the eigenvector set W_{ee} is consistent with the freedom available in the eigenvector assignment [7]. For future use, let us define

$$A_{ee1} = A_{ee} - K_{e0} C_{e0}, \quad K_{e0} = \begin{bmatrix} K_{a0}^{+} \\ K_{c0} \end{bmatrix},$$

and partition K_{e1} as

$$K_{e1} = \begin{bmatrix} K_{e11}, & K_{e12}, & \cdots, & K_{e1m_f} \end{bmatrix} \tag{20}$$

where K_{e1i} is a vector of dimension $n_e \times 1$.

Step 2 : Here, we deal with the assignment of asymptotically infinite eigenstructure (i.e. the fast time-scale structure) and apply it to m_f subsystems represented by (13) of Part 1. This step is only needed when $n_f > 0$. As discussed earlier, a complete freedom is available to specify any number r ($r \leq n_f$) of fast time scales. The simplest case is to choose $r = 1$. However, for generality, we consider the case where $r > 1$. Assignment of the fast time scales is done through selection of the set Λ_ℓ and the corresponding positive parameters μ_ℓ ($\ell = 1, \cdots, r$).

The procedure for assigning the fast time-scale eigenstructure is again accomplished in a decentralized fashion. We consider the eigenassignment problem for each i-th single-input single-output subsystem $(i = 1, \cdots, m_f)$ separately. Thus, we need to distribute elements of the specified sets Λ_ℓ and the parameters μ_ℓ $(\ell = 1, \cdots, r)$ to each of the m_f subsystems. The distribution can be done in a number of ways. Let the subsystem i be assigned r_i different time scales for some $r_i \leq q_i$ and Λ_{ij}/μ_{ij} $(1 \leq j \leq r_i)$ be the asymptotically infinite eigenvalues assigned to the subsystem i. Define n_{ij} to be the number of eigenvalues corresponding to the time scale t/μ_{ij}. That is, the set Λ_{ij} contains n_{ij} elements. As usual, the set Λ_{ij} is assumed to be closed under complex conjugation. Also, in order to have a set of well separated time scales in the subsystem i, we assume that

$$\mu_{ij}/\mu_{ij+1} \to 0 \text{ as } \mu_{ij+1} \to 0 \text{ for all } j \; (1 \leq j \leq r_i - 1). \tag{21}$$

Obviously when $r = 1$, we have applied a single time scale to all subsystems , i.e all μ_{ij} are equal to a single parameter μ and all r_i are equal to unity. In this case, the tuning parameter σ can be taken as $1/\mu$. With these preliminaries, we are now ready to design for the i-th subsystem. At first, we will find a gain matrix K_{ij} for each time-scale t/μ_{ij} $(1 \leq j \leq r_i)$. For this subsystem, we define a matrix G_{ij} of dimension $n_{ij} \times n_{ij}$ and a matrix C_{ij} of dimension $1 \times n_{ij}$ with the following structure,

$$G_{ij} = \begin{bmatrix} 0 & I_{n_{ij}-1} \\ 0 & 0 \end{bmatrix} \text{ and } C_{ij} = \begin{bmatrix} 1 & 0 \end{bmatrix}.$$

An eigenvalue assignment on this subsystem is done using a gain vector K_{ij} of dimension $n_{ij} \times 1$ such that the eigenvalues of G_{ij}^c coincide with Λ_{ij} where $G_{ij}^c = G_{ij} - K_{ij}C_{ij}$. It is clear that one can always find such a matrix K_{ij} since the pair (G_{ij}, C_{ij}) defined above is observable. Let's further partition the matrix K_{ij} as

$$K_{ij} = \begin{bmatrix} K_{ijc} \\ K_{ijd} \end{bmatrix}$$

where the last element K_{ijd} is a scalar. Moreover, since the subsystem matrix G_{ij}^c is stable, the gain K_{ijd} must also be nonzero. Next, the gains K_{ij} $(1 \le j \le r_i)$ obtained above are put together to form a gain vector that gives the desired fast time scales in the i-th subsystem. In the process, we parameterize the design solution as a function of a tuning parameter σ. For the description of this part of the ATEA design procedure, let us define the following scalars,

$$J_{i1} = 1 \ , \ J_{ij} = \prod_{\ell=1}^{j-1} K_{i\ell d} \ , \ (2 \le j \le r_i).$$

and

$$\alpha_{i0} = 0 \ , \ \alpha_{ij} = \sum_{k=1}^{j} n_{ik} \ , \ (1 \le j \le r_i).$$

Note that $\alpha_{ir_i} = q_i$. Also, for each j $(1 \le j \le r_i)$,

$$\epsilon_{i\alpha_{ij-1}+1} = \epsilon_{i\alpha_{ij-1}+2} = \cdots = \epsilon_{i\alpha_{ij-1}+n_{ij}} = \mu_{ij}$$

and

$$\eta_i = \prod_{k=1}^{q_i} \epsilon_{ik}. \tag{22}$$

Now, we are ready to give the design gain $\tilde{K}_i(\sigma)$ for the subsystem i parameterized by a variable σ as follows,

$$\tilde{K}_i(\sigma) = \left[\tilde{K}'_{i1}(\sigma), \ \tilde{K}'_{i2}(\sigma), \ \cdots, \ \tilde{K}'_{ir_i}(\sigma)\right]' \tag{23}$$

where

$$\tilde{K}_{ij}(\sigma) = \frac{1}{\eta_i} J_{ij} S_{ij} K_{ij}.$$

and

$$S_{ij} = \text{Diag}\left[\prod_{\ell=\alpha_{ij-1}+2}^{q_i} \epsilon_{i\ell} \ , \ \prod_{\ell=\alpha_{ij-1}+3}^{q_i} \epsilon_{i\ell} \ , \ \cdots, \ \prod_{\ell=\alpha_{ij-1}+n_{ij}}^{q_i} \epsilon_{i\ell}\right]. \tag{24}$$

The product $\prod_{\ell=\alpha_{ij-1}+n_{ij}}^{q_i} \epsilon_{i\ell}$ in (24) is taken to be unity when $j = r_i$. The above design formulation becomes much simpler when $r_i = 1$. For this case, let $\overline{\mu}_i$ denote the time-scale parameter, then

$$\tilde{K}_i(\sigma) = \frac{1}{(\overline{\mu}_i)^{q_i}} \left[(\overline{\mu}_i)^{q_i-1} K_{i1}, \ (\overline{\mu}_i)^{q_i-2} K_{i2}, \ \cdots, \ K_{iq_i}\right]' \tag{25}$$

where the individual gains $K_{ij}(1 \le j \le q_i)$ have been previously selected such that the eigenvalues of G_i^c are placed at the desired locations and

$$G_i^c = -\begin{bmatrix} K_{i1}, \; K_{i2}, \cdots, \; K_{iq_i-1} & K_{iq_i} \\ -I_{q_i-1} & 0 \end{bmatrix}'.$$

Here we did not discuss any eigenvector assignment. However, it turns out that our eventual design is such that the eigenvectors corresponding to the asymptotically infinite eigenvalues are naturally assigned to appropriate locations so that $M_\infty(j\omega,\sigma) \to 0$ as $\sigma \to \infty$.

Step 3 : This constitutes the last step in the ATEA design procedure. Here, various gains obtained in steps 1 and 2 are combined to form an overall observer gain for the system Σ_{yw} parameterized by a tuning parameter σ. Let's define the gain matrix $\tilde{K}_{e1}$ as

$$\tilde{K}_{e1}(\sigma) = \begin{bmatrix} \tilde{K}_{a1}^{+}(\sigma) \\ \tilde{K}_{c1}(\sigma) \end{bmatrix} = \left[\tilde{K}_{e11}(\sigma), \; \tilde{K}_{e12}(\sigma), \; \cdots, \; \tilde{K}_{e1m_f}(\sigma)\right],$$

$$\tilde{K}_{e1i}(\sigma) = \frac{1}{\eta_i} J_{ir_i} K_{ir_i d} K_{e1i}. \tag{26}$$

Note that when $r_i = 1$, the gain K_{i1d} is the same as K_{iq_i} and η_i is the same as $(\overline{\mu}_i)^{q_i}$. For the case where $n_f > 0$, the observer gain $K(\sigma)$ adjustable by a single tuning parameter σ is given by

$$K(\sigma) = \Gamma_1 \, \tilde{K}(\sigma) \, \Gamma_2^{-1} \tag{27}$$

where

$$\tilde{K}(\sigma) = \begin{bmatrix} B_{0a}^- & L_{af}^- + \tilde{H}_{af}^- & L_{ab}^- + \tilde{H}_{ab}^- \\ B_{0a}^+ + K_{a0}^+ & L_{af}^+ + \tilde{H}_{af}^+ + \tilde{K}_{a1}^+(\sigma) & L_{ab}^+ + \tilde{H}_{ab}^+ \\ B_{0b} & L_{bf} + \tilde{H}_{bf} & K_b \\ B_{0c} + K_{c0} & L_{cf} + \tilde{H}_{cf} + \tilde{K}_{c1}(\sigma) & L_{cb} + \tilde{H}_{cb} \\ B_{0f} & L_f + \tilde{K}_f(\sigma) & 0 \end{bmatrix} \tag{28}$$

and

$$\tilde{K}_f(\sigma) = \mathrm{Diag}\left[\tilde{K}_1(\sigma), \; \tilde{K}_2(\sigma), \; \cdots, \tilde{K}_{m_f}(\sigma)\right],$$

$$L_f = \left[L_1', \; L_2', \; \cdots, L_{m_f}'\right]'$$

Note that the matrices $\tilde{H}^{+}_{af}$, $\tilde{H}^{+}_{ab}$, $\tilde{H}^{-}_{af}$, $\tilde{H}^{-}_{ab}$, $\tilde{H}_{bf}$, $\tilde{H}_{cf}$ and $\tilde{H}_{cb}$ are arbitrary but finite and used to introduce additional design freedom. We can now state the following theorem.

Theorem 1. *Consider a full-order observer-based controller with gain matrix given by (27) and assume that $n_f > 0$. Then, we have the following properties:*

1. *There exists a scalar σ^* such that, for all $\sigma > \sigma^*$, the observer design is asymptotically stable. Furthermore, the observer dynamic matrix has the following time-scale structures: t, t/μ_{ij} (where $j = 1, \cdots, r_i$ and $i = 1, \cdots, m_f$). That is, its eigenvalues as $\mu_r \longrightarrow 0$ are given by*

$$\overline{\Lambda}_{-} + 0(\mu_r)\ ,\ \overline{\Lambda}_b + 0(\mu_r)\ ,\ \overline{\Lambda}_e + 0(\mu_r)\ ,$$

$$\frac{\Lambda_{ij}}{\mu_{ij}} + 0(1) \text{ for } (j = 1, \cdots, r_i) \text{ and } (i = 1, \cdots, m_f).$$

Moreover, if $\tilde{H}^{-}_{af} = 0$ and $\tilde{H}_{bf} = 0$, some finite eigenvalues of A_o are exactly equal to $\overline{\Lambda}_{-}$ and $\overline{\Lambda}_b$ for all σ rather than asymptotically tending to $\overline{\Lambda}_{-}$ and $\overline{\Lambda}_b$.

2. *CLTR is achieved in the sense that, as $\sigma \longrightarrow \infty$, $M_f(s,\sigma) \longrightarrow \overline{M}_e(s)$ pointwise in s.*

Proof : See [5]. ∎

Remark 1. *For the case when $n_f = 0$, observer gains obtained from the above ATEA procedure are independent of σ and are given by*

$$K(\sigma) = \Gamma_1 \begin{bmatrix} B^{-}_{0a} & L^{-}_{ab} \\ B^{+}_{0a} + K^{+}_{a0} & L^{+}_{ab} \\ B_{0b} & K_b \\ B_{0c} + K_{c0} & L_{cb} \end{bmatrix} \Gamma_2^{-1}. \tag{29}$$

Furthermore, the eigenvalues of A_o are precisely those of $\overline{\Lambda}_{-} \cup \overline{\Lambda}_b \cup \overline{\Lambda}_e$ and $M_f(s,\sigma) = \overline{M}_e(s)$.

Remark 2. *When $T_{zw}(s)$ is an element of $\mathbf{T}^f_{\mathrm{R}}(\Sigma)$ and due to the special structure of F in (13), $\overline{M}_e(s)$ is identically zero irrespective of the way we select the set of $n^+_a + n_c$ eigenvalues in $\Lambda_e(\sigma)$ and the set of right- and left-eigenvectors $W_e(\sigma)$ and $V_e(\sigma)$.*

Clearly, the ATEA design method presented above has the attractive feature that it is decentralized. Different time scales and eigenstructures can be assigned to the subsystems of Σ_{yw} separately. The design in each subsystem does not require an explicit value of the tuning parameter σ. The variable σ enters only in (23) or (25) when designs for different subsystems are put together to form the final observer gain with the desired time-scale structure. The variable σ will act as a tuning parameter based on the chosen time scales for the fast observer dynamics.

B. ECLTR Design Via ATEA

In the previous subsection, we have presented ATEA design methodology and utilized it for ACLTR design. The power of this method is that it explores all the degrees of freedom available and provides at the end a family of parameterized controllers $\mathbf{C}(s,\sigma)$ to achieve closed-loop transfer recovery. Depending upon the particular design requirements, one would adjust the tuning parameter σ until a desired recovery is achieved. This asymptotic procedure is no longer needed when a given target closed-loop transfer function $T_{zw}(s)$ is exactly recoverable (i.e. $T_{zw}(s) \in \mathbf{T}^f_{\text{ER}}(\Sigma)$). As discussed in section II, when $T_{zw}(s) \in \mathbf{T}^f_{\text{ER}}(\Sigma)$, F has the form given in (14). With this particular structure of F, all eigenvalues of A_o can be assigned to finite locations and the above ATEA design procedure can be simplified drastically. In fact, the design requires only finite eigenstructure assignment and does not involve fast time-scale structure assignment. The intent of this section is to describe in detail the available design freedom and provide a step-by-step design procedure in the eigenstructure assignment of A_o for exact closed-loop transfer function recovery (ECLTR).

Note that for an exactly recoverable case, the observer gain K is no longer parameterized as a function of σ and thus, dependency of σ is dropped in the following discussion. Based on the different partitions of $M_f(s)$ in section II, the design freedom available for each subsystem is as follows,

1. A set of n_a^- eigenvalues of A_o in Λ_- must be chosen to coincide

exactly with the set of stable invariant zeros of the system Σ_{yw}. The corresponding left-eigenvectors of A_o must coincide exactly with the left state-zero directions of Σ_{yw} so that $M_-(s)$ is identically zero.

2. A set of n_b eigenvalues of A_o in Λ_b can be freely assigned to any finite locations in $\mathcal{C}^-$. Moreover, the set of eigenvectors V_b corresponding to these eigenvalues must be in the null space of $(B_1 - KD_{21})'$ and satisfying the constraints defined in [7]. The resulting $M_b(s)$ will be identically zero.

3. A set of $n_a^+ + n_c$ eigenvalues of A_o in Λ_e can be freely assigned to any finite locations in $\mathcal{C}^-$ subject to the condition that any unobservable eigenvalues of Σ_{yw} must be included in the set Λ_e. Moreover, the set of eigenvectors W_{ee} corresponding to these eigenvalues can be selected freely within the constraints defined in [7]. We note that due to the specific structure of F in (14), $M_e(s)$ is zero regardless of how we select Λ_e and W_{ee}. Note that this step is not needed when $n_a^+ + n_c = 0$, i.e when the system Σ_{yw} is of minimum phase and left-invertible.

4. A set of n_f eigenvalues of A_o in Λ_d can be freely assigned to any finite locations in $\mathcal{C}^-$. (The sets Λ_∞ and V_∞ are renamed as Λ_d and V_d to highlight the fact that these eigenvalues do not need to be at infinity for exact recovery). The set of eigenvectors V_d corresponding to these eigenvalues can be selected freely within the constraints defined in [7]. With F in the form of (14), the partition $M_\infty(s)$ is identically zero irrespective of how we select Λ_d and V_d.

We now proceed to the design of an observer gain K that produces the desired finite eigenstructure to A_o for the case of exact closed-loop recovery.

Step 1a : This step deals with the assignment of finite eigenstructure to the subsystem (10) of Part 1. We choose a gain K_b such that $\lambda(A_{bb}^c)$

coincides with Λ_b, a selected set of n_b eigenvalues in $\mathcal{C}^-$ where

$$A_{bb}^c = A_{bb} - K_b C_b. \tag{30}$$

Note that existence of such a K_b is guaranteed by the property 2 of section III of Part 1 [1]. The eigenvectors of A_{bb}^c can be freely assigned within the available freedom for eigenvector assignment [7]. Under the properties of s.c.b, the above ATEA design procedure always results in a set of eigenvectors V_b for the eigenvalues Λ_b of A_o that lie in the null space of $(B_1 - KD_{21})'$; hence rendering $M_b(s) = 0$.

Step 1b : This step deals with the assignment of finite eigenstructure to the subsystems (9), (11) and (13) of Part 1. Let the matrices A_x and C_x be defined as

$$A_x = \begin{bmatrix} A_{aa}^+ & 0 & L_{af}^+ C_f \\ B_c E_{ca}^+ & A_{cc} & L_{cf} C_f \\ B_f E_a^+ & B_f E_c & A_f \end{bmatrix}, \quad C_x = \begin{bmatrix} C_{0a}^+ & C_{0c} & C_{0f} \\ 0 & 0 & C_f \end{bmatrix}. \tag{31}$$

and $\Lambda_x \equiv \Lambda_e \cup \Lambda_d$ be a set of $n_a^+ + n_c + n_f$ eigenvalues in $\mathcal{C}^-$ which must include any unobservable eigenvalues of the system Σ_{yw}. Now, we select a gain K_x such that $\lambda(A_x^c)$ coincides with Λ_x where

$$A_x^c = A_x - K_x C_x. \tag{32}$$

Note that existence of such a K_x is guaranteed by the property 2 of section III of Part 1. The eigenvectors of A_x^c can be assigned within the freedom available for eigenvector assignment [7]. Let us partition the gain matrix K_x as follows,

$$K_x = \begin{bmatrix} K_{a0}^+ & K_{a1}^+ \\ K_{c0} & K_{c1} \\ K_{f0} & K_{f1} \end{bmatrix}.$$

Step 2 : Here, the gain matrices K_b and K_x obtained in step 1 are combined to give the desired observer gain for exact closed-loop transfer recovery. It is given by

$$K = \Gamma_1 \tilde{K} \Gamma_2^{-1} \tag{33}$$

where

$$\tilde{K} = \begin{bmatrix} B_{0a}^- & L_{af}^- & L_{ab}^- \\ B_{0a}^+ + K_{a0}^+ & K_{a1}^+ & L_{ab}^+ \\ B_{0b} & L_{bf} & K_b \\ B_{0c} + K_{c0} & K_{c1} & L_{cb} \\ B_{0f} + K_{f0} & K_{f1} & 0 \end{bmatrix}. \tag{34}$$

We have the following theorem.

Theorem 2. *Consider a full-order observer-based controller with a gain given by (33). Then, this full-order observer-based controller achieves CLTR and the eigenvalues of the resulting observer design are in the set* $\Lambda_- \cup \Lambda_b \cup \Lambda_x$.

Proof : See [5]. ∎

Remark 3. *In general, the observer gain for ECLTR is not unique.*

IV. OPTIMIZATION-BASED DESIGN METHODS

Clearly from section II, the whole notion of ACLTR is to make the recovery matrix

$$M_f(s) = F(sI_n - A + KC_2)^{-1}(B_1 - KD_{21})$$

as small as possible. The previously discussed design method ATEA accomplishes this task from the perspective of asymptotic time-scale and eigenstructure assignment to the observer dynamic matrix. An alternative method is to formulate the design problem in terms of finding a gain K that minimizes some norm (e.g. H_2 or H_∞) of $M_f(s)$. That is, one can cast the ACLTR design into an optimization problem. An optimal or suboptimal solution to such problem will provide the necessary observer design gain.

From this perspective, in this section, we will cast the closed-loop transfer recovery problem into a standard H_2- or H_∞- optimization problem. To begin, we consider the following auxiliary system,

$$\Sigma_a : \begin{cases} \dot{x} = A'x + C_2{}'u + F'w, \\ y = x, \\ z = B_1{}'x + D_{21}{}'u. \end{cases} \tag{35}$$

Here, w is modeled as an exogenous disturbance input to Σ_a and u is the control input. The variables y and z represent respectively the measured system states and the controlled outputs. If we consider a state-feedback law for the control u,

$$u = -K'x. \tag{36}$$

Then it is simple to verify that the closed-loop transfer function from w to z, denoted by $T_{zw}^{au}(s)$, is indeed equal to $M_f'(s)$. Now the ACLTR design problem can be casted into a problem of designing a state-feedback gain K' such that

(1) the auxiliary system Σ_a under the control law (36) is asymptotically stable and,

(2) the norm (H_2 or H_∞) of $M_f(s)$ is minimized.

There exists a vast literature on H_2 or H_∞ minimization methods. Borrowing from such a literature, subsection IV.A discusses algorithms for H_2 minimization of $M_f(s)$ while subsection IV.B does the same for H_∞ minimization. We want to emphasize that the optimization problem is cast here in terms of minimizing an appropriate norm of recovery matrix $M_f(s)$ rather than the actual recovery error $E(s)$.

It is well known that an optimal solution for either H_2 or H_∞ minimization of $M_f(s)$ does not necessarily exist, and the infimum of $||M_f(s)||_{H_2}$ or $||M_f(s)||_{H_\infty}$ is in general nonzero. However, for the class of exactly recoverable target closed-loop transfer functions $\mathbf{T}^f_{\mathrm{ER}}(\Sigma)$, the infimum of $||M_f(s)||_{H_2}$ or $||M_f(s)||_{H_\infty}$ is in fact zero and it can be attained using a finite gain K. Also, for another class of target closed-loop transfer functions, namely the class of asymptotically recoverable target closed-loops $\mathbf{T}^f_{\mathrm{AR}}(\Sigma)$, the infimum of $||M_f(s)||_{H_2}$ or $||M_f(s)||_{H_\infty}$ is also zero, and it can only be attained in an asymptotic sense by using larger and larger gain K. Whether the infimum of $||M_f(s)||_{H_2}$ or $||M_f(s)||_{H_\infty}$ is zero or not, the recovery procedure involves generating a sequence of gains with the property that in the limit H_2- or H_∞-norms of the recovery matrices approaches the infimum of $||M_f(s)||_{H_2}$ or $||M_f(s)||_{H_\infty}$ over the set of all possible gains. One normally settles with a suboptimal solution corresponding to a particular member of the sequence. In

H_2-optimization, an observer gain is generated via the solution of an algebraic Riccati equation (called hereafter H_2-ARE) parameterized in terms of a tuning parameter σ. A sequence of suboptimal gains is generated by letting σ tend to ∞. Similarly, if we let γ^* to be the infimum of $||M_f(s)||_{H_\infty}$ over the set of all possible gains, then for given a parameter γ greater than γ^*, one generates in H_∞-optimization a gain by solving an algebraic Riccati equation (called hereafter H_∞-ARE) parameterized in terms of γ so that the resulting $||M_f(s,\gamma)||_{H_\infty}$ is strictly less than γ. By gradually reducing γ, one thereby generates a sequence of suboptimal gains.

For simplicity and without loss of generality, we assume throughout this section that the matrix D_{21} is of the form,

$$D_{21} = \begin{bmatrix} I_{m_0} & 0 \\ 0 & 0 \end{bmatrix}.$$

Also, we partition the matrices B_1 and C_2 as

$$B_1 = [\, B_{1,0}, \quad B_{1,1} \,] \quad \text{and} \quad C_2 = \begin{bmatrix} C_{2,0} \\ C_{2,1} \end{bmatrix},$$

and let $A_1 = A - B_{1,0}C_{2,0}$. In the next two sections, we examine specific algorithms for the design of observer gain in the problem of CLTR using H_2 - and H_∞-optimization methods.

A. CLTR Design Via H_2-Optimization

In this subsection, we consider H_2-norm minimization of $M_f(s)$ or equivalently $T_{zw}^{au}(s)$. At first, let us look at an elegant way of computing the infimum value of $||M_f(s)||_{H_2}$ in a recent work by Stoorvogel [15]. We first recall the following lemma.

Lemma 1. *Assume that (A, C_2) is detectable. Then the infimum of $||M_f(s)||_{H_2}$ over all the stabilizing observer gains is given by*

$$\mathit{Trace}\{F\overline{P}F'\},$$

where $\overline{P} \in \Re^{n\times n}$ is the unique positive semi-definite matrix satisfying:

1. $F(\overline{P}) = \begin{bmatrix} A\overline{P} + \overline{P}A' + B_1B_1' & \overline{P}C_2' + B_1D_{21}' \\ C_2\overline{P} + D_{21}B_1' & D_{21}D_{21}' \end{bmatrix} \geq 0,$

2. $rank\, F(\overline{P}) = normrank\, \{C_2(sI_n - A)^{-1}B_1 + D_{21}\}\ \forall s \in \mathcal{C}^+/\mathcal{C}^o,$

3. $rank \begin{bmatrix} [sI - A', \; -C_2'] \\ F(\overline{P}) \end{bmatrix} = n + normrank\, \{C_2(sI_n - A)^{-1}B_1 + D_{21}\}\ \forall s \in \mathcal{C}^+/\mathcal{C}^o.$

Here normrank{·} denotes the rank of matrix {·} over the field of rational functions.

Proof : See Stoorvogel [15]. ∎

In general, as discussed earlier, the infimum of $||M_f(s)||_{H_2}$ can only be obtained asymptotically. In what follows, we give an algorithm that produces a sequence of parameterized observer gains $K(\sigma)$ for the general system Σ_a such that the H_2-norm of the recovery matrix, which is also parameterized by σ and is denoted by $M_f'(s, \sigma) = T_{zw}^{au}(s, \sigma)$, tends to the infimum of $||M_f(s)||_{H_2}$ as $\sigma \to \infty$. The algorithm consists of the following two steps:

Step 1 : Solve the following parameterized algebraic Riccati equation (H_2-ARE) for a given value of σ,

$$A_1\tilde{P} + \tilde{P}A_1' - \tilde{P}C_{2,0}'C_{2,0}\tilde{P} - \sigma\tilde{P}C_{2,1}'C_{2,1}\tilde{P} + B_{1,1}B_{1,1}' + \frac{1}{\sigma}I_n = 0, \quad (37)$$

for a positive definite solution $\tilde{P}$. We note that a unique positive definite solution $\tilde{P}$ of (37) always exists for all $\sigma > 0$. Obviously, $\tilde{P}$ is a function of σ and is denoted by $\tilde{P}(\sigma)$.

Step 2 : Let

$$K(\sigma) = [B_{1,0} + \tilde{P}(\sigma)C_{2,0}', \; \sigma\tilde{P}(\sigma)C_{2,1}']. \quad (38)$$

We have the following theorem.

Theorem 3. *Consider a full-order observer-based controller with a gain given in (38) and let $M_f(s, \sigma)$ be the resulting recovery matrix.*

Then, we have

$$\lim_{\sigma\to\infty} \tilde{P}(\sigma) = \overline{P}$$

Moreover, $||M_f(s,\sigma)||_{H_2}$ *tends to the infimum of* $||M_f(s)||_{H_2}$ as $\sigma \to \infty$, *i.e.*

$$\lim_{\sigma\to\infty} ||M_f(s,\sigma)||_{H_2} = \text{Trace}\{F\overline{P}F'\}.$$

Proof : See [5]. ∎

In view of theorem 3, it is apparent that as σ takes on larger and larger values, the design algorithm given above generates a sequence of observer gains having the property that in the limit $||M_f(s,\sigma)||_{H_2}$ approaches the infimum of $||M_f(s)||_{H_2}$ over the set of all possible gains. A suboptimal solution would result with any chosen value of the parameter σ. However, for some particular class of systems, e.g. the well-known *regular problems* (i.e. D_{21} is surjective implying that Σ_{yw} is right-invertible and has no infinite zeros, and Σ_{yw} has no invariant zeros on the $j\omega$ axis), the infimum value of $||M_f(s)||_{H_2}$ can be achieved with the following observer gain [6],

$$K = B_{1,0} + \tilde{P}C'_{2,0}, \tag{39}$$

where $\tilde{P}$ is the positive semi-definite solution of

$$A_1\tilde{P} + \tilde{P}A'_1 - \tilde{P}C'_{2,0}C_{2,0}\tilde{P} + B_{1,1}B'_{1,1} = 0.$$

The resulting infimum value of $||M_f(s)||_{H_2}$ is given by,

$$||M_f(s)||_{H_2} = \text{Trace}\{F\tilde{P}F'\}.$$

Note that in this case, the observer gain K and thus the resulting recovery matrix is not parameterized as a function of σ. We note that for a regular problem when $||M_f(s)||_{H_2} = 0$, the observer gain K as given in (39) will achieve exact loop transfer recovery (ECLTR). However, to our knowledge, no optimization-based method exists in the literature to solve for the required gain that achieves $||M_f(s)||_{H_2} = 0$, whenever it is possible, for a general class of systems other than the class of regular systems. On the other hand, a direct design procedure based on ATEA will yield an ECLTR design gain *whenever it can be done*; the algorithm is presented in subsection III.B.

Another special case of interest is as follows. Consider a left-invertible minimum-phase system Σ_{yw} which is non-strictly proper. Let the observer gain $K(\sigma)$ be given by

$$K(\sigma) = [B_{1,0},\ \sigma \tilde{P}(\sigma) C'_{2,1}],$$

where $\tilde{P}(\sigma) := \tilde{P}$ is the positive definite solution of

$$A_1 \tilde{P} + \tilde{P} A'_1 - \sigma \tilde{P} C'_{2,1} C_{2,1} \tilde{P} + B_{1,1} B'_{1,1} = 0.$$

It is simple to show that the observer gain $K(\sigma)$ given above achieves asymptotic closed-loop transfer recovery (ACLTR), i.e. the resulting $||M_f(s,\sigma)||_{H_2}$ tends to zero asymptotically as $\sigma \to \infty$. The above result has been given earlier by Chen *et al* [3].

It is of interest to investigate what type of time-scale structure and eigenstructure is assigned to the observer dynamic matrix A_o by the gain $K(\sigma)$ obtained via the algorithm given in equations (37) and (38). Clearly, the algorithm will make $M_-(s,\sigma)$, $M_b(s,\sigma)$ and $M_\infty(s,\sigma)$ zero as $\sigma \to \infty$, while shaping $M_e(s,\sigma)$ in a particular way so that the infimum of $||M_f(s)||_{H_2}$ is attained as $\sigma \to \infty$. In so doing, among all the possible choices for the time-scale structure and eigenstructure of A_o, it selects a particular choice which can easily be deduced from the results of cheap and singular control problems in [14] (see also, [17] and [9]). We have the following results.

1. As $\sigma \to \infty$, the asymptotic limits of the set of n_a^- eigenvalues $\Lambda_-(\sigma)$ and the associated set of left eigenvectors $V_-(\sigma)$ of A_o coincide respectively with the set of stable invariant zeros and the corresponding left state zero directions of Σ_{yw}. This renders $M_-(s,\sigma)$ zero as $\sigma \to \infty$.

2. As $\sigma \to \infty$, some of the n_b eigenvalues in $\Lambda_b(\sigma)$ coincide with the stable but uncontrollable eigenvalues of Σ_{yw} while the rest of them coincide with what are called 'compromise' zeros of Σ_{yw}

[14]. Also, the asymptotic limits of the associated left eigenvectors, namely $V_b(\sigma)$, fall in the null space of matrix $[B_1 - K(\sigma)D_{21}]'$ so that $M_b(s,\sigma) \to 0$ as $\sigma \to \infty$.

3. As $\sigma \to \infty$, the set of n_f eigenvalues $\Lambda_\infty(\sigma)$ of A_o tend to asymptotically infinite locations in such a way that $M_\infty(s,\sigma) \to 0$. The time-scale structure assigned to these eigenvalues depends on the infinite zero structure of Σ_{yw} (see for details in [14]). Also, the eigenvalues assigned to each fast time-scale follow asymptotically a Butterworth pattern.

4. As $\sigma \to \infty$, the asymptotic limits of n_a^+ eigenvalues in $\Lambda_e(\sigma)$ coincide with the mirror images of unstable invariant zeros of Σ_{yw}, while the associated set of left-eigenvectors of A_o coincide with the corresponding right input-zero directions of Σ_{yw}. The rest of n_c eigenvalues of $\Lambda_e(\sigma)$, as $\sigma \to \infty$, tend to some other finite locations, while the associated left-eigenvectors follow some particular directions. In the limiting process, it shapes the recovery matrix $\overline{M}_e(s)$ in a particular way so that the infimum of $||M_f(s)||_{H_2}$ is attained as $\sigma \to \infty$.

To conclude, we note that, as in the ATEA design procedure, the algorithm of equations (37) and (38) will render $M_-(s,\sigma)$, $M_b(s,\sigma)$ and $M_\infty(s,\sigma)$ zero asymptotically as $\sigma \to \infty$. Moreover, it shapes $M_e(s)$ in a particular way so that the infimum of $||M_f(s)||_{H_2}$ is attained as $\sigma \to \infty$. In contrast to this, ATEA design procedure of section III allows complete available freedom to shape the limit of recovery matrix $\overline{M}_e(s)$ in a variety of ways within the design constraints imposed by structural properties of the given system.

B. CLTR Design Via H_∞-Optimization

In this subsection, we consider H_∞-norm minimization of $M_f(s)$ or equivalently $T_{zw}^{au}(s)$. Unlike the case of H_2 -norm minimization in previous subsection, there are in general no direct methods available of exactly computing the infimum of $||M_f(s)||_{H_\infty}$, denoted here by γ^*. However, there are iterative algorithms that can approximate γ^*, at least in principle, to an arbitrary degree of accuracy (See for example [8]). Recently, for a particular class of problems, i.e. when Σ_{yw} is left-invertible and has no invariant zeros on the $j\omega$ axis, such an infimum γ^* can be explicitly calculated [2].

We now proceed to present an algorithm for computing the observer-gain matrix K such that the resulting H_∞-norm of the recovery matrix $M_f(s,\gamma)$ is less than an a-priori given scalar $\gamma > \gamma^*$. The algorithm is as follows:

Step 0 : Choose a value $\epsilon = 1$.

Step 1 : Solve the following algebraic Riccati equation (H_∞-ARE),

$$A_1\tilde{P}+\tilde{P}A_1'-\tilde{P}C_{2,0}'C_{2,0}\tilde{P}-\frac{1}{\epsilon}\tilde{P}C_{2,1}'C_{2,1}\tilde{P}+B_{1,1}B_{1,1}'+\frac{1}{\gamma^2}\tilde{P}F'F\tilde{P}+\epsilon I_n = 0, \tag{40}$$

for $\tilde{P}$. Evidently, since (40) is parameterized in terms of γ, the solution $\tilde{P}$ will be a function of γ and is denoted by $\tilde{P}(\gamma)$.

Step 2 : If $\tilde{P}(\gamma) > 0$, then we proceed to **Step 3**. Otherwise, we decrease ϵ and go back to **Step 1**. Note that for $\gamma > \gamma^*$, it is shown in [18] that there always exists a sufficiently small scalar $\epsilon^* > 0$ such that the H_∞-ARE (40) has a unique positive definite solution $\tilde{P}(\gamma)$ for each $\epsilon \in (0, \epsilon^*)$.

Step 3 : Let

$$K(\gamma) = [B_{1,0} + \tilde{P}(\gamma)C_{2,0}', \; \frac{1}{2\epsilon}\tilde{P}(\gamma)C_{2,1}']. \tag{41}$$

We have the following theorem.

Theorem 4. *Consider a full-order observer-based controller with a gain determined from (41). Let $M_f(s,\gamma)$ be the resulting recovery matrix. Then, $\|M_f(s,\gamma)\|_{H_\infty}$ is strictly less than γ and tends to γ^* as $\gamma \to \gamma^*$.*

Proof : It follows simply from the results of [18]. ∎

Remark 4. *We note that γ acts here as a tuning parameter. Since at the beginning we do not know γ^*, it could very well happen that* *a chosen value of γ may turn out to be less than γ^*. In that case, the H_∞-ARE (40) does not have any positive definite solution even for sufficiently small ϵ. Then, one has to increase the value of γ and try to solve the H_∞-ARE once again for $\tilde{P}(\gamma) > 0$. One has to repeat this procedure as many times as necessary.*

For the special case of *regular problems* defined in subsection IV.A, there exists a method of finding the gain that does not require the additional parameter ϵ. It is given by [6],

$$K(\gamma) = B_{1,0} + \tilde{P}(\gamma)C'_{2,0}, \tag{42}$$

where $\tilde{P}(\gamma) := \tilde{P}$ is the positive semi-definite solution of

$$A_1\tilde{P} + \tilde{P}A'_1 - \tilde{P}C'_{2,0}C_{2,0}\tilde{P} + B_{1,1}B'_{1,1} + \frac{1}{\gamma^2}\tilde{P}F'F\tilde{P} = 0,$$

such that $\lambda(A'_1 - C'_{2,0}C_{2,0}\tilde{P} + \gamma^{-2}F'F\tilde{P}) \subseteq \mathcal{C}^-$. A full-order observer-based controller with gain obtained from (42) will produce a CLTR design such that $||M_f(s,\gamma)||_{H_\infty}$ is strictly less than γ.

Apparently, the gain $K(\gamma)$ obtained via the H_∞-optimization algorithm of equations (40) and (41) assigns a particular time-scale structure and eigenstructure to the observer dynamic matrix A_o. An investigation into the exact nature of time-scale structure and the eigenstructure of A_o as $\gamma \to \gamma^*$ is still an open research problem. But we like to point out that, as in the ATEA design procedure, the H_∞-optimization algorithm makes the corresponding $M_-(s,\gamma)$, $M_b(s,\gamma)$ and $M_\infty(s,\gamma)$ zero asymptotically as $\gamma \to \gamma^*$. Also, the corresponding $\overline{M}_e(s)$ is shaped in a particular way so that the infimum of $||M_f(s)||_{H_\infty}$ is attained as $\gamma \to \gamma^*$. In so doing, in addition to $\Lambda_\infty(\gamma)$, some elements of $\Lambda_e(\gamma)$ may be pushed to infinite locations in $\mathcal{C}^-$ as $\gamma \to \gamma^*$. Investigation of these and other properties of H_∞-optimization algorithm of equations (40) and (41) is outside the scope of this paper.

V. COMPARISON OF 'ATEA' AND 'ARE'-BASED DESIGN ALGORITHMS

A comparison is needed between optimal or suboptimal design schemes based on solving algebraic Riccati equations (ARE's) as described in section IV and the asymptotic time-scale and eigenstructure assignment (ATEA) design schemes of section III. In this regard, our earlier paper [9] discusses several relative advantages and disadvantages of ATEA and ARE-based designs. Here, we examine ATEA design and optimization-based designs from two different perspectives: (i) numerical simplicity and, (ii) flexibility to use all the available design freedom.

Let us first consider the numerical aspects of both design methods. It is clear that the major part of optimization-based design algorithms of section IV lies in solving positive definite solution of a parameter-dependent ARE repeatedly for different values of the parameter σ or ϵ. It is well-known that these ARE's become numerically 'stiff' when the parameter takes on values that are close to a critical value. To be specific, the H_2-ARE becomes stiff as the parameter σ takes on larger and larger values, while the H_∞-ARE becomes stiff when γ approaches γ^*. This is due to the interaction of fast and slow dynamics inherent in such equations. Thus, the numerical difficulties occur not only due to the repetitive solution of ARE's but also due to the 'stiffness' of such equations. On the other hand, as evident from section III, ATEA adopts a decentralized design approach and in so doing alleviates both the problem of stiffness and the need for repetitive solution of algebraic equations. That is, in ATEA, interaction between the slow and various fast time-scales is isolated by working with the asymptotically finite and infinite eigenstructures in the observer dynamic matrix separately. The tuning parameter σ merely adjusts the relative size of different fast time scales and is introduced only parametrically in the construction of the final gain. Hence, this procedure presents no numerical difficulties whatsoever as the parameter takes on larger and larger values.

Another factor of importance in selecting a design procedure is its flexibility in addressing all the available design freedom. As summarized in section II, there exists considerable amount of freedom to shape the recovery matrix through eigenstructure assignment to the observer dynamic matrix A_o. Such a freedom can be utilized to shape $\overline{M}_e(s)$, the limit of the recovery matrix. Any optimization-based method adopts a particular way of shaping $\overline{M}_e(s)$ as dictated by the mathematical minimization procedure. For example, as discussed earlier, in H_2 optimization $\overline{M}_e(s)$ is shaped by assigning some of the eigenvalues of A_o to the mirror images of the unstable invariant zeros of Σ_{yw}, while the associated set of left-eigenvectors of A_o coincide with the corresponding right input-zero directions of Σ_{yw}. Such a shaping obviously limits the available design freedom, and may or may not be desirable from an engineering point of view. Next, available design freedom can also be utilized to characterize appropriately the behavior of asymptotically infinite or otherwise called fast eigenvalues of A_o. What we mean by the behavior of fast eigenvalues is: (a) their asymptotic directions and, (b) the rate at which they go to infinity, i.e. the fast time-scale structure of A_o. As demonstrated in [9], the behavior of fast eigenvalues has a dramatic effect on the resulting controller bandwidth. Again, optimization-based design methods fix the behavior of fast eigenvalues in a particular way that may or may not be favorable to the designer's goals. We believe that the ability to utilize all the available design freedom is a valuable asset; in particular, exploring such a freedom in the case where complete recovery is not feasible is of dire importance. ATEA design methods of section III put all the available design freedom in the hands of designer and hence are preferable to optimization-based designs of section IV. However, a clear advantage of the optimization-based schemes is that at the onset of design, they do not require much systematic planning and hence are straightforward to apply. In fact, one simply solves the concerned ARE's repeatedly for several values of the tuning parameter until an appropriate suboptimal design is found. Admittedly, ATEA design does not have such a simplicity. One needs in ATEA design to come up with a careful utilization of the available design freedom and thus the selection of available design parameters that meet the practical design specifications. This can be done by a simple iterative adjustment. At each iteration, the required calculation in ATEA design procedure is straightforward and computationally inexpensive with added advantage that the algorithm does not involve solving any 'stiff' equations.

VI. NUMERICAL EXAMPLES

Control of flexible mechanical systems has been of interests in recent years. Presented in this section is a design example for closed-loop transfer recovery taken from the benchmark problem for robust control of a flexible mechanical system [16]. Although simple in nature, this problem will however provide us a design case where the concept of closed-loop transfer recovery can be fully illustrated. The problem is to control the displacement of the second mass by applying a force to the first mass as shown in figure 1 below. At the start, it is simple to verify that the basic open-loop system has a pair of degenerate eigenvalues at the origin and one pair of flexible modes. Equations for the dynamic

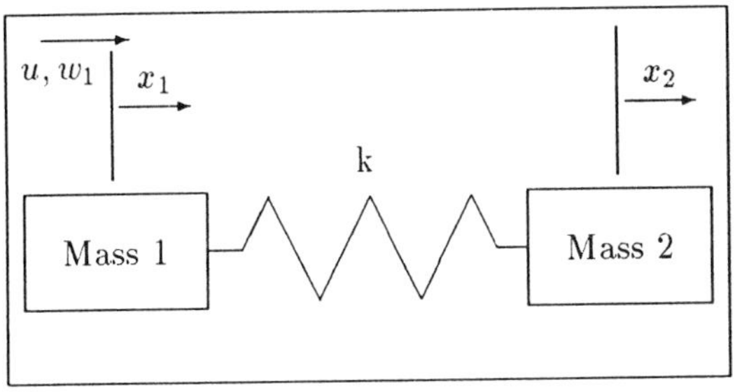

Figure 1: A Two-Mass-Spring Mass System

model are given below,

$$\begin{aligned} m_1 \ddot{x}_1 &= k(x_2 - x_1) + u + w_1 \\ m_2 \ddot{x}_2 &= k(x_1 - x_2) \end{aligned} \tag{43}$$

or

$$\begin{bmatrix} \dot{x}_1 \\ \dot{x}_2 \\ \dot{v}_1 \\ \dot{v}_2 \end{bmatrix} = \begin{bmatrix} 0 & 0 & 1 & 0 \\ 0 & 0 & 0 & 1 \\ -k/m_1 & k/m_1 & 0 & 0 \\ k/m_2 & -k/m_2 & 0 & 0 \end{bmatrix} \begin{bmatrix} x_1 \\ x_2 \\ v_1 \\ v_2 \end{bmatrix} + \begin{bmatrix} 0 \\ 0 \\ 1/m_1 \\ 0 \end{bmatrix} (u + w_1) \tag{44}$$

where $m_1 = m_2 = k_1 = k_2 = 1$, x_i, $i = 1, 2$, is the position of the i-th mass and v_i the velocity. The plant model used for robust control synthesis and closed-loop transfer recovery is given by,

$$\dot{x} = \begin{bmatrix} 0 & 0 & 1 & 0 \\ 0 & 0 & 0 & 1 \\ -1 & 1 & 0 & 0 \\ 1 & -1 & 0 & 0 \end{bmatrix} x + \begin{bmatrix} 0 & 0 \\ 0 & 0 \\ 1 & 1 \\ 0 & -1 \end{bmatrix} \begin{bmatrix} w_1 \\ w_2 \end{bmatrix} + \begin{bmatrix} 0 \\ 0 \\ 1 \\ 0 \end{bmatrix} u \tag{45}$$

$$\begin{bmatrix} z_1 \\ z_2 \end{bmatrix} = \begin{bmatrix} 1 & -1 & 0 & 0 \\ 0 & 1 & 0 & 0 \end{bmatrix} x$$

The disturbance input w_2 and the controlled output z_1 are introduced as fictitious input and output variables to represent effects of the uncertain spring parameter k in a feedback setting [19]. The second controlled output z_2 is a performance variable whose response to a unit-impulse disturbance applied at w_1 must have a settling time of 15 seconds. To begin, we synthesize a state-feedback design for robust H_∞-control to variation in the spring constant k according to the results of [19] and define it as our target closed-loop transfer function . The full-state feedback gain matrix is given as follows,

$$F = [\,1.5059 \quad -0.4942 \quad 1.7379 \quad 0.93181\,] \tag{46}$$

Singular value plot of this closed-loop transfer function $T_{zw}(j\omega)$ is shown in figure 2. Transient responses to a unit impulse input $w_1 = \delta(t)$ are given in figure 3 showing that position response of the second mass has a settling time less than 15 *sec*, as specified in the design challenge problem of [16]. This state-feedback design is extremely robust to variation in the spring constant k; namely the closed-loop system remains stable for $0 < k < \infty$. Our design goal is to recover the performance achieved under this state-feedback design using output-feedback and full-order observer-based controllers in (5).

Before we enter into the synthesis of CLTR, it is important to note that the recovery is highly dependent on the structural properties between the selected set of measurement variables y and the disturbance inputs w. First of all, the analysis given in Part 1 [1] can be used to examine different possibilities in the measurement output selection for the closed-loop recovery. Table I summarizes the results of CLTR analysis for 7 sets of measurement variables. Note that the analysis points out immediately that, with the target closed-loop transfer function being arbitrary, exact recovery (i.e ECLTR) is not possible. From the given set of sensors, only three of them (namely case c, case f and case g) can be used to achieve asymptotic recovery. That is, in these cases, one can design a full-order observer-based controller with high gain to recover the target closed-loop transfer function asymptotically. In subsection VI.A we present a CLTR design using the sensor set in case b. This set is the same as the one stated in [16] and the target closed-loop transfer function is unfortunately not recoverable in this case. Subsection 6.2 will present the results of an asymptotically recoverable design using the measurement set in case g. Notice that in the latter case, asymptotic recovery of the target transfer function $T_{zw}(s)$ can only be done with high observer gain.

A. Non-Recoverable Design Case

Let's consider the case where only measurement of position of second mass is available, i.e in case b of Table I where

$$C_2 = \begin{bmatrix} 0 & 1 & 0 & 0 \end{bmatrix} , \; D_{21} = \begin{bmatrix} 0 & 0 \end{bmatrix} , \; D_{22} = 0 \qquad (47)$$

From Table I, we recognize immediately that there are severe limitations in the recovery design imposed by the presence of the $M_e(s)$-term with $n_c = 2$. Under this circumstance, there are clearly a variety of ways one can shape the term $\overline{M}_e(s)$ in our full-order observer-based controller designs. Thus, it is critical that one examines our design goals carefully in terms of defining the limit of recovery matrix $\overline{M}_e(s)$. It should be noted that of all the three methods presented, only the ATEA procedure can be used to arbitrarily shape the term $\overline{M}_e(s)$. Optimization-based methods will produce a limit of recovery matrix $\overline{M}_e(s)$ as discussed in section V. Table II shows a sequence of design possibilities using the three methods discussed in sections III and IV. Of course, within each design method, the generated sequence is parameterized by its own scalar tuning parameter; for example, in the ATEA design we invoke the tuning parameter σ with a properly selected time-scale structure, while in H_2- and H_∞-optimization methods the parameters are σ and γ respectively. Tendency of all these sequences of designs is to make the corresponding partitions $M_-(s,c)$, $M_b(s,c)$ and $M_\infty(s,c)$ zero as the tuning parameter c (i.e. $c = \sigma$ for the ATEA design and for H_2-optimization design, and $c = \gamma$ for the H_∞-optimization design) tends to the appropriate limiting value. As seen in figure 4, the norm of the observer-gain matrix $K(c)$ in these sequences of recovery designs grows unbounded even in the case where closed-loop transfer recovery is not possible. Shown below are the corresponding limiting recovery matrix $\overline{M}_e(s)$ for each of the design methods. In this example, the limit of recovery matrix for H_2-optimization is

$$\overline{M}_e(s) = \frac{[0.1499s + 0.0376, \quad 0.0434s + 0.0684]}{s^2 + 0.7071s + 0.25} \tag{48}$$

while the one for the ATEA design is as follows,

$$\overline{M}_e(s) = \frac{[0.1082s + 0.0122, \quad 0.0434s + 0.0648]}{s^2 + 0.625s + 0.08125} \tag{49}$$

For the H_∞-optimization, the corresponding limit of recovery matrix cannot be determined since one of the eigenvalues tends to infinity. Plots of the singular values of the recovery matrix $M_f(s)$ are shown in figure 5 for the limiting case (i.e corresponding to design case number 11 of Table II) in all three design methods. Notice that the curves corresponding to the ATEA and H_2 design cases approach the limit of recovery matrix $\overline{M}_e(s)$ given in equations (48) and (49) respectively.

Due to these differences in $\overline{M}_e(s)$, we have therefore different closed-loop transfer recovery error $E(s)$. The L_∞-norm of $E(s)$ is shown in figure 6 corresponding to every design case listed in Table II. Note that none of the design methods will yield ECLTR or ACLTR as evident from the analysis summarized in Table I. It clearly shows that the error cannot be made zero regardless of what method we use for the recovery design. However, depending on the way we select $\overline{M}_e(s)$, different design considerations can be traded-off in this non-recoverable case. For example, one design consideration is the robustness to variation in the spring constant k; the results are shown in figures 7 to 9 for the ATEA, H_2 and H_∞-optimization methods respectively. Clearly, due to its complete flexibility in selecting $\overline{M}_e(s)$, the ATEA design procedure can be used to yield significantly better stability robustness. Transient responses to a unit impulse input in w_2 are shown in figures 10 to 12. Clearly, response peaks from all the output-feedback full-order observer-based controller designs are much higher than those achieved under state feedback. Results using optimization-based methods have the desired settling time of 15 sec. However, robustness of these designs are not as good as those from the ATEA design method which however exhibits a longer settling time. The example illustrates that there are many degrees of freedom in selecting $\overline{M}_e(s)$; the one shown here from the ATEA design is not meant to be the best among all possible choice of $\overline{M}_e(s)$. For example, one would need to modify $\overline{M}_e(s)$ if the settling time of 15 seconds is judged more important than the robustness consideration. An advantage of the ATEA method is that the observer design gain can be written explicitly in terms of the tuning parameter σ as follows,

$$K(\sigma) = \begin{bmatrix} 2.6562\sigma^2 - 4\sigma \\ 4\sigma \\ -3.9047\sigma^2 + 1.6 \\ 4.25\sigma^2 - 1.6 \end{bmatrix} \tag{50}$$

B. Asymptotically Recoverable Design Case

In this subsection, we consider the case where both position and acceleration of the second mass are available for feedback. The measurement output equation is given by

$$C_2 = \begin{bmatrix} 0 & 1 & 0 & 0 \\ 1 & -1 & 0 & 0 \end{bmatrix}, \quad D_{21} = \begin{bmatrix} 0 & 0 \\ 0 & 0 \end{bmatrix}, \quad D_{22} = \begin{bmatrix} 0 \\ 0 \end{bmatrix} \tag{51}$$

This corresponds to case g in Table I. Since $n_a^+ = n_c = 0$, the target closed-loop transfer function will be recoverable, but having $n_f = 4$ such a system can only recovered asymptotically. Furthermore, this implies that in all design methods the limit of recovery matrix $\overline{M}_e$ will be identically zero, and hence the H_∞-norm of the actual recovery error $E(s)$ approaches zero asymptotically as shown in figure 13. Table III shows a sequence of design possibilities from the three methods discussed in sections III and IV. Again the observer-gain in these sequences of designs for asymptotic recovery will become unbounded;

these results are shown in figure 14. Robustness results are shown in figures 15 to 17. Possibly, the significance of the ATEA design method lies in its flexibility to shape the design outcomes; for example, figure 15 shows that the ATEA design has significantly better robustness at the low-gain region where they can tolerate infinitely large change in the spring constant k. Transient responses to a unit-impulse input in w_1 are shown in figures 18 to 20. By increasing the observer gain, all the responses will approach those achieved under state feedback; this is to be expected for an asymptotically recoverable case (ACLTR). Again, in the ATEA method the observer design gain can be written explicitly in terms of the tuning parameter σ as follows,

$$K(\sigma) = \begin{bmatrix} 2\sigma & 2\sigma \\ 2\sigma & 0 \\ \sigma^2 & \sigma^2 - 1 \\ \sigma^2 & 1 \end{bmatrix} \tag{52}$$

VII. CONCLUSIONS

Presented in this paper are three design methods for closed-loop transfer recovery. Also discussed is the significance of each design method in terms of design simplicity, numerical difficulty and flexibility in utilizing the available design freedom to shape the limit of recovery matrix. All three methods of design provide explicit means of determining the appropriate observer gain matrix for closed-loop transfer recovery whenever it is possible. The design solution is usually characterized by a tuning parameter. In optimization-based methods, the gain is implicitly parameterized in terms of the solution of parameterized nonlinear algebraic Riccati equations (ARE's). On the other hand, ATEA design does not require solution of nonlinear algebraic equations. Here the tuning parameter enters the design solution only in the final step where we construct the composite observer gains from several subsystem designs. As the tuning parameter tends to its limiting value, a sequence of observer-based controllers is generated that yields in the limit a certain recovery matrix. For optimization-based methods, the limiting norm of the recovery matrix is simply the respective infimum under H_2 (or H_∞)-optimization over all possible stabilizing observer gains. Optimization-based methods usually shape the recovery matrix in a particular way which may or may not be meaningful in an engineering point of view. Our experiences indicate that these designs have unnecessarily high gain than those achieved under the ATEA methods for a comparable size of the recovery error. Fundamentally, the ATEA method offers complete flexibility in shaping the recovery error utilizing all the available design freedom within the constraints imposed by the structural properties of the given system. In the case of exact closed-loop transfer recovery (ECLTR), the ATEA method can in all circumstances be used to obtain the appropriate observer design. By contrast optimization-based design approaches are simpler to use; however they are prone to numerical problems associated with solving "stiff" algebraic equations when the tuning parameters approach a certain critical limit. Different aspects of the CLTR design are illustrated in a numerical example for both a non-recoverable and a recoverable situations. All the design methods discussed in the paper have been implemented in a "Matlab" software package.

Acknowledgement

The work of B. M. Chen and A. Saberi is supported in part by Boeing Commercial Airplane Group and in part by NASA Langley Research Center under grant contract NAG-1-1210. The work of U. Ly is supported in part by NASA Langley Research Center under grant contract NAG-1-1210.

References

[1] B. M. Chen, A. Saberi and U. Ly, "Closed-Loop Transfer Recovery with Observer-Based Controllers – Part 1: Analysis," To appear in *Control and Dynamic Systems: Advances in Theory and Applications*, Academic Press, Inc. (1991).

[2] B. M. Chen, A. Saberi and U. Ly, "Exact Computation of the Infimum in H_∞-Optimization Via State Feedback," *Proceedings*

of The Twenty-Eighth Annual Allerton Conference, Monticello, Illinois, pp. 745-754 (1990).

[3] B. M. Chen, A. Saberi, S. Bingulac, and P. Sannuti, "Loop Transfer Recovery for Non-Strictly Proper Plants," *Control–Theory and Advanced Technology*, Vol. 6, No. 4, pp. 573-594 (1990).

[4] B. M. Chen, A. Saberi and P. Sannuti, "Loop Transfer Recovery for General Nonminimum Phase Non-strictly Proper Systems, Part 1 — Analysis," submitted for publication (1991).

[5] B. M. Chen, A. Saberi and P. Sannuti, "Loop Transfer Recovery for General Nonminimum Phase Non-strictly Proper Systems, Part 2 — Design," submitted for publication (1991).

[6] J. Doyle, K. Glover, P. P. Khargonekar and B. A. Francis, "State Space Solutions to Standard H_2 and H_∞ Control Problems", *IEEE Transactions on Automatic Control*, Vol. 34, No. 8, pp. 831-847 (1989).

[7] B. C. Moore, "On the Flexibility Offered by State Feedback in Multivariable Systems Beyond Closed Loop Eigenvalue Assignment," *IEEE Transactions on Automatic Control*, AC-21, pp. 689-692 (1976).

[8] P. Pandey, C. Kenney, A. Laub and A. Packard, "Algorithms for Computing the Optimal H_∞-Norm," *Proceedings of the 29th CDC*, Honolulu, Hawaii, pp. 2628-2689 (1990).

[9] A. Saberi, B. M. Chen and P. Sannuti, " Theory of ALTR for Nonminimum Phase Systems, Recoverable Target Loops, Recovery in a Subspace—Part 2: Design," *International Journal of Control*, Vol. 53, No. 5, pp. 1117-1160 (1991).

[10] P. Sannuti and A. Saberi," A Special Coordinate Basis of Multivariable Linear Systems - Finite and Infinite Zero Structure, Squaring Down and Decoupling," *International Journal of Control*, Vol.45, No. 5, pp. 1655-1704 (1987).

[11] A. Saberi and P. Sannuti, "Squaring Down of Non-Strictly Proper Systems," *International Journal of Control*, Vol. 51, No. 3, pp. 621-629 (1990).

[12] A. Saberi and P. Sannuti, "Time-Scale Structure Assignment in Linear Multivariable Systems Using High-gain Feedback," *International Journal of Control*, Vol.49, No. 6, pp. 2191-2213 (1989).

[13] A. Saberi and P. Sannuti, "Observer Design for Loop Transfer Recovery and for Uncertain Dynamical Systems," *IEEE Transactions on Automatic Control*, Vol. 35, No. 8, pp. 878-897 (1990).

[14] A. Saberi and P. Sannuti, "Cheap and Singular Controls for Linear Quadratic Regulators," *IEEE Transactions on Automatic Control*, AC-32, pp.208-219 (1987).

[15] A. A. Stoorvogel, "The Singular Linear Quadratic Gaussian Control Problem," Preprint, (1990).

[16] B. Wie and D. Bernstein,"A Benchmark Problem for Robust Control Design," *Proceedings of the 1990 American Control Conference*, May 23-25 (1990).

[17] Z. Zhang and J. S. Freudenberg, "Loop Transfer Recovery for Nonminimum Phase Plants," *IEEE Transactions on Automatic Control*, Vol.35, No.5, pp. 547-553 (1990).

[18] K. Zhou and P. Khargonekar, "An Algebraic Riccati Equation Approach to H_∞-Optimization," *Systems & Control Letters*, Vol. 11, pp. 85-91 (1988).

[19] B. Wie, Q. Liu and K-W Byun, "Robust H_∞ Control Synthesis Method and its Application to a Benchmark Problem," presented at the *1990 American Control Conference*, San Diego, California, May 23-25, 1990.

Table I: Analysis of Closed-Loop Transfer Recovery

Case	Measurement	n_a^-	n_a^+	n_b	n_c	n_f	CLTR
a	x_1	0	0	0	2	2	Non-recoverable
b	x_2	0	0	0	2	2	Non-recoverable
c	x_1 , x_2	0	0	0	0	4	Asymptotic Recoverable
d	x_1 , v_1	0	0	1	2	1	Non-recoverable
e	x_2 , v_2	0	0	1	2	1	Non-recoverable
f	x_1 , a_1	2	0	0	0	2	Asymptotic Recoverable
g	x_2 , a_2	0	0	0	0	4	Asymptotic Recoverable

Table II: Values of the Tuning Design Parameters (case b)

Design No.	ATEA	H_2-Optimization	H_∞-Optimization
	σ	σ	γ
1	0.75	1	1
2	1	10	0.9
3	2	10^2	0.8
4	3	10^3	0.7
5	4	10^4	0.6
6	5	10^5	0.5
7	6	10^6	0.4
8	7	10^7	0.3
9	8	10^8	0.275
10	9	10^9	0.25
11	10	10^{10}	0.225

Table III: Values of the Tuning Design Parameters (case g)

Design No.	ATEA	H_2-Optimization	H_∞-Optimization
	σ	σ	γ
1	0.25	1	1
2	0.5	10	0.9
3	1	10^2	0.8
4	2	10^3	0.7
5	5	10^4	0.6
6	10	10^5	0.5
7	15	10^6	0.4
8	20	10^7	0.3
9	30	10^8	0.275
10	40	10^9	0.25
11	50	10^{10}	0.225

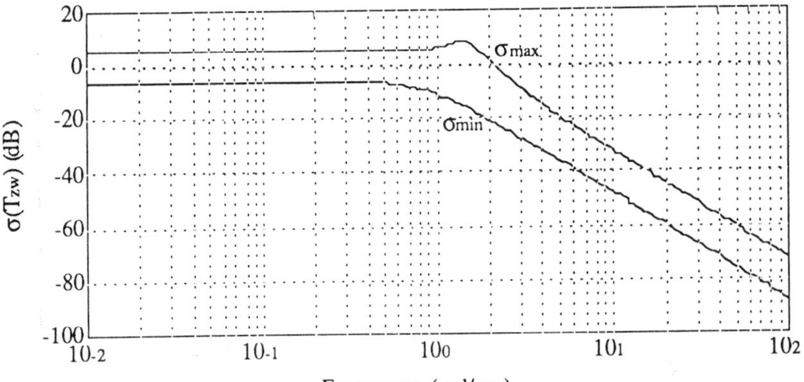

Figure 2: Singular Value Plot of Target Closed-Loop $T_{zw}(s)$

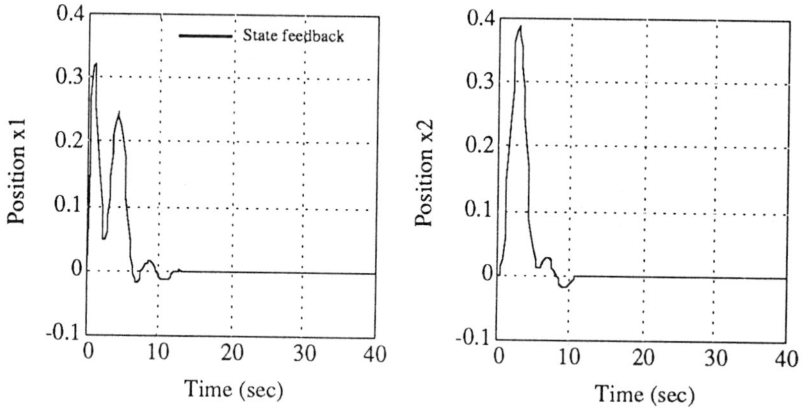

Figure 3: Impulse Responses in the Full-State Feedback Design

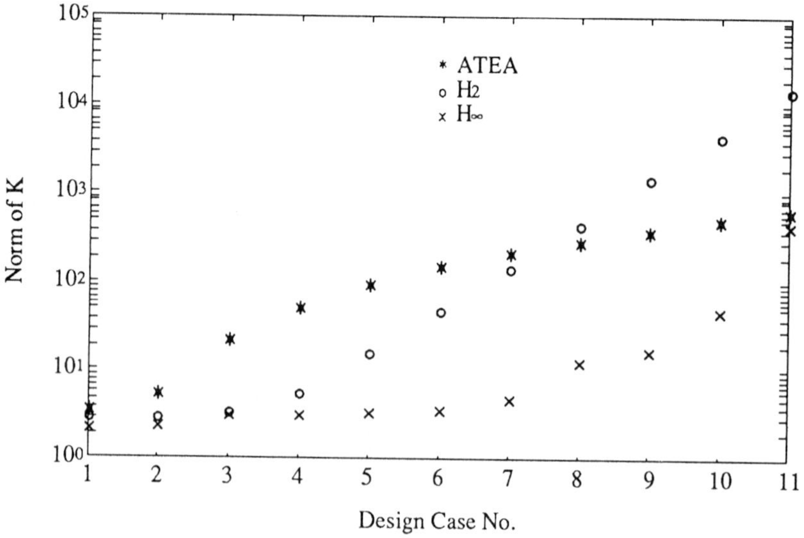

Figure 4: Norm of Observer Gain K for Designs in case b

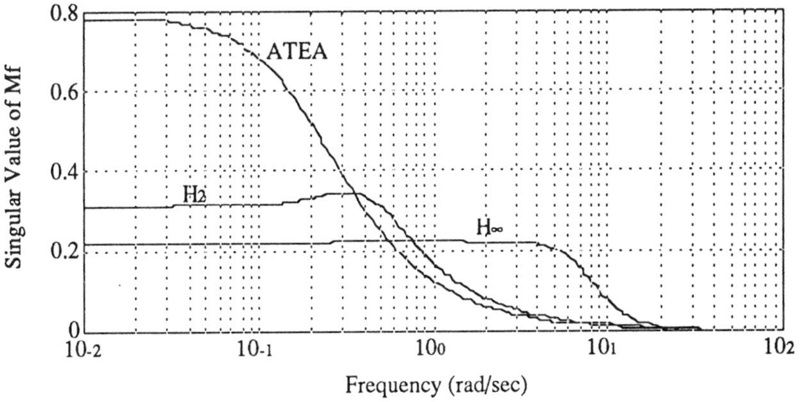

Figure 5: Singular Values of $M_f(s)$ for Design No. 11 (case b)

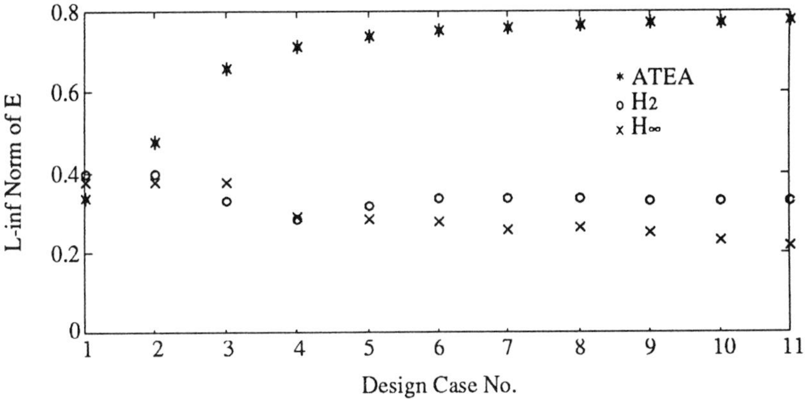

Figure 6: L_∞-Norm of $E(s)$ for Designs in case b

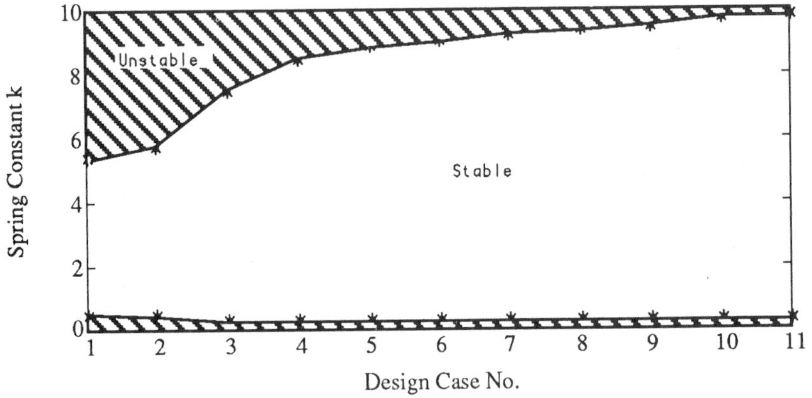

Figure 7: Robustness of ATEA Designs (case b) to Variation in k

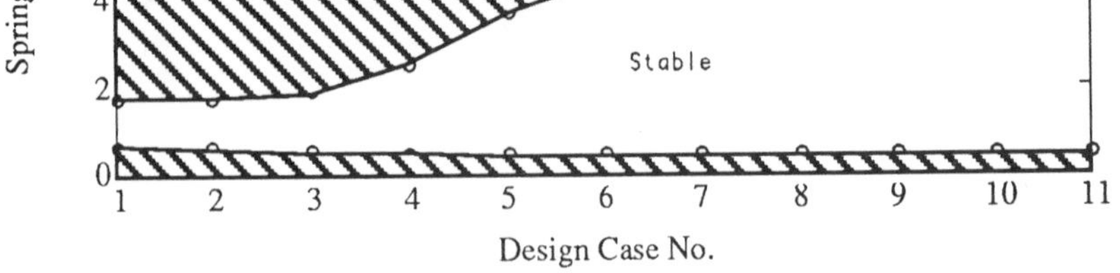

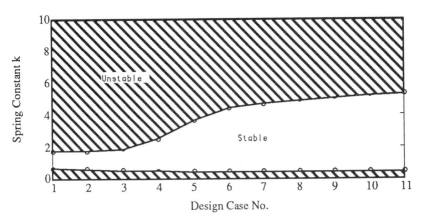

Figure 8: Robustness of H_2 Designs (case b) to Variation in k

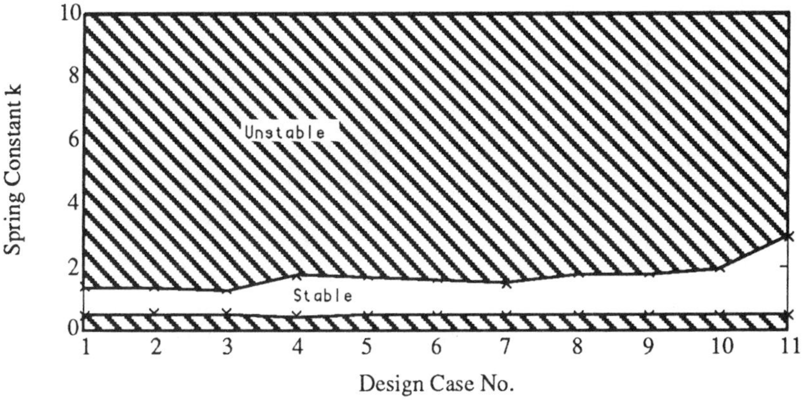

Figure 9: Robustness of H_∞ Designs (case b) to Variation in k

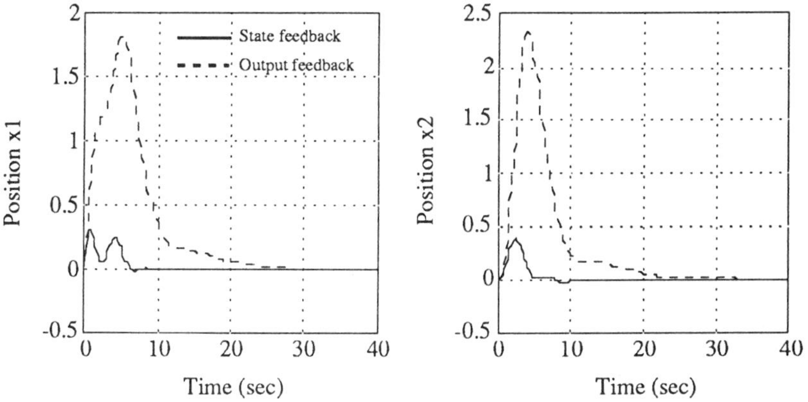

Figure 10: Impulse Responses for ATEA Design No. 1 (case b)

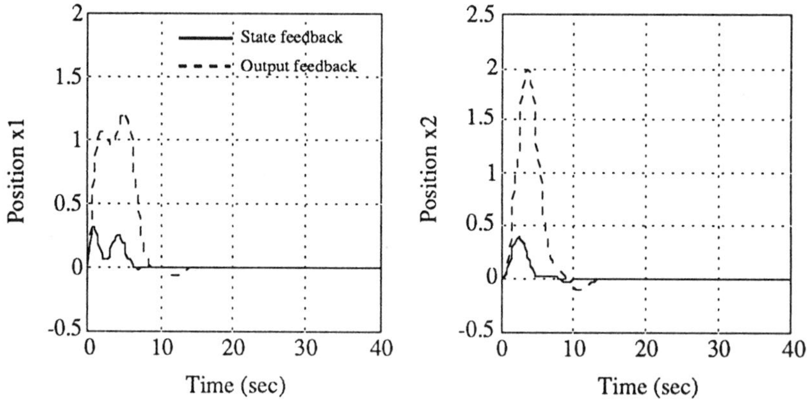

Figure 11: Impulse Responses for H_2 Design No. 3 (case b)

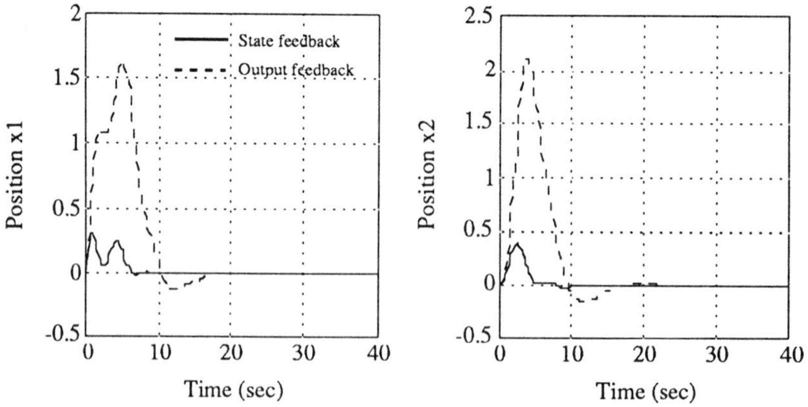

Figure 12: Impulse Responses for H_∞ Design No. 5 (case b)

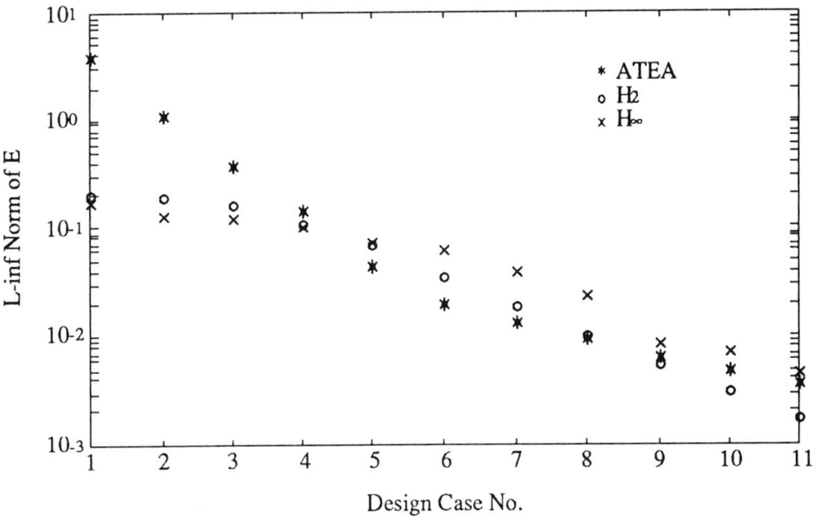

Figure 13: L_∞-Norm of $E(s)$ for Designs in case g

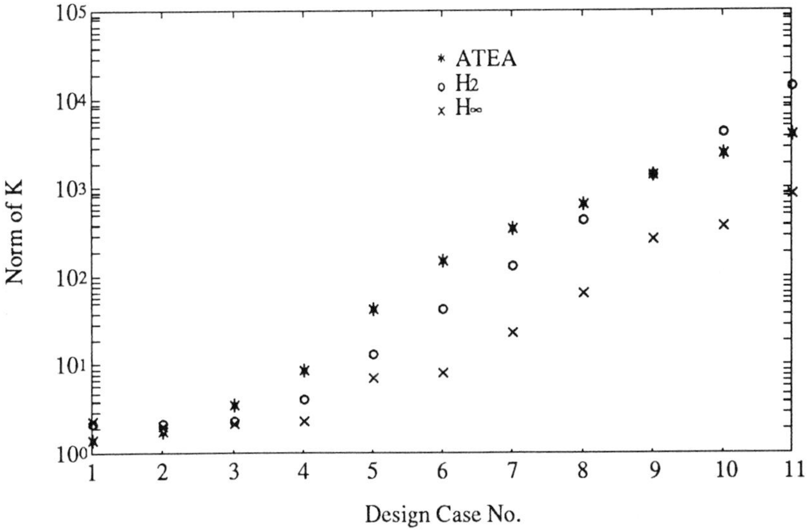

Figure 14: Norm of Observer Gain K for Designs in case g

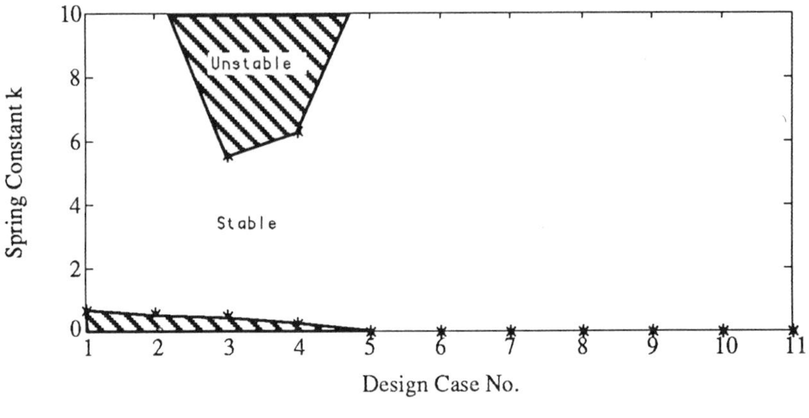

Figure 15: Robustness of ATEA Designs (case g) to Variation in k

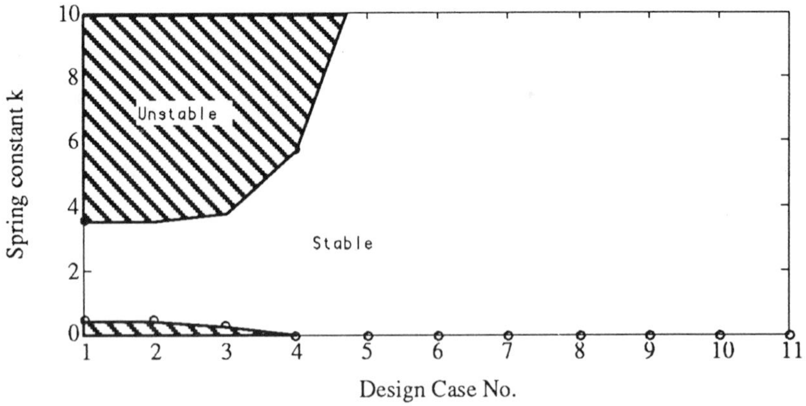

Figure 16: Robustness of H_2 Designs (case g) to Variation in k

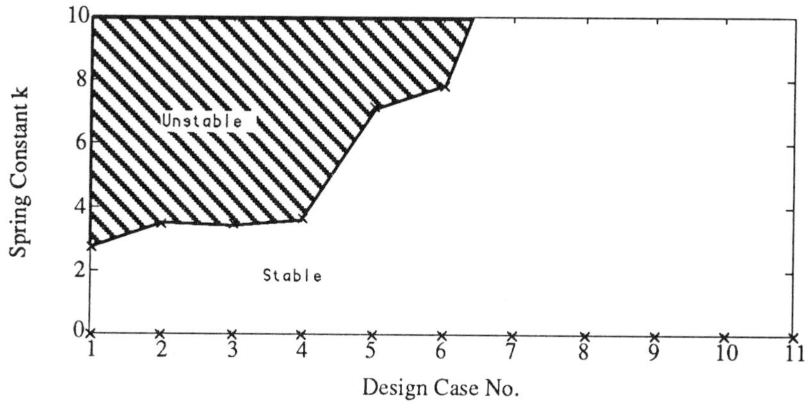

Figure 17: Robustness of H_∞ Designs (case g) to Variation in k

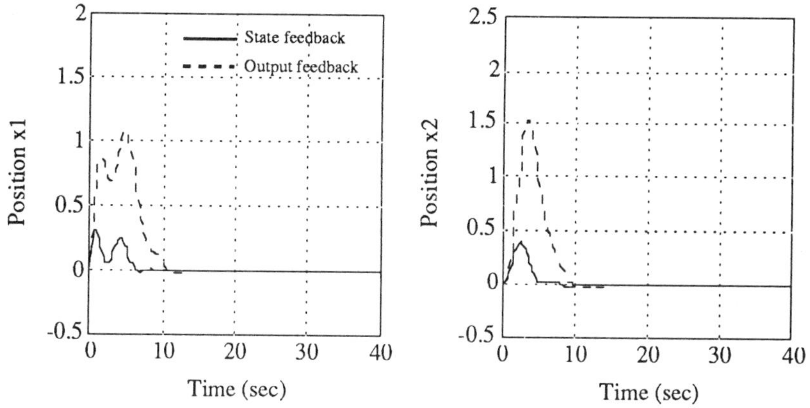

Figure 18: Impulse Responses for ATEA Design No. 3 (case g)

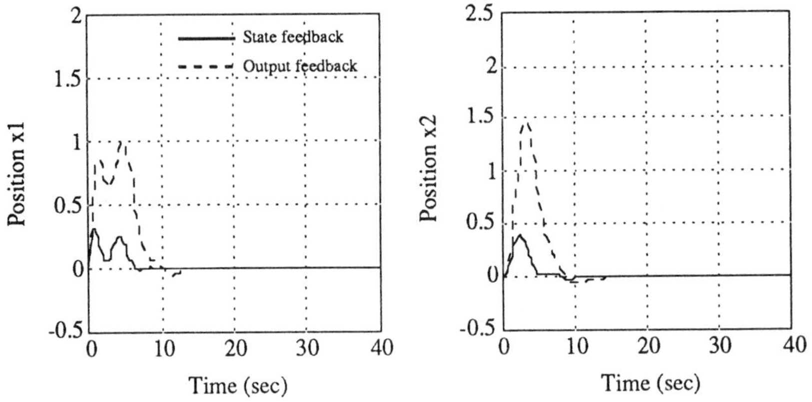

Figure 19: Impulse Responses for H_2 Design No. 3 (case g)

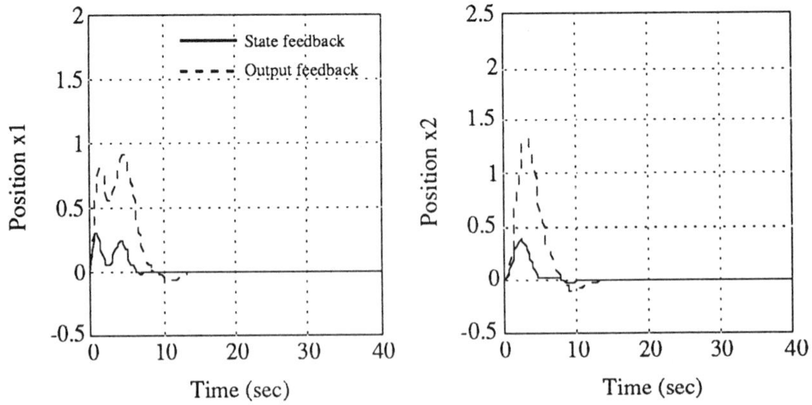

Figure 20: Impulse Responses for H_∞ Design No. 4 (case g)

Robust Adaptation in Slowly Time-Varying Systems: Double-Algebra Theory *

Le Yi Wang
Department of Electrical and Computer Engineering,
Wayne State University,
5050 Anthony Wayne Drive,
Detroit, Michigan 48202, U.S.A.

George Zames
Department of Electrical Engineering,
McGill University,
3480 University Street,
Montreal, Quebec,
Canada, H3A 2A7

December 27, 1991

Abstract

In this paper, a common algebraic framework is introduced for the frozen-time analysis of stability and H^∞ optimization in slowly time-varying systems, based on the notion of a normed algebra on which local and global products are defined. Relations between local stability, local (near) optimality, local coprime factorization, and global versions of these properties are sought.

The framework is valid for time-domain disturbances in ℓ^∞. H^∞-behavior is related to ℓ^∞ input-output behavior via the device of an approximate isometry between frequency and time-domain norms.

The double-algebra concept is first elaborated for Volterra operators which approximately commute with the shift. The main algebraic properties and norm inequalities are summarized. Local conditions for global invertibility are obtained. Classical frozen-time stability conditions are incorporated in relations between local and global spectra.

*Based on [33] [28].

An explicit formula is established linking global and local sensitivity for systems with stable plants, where local sensitivity is a Lipschitz continuous function of data. Frequency-domain estimates of time-domain sensitivity norms, which become accurate as rates of time-variation approach zero, are obtained. Notions of adaptive versus nonadaptive (robust) control are introduced. It is shown that adaptive control can achieve better sensitivity than optimal nonadaptive control.

It is demonstrated by an example that, in general, H^∞-optimal interpolants do not depend Lipschitz continuously on data. However, δ-suboptimal interpolants of the AAK central (maximal entropy) type are shown to satisfy a tractable Lipschitz condition.

1 INTRODUCTION

This paper considers the problem of adaptive optimization using H^∞ and frozen-time methods, continuing the work started in [23].

The ability to adapt or learn from experience is limited to those aspects of data which persist or, at most, vary slowly with time. It seems worthwhile, therefore, to single out features of optimization which are peculiar to slowly time-varying systems[1]. There are results from the 1960's on the stability of such systems [13] [5], which were obtained using contraction and monotonicity ideas and doing frozen-time analysis on a case by case basis. In the course of trying to extend those ideas to H^∞ optimization, it became apparent that a common framework for these problems, extracting their recurrent algebraic properties, is possible.

The framework introduced here involves a space of input-output maps which approximately commute with the shift, and which are represented by Volterra operators in the time-domain and "local" transfer functions in the frequency domain. Two products are defined on this space, one global and the other local, creating the notion of a double algebra of operators. Local and global versions of properties such as spectrum, stability, optimal sensitivity, and coprime factorization can be defined in terms of these two products, and relations established between them, enabling global properties to be deduced from the more easily computed local ones, in systems that vary slowly.

[1]Despite the existence of more general theories of optimization for time-varying systems, e.g. [10], which have yet to be exploited to produce controllers causally dependent on data.

In a nontrivial theory of adaptive optimization, the effects of persistent disturbances, say in ℓ^∞, have to be considered. H^∞ interpolation theory is usually set in a Hilbert space and in that setting is restricted to transient disturbances. One possibility is to abandon H^∞ altogether and turn to ℓ^1 kernel optimization, but then most spectral information is lost, together with the extensive qualitative insight which spectral data provides. The alternative chosen here is to relate H^∞ frequency-domain properties with ℓ^1 kernel behavior via the device of an approximate isometry between certain frequency and exponentially weighted time-domain norms. (The results obtained here for ℓ^∞ input-output disturbances can be modified for the simpler case of ℓ^2 disturbances which have been treated elsewhere [27].)

The basic notions of a double algebra are set up in the time and frequency domains, and applied to deduce relationships between local and global stability for feedback systems with stable components. Classical frozen-time stability results, including a Nyquist type frequency-domain condition [13] and an example of Desoer's [5], are reinterpreted in terms of a relationship between local and global spectra.

Adaptive sensitivity reduction is investigated for stable plants, based on local H^∞ interpolation. An explicit (double) algebraic expression for adaptive design is obtained and is applied to an example involving rejection of narrowband disturbances of uncertain frequency in which the local optimum depends Lipschitz continuously on the data. Notions of adaptive and robust (non-adaptive) sensitivity minimization are introduced, and adaptive minimization is shown to give better sensitivity than H^∞ optimal robust minimization.

However, it is shown that, in general, H^∞ optimal interpolants do not depend Lipschitz continuously on data. On the other hand, δ-suboptimal interpolants of the AAK central (maximal entropy) type [1] are shown to have a Lipschitz dependence on data, and can serve as a basis for frozen-time design.

Relations between local and global coprime-factorizations are summarized in [26] and will be reported in a separate article.

Previous Results. Stability conditions for systems with slowly time-varying parameters go back to the 1960's-70's. The notion of frozen-time stability in an input-output setting, using moving window averaging, was introduced in [13]. Closely related conditions in a state-space setting were introduced by Desoer and students, (see, e.g., [5]) and are described in [9]. Algebraic approaches to input-output feedback go back to the same period, culminating in generalized coprime-factorization [8] and operator-norm sensitivity minimization [31]. Feintuch and Francis obtained an early result on optimization of time-varying systems based on Arveson's theory [10]. The sensitivity and well-posedness of H^∞ interpolants with respect to data is discussed in a recent paper by Smith [18], from a different point of view.

For contemporaneous approaches to time-varying systems, see Khargonekar and Poolla, who have introduced a variety of algebraic and optimization results, e.g. [16] [17]; and Verma [19] who has studied relations between robustness and coprime factorization; and Ioannou and Tsakalis [15]. Ball, Foias, Helton and Tannenbaum extended optimal interpolation to nonlinear systems, e.g., in [3]. Recently, Dahleh and Dahleh [6] have started using the notion of frozen-time system introduced in [23] [24].

Parts of this paper were included in [23] [24] [25] [26].

NOTATION

$\mathbb{R}, \mathbb{C}, \mathbb{Z}$ denote the reals, complex numbers and integers. The complex conjugate of any $x \in \mathbb{C}$ is $\overline{x}$.

$\mathbb{K}^n$ and $\mathbb{K}^{n\times n}$ denote n-tuples and $n \times n$ matrices over a ring $\mathbb{K}$. $\mathbb{C}^n$ is viewed as a Euclidean space; for $x \in \mathbb{C}^n$ the conjugate transpose is x^* and norm $|x| = (x^*x)^{\frac{1}{2}}$. For a matrix $K \in \mathbb{C}^{n\times n}$, $|K|$ is its largest singular value.

$\ell^p_\sigma[a, b]$, $1 \le p \le \infty$, $\sigma \ge 1$, denotes the space of sequences $u(t)$, $t = a, a+1, \cdots, b$, $t \in \mathbb{Z}$, either of vectors in $\mathbb{C}^n$ or matrices in $\mathbb{C}^{n\times n}$ for which

$$\|u\|_{\ell^p_\sigma} := \begin{cases} \left[\sum_{t=a}^{b}(|u(t)|\sigma^t)^p\right]^{\frac{1}{p}} < \infty & \text{for } 1 \le p < \infty, \\ \sup_{t\in[a,b]} |u(t)|\sigma^t < \infty & \text{for } p = \infty. \end{cases}$$

The dimension n will usually be suppressed in notation. Occasionally, it will be emphasized by the symbols $(\ell^p_\sigma[a, b])^n$ or $(\ell^p_\sigma[a, b])^{n\times n}$.

H^p_σ, $1 \le p \le \infty$, $\sigma \ge 1$, denotes the H^p space either of $\mathbb{C}^n$-vector or $\mathbb{C}^{n\times n}$-matrix functions $K(z)$ on the disc $|z| < \sigma$ for which $\|K(\cdot)\|_{H^p_\sigma} := \|K(\sigma(\cdot))\|_{H^p}$. H^p_σ is viewed as a subspace of L^p_σ, the space of L^p functions of the circle of radius σ.

$\mathcal{F}(u)$ denotes the discrete Fourier transform of any $u \in \ell^2_\sigma(-\infty,\infty)$,

$$\mathcal{F}(u)(z) := \sum_{t=-\infty}^{\infty} u(t)z^t, \ |z| = \sigma.$$

$\mathcal{F}(u)$ will also be represented by $\hat{u}$.

$\mathcal{F}^{-1}(K) \in \ell^2_\sigma$ denotes the inverse transform of any $K \in L^2_\sigma$ defined for $t \in \mathbb{Z}$ by

$$\mathcal{F}^{-1}(K)(t) = \frac{1}{2\pi}\int_0^{2\pi} K(\sigma e^{i\theta})(\sigma e^{i\theta})^{-t} d\theta.$$

Functions in ℓ^p_σ will be denoted by lower case letters, in H^p_σ by capitals, and operators in either space by boldface capitals.

$\mathbf{\Pi}_t$, $t \in \mathbb{Z}$ denotes the truncation operator which maps any linear-space sequence $f(\tau)$, $\tau \in \mathbb{Z}$, into $f_t(\tau)$, where $f_t(\tau) = f(\tau)$ for $\tau \le t$, $f_t(\tau) = 0$ elsewhere.

2 LOCAL-GLOBAL ALGEBRAS IN THE TIME-DOMAIN

Freezing will be applied to operations as diverse as inversion, factorization, optimization, etc. If performed on a case-by-case basis, freezing can become extremely unwieldy. The alternative taken here will be to extract the common features of the various freezing operations, albeit at the cost of some extra abstraction and notation at the outset.

The standard setup for input-output stability [30] [9] will be used. Stable systems will be represented by elements of a normed linear space of causal bounded linear operators denoted by $\mathbf{B}$. Unstable systems will lie in a (larger) "extended" space $\mathbf{B}_e$. These spaces will be specialized to convolution operators on $\ell^\infty(-\infty,\infty)$ as follows.

Let A denote the Banach space of $\mathbb{C}^n$-valued functions $\ell^\infty(-\infty,\infty)$. (Later, in Section 3, A will be equipped with certain equivalent auxiliary norms.) Stable systems will belong to the Banach space $\mathbf{B}$ of bounded causal linear operators $\mathbf{K} : A \to A$ which have *convolution sum* representations,

$$(\mathbf{K}u)(t) = \sum_{\tau=-\infty}^{t} k(t,\tau)u(\tau), \quad t \in \mathbb{Z} \tag{1}$$

where the *kernel* $k : \mathbb{Z}^2 \to \mathbb{C}^{n\times n}$ is assumed, for each $t \in \mathbb{Z}$, to satisfy $k(t,\cdot) \in \ell^1(-\infty,\infty)$; $\sup_{t\in\mathbb{Z}} \|k(t,\cdot)\|_{\ell^1} =: \|\mathbf{K}\|_{\mathbf{B}} < \infty$; and $k(t,\tau) = 0$ whenever $t < \tau$. The norm on $\mathbf{B}$ will be $\|\cdot\|_{\mathbf{B}}$.

Unbounded systems will belong to a linear extension of $\mathbf{B}$, denoted by $\mathbf{B}_e$, defined as follows: Let A° be the subspace of functions of finite support in A,

$$A^\circ := \{u \in A : u(t) = 0 \text{ for } t > t_u, \quad t < t'_u\}$$

where $t'_u, t_u \in \mathbb{Z}$ depend on u. $(A^\circ)_e$ is the linear space of sequences whose truncations $\mathbf{\Pi}_t u$ lie in A° for each $t \in \mathbb{Z}$. Then $\mathbf{B}_e$ is the linear space of operators in $(A^\circ)_e$ which have convolution sum representations of the form (1).

$\mathbf{B}$ can be viewed as a subspace of $\mathbf{B}_e$ modulo the following *equivalence*: To each $\mathbf{K} \in \mathbf{B}$ assign the unique bounded operator $\mathbf{K}_e \in \mathbf{B}_e$ obtained by first restricting $\mathbf{K}$ from A down to A°, and then extending to $(A^\circ)_e$; the map $\mathbf{K} \to \mathbf{K}_e$ is an equivalence between bounded operators in $\mathbf{B}$ and $\mathbf{B}_e$.

2.1 Local Systems

If an operator $\mathbf{K} \in \mathbf{B}_e$ changes greatly over time, it can not be approximated globally by a single time-invariant system. Instead, frozen time analysis employs a sequence of time-invariant systems, $\mathbf{K}_t$, $t \in \mathbb{Z}$, each of which approximates a slowly varying $\mathbf{K}$ in a neighbourhood of a particular t. Coupling between the $\mathbf{K}_t$ complicates the global perturbation analysis of $\mathbf{K}$. Our formalism seeks to bring out the main features of this analysis.

The approach here is to chose each time-invariant $\mathbf{K}_t$ to produce outputs which coincide with those of $\mathbf{K}$ at $t \in \mathbb{Z}$. [2] If $\mathbf{K}$ is any linear operator defined by a convolution sum

$$(\mathbf{K}u)(t) = \sum_{\theta=-\infty}^{\infty} k(t,\theta)u(\theta), \qquad t \in \mathbb{Z} \tag{2}$$

then the *local* system of $\mathbf{K}$ at $\tau \in \mathbb{Z}$ is the (time-invariant) operator $\mathbf{K}_\tau$ with the same domain as $\mathbf{K}$ satisfying

$$(\mathbf{K}_\tau u)(t) := \sum_{\theta=-\infty}^{\infty} k(\tau, \tau - (t-\theta))u(\theta), \quad t \in \mathbb{Z},$$

[2]Other choices of $\mathbf{K}_t$, involving combinations of averaging and freezing are possible (see Remarks 2.1, 2.2), but this one seems to give the simplest theory where the sole information about $\mathbf{K}$ is that it varies slowly.

i.e., the kernel of $\mathbf{K}_\tau$ is a Toeplitz matrix determined by a single row of the matrix $k(\tau, \theta)$. The terms local and frozen-time will be used interchangeably.

Although in general the notation $\mathbf{K}_\tau$ will denote the local system of $\mathbf{K} \in \mathbf{B}_e$ at time $\tau \in \mathbb{Z}$, an exception will be made in the case of $\mathbf{\Pi}_\tau$, which denotes a truncation operator at τ.

2.2 The Normed Double Algebra $\mathbb{B}$

We define two products on the space $\mathbf{B}_e$: (1) The usual composition product, which will be called the *global product*, and denoted explicitly by $\cdot$, although that symbol, as usual, will mostly be suppressed in notation, i.e., $\mathbf{F} \cdot \mathbf{K} = \mathbf{FK}$; and (2) a *local product*, denoted by $\otimes$ and defined as follows: For any $\mathbf{F}, \mathbf{K} \in \mathbf{B}_e$, $\mathbf{F} \otimes \mathbf{K}$ is the unique operator in $\mathbf{B}_e$ whose local operators satisfy

$$(\mathbf{F} \otimes \mathbf{K})_t = \mathbf{F}_t \mathbf{K}_t \qquad \forall t \in \mathbb{Z}. \tag{3}$$

A *double algebra* is any subspace of $\mathbb{B}_e$ which is equipped with both products and is an algebra with respect to each one. In particular, the space $\mathbb{B}_e$ equipped with both products is clearly a double algebra. Henceforth, the symbol $\mathbb{B}_e$ will denote this double algebra.

A *normed double algebra* (NDA) is a double algebra on which local and global norms, $\|\cdot\|^\ell$ and $\|\cdot\|^g$, are defined satisfying the inequalities

$$\|\mathbf{FK}\|^g \leq \|\mathbf{F}\|^g \|\mathbf{K}\|^g, \tag{4}$$
$$\|\mathbf{F} \otimes \mathbf{K}\|^\ell \leq \|\mathbf{F}\|^\ell \|\mathbf{K}\|^\ell. \tag{5}$$

In particular, the space $\mathbb{B}$ equipped with both products, and with local and global norms taken to be equal to $\|\cdot\|_\mathbf{B}$ is an NDA. In the following text assume $\mathbb{B}$ to be an NDA so equipped.

(In subsequent developments, the local norm may be distinct from and superior to the global norm, i.e., $\|\cdot\|^g \leq \|\cdot\|^l$, or vice versa.)

If $\mathbf{A}$ is any (normed) double algebra, its restriction to one of its products (and norms) will be denoted by the prefixes $\mathbb{L}$ for local and $\mathbf{G}$ for global. $\mathbb{L}\mathbf{A}$ and $\mathbf{GA}$ will be called the *local (normed) algebra* and *global (normed) algebra*, respectively. If $\mathbb{L}\mathbf{A}$ and $\mathbf{GA}$ are both complete normed algebras then $\mathbf{A}$ is a Banach double algebra. $\mathbb{B}$ is an example.

$\mathbf{K} \in \mathbf{A}$ has a *local inverse* in $\mathbf{A}$, denoted by $\mathbf{K}^{\text{(-1)}}$, if $\mathbf{K}^{\text{(-1)}}$ is an inverse of $\mathbf{K}$ in $\mathbb{L}\mathbf{A}$, i.e.

$$\mathbf{K} \otimes \mathbf{K}^{\text{(-1)}} = \mathbf{K}^{\text{(-1)}} \otimes \mathbf{K} = \mathbf{I}$$

and a *global inverse* denoted by $\mathbf{K}^{-1}$ if $\mathbf{K}^{-1}$ is an inverse of $\mathbf{K}$ in $\mathbb{G}\mathbf{A}$. Similarly, any object defined in $\mathbb{L}\mathbf{A}$ (in $\mathbb{G}\mathbf{A}$) will be termed the local (global) object in $\mathbf{A}$.

Relations between local and global properties will depend upon the following.

2.3 The Local-Global Coupling

This is expressed by the *product difference* binary operator $\nabla : \mathbb{B}_e \times \mathbb{B}_e \to \mathbb{B}_e$ defined by

$$\mathbf{F}\nabla\mathbf{K} = \mathbf{F}\mathbf{K} - \mathbf{F} \otimes \mathbf{K}. \tag{6}$$

We seek a relation between local and global invertibility in a normed double algebra $\mathbf{A}$, as this determines stability. Observe first, that $\mathbf{K} \in \mathbb{B}_e$ has a global inverse in $\mathbb{B}_e$ if and only if $k(t,t)$ is invertible in $\mathbb{C}^{n\times n}$ for each $t \in \mathbb{Z}$ (for then $\mathbf{K}$ decomposes into the sum of a memoryless invertible operator and a strictly causal one). Conditions for local invertibility in $\mathbb{B}_e$ are identical to global ones. However, in a general normed double algebra $\mathbf{A}$ this will not be true, and we get the following inversion results.

Let $\mathbf{A}$ be any normed double subalgebra of $\mathbb{B}_e$ (not necessarily complete) with global norm $\|\cdot\|$ subject to the *Norm Characterization Property*

$$P1: \qquad \mathbf{K} \in \mathbf{A} \Leftrightarrow (\mathbf{\Pi}_t\mathbf{K}) \in \mathbf{A} \ \ \forall t \in \mathbb{Z} \text{ and } \sup_t \|\mathbf{\Pi}_t\mathbf{K}\| < \infty,$$

where $\{\mathbf{\Pi}_t\}_{t\in\mathbb{Z}}$ is the family of truncation operators, e.g., $\mathbb{B}$ is certainly so characterized by its norm $\|\cdot\|_{\mathbb{B}}$.

Inversion Lemma 2.1.

(a) *If* $\mathbf{K} \in \mathbf{A}$ *has a local inverse* $\mathbf{K}^{\text{(-1)}} \in \mathbf{A}$, *then it has a global inverse* $\mathbf{K}^{-1} \in \mathbf{A}$ *whenever either (1)* $\|\mathbf{K}^{\text{(-1)}}\nabla\mathbf{K}\| < 1$, *in which case*

$$\mathbf{K}^{-1} = (\mathbf{K}^{\text{(-1)}}\mathbf{K})^{-1}\mathbf{K}^{\text{(-1)}} = (\mathbf{I} + \mathbf{K}^{\text{(-1)}}\nabla\mathbf{K})^{-1}\mathbf{K}^{\text{(-1)}} \tag{7}$$

and

$$\|\mathbf{K}^{-1}\| \le \|\mathbf{K}^{\text{(-1)}}\|\{1 - \|\mathbf{K}^{\text{(-1)}}\nabla\mathbf{K}\|\}^{-1}, \tag{8}$$

or (2) $\|\mathbf{K}\nabla\mathbf{K}^{\textcircled{\scriptsize -1}}\| < 1$, *in which case*

$$\mathbf{K}^{-1} = \mathbf{K}^{\textcircled{\scriptsize -1}}(\mathbf{K}\mathbf{K}^{\textcircled{\scriptsize -1}})^{-1} = \mathbf{K}^{\textcircled{\scriptsize -1}}(\mathbf{I} + \mathbf{K}\nabla\mathbf{K}^{\textcircled{\scriptsize -1}})^{-1} \tag{9}$$

and

$$\|\mathbf{K}^{-1}\| \leq \|\mathbf{K}^{\textcircled{\scriptsize -1}}\|\{1 - \|\mathbf{K}\nabla\mathbf{K}^{\textcircled{\scriptsize -1}}\|\}^{-1}. \tag{10}$$

(b) *Part (a) remains valid under an interchange of global norms, products and inverses with their local counterparts.* □

(For an extension of this result to certain norm-open algebras, see Lemma 2.2.)

Proof:

(a) If $\mathbf{K}^{\textcircled{\scriptsize -1}}$ is in $\mathbf{A}$, each matrix $k(t,t)$, $t \in \mathbb{Z}$, has an inverse in $\mathbb{C}^{n\times n}$, where k is the kernel of $\mathbf{K}$. Therefore, $\mathbf{K}^{-1}$ exists in $\mathbf{B}_e$. Furthermore, from the identities

$$\mathbf{K}^{\textcircled{\scriptsize -1}}\mathbf{K} = \mathbf{I} + \mathbf{K}^{\textcircled{\scriptsize -1}}\mathbf{K} - \mathbf{K}^{\textcircled{\scriptsize -1}} \otimes \mathbf{K} = \mathbf{I} + \mathbf{K}^{\textcircled{\scriptsize -1}}\nabla\mathbf{K} \tag{11}$$

we get, after multiplication by $\mathbf{K}^{-1}$ on the right, $\mathbf{K}^{-1} = \mathbf{K}^{\textcircled{\scriptsize -1}} - (\mathbf{K}^{\textcircled{\scriptsize -1}}\nabla\mathbf{K})\mathbf{K}^{-1}$. Subject to the norm characterization property (P1) and causality of $\mathbf{K}$, the usual "small gain" argument applied in the global algebra $\mathbf{GA}$ gives

$$\begin{aligned}\|\mathbf{\Pi}_t\mathbf{K}^{-1}\| &\leq \{1 - \|\mathbf{K}^{\textcircled{\scriptsize -1}}\nabla\mathbf{K}\|\}^{-1}\|\mathbf{\Pi}_t\mathbf{K}^{\textcircled{\scriptsize -1}}\| \\ &\leq \{1 - \|\mathbf{K}^{\textcircled{\scriptsize -1}}\nabla\mathbf{K}\|\}^{-1}\|\mathbf{K}^{\textcircled{\scriptsize -1}}\|, \quad t \in \mathbb{Z}\end{aligned} \tag{12}$$

provided $\|\mathbf{K}^{\textcircled{\scriptsize -1}}\nabla\mathbf{K}\| < 1$ in which case, since the bound (12) holds for all $t \in \mathbb{Z}$, $\mathbf{K}^{-1}$ is in $\mathbf{A}$. Expressions (7, 8) now follow from (11, 12). The proof of (9, 10) is obtained similarly by multiplying $\mathbf{K}\mathbf{K}^{\textcircled{\scriptsize -1}}$ by $\mathbf{K}^{-1}$ on the left.

(b) The proof remains valid under the specified interchange. □

The condition that $\mathbf{K}\nabla\mathbf{G}$ is small will be related to the smallness of the commutant of $\mathbf{G}$ with the shift, i.e., to slow variation of the kernel in the time domain, or, in Section 3, of the transfer function in the frequency domain. First, however, we summarize some elementary algebraic identities involving shift invariant and memoryless operators.

2.4 Algebraic Identities

The ensuing identities are obtained more or less immediately from the definitions of $\otimes$ and ∇.

Let $\mathbf{T} \in \mathbb{B}$ denote the shift, $(\mathbf{T}u)(t) = u(t-1)$, $t \in \mathbb{Z}$. An operator $\mathbf{G} \in \mathbf{B}_e$ is *shift-invariant* if its *commutant* $\mathbf{TG} - \mathbf{GT}$ (i.e., with the shift) vanishes. From the definition of $\otimes$,

$$\mathbf{K} \otimes \mathbf{T} - \mathbf{T} \otimes \mathbf{K} = 0 \tag{13}$$

for all $\mathbf{K} \in \mathbb{B}_e$, implying that all operators are locally shift-invariant. The term shift-invariant will be reserved for the global property. If $\mathbf{G}$ is shift invariant then

$$\mathbf{KG} = \mathbf{K} \otimes \mathbf{G}, \qquad \mathbf{K} \nabla \mathbf{G} = 0. \tag{14}$$

From these identities, the commutant of $\mathbf{K}$ is precisely the difference between the local and global products of $\mathbf{K}$ and the shift,

$$\mathbf{TK} - \mathbf{KT} = \mathbf{T} \nabla \mathbf{K}. \tag{15}$$

The simplicity of this relation of the ∇ operator to the commutant is noteworthy.

For any $\mathbf{F}, \mathbf{K}$, and shift-invariant $\mathbf{G}$ in $\mathbb{B}_e$,

$$(\mathbf{K} \nabla \mathbf{F})\mathbf{G} = \mathbf{K} \nabla (\mathbf{FG}). \tag{16}$$

Let $(\Delta \mathbf{\Pi}_\tau)$, $\tau \in \mathbb{Z}$, denote the family of projection operators, $(\Delta \mathbf{\Pi}_\tau) := \mathbf{\Pi}_\tau - \mathbf{\Pi}_{\tau-1}$. An operator $\mathbf{K} \in \mathbb{B}_e$ has *no memory* if

$$(\Delta \mathbf{\Pi}_\tau)\mathbf{K} = (\Delta \mathbf{\Pi}_\tau)\mathbf{K}(\Delta \mathbf{\Pi}_\tau)$$

i.e., $(\Delta \mathbf{\Pi}_\tau)(\ell_e^\infty)$ is an invariant subspace of $\mathbf{K}$. The identities (14), (16) which were previously shown to hold for $\mathbf{G}$ shift invariant, also hold for $\mathbf{G}$ arbitrary if $\mathbf{K}$ has no memory. In the last case, if $\mathbf{F}$ is arbitrary

$$(\mathbf{KF}) \nabla \mathbf{G} = \mathbf{K}(\mathbf{F} \nabla \mathbf{G}).$$

Any $\mathbf{F} \in \mathbb{B}$ can be expressed as a Volterra sum or, equivalently, as a linear combination of powers of $\mathbf{T}$,

$$\mathbf{F} \sim \sum_{r=0}^{\infty} \mathbf{F}^{(r)} \mathbf{T}^r \tag{17}$$

in which $\mathbf{F}^{(r)} \in \mathbb{B}$, $r = 1, 2, \ldots$, are operators without memory[3] whose kernels satisfy: $f^{(r)}(t, \tau) = f(t, t-r)$ when $\tau = t$, $f^{(r)}(t, \tau) = 0$ elsewhere. Series convergence here is weak ℓ^1 convergence, defined as follows:

[3]The expression (17) means that $\mathbb{B}$ is a module spanned by powers of $\mathbf{T}$.

Definition.[4] *A sequence of operators* $\mathbf{K}_n \in \mathbf{B}$ *weakly* (ℓ^1) *converges to* $\mathbf{K} \in \mathbf{B}$ *iff given any* $u \in \ell^\infty(-\infty, \infty)$ *and any functional* f *(with kernel) in* $\ell^1(-\infty, \infty)$, $f : \ell^\infty(-\infty, \infty) \to \mathbb{C}$, *the sequence* $f(\mathbf{K}_n u)$ *converges to* $f(\mathbf{K}u)$.

$\mathbf{F}\nabla\mathbf{K}$ can now be expressed as a linear combination of commutants $\mathbf{T}^r\nabla\mathbf{K}$ $(= \mathbf{T}^r\mathbf{K} - \mathbf{K}\mathbf{T}^r)$ with coefficients in $\mathbf{B}$.

Proposition 2.1. $\mathbf{T}\nabla\mathbf{K}$ coincides with the commutant (15) and

$$\mathbf{F}\nabla\mathbf{K} = \sum_{r=1}^{\infty} \mathbf{F}^{(r)}(\mathbf{T}^r\nabla\mathbf{K}) \tag{18}$$

where $\mathbf{F}^{(r)}$ has no memory, and the series weakly (ℓ^1) converges in $\mathbf{B}$.

Proof: Equation (15) follows from the identities $\mathbf{T}\nabla\mathbf{K} = \mathbf{T}\mathbf{K} - \mathbf{T}\otimes\mathbf{K}$ (by definition of ∇) and $\mathbf{T}\otimes\mathbf{K} = \mathbf{K}\otimes\mathbf{T} = \mathbf{K}\mathbf{T}$ (from the definition of $\otimes$ and the shift-invariance of $\mathbf{T}$). The expression (17) for $\mathbf{F}$ gives

$$\mathbf{F}\nabla\mathbf{K} = (\sum_r \mathbf{F}^{(r)}\mathbf{T}^r)\mathbf{K} - (\sum_r \mathbf{F}^{(r)}\mathbf{T}^r)\otimes\mathbf{K}.$$

As $\mathbf{F}^{(r)}$ has no memory, $(\mathbf{F}^{(r)}\mathbf{T}^r)\otimes\mathbf{K} = \mathbf{F}^{(r)}(\mathbf{T}^r\otimes\mathbf{K})$; also, weak (ℓ^1) convergence in $\mathbf{B}$ of a sequence $\mathbf{G}_n$ implies that of $\mathbf{G}_n\otimes\mathbf{K}$; therefore

$$\mathbf{F}\nabla\mathbf{K} = \sum_r \mathbf{F}^{(r)}(\mathbf{T}^r\mathbf{K} - \mathbf{T}^r\otimes\mathbf{K})$$

which implies (18). □

Equation (18) suggests that $\mathbf{F}\nabla\mathbf{K}$ is small whenever $\mathbf{K}$ has a small commutant with $\mathbf{T}$ and the memory of $\mathbf{F}$ decays quickly enough, motivating the following development.

2.5 Slowly Time-Varying Systems and Exponential Memories

The (time) *variation-rate* of $\mathbf{K} \in \mathbf{B}_e$ in any given norm $\|\cdot\|$, denoted by $d_{\|\cdot\|}(\mathbf{K})$ is

$$d_{\|\cdot\|}(\mathbf{K}) := \|\mathbf{K}\mathbf{T} - \mathbf{T}\mathbf{K}\|. \tag{19}$$

$\mathbf{K}$ will be said to *commute approximately with the shift if* $d_{\|\cdot\|}(\mathbf{K}) < \|\mathbf{K}\|$. Actually, $d_{\|\cdot\|}(\mathbf{K})$ will also be required to be small in relation to certain

[4]This coincides with the usual weak operator convergence provided the domain of operators in $\mathbf{B}$ is taken to be A° (i.e., modulo the equivalence mentioned in Section 1), the dual-space of A° being $\ell^1(-\infty, \infty)$.

other constants as yet to be specified, and the term *"approximately"* is included mainly as a reminder of intent.

(An estimate of the form $\|\mathbf{F}\nabla\mathbf{K}\|_{\mathbf{B}} \le \gamma\, d_{\mathbf{B}}(\mathbf{K})$ is obtained[5] whenever the kernel of $\mathbf{F}$ satisfies the summability assumption $\gamma := \sup_t \sum_{r=0}^{\infty} |rf(t,t-r)| < \infty$. However, this kind of summability is not preserved under both $(\cdot, \otimes)$ products. Instead, we focus on the stronger assumption that operators have exponentially decaying ℓ^1 memories.)

For any $\sigma > 1$, let $\|\cdot\|_{(\sigma)}$ be the function from operators $\mathbf{K} \in \mathbf{B}_e$ to $\mathbf{R}$ defined by

$$\|\mathbf{K}\|_{(\sigma)} = \sup_{t\in\mathbf{Z}} \sum_{\tau=-\infty}^{t} |k(t,\tau)\sigma^{(t-\tau)}| \tag{20}$$

where k is the kernel of $\mathbf{K}$. Equation (20) equals $\sup_{t\in\mathbf{Z}} \|k(t,t-(\cdot))\|_{\ell^1_\sigma}$. Let $\mathbf{E}_\sigma$ be the subspace of $\mathbf{B}$

$$\mathbf{E}_\sigma := \{\mathbf{K} \in \mathbf{B} : \|\mathbf{K}\|_{(\sigma)} < \infty\},$$

which is a Banach space under $\|\cdot\|_{(\sigma)}$ as norm. The variation rate of $\mathbf{K}$ in this norm is $d_\sigma(\mathbf{K})$.

Proposition 2.2. *The Banach space $\mathbf{E}_\sigma$ under the products $(\cdot, \otimes)$ is a Banach double algebra. For any $\mathbf{F} \in \mathbf{E}_{\sigma_0}, \mathbf{K} \in \mathbf{E}_\sigma$, $\sigma_0 > \sigma$,*

$$\|\mathbf{F}\nabla\mathbf{K}\|_{(\sigma)} \le \sigma^{-1}(e\,\ell n(\sigma_0/\sigma))^{-1}\|\mathbf{F}\|_{(\sigma_0)} d_\sigma(\mathbf{K}). \tag{21}$$

Proof: It is shown in Appendix 1, Proposition A1.1 that $\mathbf{E}_\sigma$ is a normed algebra under the local and global products, and therefore a normed double algebra (with local and global norms equal to $\|\cdot\|_{(\sigma)}$). From the Volterra sum representation of operators in $\mathbf{E}_\sigma$ by kernels in the complete space ℓ^1_σ, it can be deduced that $\mathbf{E}_\sigma$ is complete, and is therefore a Banach double algebra.

From (18, 17) we get for the kernel g of $\mathbf{F}\nabla\mathbf{K}$,

$$\begin{aligned} &|g(t,\varsigma)\sigma^{(t-\varsigma)}| \\ &= \left|\sum_{r=-\infty}^{\infty} f(t,t-r)[k(t-r,\varsigma) - k(t,r+\varsigma)]\sigma^{(t-\varsigma)}\right| \\ &\le \sum_{r=-\infty}^{\infty} |f(t,t-r)\sigma_0^r| \cdot |k(t-r,\varsigma) - k(t,r+\varsigma)|\sigma^{(t-r-\varsigma)}(\sigma_0/\sigma)^{-r}, \end{aligned}$$

[5] using the facts that $\|\mathbf{F}^{(r)}\|_{\mathbf{B}} = \sup_t |f(t,t-r)|$, and $\|\mathbf{T}^r\nabla\mathbf{K}\|_{\mathbf{B}} \le r\|\mathbf{T}\nabla\mathbf{K}\|_{\mathbf{B}}$.

$$\begin{aligned}
&\|g(t,t-(\cdot))\|_{\ell^1_\sigma} \\
\leq\ & \|f(t,t-(\cdot))\|_{\ell^1_{\sigma_0}} \sup_r \|(k(t,r+(\cdot)) - k(t-r,\cdot))\|_{\ell^1_\sigma} (\sigma_0/\sigma)^{-r} \\
\leq\ & \|\mathbf{F}\|_{(\sigma_0)} \sigma^{-1} d_\sigma(\mathbf{K}) \sup_{r\geq 0} r(\sigma_0/\sigma)^{-r}
\end{aligned}$$

and, as $\sup_{r\geq 0} r\theta^{-r} \leq (e\, \ell n\, \theta)^{-1}$ for $\theta > 1$, the conclusion (21) follows. □

2.6 Frozen-Time Stability in the Time Domain

Most classical frozen-time stability conditions can be encompassed in a statement relating the existence of local and global inverses. Time-domain conditions are contained in the following.

Corollary 2.1. *If* $\mathbf{G}$ *and* $\mathbf{K}$ *are in* $\mathbf{E}_{\sigma_0}$, $\sigma_0 > 1$, *and either* $\mathbf{G}$ *has no memory or* $\mathbf{K}$ *is shift-invariant, then existence of the local inverse* $(\mathbf{I}+\mathbf{G}\otimes\mathbf{K})^{\circledast}$ *in* $\mathbf{E}_{\sigma_0}$ *implies that of the global inverse* $(\mathbf{I}+\mathbf{GK})^{-1}$ *in* $\mathbf{B}$, *provided that*

$$d_{(1)}(\mathbf{G}\otimes\mathbf{K}) < (e\,\ell n\,\sigma_0)\|(\mathbf{I}+\mathbf{G}\otimes\mathbf{K})^{\circledast}\otimes[(1-\alpha)\mathbf{I}-\alpha\mathbf{G}\otimes\mathbf{K}]\|^{-1}_{(\sigma_0)} \quad (22)$$

for some $\alpha \in \mathbf{R}$.

Proof: The assumption that $\mathbf{G}$ is memoryless or $\mathbf{K}$ is shift-invariant implies that $\mathbf{G}\otimes\mathbf{K} = \mathbf{GK}$ and, by Inversion Lemma 2.1, Corollary 2.1 is true provided that

$$\|(\mathbf{I}+\mathbf{G}\otimes\mathbf{K})^{\circledast}\nabla(\mathbf{I}+\mathbf{G}\otimes\mathbf{K})\|_{\mathbf{B}} < 1. \quad (23)$$

As $\mathbf{I}$ has no memory and is shift invariant, it is true, in general, that $\mathbf{A}\nabla\mathbf{B} = (\mathbf{A}-\alpha\mathbf{I})\nabla(\mathbf{B}-\mathbf{I})$; therefore (23) is equivalent to

$$\|[(\mathbf{I}+\mathbf{G}\otimes\mathbf{K})^{\circledast}\otimes((1-\alpha)\mathbf{I}-\alpha\mathbf{G}\otimes\mathbf{K})]\nabla(\mathbf{G}\otimes\mathbf{K})\|_{\mathbf{B}} < 1.$$

By (21), (22) is sufficient for (23). □

Unfortunately, Corollary 2.1 involves the estimation of the ℓ^1-kernel norm of an inverse, which is seldom an analytically tractable object, and we will therefore emphasize alternative methods in the frequency domain. First, however, an example of Desoer [5], which is nicely tractable, is included to illustrate the symbolism.

Example 2.1. Stability of the difference equation

$$x(t) = G_t x(t-1) + F_t u(t), \qquad t \in \mathbf{Z}, \quad (24)$$

$x(t), u(t) \in \mathbf{R}^n$; $G_t, F_t \in \mathbf{R}^{n\times n}$, is to be deduced from its local properties. If $u \in \ell^\infty(-\infty, \infty)$ and G_t, F_t are bounded functions of $t \in \mathbf{Z}$, (24) can be expressed in operator form,

$$x = \mathbf{G}\mathbf{T}x + \mathbf{F}u$$

which has the solution $x = (\mathbf{I} - \mathbf{G}\mathbf{T})^{-1}\mathbf{F}u$, where $(\mathbf{I} - \mathbf{G}\mathbf{T})^{-1} \in \mathbf{B}_e$. The a priori assumption that the frozen-time system is exponentially stable means that $(\mathbf{I} - \mathbf{G} \otimes \mathbf{T})^{\ominus 1}$ is in $\mathbf{E}_{\sigma_0}$ whenever $1 < \sigma_0 < \nu^{-1}$, where ν is the supremum of the spectral radii of the matrices G_t, $t \in \mathbf{Z}$. The actual system is ℓ^∞-stable if the global inverse $(\mathbf{I} - \mathbf{G}\mathbf{T})^{-1}$ is in $\mathbf{B}$ which, by Corollary 2.1 ($\alpha = 1$), is ensured whenever the variation rate of $\mathbf{G}$ satisfies

$$d_{(1)}(\mathbf{G}) < (e\ \ell n\ \sigma_0)\|(\mathbf{I} - \mathbf{G} \otimes \mathbf{T})^{\ominus 1} \otimes \mathbf{G}\|_{(\sigma_0)}^{-1} \tag{25}$$

for some $\sigma_0 \in (1, \nu^{-1})$. The norm in (25) can be estimated as in [5], where it is shown that for any β, $\nu < \beta < 1$, (as $\mathbf{G}$ has no memory and $\mathbf{G}_t$ is finite-dimensional), $\sup_{r\in\mathbf{Z}} \|(\mathbf{G}/\beta)^r\|_{(1)} =: \gamma$ is a finite constant depending on β. Therefore

$$\begin{aligned}\|(\mathbf{I} - \mathbf{G} \otimes \mathbf{T})^{\ominus 1} \otimes \mathbf{G}\|_{(\sigma_0)} &\leq \sup_{t\in\mathbf{Z}} \sum_{i=0}^{\infty} |g(t,t)^{i+1}|\sigma_0^i \\ \leq \gamma \sum_{i=0}^{\infty} \beta^{i+1}\sigma_0^i &\leq \gamma\beta(1 - \sigma_0\beta)^{-1}, \quad \sigma_0\beta < 1,\end{aligned}$$

where $g(t,t)$ equals G_t (i.e., the matrix representing the no memory system $\mathbf{G}_t$). The choice $\beta = \frac{1}{2}(1 + \nu)$, $\sigma_0 = \frac{1}{2}(1 + \beta^{-1})$ and observation that $\ell n\ \sigma_0 > 1 - \sigma_0^{-1}$ gives a sufficient condition for stability, $d_{(1)}(\mathbf{G}) < \frac{e(1-\beta)^2}{2\gamma\beta(1+\beta)}$. If $\nu =: 1 - 2\varepsilon$, the rate bound is better than $e\varepsilon^2/4\gamma$.

Remark 2.1. It would appear that frozen-time analysis is distinguished from usual perturbation analysis in that the object perturbed is not any particular operator but rather a method of computing products. Double algebras capture this fact.

Remark 2.2. Our particular choice of local system $\mathbf{K}_t$ seems natural if $\mathbf{K}$ is assumed to be slowly varying in the sense of having a commutant norm (19) near zero. Other choices of $\mathbf{K}_t$ might be better under other assumptions. For example, consider the map $\{\cdot\}_t : \mathbf{B} \to \mathbf{B}$, $\mathbf{K} \to \{\mathbf{K}\}_t$, where the kernel $k_t(\cdot)$ of $\{\mathbf{K}\}_t$ is an average of the past of the kernel $k(\cdot,\cdot)$

of $\mathbf{K}$, namely

$$k_t(\delta) = \lambda \sum_{\tau=-\infty}^{t} \varsigma^{(\tau-t)} k(\tau, \tau-\delta),$$

where $0 < \varsigma < 1$, $\lambda \sum_{\tau=-\infty}^{t} \varsigma^{(\tau-t)} = 1$. The choice of local system $\{\mathbf{K}\}_t$ would seem more appropriate if the slow variation assumption were valid for the averaged operator as opposed to the original $\mathbf{K}$. Whatever the choice of local system, the local product (3) and double algebra framework would be preserved.

2.7 Nests of NDA's

In preparation for the frequency domain, let us axiomatically introduce a recurrent concept which is exemplified by the parametrized family $\{\mathbf{E}_\sigma\}$ of double algebras.

Definition. A *continuous nest of NDAs* is a set $\{\mathbf{A}_\sigma\}$ of nontrivial normed double subalgebras of $\mathbb{B}$ depending on a parameter σ, $\sigma \geq 1$, with the following properties:

(1) $\{\mathrm{A}_\sigma\}$ is monotone by containment, i.e., $\mathbf{A}_{\sigma_0} \subseteq \mathbf{A}_\sigma \subseteq \mathbb{B}$, if $1 \leq \sigma \leq \sigma_0$, containment being strict whenever $1 \neq \sigma \neq \sigma_0$.

(2) For $\mathbf{K} \in \mathbf{A}_{\sigma_0}$, the local and global norms, $\|\mathbf{K}\|_\sigma^\ell$ and $\|\mathbf{K}\|_\sigma^g$, depend continuously on σ and are monotone in σ, i.e., if $1 < \sigma < \sigma_0$ then

$$\|\mathbf{K}\|_{\mathbf{B}}^a \leq const. \|\mathbf{K}\|_\sigma^b \leq const. \|\mathbf{K}\|_{\sigma_0}^c$$

for every combination of local / global norm superscripts (a,b,c) chosen from (g, ℓ), the constants being independent of $\mathbf{K}$.

(3) Each NDA $\mathbf{A}_\sigma$ is characterized by the global norm $\|\cdot\|_\sigma^g$ according to Property (P1) of Section 2.3.

An NDA is *continuously nested* if it is in the nest or is a union of nest elements of the form $\mathbf{A} := \cup_{\sigma > \sigma_1} \{\mathbf{A}_\sigma\}$.

The NDA's in this paper all satisfy an additional inequality, linking local-norm rates of change to global behavior, which, however, is not part of the nest definition. For $\sigma < \sigma_0$,

$$\|\mathbf{K}\nabla\mathbf{F}\|_\sigma^g \leq const. \|\mathbf{K}\|_{\sigma_0}^\ell \quad d_{\|\cdot\|_{\sigma_0}^\ell}(\mathbf{F}).$$

An extension of the Inversion Lemma to certain open normed double algebras will be required.

Extended Inversion Lemma 2.2. *If* $\mathbf{A} = \cup_{\sigma>\sigma_1}\{\mathbf{A}_\sigma\}$ *is a continuously nested NDA, then Inversion Lemma 2.1 holds provided* $\|\cdot\|$ *is identified with* $\|\cdot\|_{\sigma_1}$.

Proof: If $\mathbf{K}$ and $\mathbf{K}^{\ominus 1}$ are in $\mathbf{A}$, they are certainly in $\mathbf{A}_{\sigma_1}$. By Lemma 2.1, $\mathbf{K}^{-1}$ is in $\mathbf{A}_{\sigma_1}$ and satisfies the inequalities (7, 9). All that remains to be shown is that $\mathbf{K}^{-1}$ is actually in $\mathbf{A}$. Under our hypothesis, $\mathbf{K}$ and $\mathbf{K}^{\ominus 1}$ are in some $\mathbf{A}_\sigma$, $\sigma > \sigma_1$, where $\|\mathbf{K}\nabla\mathbf{K}^{\ominus 1}\|^g_{\sigma_1} < 1$ implies $\|\mathbf{K}\nabla\mathbf{K}^{\ominus 1}\|^g_\sigma < 1$ by continuity of $\|\cdot\|^g_\sigma$. Lemma 2.1 now implies that $\mathbf{K}^{-1} \in \mathbf{A}_\sigma$, $\sigma > \sigma_1$, and therefore $\mathbf{K}^{-1} \in \mathbf{A}$. □

3 LOCAL-GLOBAL ALGEBRAS IN THE FREQUENCY DOMAIN

For any $\mathbf{K} \in \mathbb{B}$ and $\tau \in \mathbb{Z}$, the convolution kernel $k(\tau, \tau - (\cdot))$ of $\mathbf{K}_\tau$ has a well defined transform $\mathcal{F}(\mathbf{K}_\tau)$

$$\mathcal{F}(\mathbf{K}_\tau)(z) := \sum_{\theta=0}^{\infty} k(\tau, \tau - \theta)z^\theta, \quad |z| < 1 \tag{26}$$

in H^∞ called the *transfer function* of $\mathbf{K}_\tau$ and denoted by $\widehat{\mathbf{K}}_\tau$. $\widehat{\mathbf{K}}_\tau$ will be called the *local transfer function* (*local transform*) of $\mathbf{K}$ (of $k(\tau, \tau - (\cdot))$ at τ. For $\mathbf{K} \in \mathbb{E}_\sigma$, $\widehat{\mathbf{K}}_\tau$ is in H^∞_σ.

Notation: $\widehat{\mathbf{K}}$ will denote the sequence $\{\mathcal{F}(\mathbf{K}_\tau)\}, \tau \in \mathbb{Z}$, of local transfer functions of $\mathbf{K}$. In the case of shift-invariant $\mathbf{K}$, all elements of that sequence are identical. In that special case, to keep the notation simple, $\widehat{\mathbf{K}}$ will also denote a particular element of the sequence in H^∞. The context will be relied upon to resolve the ambiguity.

The local product is equivalent to a sequence of transfer function products, $(\widehat{\mathbf{G}\otimes\mathbf{K}})_\tau = (\widehat{\mathbf{G}_\tau})(\widehat{\mathbf{K}_\tau})$.

Although operators in $\mathbb{B}$ (in $\mathbb{E}_\sigma$) have local transfer functions in H^∞ (in H^∞_σ), the reverse is not true. $\mathbb{B}$ and $\mathbb{E}_\sigma$ have no precise characterizations in terms of transfer functions. To deal with operators initially specified in the frequency domain, we turn instead to subalgebras of $\mathbb{B}$ which have such a specification.

Define the function μ of operators $\mathbf{K} \in \mathbb{B}$

$$\mu^{(p)}_\sigma(\widehat{\mathbf{K}}) := \sup_{\tau\in\mathbb{Z}} \|\widehat{\mathbf{K}}_\tau\|_{L^p_\sigma},$$

$2 \le p \le \infty$, (L^p_σ = the L^p space of the circle $|z| = \sigma$). In the case $p = \infty$, omit[6] the superscript, i.e., $\mu_\sigma(\hat{\mathbf{K}}) = \mu_\sigma^{(\infty)}(\hat{\mathbf{K}})$.

Introduce two subspaces of operators in $\mathbb{B}$, one specified (in the time domain) by

$$\underline{\mathbf{E}}_\sigma = \{\mathbf{K} \in \mathbb{B} : \|\mathbf{K}\|_{(\sigma_0)} < \infty \text{ for some } \sigma_0 > \sigma\},$$

the second (in the frequency domain) by

$$\underline{\hat{\mathbf{E}}}_\sigma = \{\mathbf{K} \in \mathbb{B} : \mu_{\sigma_0}(\hat{\mathbf{K}}) < \infty \text{ for some } \sigma_0 > \sigma\},$$

and equip each with local and global products.

Proposition 3.1. *$\underline{\mathbf{E}}_\sigma$ and $\underline{\hat{\mathbf{E}}}_\sigma$ are identical double algebras.* □

Proof: In Appendix I.

(Henceforth, we will often not distinguish between $\underline{\mathbf{E}}_\sigma$ and $\underline{\hat{\mathbf{E}}}_\sigma$). Equip $\underline{\mathbf{E}}_\sigma$ with local and global norms,

$$\begin{aligned} \|\mathbf{K}\|^g_\sigma &= \|\mathbf{K}\|_{(\sigma)}, \\ \|\mathbf{K}\|^\ell_\sigma &= \mu_\sigma(\hat{\mathbf{K}}). \end{aligned}$$

$\underline{\mathbf{E}}_\sigma$ is now a normed double algebra, and conforms to our definition of being continuously nested, as $\underline{\mathbf{E}}_\sigma = \cup_{\sigma_0 > \sigma}\{\mathbf{E}_{\sigma_0}\}$. Although the space $\underline{\mathbf{E}}_\sigma$ is not closed in either norm, the extended version of the Inversion Lemma, Lemma 2.2, will apply nevertheless.

$\underline{\mathbf{E}}_\sigma$ will be the algebra of choice for problems initially specified in the frequency domain. The two norms will serve to compute local-norm approximants to global-norm behavior.

Remark 3.1. The closure of the space $\underline{\mathbf{E}}_\sigma$ in the kernel norm $\|\cdot\|_{(\sigma)}$ is $\mathbf{E}_\sigma$. Let $\overline{\mathbf{E}}_\sigma$ denote the space

$$\overline{\mathbf{E}}_\sigma := \{\mathbf{K} \in \mathbb{B} : \quad \mu_\sigma(\hat{\mathbf{K}}) < \infty\}. \tag{27}$$

These three spaces are related by the inclusions

$$\underline{\mathbf{E}}_\sigma \subset \mathbf{E}_\sigma \subset \overline{\mathbf{E}}_\sigma. \tag{28}$$

Also, $\overline{\mathbf{E}}_{\sigma_0} \subset \mathbf{E}_\sigma$ for $\sigma < \sigma_0$. Of the three spaces (28), only $\underline{\mathbf{E}}_\sigma$ is a normed double algebra with both time and frequency domain characterizations.

[6]Note that for time invariant $\mathbf{K}$, the notations $\mu_\sigma^{(p)}(\hat{\mathbf{K}})$ and $\|\hat{\mathbf{K}}\|_{L^p_\sigma}$ coincide, and will be used interchangeably.

3.1 An Auxiliary Time-Domain Norm

There is now a double algebra $\underline{\mathbf{E}}_\sigma$, which (viewed as a space) has equivalent descriptions in the time-domain via the kernel norm $\|\cdot\|_{(\sigma)}$, and the frequency domain via the transfer-function norm $\mu_\sigma(\cdot)$. However, these norms are incommensurate, and inconvenient for the estimation of $\ell^\infty(-\infty,\infty)$ time-domain behavior from local frequency-domain properties, unlike, e.g., the time-invariant situation in $\ell^2(-\infty,\infty)$ where Parseval's Theorem provides an isometry between kernel and transform representations. Instead, we introduce an auxiliary time-domain norm on $\mathbf{B}$, denoted by $\|\cdot\|_{a(\sigma)}$, which is equivalent to the $\ell^\infty(-\infty,\infty)$-induced operator norm on $\mathbf{B}$, and produces a kind of approximate isometry for slowly varying operators; i.e., subject to certain assumptions on uniformity of growth, $\|\cdot\|_{a(\sigma)}$ will have the properties

$$\mu_\sigma(\widehat{\mathbf{K}}) - \alpha \le \|\mathbf{K}\|_{a(\sigma)} \le \mu_\sigma(\widehat{\mathbf{K}}) + \beta$$

where $\beta \to 0$ as the variation rate ρ of $\widehat{\mathbf{K}}$ approaches[7] 0, and $\alpha \to 0$ as $\rho \to 0$ and $\sigma \to 1$. By making σ depend on ρ and approach 1 as $\rho \to 0$, it is possible to ensure that $\|\mathbf{K}\|_{a(\sigma)} \to \mu_\sigma(\widehat{\mathbf{K}})$ as $\rho \to 0$. This device of an asymptotic isometry formalizes an idea originally used to obtain a BIBO version of the circle criterion in [29].

Equip the space $\ell^\infty(-\infty,\infty) =: A$ with the family of auxiliary norms depending on the parameter σ, $1 < \sigma \le \infty$,

$$\|u\|_{a(\sigma)} := \begin{cases} \kappa_\sigma^{-1} \sup_{t\in\mathbf{Z}} (\sum_{\tau=-\infty}^{t} |u(\tau)\sigma^{-(t-\tau)}|^2)^{1/2}, & \sigma < \infty, \\ \|u\|_{\ell^\infty}, & \sigma = \infty, \end{cases}$$

where $\kappa_\sigma := (1-\sigma^{-2})^{\frac{1}{2}}$, $(= (\sum_{r=0}^{\infty} \sigma^{-2r})^{\frac{1}{2}})$, i.e., $\|u\|_{a(\sigma)}$ is the ℓ^∞ norm of a moving window average of u, with an exponential window normalized to have unit ℓ^2 norm.

The norms in this family are equivalent to each other,

$$\|u\|_{a(\sigma_1)} \le const.\|u\|_{a(\sigma_2)} \le const.\|u\|_{a(\sigma_1)}, \quad 1 < \sigma_1 < \sigma_2$$

and to the ℓ^∞ norm,

$$\|u\|_{a(\sigma)} \le \|u\|_{\ell^\infty} \le \kappa_\sigma \|u\|_{a(\sigma)},$$

[7]The term "approaches" must be suitably interpreted; see Section 3.4.

the constants being independent of u.

Each auxiliary norm on A induces an *auxiliary operator* norm on the *space* of operators; for $\mathbf{K} \in \mathbf{B}$

$$\|\mathbf{K}\|_{a(\sigma)} := \sup\{\|\mathbf{K}u\|_{a(\sigma)} : u \in A, \ \|u\|_{a(\sigma)} \leq 1\}.$$

Assume the space $\mathbf{B}$ and global algebra $\mathbf{GB}$ to be equipped with the family of auxiliary operator norms $\|\cdot\|_{a(\sigma)}$, $1 < \sigma \leq \infty$. The norms in this family inherit from A equivalence to each other and to the ℓ^∞-induced operator norm on any $\mathbf{K} \in \mathbf{B}$. That last norm equals the ℓ^1 norm of the kernel k of $\mathbf{K}$, i.e.,

$$\|\mathbf{K}\|_{\mathbf{B}} = \|\mathbf{K}\|_{a(\infty)} = \sup_t \|k_t\|_{\ell^1}, \qquad \mathbf{K} \in \mathbf{B}. \tag{29}$$

Since $\mathbf{GB}$ is a Banach algebra under the $\|\cdot\|_{\mathbf{B}}$ norm, we obtain the following proposition.

Proposition 3.2. *The global algebra* $\mathbf{GB}$ *under the auxiliary norm* $\|\cdot\|_{a(\sigma)}$ *is a Banach algebra.* (However, as the constant in the inequality $\|\mathbf{K} \otimes \mathbf{G}\|_{a(\sigma)} \leq const. \|\mathbf{K}\|_{a(\sigma)} \|\mathbf{G}\|_{a(\sigma)}$ differs from unity, the local algebra $\mathbf{LB}$ is not a normed algebra under $\|\cdot\|_{a(\sigma)}$.)

In view of the equivalence of $\|\cdot\|_{\mathbf{B}}$ and $\|\cdot\|_{a(\sigma)}$, we can employ $\|\cdot\|_{a(\sigma)}$ as the global norm for $\mathbf{B}$ in place of $\|\cdot\|_{(\sigma)}$, and will do so in the rest of Section 3.

The auxiliary norms are bounds on an operator which are uniform in time. Occasionally, we shall relate these to certain finer bounds emphasizing particular times. The (exponentially weighted) *recent past seminorms* $\|\mathbf{K}\|_{a(\sigma;t)}$ of $\mathbf{K} \in \mathbf{B}$ are defined by the equation

$$\|\mathbf{K}\|_{a(\sigma;t)} := \kappa_\sigma^{-1} \sigma^{-t} \sup\left\{ \|\mathbf{\Pi}_t \mathbf{K}u\|_{\ell^2_\sigma} : u \in A, \ \|u\|_{a(\sigma)} \leq 1 \right\}.$$

The auxiliary norm is the supremum of these seminorms,

$$\|\mathbf{K}\|_{a(\sigma)} = \sup_t \|\mathbf{K}\|_{a(\sigma;t)}.$$

3.2 Slowly Time-Varying Transfer Functions

Definition. An operator $\mathbf{K} \in \overline{\mathbf{E}}_\sigma$ has a *slowly varying local transfer function with rate* $\partial_\sigma^{(p)}(\widehat{\mathbf{K}})$, $2 \leq p \leq \infty$, if

$$\partial_\sigma^{(p)}(\widehat{\mathbf{K}}) := \sup_{t \in \mathbf{Z}} \|\widehat{\mathbf{K}}_{t+1} - \widehat{\mathbf{K}}_t\|_{L^p_\sigma} < \mu_\sigma^{(p)}(\widehat{\mathbf{K}}), \tag{30}$$

and in case $p = \infty$, *write* $\partial_\sigma(\widehat{\mathbf{K}}) := \partial_\sigma^{(\infty)}(\widehat{\mathbf{K}})$. ($\partial_\sigma^{(p)}(\widehat{\mathbf{K}})$ will later be assumed small in relation to certain additional constants.)

Proposition 3.3. *For* $\sigma_0 > \sigma > 1$, *the transfer function and operator (commutant) variation rates are related,*

$$d_\sigma(\mathbf{K}) \leq \sigma\kappa_{(\sigma_0/\sigma)}\partial_{\sigma_0}(\widehat{\mathbf{K}}),$$

and $\mathbf{K}$ *commutes approximately with the shift whenever*

$$\partial_{\sigma_0}(\widehat{\mathbf{K}}) < \sigma^{-1}\kappa^{-1}_{(\sigma_0/\sigma)}\|\mathbf{K}\|_{(\sigma)}.$$

Proof: Follows (101) of Appendix I.

The time and frequency-domain norms are related by inequalities. The main ones are listed in the next two propositions whose proofs, together with subsidiary inequalities are in Appendix II.

In Propositions 3.4-3.5, $1 < \sigma \leq \infty$, $t \in \mathbf{Z}$, and $2 \leq p \leq \infty$. The constant $\kappa_\sigma^{(p)}$ is defined by

$$\kappa_\sigma^{(p)} = \begin{cases} (\sigma - 1)^{-1} & if\ p = \infty, \\ \sqrt{n}\{\sum_{r=0}^{\infty}(r\sigma^{-r})^2\}^{\frac{1}{2}} & if\ p < \infty. \end{cases}$$

Proposition 3.4. Transfer-function Bounds.

(a) **Bounds on** $\mathbf{K} \in \mathbf{E}_\sigma$.

$$\|\mathbf{K}\|_{a(\sigma;t)} \leq \|\widehat{\mathbf{K}}\|_{H_\sigma^\infty} \quad if\ \mathbf{K}\ is\ shift\ invariant, \tag{31}$$

$$\|\mathbf{K}\|_{a(\sigma;t)} \leq \mu_\sigma(\widehat{\mathbf{K}}_t) + \kappa_\sigma^{(p)}\partial_\sigma^{(p)}(\widehat{\mathbf{K}}) \quad if\ \widehat{\mathbf{K}}\ is\ slowly\ varying, \tag{32}$$

$$\|\mathbf{K}\|_{a(\sigma;t)} \leq \kappa_\sigma\sqrt{n}\mu_\sigma^{(p)}(\widehat{\mathbf{K}}) \leq \kappa_\sigma\sqrt{n}\mu_\sigma(\widehat{\mathbf{K}}) \quad always. \tag{33}$$

(b) **Local Bounds on** $(\mathbf{GK}+\mathbf{F})$ **in** $\mathbf{E}_\sigma$. *For* $\mathbf{F}, \mathbf{K}$ *with slowly varying transfer functions,* $\mathbf{S} := \mathbf{GK} + \mathbf{F}$, $\mathbf{S}^\ell := \mathbf{G} \otimes \mathbf{K} + \mathbf{F}$,

$$\|\mathbf{S}\|_{a(\sigma;t)} - \mu_\sigma(\widehat{\mathbf{S}}_t^\ell) \leq (\sigma - 1)^{-1}\{\mu_\sigma(\widehat{\mathbf{G}})\partial_\sigma(\widehat{\mathbf{K}}) + \partial_\sigma(\widehat{\mathbf{G}})[\mu_\sigma(\widehat{\mathbf{K}}) + (\sigma - 1)^{-1}\partial_\sigma(\widehat{\mathbf{K}})] + \partial_\sigma(\widehat{\mathbf{F}})\} \quad if\ \widehat{\mathbf{G}}\ is\ slowly\ varying, \tag{34}$$

$$\|\mathbf{S}\|_{a(\sigma;t)} - \mu_\sigma(\widehat{\mathbf{S}}_t^\ell) \leq \kappa_\sigma^{(p)}\{\kappa_\sigma\mu_\sigma(\widehat{\mathbf{G}})\partial_\sigma^{(p)}(\widehat{\mathbf{K}}) + \partial_\sigma^{(p)}(\widehat{\mathbf{S}}^\ell)\} \quad always. \tag{35}$$

(c) **Local Bounds on $\mathbf{G}\nabla\mathbf{K}$ in $\mathbf{E}_\sigma$.** *For* $\mathbf{K}$ *with slowly varying transfer functions,*

$$\|\mathbf{G}\nabla\mathbf{K}\|_{a(\sigma)} \leq (\sigma-1)^{-1}\{\mu_\sigma(\widehat{\mathbf{G}})\partial_\sigma(\widehat{\mathbf{K}}) + \partial_\sigma(\widehat{\mathbf{G}})\|\mathbf{K}\|_{a(\sigma)} + \partial_\sigma(\widehat{\mathbf{G}\otimes\mathbf{K}})\} \quad \text{if } \widehat{\mathbf{G}} \text{ is slowly varying,} \tag{36}$$

$$\|\mathbf{G}\nabla\mathbf{K}\|_{a(\sigma)} \leq \kappa_\sigma\kappa_\sigma^{(p)}\mu_\sigma(\widehat{\mathbf{G}})\partial_\sigma^{(p)}(\widehat{\mathbf{K}}) \qquad \text{always.} \tag{37}$$

(d) **Bounds[8] on $\partial_\sigma^{(p)}(\cdot)$.**

$$\partial_\sigma^{(p)}(\widehat{\mathbf{G}\otimes\mathbf{K}}) \leq \mu_\sigma(\widehat{\mathbf{G}})\partial_\sigma^{(p)}(\widehat{\mathbf{K}}) + \partial_\sigma^{(p)}(\widehat{\mathbf{G}})\mu_\sigma(\widehat{\mathbf{K}}), \tag{38}$$

$$\partial_\sigma^{(p)}(\widehat{\mathbf{G}^{\ominus}}) \leq [\mu_\sigma(\widehat{\mathbf{G}^{\ominus}})]^2\partial_\sigma^{(p)}(\widehat{\mathbf{G}}). \tag{39}$$

3.3 Radial Growth Inequalities

If $K \not\equiv 0$ is in $H^\infty_{\sigma_0}$, $\sigma_0 > 1$, Hardy's Convexity Theorem (Duren [7]) implies the radial growth condition,

$$\|K\|_{H^\infty_\sigma}/\|K\|_{H^\infty} \leq \nu^{\left(\frac{\ln\sigma}{\ln\sigma_0}\right)}, \quad \nu := \|K\|_{H^\infty_{\sigma_0}}/\|K\|_{H^\infty}, \quad 1 \leq \sigma < \sigma_0.$$

$\mathbf{K} \in \mathbf{E}_{\sigma_0}$ has *uniform* (i.e., in t) *radial growth* with constant $\nu_{\sigma_0}(\widehat{\mathbf{K}})$ iff

$$\nu_{\sigma_0}(\widehat{\mathbf{K}}) := \sup_{t\in\mathbf{Z}} \|\widehat{\mathbf{K}}_t\|_{H^\infty_{\sigma_0}}/\|\widehat{\mathbf{K}}_t\|_{H^\infty} < \infty$$

in which case

$$\mu_\sigma(\widehat{\mathbf{K}}) \leq \mu_1(\widehat{\mathbf{K}})\nu_{\sigma_0}(\widehat{\mathbf{K}})^{\left(\frac{\ln\sigma}{\ln\sigma_0}\right)}.$$

Proposition 3.5. *If the operators* $\mathbf{S}$, $\mathbf{S}^\ell$ *defined in (34) are in* $\mathbf{E}_{\sigma_0}$, $\sigma_0 > 1$, *and* $\mathbf{S}^\ell$ *has uniform radial growth, then*

$$\left|\|\widehat{\mathbf{S}}\|_{a(\sigma;t)} - \mu_\sigma(\widehat{\mathbf{S}}^\ell_t)\right| \leq \mu_1(\widehat{\mathbf{S}}^\ell)\{\nu_{\sigma_0}(\widehat{\mathbf{S}}^\ell)^{\left(\frac{\ln\sigma}{\ln\sigma_0}\right)} - 1\} + \kappa_\sigma^{(p)}\partial_\sigma^{(p)}(\widehat{\mathbf{S}}^\ell) + \|\mathbf{G}\nabla\mathbf{K}\|_{a(\sigma)}. \tag{40}$$

Proof. In Appendix II.

---

[8] The bounds (39) remain valid for certain noncausal operators $\mathbf{G}^{\ominus}$; see the definition 4.1.

3.4 Remarks on Variable Rates

In adaptation problems, the variation rate of a system is often adjustable, e.g., by reduction of the rate of change of its (memoryless) gain parameters, in a way which distorts time-scales but in a certain sense preserves local values and, especially, preserves operator norms. We wish to describe the behavior of such a system as rates approach zero. The details of how rate variation is achieved can be complex in discrete time, and can be disregarded for our purposes except as follows.

Definition. *The rate $d_\sigma(\cdot)$ of an operator $\mathbf{K} \in \mathbf{E}_\sigma$ is variable towards* 0 *if* $\mathbf{K}$ *is an element of a set of operators $\mathbf{K}(\rho)$ depending on a parameter ρ such that*[9]

(1) ρ has values in $\mathbf{R}_+$ with zero as limit point; and for some ρ_1, $\mathbf{K}(\rho_1) = \mathbf{K}$;

(2) $d_\sigma(\mathbf{K}(\rho)) \leq \rho$;

(3) The local norm $\|\mathbf{K}(\rho)\|_{(\sigma)}$ is invariant with ρ.

A similar definition is made for the transfer function rate $\partial_\sigma(\widehat{\mathbf{K}})$ under local norm $\mu_\sigma(\cdot)$ (replacing $d_\sigma(\mathbf{K})$ and $\|\cdot\|_{(\sigma)}$).

Remark 3.2. In (40) if the operators $\mathbf{G}, \mathbf{K}, \mathbf{F}$ have rates $\partial_\sigma(\cdot)$ variable towards 0 and approaching 0, and $\sigma \to 1$, then (40) implies that the auxiliary time-domain norm $\|\mathbf{S}\|_{a(\sigma)}$ approaches the transfer function norm $\mu_\sigma(\widehat{\mathbf{S}}^\ell)$; in this sense, the two norms are asymptotically isometric.

Remark 3.3. This asymptotic isometry can be deduced without assuming uniform radial growth or using (40). An alternative inequality based on the continuity of $\mu_\sigma(\cdot)$ versus σ can be obtained as follows. For $\mathbf{S}, \mathbf{S}^\ell$ defined as in (34) and in some $\overline{\mathbf{E}}_{\sigma_0}$, $\sigma_0 > \sigma$, the inequalities

$$\begin{aligned} \mu_1(\widehat{\mathbf{S}}_t^\ell) &- \kappa_\sigma^{(p)}\partial_\sigma^{(p)}(\widehat{\mathbf{S}}^\ell) - \|\mathbf{G}\nabla\mathbf{K}\|_{a(\sigma)} \\ &\leq \|\mathbf{S}\|_{a(\sigma;t)} \\ &\leq \mu_\sigma(\widehat{\mathbf{S}}_t^\ell) + \kappa_\sigma^{(p)}\partial_\sigma^{(p)}(\widehat{\mathbf{S}}^\ell) + \|\mathbf{G}\nabla\mathbf{K}\|_{a(\sigma)} \end{aligned} \tag{41}$$

hold. However, if s_t^ℓ denotes the kernel of $\mathbf{S}_t^\ell$, we have

$$\mu_\sigma(\widehat{\mathbf{S}}_t^\ell) - \mu_1(\widehat{\mathbf{S}}_t^\ell) \leq \sup_t \sum_{\tau=0}^{\infty} |s_t^\ell(\tau)|\sigma_0^\tau\left(\frac{\sigma}{\sigma_0}\right)^\tau(1-\sigma^{-\tau})$$

[9] An example would be the scale expansion: ρ assumes the values $\rho = r^{-1}\partial_\sigma(\mathbf{K}), r = 1, 2, \cdots$; $\mathbf{K}(\rho) = \mathbf{K}$ when $r = 1$; when $r > 1$, $\mathbf{K}_{rt} = \mathbf{K}_t$ and $\mathbf{K}_t$ is affine in t in any interval $[rt, r(t+1)]$.

$$\leq \; (1-\sigma^{-1})\kappa^{(2)}_{(\sigma_0/\sigma)}\mu_{\sigma_0}(\widehat{\mathbf{S}}^{\ell}). \tag{42}$$

Equations (42) and (41) give

$$\left| \|\mathbf{S}\|_{a(\sigma;t)} - \mu_1(\widehat{\mathbf{S}}^{\ell}_t) \right|$$

$$\leq \; \left[(1-\sigma^{-1})\kappa^{(2)}_{(\sigma_0/\sigma)}\right]\mu_{\sigma_0}(\widehat{\mathbf{S}}^{\ell}) + \kappa^{(p)}_{\sigma}\sigma^{(p)}_{\sigma}(\widehat{\mathbf{S}}^{\ell}) + \|\mathbf{G}\nabla\mathbf{K}\|_{a(\sigma)}. \tag{43}$$

The coefficient of $\mu_{\sigma_0}(\widehat{\mathbf{S}}^{\ell})$ in (43) approaches 0 as $\sigma \to 1$. If, moreover, $\mathbf{G}, \mathbf{K}, \mathbf{F}$ have rates $\partial^{(p)}_{\sigma}(\cdot) \leq \rho$ variable towards 0, (43) implies that

$$\lim_{\sigma\to 1}\lim_{\rho\to 0}\left| \|\mathbf{S}(\rho)\|_{a(\sigma;t)} - \mu_{\sigma}(\widehat{\mathbf{S}}^{\ell}_t(\rho)) \right| = 0.$$

3.5 Frozen-Time Stability in the Frequency Domain

Although the initial reason for introducing double algebras was to deal with adaptive optimization, their chief application in Section 3 is to obtain frozen-time conditions for stability in the frequency domain. The main idea here is to introduce notions of local spectrum and local resolvent to parallel the global (i.e., usual) ones. Most classical frequency domain conditions of the frozen-time type can then be incorporated in a statement linking local and global resolvents.

The *spectrum* $Spec_{\mathbf{A}}(\mathbf{K})$ of an operator $\mathbf{K}$ in a normed algebra $\mathbf{A}$ is the set $\{\lambda \in \mathbb{C} : (\lambda\mathbf{I} - \mathbf{K})$ has no inverse in $\mathbf{A}\}$. The *resolvent* set $Res_{\mathbf{A}}(\mathbf{K})$ is the complement of the spectrum, and the *γ-sublevel set* of that resolvent is

$$Res_{\mathbf{A};\gamma}(\mathbf{K}) = \{\lambda \in \mathbb{C} : \|(\lambda\mathbf{I} - \mathbf{K})^{-1}\|_{\mathbf{A}} \leq \gamma\}.$$

The *local (global) spectrum* of an operator in a normed double algebra $\mathbf{A}$ is its spectrum in the local algebra $\mathbb{L}\mathbf{A}$ (global algebra $\mathbf{GA}$); the *local (global) resolvent set and its γ-sublevel sets* are similarly defined.

Let $\mathbf{G}, \mathbf{K}$ be operators in $\overline{\mathbf{E}}_{\sigma}$, $\sigma > 1$, where $\mathbf{G}$ has no memory and $\mathbf{K}$ is shift invariant. ($\overline{\mathbf{E}}_{\sigma}$ is defined in (27)). The local and global norms on $\overline{\mathbf{E}}_{\sigma}$ are $\mu_{\sigma}(\cdot)$ and $\|\cdot\|_{a(\sigma)}$, and $\mathbf{G}$ is locally normalized, $\mu_{\sigma}(\widehat{\mathbf{G}}) = 1$. The γ-sublevel sets of $\mathbf{G} \otimes \mathbf{K}$ in the local algebra $\mathbb{L}\overline{\mathbf{E}}_{\sigma}$ can be determined from a plot of the spectrum of $\mathbf{K}$ on a Nichols or Hall chart. The main result of Section 3 is the following.

Corollary 3.1. *The local sublevel sets of* $\widehat{\mathbf{G}} \otimes \widehat{\mathbf{K}}$ *are related to the global resolvent of* **GK** *by the inclusion*

$$Res_{\mathbf{L}\overline{\mathbf{E}}_{\sigma;\gamma}}(\widehat{\mathbf{G}} \otimes \widehat{\mathbf{K}}) \subset Res_{\,\mathbf{G}\mathbf{B}}(\mathbf{GK})$$

provided

$$\partial_\sigma(\widehat{\mathbf{G}}) < (\sigma - 1)/\gamma\mu_\sigma(\widehat{\mathbf{K}})(1 + \gamma\mu_\sigma(\widehat{\mathbf{K}})). \tag{44}$$

Proof: In Appendix II.

Note that for **G** with memory, Corollary 3.1 still holds with (44) replaced by

$$\partial_\sigma(\widehat{\mathbf{G}}) < (\sigma - 1)/\gamma\mu_\sigma(\widehat{\mathbf{K}})(1 + \gamma\|\mathbf{G}\|_{a(\sigma)}\mu_\sigma(\widehat{\mathbf{K}})).$$

Example 3.1. Corollary 3.1 contains a frequency domain stability test. For scalar systems and **G** memoryless, the following prescription is obtained. Plot the σ-shifted discrete Nyquist diagram of the time-invariant operator **K**, i.e., $\widehat{\mathbf{K}}(\sigma e^{j\theta})$ versus θ, $\theta \in [-\pi, \pi)$. (See [13] for details in the case of continuous-time systems). The local spectrum of $\mathbf{G} \otimes \mathbf{K}$ in $\overline{\mathbf{E}}_\sigma$ is the union of the interiors of such diagrams of $c\mathbf{K}$, evaluated for each constant c in the range of the kernel of the memoryless operator **G**, i.e., $c \in \{g(t,t) : t \in \mathbf{Z}\}$. Let γ, $0 < \gamma < \infty$ denote the supremal Nichols contour intersected by this local spectrum. Then the 1 point (is not encircled by any of the Nyquist diagrams and) belongs to the γ-sublevel resolvent set of $\mathbf{G} \otimes \mathbf{K}$. Therefore the closed-loop operator $(\mathbf{I} - \mathbf{GK})^{-1}$ is ℓ^∞-bounded (i.e., globally in **B**) by the Corollary, provided the rate of variation $\partial_\sigma(\widehat{\mathbf{G}})$ of $\widehat{\mathbf{G}}$ satisfies inequality (44).

4 ADAPTIVE DESIGN BY LOCAL INTERPOLATION

One of the aims of this work is to obtain a paradigm of adaptive feedback in the H^∞ context, and a comparison of adaptive versus nonadaptive feedback. A prerequisite for such a comparison would appear to be some means of computing optimal or nearly optimal performance under time-varying weightings, in order to determine whether the best that can be achieved with updated information is better than the best that can be achieved without. Frozen-time analysis provides a way of obtaining approximate optimization, which can be used to construct an elementary paradigm. However, as freezing is involved in several operations, including inner-outer factorization and optimal H^∞ interpolation, it can involve some messy bookkeeping. Our formalism seeks to tidy up this process.

The main objective here is to synthesize a global sensitivity from a prescribed local one, which may be locally optimal or nearly optimal, and to determine how well the global (optimum) solution approximates the local (optimum) one. The double-algebra symbolism allows an explicit description of this problem.

In this section, the double algebra will be $\underline{\mathbf{E}}_\sigma$, with local norm $\mu_\sigma(\cdot)$ and global norm $\|\cdot\|_{a(\sigma)}$.

Suppose that $\mathbf{W}_1, \mathbf{W}_2 \in \underline{\mathbf{E}}_\sigma$ represent two weightings, and $\mathbf{G} \in \underline{\mathbf{E}}_\sigma$ represents a strictly causal plant. It is standard that feedbacks which are globally stabilizing in $\underline{\mathbf{E}}_\sigma$, i.e., maintain all closed-loop operators in $\underline{\mathbf{E}}_\sigma$, can be parametrized by a compensator $\mathbf{Q} \in \underline{\mathbf{E}}_\sigma$ which gives a sensitivity $(\mathbf{I} - \mathbf{G}\mathbf{Q})$, and a *weighted sensitivity* $\mathbf{S} \in \underline{\mathbf{E}}_\sigma$,

$$\mathbf{S} = \mathbf{W}_2(\mathbf{I} - \mathbf{G}\mathbf{Q})\mathbf{W}_1. \tag{45}$$

Denote $\mathbf{W}_2\mathbf{G}$ by $\mathbf{G}_W$ and suppose that it has a local factorization,

$$\mathbf{G}_W = \mathbf{U} \otimes \mathbf{G}^{out}$$

where $\mathbf{U}$ and $\mathbf{G}^{out}$ are *locally inner* and *locally outer* in $\mathbf{E}_{\sigma_0}$ for some $\sigma_0 > \sigma$, i.e., for each $t \in \mathbf{Z}$, $\widehat{\mathbf{U}}_t(\sigma_0(\cdot)) \in H^\infty$ is inner and $(\mathbf{G}^{out})_t(\sigma_0(\cdot)) \in H^\infty$ is outer.

We are given a sensitivity $\mathbf{S}^\ell \in \underline{\mathbf{E}}_\sigma$ which locally interpolates $\mathbf{W}_2\mathbf{W}_1$ at $\mathbf{U}$, i.e., for which there exists $\mathbf{Q}_1 \in \underline{\mathbf{E}}_\sigma$ such that

$$\mathbf{S}^\ell = \mathbf{W}_2\mathbf{W}_1 - \mathbf{U} \otimes \mathbf{Q}_1 \tag{46}$$

where $\widehat{\mathbf{S}}^\ell$ is smaller than $\widehat{\mathbf{W}_2\mathbf{W}_1}$ in $\mu_\sigma(\cdot)$. $\mathbf{Q}$ is now chosen to locally realize $\mathbf{S}^\ell$. The simplest case[10] will be considered here, in which $\mathbf{Q}$ is chosen to satisfy.

$$\mathbf{S}^\ell = \mathbf{W}_2\mathbf{W}_1 - \mathbf{U} \otimes \mathbf{G}^{out} \otimes (\mathbf{Q}\mathbf{W}_1). \tag{47}$$

This choice of $\mathbf{Q}$ can be described as a local product of noncausal operators, provided the domains of definition of the various local functions are first

[10] This case occurs when the global products in (46) are accessible to computation. Alternatively, the global products can be replaced by local ones. The ensuing inequalities then involve more terms, generated by the additional local factors, but the essentials remain similar.

extended to noncausal operators. Previously, these domains included the space $\underline{\mathbf{E}}_\sigma$ of operators with kernels $k \in \ell^1_\sigma(-\infty,\infty)$ satisfying the causality constraint $k(t,\tau) = 0$ for $t < \tau$.

Definition 4.1. *Henceforth, the definitions of local product* $\mathbf{K} \otimes \mathbf{F}$*, transform norm* $\mu_\sigma(\widehat{\mathbf{K}})$*, and rate* $\partial_\sigma^{(p)}(\widehat{\mathbf{K}})$ *are extended to operators with (possibly noncausal) kernels* $k(t, t-(\cdot)) \in \ell^1_\sigma(-\infty,\infty)$. (The definitions as originally stated can be extended intact.)

$\mathbf{Q}$ satisfying (47) is now explicitly given by

$$\mathbf{Q} := [\mathbf{G}_W^{(-1)} \otimes (\mathbf{W}_2\mathbf{W}_1 - \mathbf{S}^\ell)]\mathbf{W}_1^{-1}, \tag{48}$$

provided the inverses in (48) exist, where $\mathbf{G}_W^{(-1)}$ may be noncausal with Fourier transform in L^∞_σ. The problem is to determine whether (48) is stabilizing and makes the (true global) sensitivity $\mathbf{S}$ a good approximant to $\mathbf{S}^\ell$ for slowly-varying $\mathbf{G}$, $\mathbf{W}_i$, and $\mathbf{S}^\ell$.

Assumption 4.1 for Theorem 4.1. *($2 \le p \le \infty$, $\sigma > 1$)*

(a) $\mathbf{W}_1^{-1}$ *and* $\mathbf{W}_2^{-1}$ *are in* $\underline{\mathbf{E}}_\sigma$*;*

(b) $(\mathbf{G}^{out})^{(-1)} \in \underline{\mathbf{E}}_\sigma$.

Sufficient conditions for the local interpolation (46) and assumption (b) to hold are that there exists $\sigma_0 > \sigma$ *for which*

(a′) The zero vectors of $(\widehat{\mathbf{S}^\ell} - \widehat{\mathbf{W}_2\mathbf{W}}_1)_t$ *in* $|z| \le \sigma_0$ *contain the zero vectors of* $\widehat{\mathbf{U}}_t(z)$ *for each* $t \in \mathbb{Z}$ *(where* $\varsigma \in \mathbb{C}^n$ *is a zero vector of* $\widehat{\mathbf{K}}(z)$ *at* z *if* $\widehat{\mathbf{K}}(z)\varsigma = 0$*);*

(b′) $|\widehat{\mathbf{U}}_t(z)^{-1}|$, $|\widehat{\mathbf{G}}_t^{out}(z)^{-1}|$ *are bounded in an annulus* $\sigma \le |z| \le \sigma_0$*, uniformly in* t *and* z.

It will be recalled that the following constants were defined in Section 3,

$$\kappa_\sigma = \frac{\sigma}{\sqrt{\sigma^2-1}}, \qquad \kappa_\sigma^{(p)} = \begin{cases} \frac{1}{\sigma-1}, & \text{if } \ p=\infty, \\ \sqrt{n}\left[\sum_{r=0}^{\infty}(r\sigma^{-r})^2\right]^{1/2}, & \text{if } \ 2<p<\infty. \end{cases} \tag{49}$$

Theorem 4.1. $\mathbf{Q}$ *defined by (48) stabilizes* $\mathbf{G}$ *in* $\mathbf{B}$. *If* $\mathbf{G}_W, \mathbf{G}_W^{(-1)}$ *and* $\mathbf{S}^\ell$ *are slowly varying, then the weighted sensitivity* $\mathbf{S} \in \underline{\mathbf{E}}_\sigma$ *is explicitly given by*

$$\mathbf{S} = \mathbf{S}^\ell + \mathbf{G}_W \nabla(\mathbf{G}_W^{(-1)} \otimes \mathbf{F}) \tag{50}$$

where

$$\mathbf{F} := \mathbf{W}_2\mathbf{W}_1 - \mathbf{S}^\ell \ \textit{and}\ \mathbf{G}_W = \mathbf{U} \otimes \mathbf{G}^{out}.$$

Moreover, the difference $\|\mathbf{S}\|_{a(\sigma;t)} - \mu_\sigma(\widehat{\mathbf{S}}_t^\ell) =: \Delta(t)$ *has the upper bound*

$$\Delta(t) \leq \kappa_\sigma^{(p)}\{\kappa_\sigma\theta(\rho,\sigma) + \partial_\sigma^{(p)}(\widehat{\mathbf{S}}^\ell)\}, \tag{51}$$

and $\|\mathbf{S}\|_{a(\sigma;t)}$ *also has the lower bound*

$$\mu_1(\widehat{\mathbf{S}}_t^\ell) - \kappa_\sigma^{(p)}\partial_\sigma^{(p)}(\widehat{\mathbf{S}}^\ell) - \kappa_\sigma\kappa_\sigma^{(p)}\theta(p,\sigma) \leq \|\mathbf{S}\|_{a(\sigma;t)}, \tag{52}$$

where

$$\theta(p,\sigma) := \mu_\sigma(\widehat{\mathbf{G}}_W)\partial_\sigma^{(p)}(\widehat{\mathbf{G}}_W^{\textcircled{-1}} \otimes \widehat{\mathbf{F}}) \tag{53}$$

$$\partial_\sigma^{(p)}(\mathbf{F}) \leq \partial_\sigma(\widehat{\mathbf{W}}_2\widehat{\mathbf{W}}_1) + \partial_\sigma(\widehat{\mathbf{S}}^\ell). \tag{54}$$

Remark 4.1. It follows from (51) that in a variable rate situation (see Section 3.4), if the rates approach 0, the frequency-domain norm $\mu_\sigma(\widehat{\mathbf{S}}_t^\ell)$ of the local sensitivity $\mathbf{S}_t^\ell$ approaches an upper bound on the time-domain norm $\|\mathbf{S}\|_{a(\sigma;t)}$ of the global sensitivity $\mathbf{S}$.

If $\mathbf{S}^\ell$ actually satisfies the uniform radial growth condition defined in Section 3.3, then it follows from (51), (52) and the inequality

$$\mu_\sigma(\widehat{\mathbf{S}}_t^\ell) - \mu_1(\widehat{\mathbf{S}}_t^\ell) \leq \mu_1(\widehat{\mathbf{S}}_t^\ell)\left(\nu_{\sigma_0}(\widehat{\mathbf{S}}^\ell)^{\left(\frac{\ln\sigma}{\ln\sigma_0}\right)} - 1\right)$$

that $\Delta(t)$ is bounded by

$$|\Delta(t)| \leq \mu_1(\widehat{\mathbf{S}}_t^\ell)\left(\nu_{\sigma_0}(\widehat{\mathbf{S}}^\ell)^{\left(\frac{\ln\sigma}{\ln\sigma_0}\right)} - 1\right) + \kappa_\sigma^{(p)}\partial_\sigma^{(p)}(\widehat{\mathbf{S}}^\ell) + \kappa_\sigma\kappa_\sigma^{(p)}\theta(p,\sigma); \tag{55}$$

and if, in addition, $\sigma \to 1$, the frequency and time-domain norms mentioned at the begining of this remark approach each other, by (55).

Proof of Theorem 4.1. As $\mathbf{Q}$ satisfies (48), it has the form

$$\mathbf{Q} = \left\{(\mathbf{G}^{out})^{\textcircled{-1}} \otimes \left(\mathbf{U}^{\textcircled{-1}} \otimes (\mathbf{W}_2\mathbf{W}_1 - \mathbf{S}^\ell)\right)\right\}\mathbf{W}_1^{-1}.$$

Since $(\mathbf{G}^{out})^{\textcircled{-1}}$, $\mathbf{U}^{\textcircled{-1}} \otimes (\mathbf{W}_2\mathbf{W}_1 - \mathbf{S}^\ell)$, and $\mathbf{W}_1^{-1}$ are in $\underline{\mathbf{E}}_\sigma$ by assumptions (a,b) and the local interpolation condition (46), and $\underline{\mathbf{E}}_\sigma$ is closed under the $\otimes$ and $\cdot$ products, $\mathbf{Q}$ is in $\underline{\mathbf{E}}_\sigma$, i.e., $\mathbf{Q}$ stabilizes $\mathbf{G}$ in $\underline{\mathbf{E}}_\sigma$.

The bounds (51, 52) are obtained from the inequalities (35, 40), respectively, applied to the expressions $\mathbf{S}^\ell = \mathbf{W}_2\mathbf{W}_1 - \mathbf{G}_W \otimes (\mathbf{G}_W^{\textcircled{-1}} \otimes \mathbf{F})$ and $\mathbf{S} = \mathbf{W}_2\mathbf{W}_1 - \mathbf{G}_W(\mathbf{G}_W^{\textcircled{-1}} \otimes \mathbf{F})$ where the identifications $\mathbf{G} \to -\mathbf{G}_W$, $\mathbf{K} \to \mathbf{G}_W^{\textcircled{-1}} \otimes \mathbf{F}$ and $\mathbf{F} \to \mathbf{W}_2\mathbf{W}_1$ are made. □

4.1 Local H^∞ Adaptive Optimization

A natural idea for adaptive compensation is to make the weighted sensitivity $\mathbf{S}_t$ at time $t \in \mathbf{Z}$ depend on the local behavior of weighted plant inner and outer factors $\mathbf{U}_t, \mathbf{G}_t^{out}$, and[11] weighting $\mathbf{W}_t := (\mathbf{W}_2\mathbf{W}_1)_t$. That triple of local operators constitutes the *data* available at t about the plant in the form of a nominal model and a band of uncertainty. The data may be available a priori or acquired through identification or a combination of both. In frozen-time adaptive design the controller generates a local approximation $\mathbf{S}_t^\ell$ to $\mathbf{S}_t$ based on that data. The adaptation law can be represented by a map $\mathcal{S}$,

$$\mathcal{S} : \mathbf{Z} \times (H_{\sigma_0}^\infty)^2 \to (H_{\sigma_0}^\infty), \quad \widehat{\mathbf{S}}_t^\ell = \mathcal{S}(\widehat{\mathbf{U}}_t, \widehat{\mathbf{W}}_t), \quad \sigma_0 > \sigma.$$

If the data varies slowly, Theorem 4.1 provides a basis for frozen-time adaptation provided $\widehat{\mathbf{S}}^\ell$ also varies slowly. A sufficient condition for this is that $\mathcal{S}$ be Lipschitz in its variables, i.e., that there be constants $\lambda_U^{(p)}, \lambda_W^{(p)}$ for which the inequality

$$\|\widehat{\mathbf{S}}_t^\ell - \widehat{\mathbf{S}}_{t-1}^\ell\|_{H_{\sigma_0}^p} \leq \lambda_U^{(p)} \|\widehat{\mathbf{U}}_t - \widehat{\mathbf{U}}_{t-1}\| + \lambda_W^{(p)} \|\widehat{\mathbf{W}}_t - \widehat{\mathbf{W}}_{t-1}\|$$

holds with $\|\cdot\|$ representing $\|\cdot\|_{H_{\sigma_0}^\infty}$ or, in more compact matrix notation,

$$\partial_{\sigma_0}^{(p)}(\widehat{\mathbf{S}}^\ell) \leq \Lambda_\sigma^{(p)} \partial_{\sigma_0}([\widehat{\mathbf{U}}, \widehat{\mathbf{W}}]^{tr}) \tag{56}$$

where $\partial_\sigma([\widehat{\mathbf{K}}_{ij}])$ means $[\partial_\sigma(\widehat{\mathbf{K}}_{ij})]$ and $\Lambda_\sigma^{(p)}$ is the row matrix $[\lambda_U^{(p)}, \lambda_W^{(p)}]$ of Lipschitz constants. In the case of variable rates, it will be assumed the Lipschitz constants hold independently of rate.

One can try to design $\mathbf{S}^\ell$ by local $H_{\sigma_0}^\infty$ optimization, which gives an optimal weighted sensitivity $\mathbf{S}_{opt(\sigma_0)}^\ell$ satisfying

$$\|(\widehat{\mathbf{S}}_{opt(\sigma_0)}^\ell)_t\|_{H_{\sigma_0}^\infty} = \inf_{Q \in H_{\sigma_0}^\infty} \|\widehat{\mathbf{W}}_t - \widehat{\mathbf{U}}_t Q\|_{H_{\sigma_0}^\infty}, \quad t \in \mathbf{Z},$$

which would imply

$$\mu_{\sigma_0}(\widehat{\mathbf{S}}_{opt(\sigma_0)}^\ell) = \inf_{\mathbf{Q} \in \overline{\mathbf{E}}_{\sigma_0}} \mu_{\sigma_0}(\widehat{\mathbf{W}} - \widehat{\mathbf{U} \otimes \mathbf{Q}}).$$

In the case in which $\widehat{\mathbf{S}}_{opt(\sigma_0)}^\ell$ is a Lipschitz function of data, Theorem 4.1 gives a global sensitivity $\mathbf{S}$ which approximates the local optimum. These cases are illustrated by the following example and discussion of robust versus adaptive design.

[11]For simplicity, assume that $\mathbf{S}^\ell$ does not depend on $\mathbf{W}_1, \mathbf{W}_2$ separately.

5 ROBUST VS. ADAPTIVE SENSITIVITY MINIMIZATION

Metric information about uncertain perturbations or disturbances is represented by a weighting operator $\mathbf{W} \in \underline{\mathbf{E}}_{\sigma}$, $\sigma > 1$. At time t, disturbances are assumed to lie in the image under $\mathbf{W}_t$ of the unit ball of $\ell^2_{\sigma_0}(-\infty, t)$, $\sigma_0 > \sigma$, in the case of noise, (or of $H^\infty_{\sigma_0}$ in the case of transfer function uncertainty). The smaller the weighting, the more tightly the uncertainty is confined in $\mathbb{C}^n$ (or $\mathbb{C}^{n\times n}$) at time t, and therefore the greater the information pertaining to that time. (Information can be measured by ε-entropy or ε-dimension [30]; although quantitative information measures will not be estimated here, we note that they depend monotonely on the weighting). We distinguish *a priori* information at some starting time t_o, and *a posteriori* information at time $\tau \geq t_o$ represented by operators $\mathbf{W}^o$ and $\mathbf{W}^\tau$. The difference between $\mathbf{W}^o$ and $\mathbf{W}^\tau$ represents a reduction of uncertainty or acquisition of information in the interval $[t_o, \tau]$, and this reduction is reflected in a shrinkage of weighting, $|(\mathbf{W}^\tau)_t(z)| \leq |(\mathbf{W}^o)_t(z)|$ for at least some $t \geq \tau$ and z in some subset of the circle $|z| = \sigma_0$ of non-zero length. A sensitivity reduction scheme will be called *robust* or *adaptive* if based on a priori or a posteriori information, respectively. A controller which achieves a sensitivity which is better than an optimal robust one is necessarily adaptive, and the question arises how much advantage adaptation provides. For slowly time-varying systems, this can be answered independently of how the information was obtained.

Example 5.1

We will introduce a family of "narrow band" disturbance weighting functions whose center frequencies become known with increasing accuracy, and whose envelope is easy to compute.

Let $f(\cdot) : [0, \pi] \to \mathbf{R}$ be a differentiable monotone decreasing function satisfying $f(0) = 1, f(\theta) = \xi$ for $\theta \geq \frac{\alpha}{2}\pi$, where $0 < \xi \ll 1, 0 < \alpha \ll 1$ are constants. $f(\cdot)$ will be fixed.

Let $\sigma_0 > \sigma > 1$ be fixed. A narrowband weighting $V_{(\theta_0)} \in H^\infty_{\sigma_0}$, with center θ_0, $\frac{\alpha}{2}\pi \leq \theta_0 \leq (1 - \frac{\alpha}{2})\pi$, is a function such that $V_{(\theta_0)}(\sigma_0(\cdot))$ is outer in H^∞, defined in terms of its boundary magnitude by

$$|V_{(\theta_0)}(\sigma_0 e^{i\theta})| = \begin{cases} f(|\theta - \theta_0|) & \text{for } 0 \le \theta \le \pi \\ \left|V_{(\theta_0)}(\sigma_0 e^{-i\theta})\right| & \text{for } -\pi \le \theta \le 0 \end{cases}$$

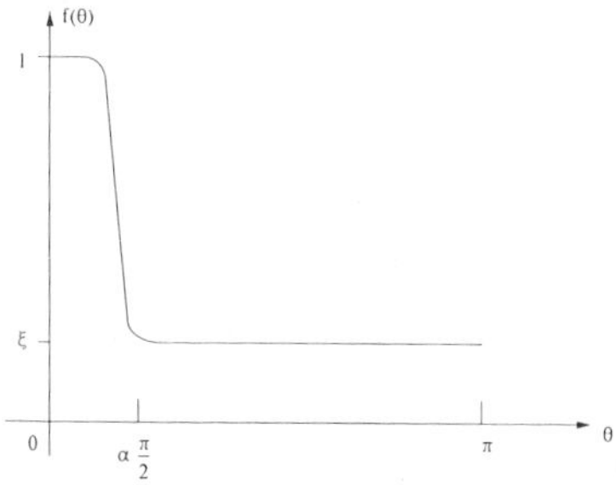

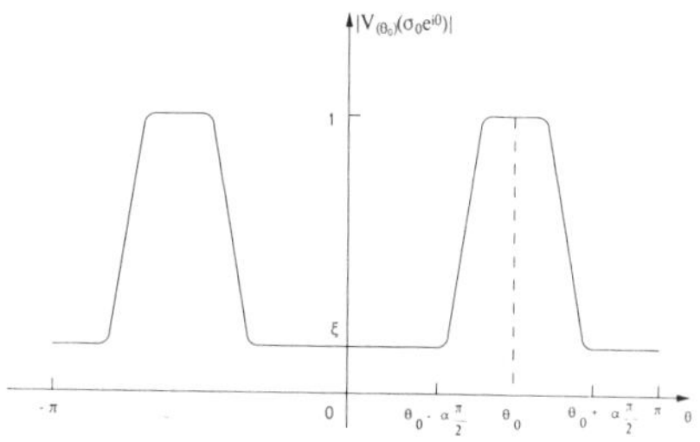

Narrowband disturbances with uncertain center frequencies will be represented as elements of a family of such narrowband weightings,

$$\mathcal{N}(\beta, c) = \left\{ V_{(\theta_0)} \in H^\infty_{\sigma_0} : |\frac{2}{\pi}\theta_0 - c| \le \beta \le (2 - \alpha) \right\}.$$

The center frequencies lie in an interval with midpoint c and width β; β is a measure of uncertainty about center frequencies.

Let $\tilde{V}_{(\beta,c)} \in H^\infty_{\sigma_0}$ denote the *envelope* weighting of the family, $\tilde{V}_{(\beta,c)}(\sigma_0(\cdot))$ outer in H^∞, and satisfying

$$\left|\tilde{V}_{(\beta,c)}(\sigma_0 e^{i\theta})\right| = \sup_{V \in \mathcal{N}(\beta,c)} \left|V(\sigma_0 e^{i\theta})\right|, \quad -\pi \le \theta \le \pi.$$

A priori information about the disturbances is that they belong to the family $\mathcal{N}(\beta_0, c_0)$. (The a priori weighting is assumed to be time invariant.)

Sensitivity is to be minimized in a SISO time-invariant plant $\mathbf{G} \in \mathbf{E}_{\sigma_0}$, whose inner part consists of one zero at the origin, i.e., the inner factor in $\mathbf{E}_{\sigma_0}$ is $U(z) = \sigma_0^{-1} z$. A robust control based on the a priori envelope, $\widehat{\mathbf{W}}_0 = \tilde{V}_{(\beta_0, c_0)}$ achieves

$$\mu_{\sigma_0}(\widehat{\mathbf{S}}_{rbst}) = \inf_{\mathbf{Q} \in \overline{\mathbf{E}}_{\sigma_0}} \mu_{\sigma_0}(\widehat{\mathbf{W}}_0 - \widehat{\mathbf{U}}\widehat{\mathbf{Q}}) = \widehat{\mathbf{W}}_0(0). \tag{57}$$

In an interval $[0, t]$, additional information is received about the disturbances, and results in a shrinkage in a posteriori uncertainty about the center frequency parameter, i.e., β_t is monotone decreasing as $t \to \infty$. An adaptive local optimization of the worst case sensitivity, based on the a posteriori envelope

$$\widehat{\mathbf{W}}_t := \begin{cases} \widetilde{V}_{(\beta_t,c_t)} & \text{for } t \geq 0, \\ \widetilde{V}_{(\beta_0,c_0)} & \text{for } t \leq 0, \end{cases}$$

based on Theorem 4.1, achieves

$$\mu_{\sigma_0}\left[\left(\widehat{\mathbf{S}}^{\ell}_{adpt}\right)_t\right] = \inf_{\widehat{\mathbf{Q}}_t \in H^\infty_{\sigma_0}} \|\widehat{\mathbf{W}}_t - \widehat{\mathbf{G}}_t\widehat{\mathbf{Q}}_t\|_{H^\infty_{\sigma_0}} = \widehat{\mathbf{W}}_t(0) \tag{58}$$

and the resulting adaptive sensitivity achieved is

$$\mathbf{S}_{adpt} = \mathbf{S}^{\ell}_{adpt} + \mathbf{U}\nabla\left(\mathbf{U}^{\ominus 1} \otimes (\mathbf{W} - \mathbf{S}^{\ell})\right).$$

The constants in (57, 58) can be expressed in terms of the logarithmic bandwidth $\chi(t)$ of the envelope at time t, defined by

$$\log \chi(t) = \frac{1}{2\pi}\int_{-\pi}^{\pi} \log\left|\widetilde{V}_{(\beta_t,c_t)}(\sigma_0 e^{i\theta})\right| d\theta.$$

From the assumption that $f(\theta) = \xi$ for $\theta \geq \frac{\alpha}{2}\pi$ and the fact that $|\widetilde{V}_{(\beta_0,c_0)}(\cdot)|$ is a widening of $|\widetilde{V}_{(\beta_t,c_t)}(\cdot)|$ by $(\beta_0 - \beta_t)$, we deduce that

$$\log \chi(t) = \log \chi(0) + (\beta_0 - \beta_t)\log \xi.$$

By Jensen's Theorem,

$$\widetilde{V}_{(\beta_t,c_t)}(0) = e^{\log \chi(t)} = \chi(0)\xi^{(\beta_0-\beta_t)}. \tag{59}$$

Let us evaluate the recent past norms $\|\cdot\|_{a(\sigma_0;t)}$ of the sensitivity for the robust and adaptive controllers. In the robust case

$$\mu_1(\widehat{\mathbf{S}}_{rbst}) \leq \|\mathbf{S}_{rbst}\|_{a(\sigma_0)} \leq \mu_{\sigma_0}(\widehat{\mathbf{S}}_{rbst}) = \chi(0).$$

In this simple example, the $\mu_{\sigma_0}(\widehat{\mathbf{S}}_{rbst})$ norm is independent of σ_0 and we get, from (57) and (59)

$$\|\mathbf{S}_{rbst}\|_{a(\sigma_0)} = \chi(0). \tag{60}$$

In the adaptive case, (58) and (59) give

$$\mu_{\sigma_0}\left[\left(\widehat{\mathbf{S}}^{\ell}_{adpt}\right)_t\right] = \chi(0)\xi^{(\beta_0-\beta_t)}. \tag{61}$$

Suppose now that β_t and c_t change slowly, $|\beta_t - \beta_{t-1}| \leq \rho_\beta, |c_t - c_{t-1}| \leq \rho_c$, and $|\frac{df(\theta)}{d\theta}| \leq \rho_f$. The rates of $\mathbf{W}$ and $\mathbf{S}^\ell_{adpt}$ are

$$\begin{aligned} \partial_{\sigma_0}(\widehat{\mathbf{W}}) &\leq \rho := \rho_f(\rho_\beta + \rho_c) \\ \partial_{\sigma_0}(\widehat{\mathbf{S}}^\ell_{adpt}) &= \partial_{\sigma_0}(\tilde{V}_{(\beta_t, c_t)}(0)) \leq \partial_{\sigma_0}(\widehat{\mathbf{W}}). \end{aligned}$$

As $\mathbf{S}^\ell_{adpt}$ depends Lipschitz continuously ($L^\infty \to L^\infty$) on $\mathbf{W}$, the rate of the local optimal sensitivity becomes small as $\rho \to 0$, and we can base our solution on it. To evaluate the upper bound (51) of Theorem 4.1, we note that $\widehat{\mathbf{G}}_W(z) = \widehat{\mathbf{U}}(z) = \sigma_0^{-1}z$, $\mu_{\sigma_0}(\widehat{\mathbf{G}}_W) = 1$,

$$\begin{aligned} \theta(\infty, \sigma_0) = \partial_{\sigma_0}(\widehat{\mathbf{F}}) &\leq \partial_{\sigma_0}(\widehat{\mathbf{W}}) + \partial_{\sigma_0}(\widehat{\mathbf{S}}^\ell_{adpt}) \\ &\leq 2\rho \end{aligned}$$

which gives

$$\|\mathbf{S}_{adpt}\|_{a(\sigma_0;t)} \leq \mu_{\sigma_0}(\widehat{\mathbf{S}}^\ell_{adpt}) + \rho\kappa^{(\infty)}_{\sigma_0}[1 + 2\kappa_{\sigma_0}]. \tag{62}$$

A lower bound is computed using (41) of Section 3, giving

$$\|\mathbf{S}_{adpt}\|_{a(\sigma_0;t)} \geq \mu_1(\widehat{\mathbf{S}}^\ell_{adpt}) - \rho\kappa^{(\infty)}_{\sigma_0}[1 + 2\kappa_{\sigma_0}]. \tag{63}$$

As μ_{σ_0} is independent of σ_0 in (62, 63), and using (61),

$$\left| \|\mathbf{S}_{adpt}\|_{a(\sigma_0;t)} - \chi(0)\xi^{(\beta_0 - \beta_t)} \right| \leq \rho\kappa^{(\infty)}_{\sigma_0}[1 + 2\kappa_{\sigma_0}]$$

and by (60),

$$\frac{\|\mathbf{S}_{adpt}\|_{a(\sigma_0;t)}}{\|\mathbf{S}_{rbst}\|_{a(\sigma_0;t)}} \leq \xi^{(\beta_0 - \beta_t)} + \rho\kappa^{(\infty)}_{\sigma_0}[1 + 2\kappa_{\sigma_0}]\chi^{-1}(0). \tag{64}$$

In the limit of slow time variation, as $\rho \to 0$, (64) shows that adaptive sensitivity is better than optimal robust sensitivity by a factor $\xi^{(\beta_0 - \beta_t)}$, where $(\beta_0 - \beta_t)$ is the reduction in log-bandwidth of the disturbance weighting resulting from extra information about disturbances acquired in the intervening interval $[0, t]$.

Remark 5.1. It should be emphasized that the point of this example is not the trivial one that the smaller the weighting the smaller the time invariant sensitivity. Rather, it is that if a way is found of computing sensitivity optimally under time-varying weightings, then that opens up a means of determining whether a black box is adaptive or merely robust, provided that weighting can be interpreted as a measure of information (i.e., decay of the former is equivalent to growth of the latter). It is worth noting that the internal structure or parametrization of the system, or the manner in which information is acquired, are irrelevant here.

6 δ-SUBOPTIMAL SENSITIVITY

However, it will be shown in Section 7 that the optimal local sensitivity $\widehat{\mathbf{S}}^{\ell}_{opt}$ is not always Lipschitz, and is therefore not always a suitable candidate for frozen-time design. On the other hand, it will be shown that the AAK maximal-entropy (local) interpolation, selected to be δ-suboptimal at each $t \in \mathbf{Z}$, $\delta > 0$, produces a local sensitivity $\mathbf{S}^{\ell}_{(\sigma_0,\delta)}$ which is *δ-suboptimal* in the sense that

$$\mu_{\sigma_0}(\widehat{\mathbf{S}}^{\ell}_{(\sigma_0,\delta)}) \leq \mu_{\sigma_0}(\widehat{\mathbf{S}}^{\ell}_{opt(\sigma_0)}) + \delta \tag{65}$$

and satisfies a Lipshitz continuity condition (56), (see Corollary 7.2).

Subject to the Lipschitz condition (56), the global sensitivity $\mathbf{S}_{(\sigma_0,\delta,\rho)}$ achieved using such a locally δ-suboptimal adaptation scheme (specified by (47, 45)) is δ'-suboptimal as follows. Denote $[\widehat{\mathbf{S}}^{\ell}_{opt(\sigma_0)}]_t$ by $\widehat{\mathbf{S}}^{\ell}_{opt(\sigma_0;t)}$.

Let $\sigma_0 > \sigma \geq 1$, and suppose Assumption 4.1 holds.

Corollary 6.1. *(a) Given any $\delta' > \delta$, $\mathbf{S}_{(\sigma_0,\delta,\rho)}$ has the bound*

$$\|\mathbf{S}_{(\sigma_0,\delta,\rho)}\|_{a(\sigma;t)} \leq \mu_{\sigma_0}(\widehat{\mathbf{S}}^{\ell}_{opt(\sigma_0;t)}) + \delta', \tag{66}$$

provided the variation rates of $\mathbf{U}$ *and* $\mathbf{W}$ *satisfy the inequality*

$$C^{(p)}_{\sigma_0}\partial^{(p)}_{\sigma_0}([\widehat{\mathbf{U}}, \widehat{\mathbf{W}}]^{\mathrm{tr}}) \leq \delta' - \delta \tag{67}$$

where $C^{(p)}_{\sigma_0}$ is the row matrix

$$C^{(p)}_{\sigma_0} := \alpha^{(p)}_{\sigma_0}\left(\kappa_{\sigma_0}\Lambda^{(p)}_{\sigma_0} + \begin{bmatrix} \alpha^{(p)}_{\sigma_0}\mu_{\sigma_0}(\widehat{\mathbf{S}}^{\ell}_{opt(\sigma_0)} + \widehat{\mathbf{W}}) \\ 1 \end{bmatrix}^{\mathrm{tr}}\right)$$

with $\alpha^{(p)}_{\sigma_0} := \kappa_{\sigma_0}\kappa^{(p)}_{\sigma_0}\mu_{\sigma_0}(\widehat{\mathbf{G}}_W)\mu_{\sigma_0}(\widehat{\mathbf{G}}^{(-1)}_W)$, and $\Lambda^{(p)}_{\sigma_0}$ is the Lipschitz constants defined in (56).

If the data operators $\mathbf{U}, \mathbf{W}$, *have variable rates towards* 0, $\partial^{(p)}_{\sigma_0}(\cdot) \leq \rho$, *then*

$$\lim_{\rho\to 0} \|\mathbf{S}_{(\sigma_0,\delta,\rho)}\|_{a(\sigma;t)} \leq \mu_{\sigma_0}(\widehat{\mathbf{S}}^{\ell}_{opt(\sigma_0;t)}) + \delta. \tag{68}$$

(b)
$$\lim_{\sigma_0\to 1}\lim_{\delta\to 0}\lim_{\rho\to 0}\|\mathbf{S}_{(\sigma_0,\delta,\rho)}\|_{a(\sigma;t)} = \mu_1(\widehat{\mathbf{S}}^{\ell}_{opt(1;t)}). \tag{69}$$

Proof: (a) (66, 67) are obtained from Theorem 4.1, by substituting (56) for $\partial^{(p)}_{\sigma_0}(\widehat{\mathbf{S}}^{\ell})$ into (51), and noting that $\mu_\sigma(\cdot)$ and $\partial^{(p)}_{\sigma}(\cdot)$ are monotone increasing in σ for $1 \le \sigma \le \sigma_0$.

If $\mathbf{U}$ and $\mathbf{W}$ have variable rates approaching 0, as the Lipschitz constants $\Lambda^{(p)}_{\sigma_0}$ are not dependent on ρ, $\mathbf{S}^{\ell}_{(\sigma_0,\delta)}$ has variable rate approaching 0, and $\mu_{\sigma_0}(\widehat{\mathbf{S}}^{\ell}_{(\sigma_0,\delta)}) \le \mu_{\sigma_0}(\widehat{\mathbf{S}}^{\ell}_{opt(\sigma_0)}) + \delta$ by hypothesis; (68) follows by taking the indicated limits of (51).

(b) Since $\mu_1(\widehat{\mathbf{S}}^{\ell}_{opt(1;t)}) \le \mu_1(\widehat{\mathbf{S}}^{\ell}_{opt(\sigma_0;t)})$ by optimality, after taking the limits $\rho \to 0$ and $\delta \to 0$ in (52) and (65), we obtain

$$\mu_1(\widehat{\mathbf{S}}^{\ell}_{opt(1;t)}) \le \lim_{\delta\to 0}\lim_{\rho\to 0}\|\mathbf{S}_{(\sigma_0,\delta,\rho)}\|_{a(\sigma;t)} \le \mu_{\sigma_0}(\widehat{\mathbf{S}}^{\ell}_{opt(\sigma_0;t)}).$$

To emphasize the dependence of the inner-outer factorization on σ, in the rest of the proof write $\mathbf{U}_{(\sigma;t)}$ for $\mathbf{U}_t$ and $\mathbf{G}^{out}_{(\sigma;t)}$ for $\mathbf{G}^{out}_t$, i.e., in this notation $(\mathbf{G}_W)_t = \mathbf{U}_{(\sigma;t)}\mathbf{G}^{out}_{(\sigma;t)}$ where σ is a variable. It will be shown below that $\mu_\sigma(\widehat{\mathbf{S}}^{\ell}_{opt(\sigma;t)})$, viewed as a function of σ, satisfies the Lipschitz condition

$$\begin{aligned}|\mu_{\sigma_0}(\widehat{\mathbf{S}}^{\ell}_{opt(\sigma_0;t)}) - \mu_1(\widehat{\mathbf{S}}^{\ell}_{opt(1;t)})| \;&\le\; \mu_{\sigma_0}(\widehat{\mathbf{W}}_t)\mu_1(\widehat{\mathbf{U}}_{(\sigma_0;t)}(\sigma_0(\cdot)) - \widehat{\mathbf{U}}_{(1;t)}(\cdot)) \\ &\quad + \mu_1(\widehat{\mathbf{W}}_t(\sigma_0(\cdot)) - \widehat{\mathbf{W}}_t(\cdot)).\end{aligned} \tag{70}$$

It can be shown (see [22]) that for $\mathbf{G}_W$ satisfying Assumption 4.1 there exists a unitary matrix $B \in \mathbb{C}^{n\times n}$ such that

$$\mu_1(\widehat{\mathbf{U}}_{(\sigma_0;t)}(\sigma_0(\cdot))B - \widehat{\mathbf{U}}_{(1;t)}(\cdot)) \le Const.\|(\mathbf{G}_W)_t(\sigma_0(\cdot)) - (\mathbf{G}_W)_t(\cdot)\|_{(1)}. \tag{71}$$

Since $\mu_{\sigma_0}(\widehat{\mathbf{S}}^{\ell}_{opt(\sigma_0;t)})$ is not affected by the multiplication of the inner factor $\widehat{\mathbf{U}}_{(\sigma_0;t)}(\sigma_0(\cdot))$ by a unitary constant matrix B, we obtain from (70) and (71) that

$$\begin{aligned}|\mu_{\sigma_0}(\widehat{\mathbf{S}}^{\ell}_{opt(\sigma_0;t)}) - \mu_1(\widehat{\mathbf{S}}^{\ell}_{opt(1;t)})| \;&\le\; Const.\|(\mathbf{G}_W)_t(\sigma_0(\cdot)) - (\mathbf{G}_W)_t(\cdot)\|_{(1)} \\ &\quad + \mu_1(\widehat{\mathbf{W}}_t(\sigma_0(\cdot)) - \widehat{\mathbf{W}}_t(\cdot)).\end{aligned} \tag{72}$$

It is easy to show that under Assumption 4.1,

$$\|(\mathbf{G}_W)_t(\sigma_0(\cdot)) - (\mathbf{G}_W)_t(\cdot)\|_{(1)} \to 0,$$

$$\mu_1(\widehat{\mathbf{W}}_t(\sigma_0(\cdot)) - \widehat{\mathbf{W}}_t(\cdot)) \to 0,$$

as $\sigma_0 \to 0$, which implies

$$\mu_{\sigma_0}(\widehat{\mathbf{S}}^{\ell}_{opt(\sigma_0;t)}) \to \mu_1(\widehat{\mathbf{S}}^{\ell}_{opt(1;t)}),$$

and (69) follows.

It remains only to establish (70). This will be done in Proposition 7.1 of the next section. Inequality (70) will follow from (76) with the following substitution: $U_1 = \widehat{\mathbf{U}}_{(\sigma_0;t)}(\sigma_0(\cdot))$, $U_2 = \widehat{\mathbf{U}}_{(1;t)}(\cdot)$, $V_1 = \widehat{\mathbf{W}}_t(\sigma_0(\cdot))$, $V_2 = \widehat{\mathbf{W}}_t(\cdot)$, $S_1 = \widehat{\mathbf{S}}^{\ell}_{opt(\sigma_0;t)}(\sigma_0(\cdot))$, $S_2 = \widehat{\mathbf{S}}^{\ell}_{opt(1;t)}(\cdot)$. □

Remark 6.1. At each instant $t \in \mathbf{Z}$, the global time-domain sensitivity (under persistent disturbances) $\|\mathbf{S}_{(\sigma_0,\delta,\rho)}\|_{a(\sigma;t)}$ is at most δ'-inferior to the optimum $\mu_{\sigma_0}(\widehat{\mathbf{S}}^{\ell}_{opt(\sigma_0;t)})$ achieved by local frequency domain interpolation; it is δ-inferior in the limit of zero variation and, since δ is arbitrary, actually approaches the μ_1 norm of the H^∞ optimum as $\sigma_0 \to 1$. So long as that norm depends continuously on σ_0 near $\sigma_0 = 1$, i.e.,

$$\lim_{\sigma_0 \to 1} \mu_{\sigma_0}(\widehat{\mathbf{S}}^{\ell}_{opt(\sigma_0;t)}) = \mu_1(\widehat{\mathbf{S}}^{\ell}_{opt(1;t)}), \tag{73}$$

this means that in a certain sense the best that can be achieved with slowly-varying data and slowly-varying compensators is near the time-invariant best, but does not preclude the possibility that quickly varying compensators might do better still.

7 CONTINUITY PROPERTIES OF H^∞ INTERPOLANTS

The continuity properties used in local interpolation will be developed in this section. As the normed spaces H^∞_σ and H^∞ are isometrically related by the change of variable $z \Leftrightarrow \sigma z$, there is no loss of generality in considering H^∞ only. *Throughout Section 7,* $\|\cdot\|_{L^p}$ *will be abbreviated to* $\|\cdot\|_p$*, for* $p = 2, \infty$*, and the unsubscripted norm symbol* $\|\cdot\|$ *will denote the operator norm of an operator between two Hilbert spaces.*

$S \in L^\infty$ will be said to *interpolate a pair of data functions* (V, U), $V \in L^\infty$, $U \in H^\infty$, U inner, if there exists $Q \in H^\infty$ satisfying

$$S = V - UQ. \tag{74}$$

(In Section 6, both data functions were locally in H^∞, so that this interplolation is more general.) If the data (V, U) is variable, and $\mathcal{S}(\cdot,\cdot)$ is an interpolation rule which assigns interpolants to data pairs, the interpolation rule will be called *Lipschitz continuous* $(L^\infty \to L^2)$ if there are constants $\gamma_V^o \geq 0$, $\gamma_U^o \geq 0$, for which any two assignments $S_i = \mathcal{S}(V_i, U_i)$, $i = 1, 2$, satisfy the inequalities,

$$\|S_2 - S_1\|_2 \leq \gamma_V^o \|V_2 - V_1\|_\infty + \gamma_U^o \|U_2 - U_1\|_\infty \tag{75}$$

or, in incremental notation which will be used in the proofs,

$$\|\Delta S\|_2 \leq \gamma_V^o \|\Delta V\|_\infty + \gamma_U^o \|\Delta U\|_\infty.$$

The rule will be said to yield *Lipschitz continuous norm* if for constants $\gamma_V' \geq 0, \gamma_U' \geq 0$,

$$\big|\|S_2\|_\infty - \|S_1\|_\infty\big| \ \leq \ \gamma_V' \|V_2 - V_1\|_\infty + \gamma_U' \|U_2 - U_1\|_\infty. \tag{76}$$

Now consider the optimal interpolation rule, $(V, U) \to S_{opt}$, where

$$\|S_{opt}\|_\infty = \inf_{Q \in H^\infty} \|V - UQ\|_\infty =: \mu_0.$$

Let $\mathbf{\Gamma}_{U^*V} : \ell_+^2 \to \ell_+^2$ (where $\ell_+^2 := \ell^2[0,\infty)$) denote the Hankel operator with symbol U^*V. By Nehari's Theorem, $\mu_0 = \|\mathbf{\Gamma}_{U^*V}\|$.

Proposition 7.1. *The norm of the optimal interpolant depends Lipschitz continuously on the data (V, U); indeed, (76) is satisfied with constants $\gamma_V' = 1, \gamma_U' = \min_{i=1,2} \|V_i\|_\infty$.*

Proof: Write $\mathbf{\Gamma}_i := \mathbf{\Gamma}_{U_i^* V_i}$. By Nehari's Theorem,

$$\big|\ \|S_2\|_\infty - \|S_1\|_\infty\ \big| = \big|\ \|\mathbf{\Gamma}_2\| - \|\mathbf{\Gamma}_1\|\ \big| \leq \|\mathbf{\Gamma}_2 - \mathbf{\Gamma}_1\|.$$

As $\mathbf{\Gamma}_K$ is unitarily equivalent to a projection of K, $\mathbf{\Gamma}_K$ is linear in K and $\|\mathbf{\Gamma}_K\| \leq \|K\|_\infty$. Therefore,

$$\begin{aligned}\|\mathbf{\Gamma}_2 - \mathbf{\Gamma}_1\| &\leq \|U_2^* V_2 - U_1^* V_1\|_\infty \\ &\leq \|U_2^* - U_1^*\|_\infty \|V_2\|_\infty + \|U_1^*\|_\infty \|V_2 - V_1\|_\infty.\end{aligned}$$

Without loss of generality we can assume V_2 to be the smaller of V_i, $i = 1, 2$, which implies (76) with $\gamma_U' = \|V_2\|_\infty = \min_{i=1,2} \|V_i\|_\infty$, and $\gamma_V' = \|U_1^*\|_\infty = 1$. □

However, the optimal interpolant itself does not depend Lipschitz continuously on the data, and in fact can have an infinite modulus of continuity, as the following counterexample shows.

Example 7.1. Consider the problem of optimally interpolating (W_ω, U) in H^∞, where $U \in H^\infty$ is fixed, $U(z) = \frac{\beta_1 - z}{\beta_1 z - 1}\frac{\beta_2 - z}{\beta_2 z - 1}$, $0 < \beta_1 < \beta_2 < 1$, $(i = 1, 2)$, and $W_\omega \in H^\infty$ is variable depending on a parameter $\omega > 0$, $W_\omega(e^{j\theta}) = \overline{W_\omega(e^{-j\theta})}$. By the Nevanlinna-Pick theory, the optimal interpolant of (W_ω, U) has the form

$$S_\omega = \mu_\omega \frac{(\alpha_\omega - z)}{(\alpha_\omega z - 1)}, \ |\alpha_\omega| < 1, \ \mu_\omega \in \mathbf{R},$$

where S_ω satisfies the interpolation constraints $S_\omega(\beta_i) = W_\omega(\beta_i)$, $i = 1, 2$. Consider any W_ω for which the ratio $W_\omega(\beta_2)/W_\omega(\beta_1) =: \rho_\omega$ approaches 1 as $\omega \to 0$, and which satisfies the inequality

$$\left|\frac{d\rho_\omega}{d\omega}\right| \geq \theta \left\|\frac{dW_\omega}{d\omega}\right\|_\infty, \qquad \theta > 0.$$

For example, $W_\omega := 1 + \omega W'$, where $W' \in H^\infty$, $\|W'\|_\infty < 1$, $W'(\beta_1) = 0$, $W'(\beta_2) > 0$ will have these properties. Let $\|dW_\omega\|_\infty$ denote $\|W' d\omega\|_\infty$. We will show that as $\omega \to 0$, $\|dS_\omega\|_2/\|dW_\omega\|_\infty \to \infty$, implying that the optimal interpolant has an infinite modulus of continuity as a function of W and is certainly not Lipschitz.

As $\omega \to 0$ we get

$$\frac{dS}{\|dW\|_\infty} = \mu \frac{(z^2 - 1)}{(\alpha z - 1)^2} \frac{d\alpha}{\|dW\|_\infty} + \frac{(\alpha - z)}{(\alpha z - 1)} \frac{d\mu}{\|dW\|_\infty} \tag{77}$$

where S, W, α, μ, all depend on ω. The term proportional to $d\mu$ is ≤ 1 for $|z| \leq 1$ by Proposition 7.1, so it is enough to establish the unboundedness of the term proportional to $d\alpha$. Now $\omega \to 0$ implies that $\rho_\omega \to 1$ which, it is not hard to show, implies that $\alpha \to -1$ from the right, and $|d\alpha/d\omega| \to (\beta_2 - \beta_1)^{-1}\frac{(1+\beta_2)^2(1+\beta_1)^2}{2(\beta_1+\beta_2+\beta_1\beta_2)} W'(\beta_2) > 0$. Therefore, for ω small enough

$$\begin{aligned} \frac{|dS(z)|}{\|dW\|_\infty} &\geq \left|\mu \frac{(z^2 - 1)}{(\alpha z - 1)^2} \frac{d\alpha}{d\omega} \frac{d\omega}{\|dW\|_\infty}\right| - 1 \\ &\geq const. \frac{|z^2 - 1|}{|\alpha z - 1|^2} - 1, \qquad |z| \leq 1, \end{aligned} \tag{78}$$

as $d\omega/\|dW\|_\infty = (\|W'\|_\infty)^{-1}$. Contour integration now gives

$$\|dS\|_2/\|dW\|_\infty \geq C_1(1 - \alpha^2)^{-\frac{1}{2}} + C_2,$$

(where C_1, C_2 are constants) which grows without bound as $\alpha \to -1$, and therefore as $\omega \to 0$. □

This means that the optimal interpolant is not a suitable candidate for the local interpolation outlined in Section 4. Instead, we turn to a δ-suboptimal interpolant based on the AAK parametrization, which has the requisite Lipschitz continuity.

7.1 Lipschitz Continuity of AAK Central Interpolants

For $V \in L^\infty$, consider $S \in L^\infty$ which interpolates (V, I) and satisfies

$$\|S\|_\infty \leq \rho. \tag{79}$$

(Eventually, V will be identified with U^*W of Section 6, and US with the resulting sensitivity). Let $\mathbf{\Gamma}_V : \ell^2_+ \to \ell^2_+$ be the Hankel operator with symbol V. By Nehari's Theorem, interpolants satisfying (79) exist if $\rho \geq \|\mathbf{\Gamma}_V\|$ and will be called *δ-suboptimal* if $\rho - \|\mathbf{\Gamma}_V\| \leq \delta$, $\delta > 0$. The AAK parametrization provides two pairs of functions in H^2, $P_\pm$ and $Q_\pm$ with the property that every δ-suboptimal interpolant can be expressed in the form

$$S(z) = \rho(Q_-(z) + P_-(z)\mathrm{E}(z))(P_+(z) + Q_+(z)\mathrm{E}(z))^{-1}, \quad |z| = 1, \tag{80}$$

for some $\mathrm{E} \in H^\infty$, $\|\mathrm{E}\|_\infty \leq 1$. The *central* (or *maximal entropy*) *δ-suboptimal* interpolant is obtained when $\mathrm{E} = 0$,

$$S(z) = \rho Q_-(z)P_+^{-1}(z) \tag{81}$$

and is unique subject to the constraint that $\rho = \|\mathbf{\Gamma}_V\| + \delta$, which will be assumed to hold. The main objective of Section 7 will be to prove Theorem 7.1, establishing the Lipschitz continuity ($L^\infty \to L^2$) of this interpolant, expressed by the inequality

$$\|S_2 - S_1\|_2 \leq \gamma_V \|V_2 - V_1\|_\infty, \tag{82}$$

where S_1, S_2 are any two such interpolants of the respective data pairs (V_1, I), (V_2, I).

In Theorem 7.1, let $\mu_i := \|\mathbf{\Gamma}_{V_i}\|$, $\alpha_i := (\mu_i/\rho_i)^2$, $\mu := \max(\mu_1, \mu_2)$, $\alpha := \max(\alpha_1, \alpha_2)$.

Theorem 7.1. *The AAK central δ-suboptimal interpolant satisfies the Lipschitz condition (82) with constant*

$$\gamma_v = n^{\frac{1}{2}} \frac{\sqrt{1+\alpha}}{(1-\sqrt{\alpha})^2}\left[1+\left(\frac{1+\alpha}{1+\sqrt{\alpha}}\right)\right] \tag{83}$$

$$\leq 2\sqrt{2} n^{\frac{1}{2}} \left(\frac{\mu+\delta}{\delta}\right)^2. \tag{84}$$

7.2 The AAK Construction and the Proof of Theorem 7.1

A Hankel operator $\mathbf{\Gamma}_V : \ell^2_+ \to \ell^2_+$ with symbol $V \in L^\infty$ has the infinite matrix representation $[v_{j+k-1}]$, $j,k \geq 1$, where $v_k \in \mathbb{C}^{n\times n}$ are negative matrix Fourier coefficients

$$v_k = \frac{1}{2\pi}\int_{-\pi}^{\pi} e^{ik\theta} V(e^{i\theta}) d\theta.$$

The following construction was introduced by Adamjan, Arov and Krein in [1]. Let $\mathbf{T}_+ : \ell^2_+ \to \ell^2_+$ denote the right shift operator in ℓ^2_+, $(\mathbf{T}_+ u)(t) = u(t-1)$ if $t \geq 1$, $= 0$ if $t = 0$. View $\mathbb{C}^n$ as the subspace of ℓ^2_+ consisting of sequences of the form $\{x_1, 0, 0, \cdots\}$, $x_1 \in \mathbb{C}^n$. Denote $\mathbf{\Gamma}_V$ by $\mathbf{\Gamma}$. For any $\rho > \|\mathbf{\Gamma}\|$, assumed to be fixed, introduce the following operators, with domains and codomains as shown

$$\begin{array}{lll}
\mathbf{R} := (\rho^2\mathbf{I} - \mathbf{\Gamma}^*\mathbf{\Gamma})^{-1} & \tilde{\mathbf{R}} := (\rho^2\mathbf{I} - \mathbf{\Gamma}\mathbf{\Gamma}^*)^{-1} & : \ell^2_+ \to \ell^2_+ \\
\mathbf{G} := (\mathbf{\Pi}_{\mathbb{C}^n}\mathbf{R}|\mathbb{C}^n)^{-\frac{1}{2}} & \tilde{\mathbf{G}} := (\mathbf{\Pi}_{\mathbb{C}^n}\tilde{\mathbf{R}}|\mathbb{C}^n)^{-\frac{1}{2}} & : \mathbb{C}^n \to \mathbb{C}^n \\
\mathbf{P} := \rho\mathbf{R}\mathbf{G} & \tilde{\mathbf{P}} := \rho\tilde{\mathbf{R}}\tilde{\mathbf{G}} & : \mathbb{C}^n \to \ell^2_+ \\
\mathbf{Q} := \mathbf{T}_+\mathbf{\Gamma}\mathbf{R}\mathbf{G} & \tilde{\mathbf{Q}} := \mathbf{T}_+\mathbf{\Gamma}^*\tilde{\mathbf{R}}\tilde{\mathbf{G}} & : \mathbb{C}^n \to \ell^2_+
\end{array} \tag{85}$$

All operators defined in (85) depend on $(\mathbf{\Gamma}, \rho)$.

The operator $\mathbf{P}$ is isomorphic to a multiplication operator with multiplicand $P_+(\cdot)$ in $(H^2)^{n\times n}$ satisfying

$$P_+(z)h = \mathcal{F}[\mathbf{P}h](z), \qquad |z| = 1,$$

for $h \in \mathbb{C}^n$, where $\mathcal{F} : \ell^2_+ \to H^2$ denotes the z-transform. $P_+(\cdot)$ is determined by $\mathbf{P}$ uniquely up to L^2 equivalence. If $\{\varsigma_1, \varsigma_2, \cdots, \varsigma_n\}$ denotes the basis $\varsigma_1 = [1, 0, 0, \cdots, 0]^{tr}$, $\varsigma_2 = [0, 1, 0, 0, \cdots]^{tr}$ for $\mathbb{C}^n$, then P_+ is expressed as

$$P_+(z) = [\mathcal{F}[P\varsigma_1](z), \cdots, \mathcal{F}[P\varsigma_n](z)], \qquad |z| = 1.$$

Similarly, we define functions $Q_+ \in (H^2)^{n\times n}$, and $P_-, Q_- \in (L^2 \ominus H^2)^{n\times n}$ to satisfy

$$\begin{aligned} Q_+(z)h &= \mathcal{F}[\tilde{\mathbf{Q}}h](z) \\ P_-(z)h &= \mathcal{F}[\tilde{\mathbf{P}}h](\bar{z}) \\ Q_-(z)h &= \mathcal{F}[\mathbf{Q}h](\bar{z}) \end{aligned}$$

for $|z| = 1$ and $h \in \mathbb{C}^n$. $P_\pm$ and $Q_\pm$ satisfy the identity (see [1], p.150),

$$P_+^*(z)P_+(z) - Q_+^*(z)Q_+(z) = I, \quad |z| = 1, \tag{86}$$

(a.e. in z, but we shall cease distinguishing between functions equal everywhere and a.e.), which implies that for any $\mathrm{E} \in H^\infty$, $\|\mathrm{E}\|_\infty \le 1$, $P_+(z)$ and $[P_+(z) + Q_+(z)\mathrm{E}(z)]$ are invertible and

$$|P_+^{-1}(z)| \le 1, \quad |z| = 1. \tag{87}$$

(Recall that $\|\cdot\|$ denotes an operator norm, which of course depends on the domain and codomain.)

Lemma 7.1. *If* $\mathbf{K} : \mathbb{C}^n \to \ell_+^2$, *and* $K(z)h = \mathcal{F}(\mathbf{K}h)(z)$ *for each* $h \in \mathbb{C}^n, |z| = 1$, *then*

$$\|K\|_2 \le n^{\frac{1}{2}} \|\mathbf{K}\|. \tag{88}$$

Proof: The hypotheses imply that $K \in H^2$, and K satisfies the inequalities

$$\begin{aligned} \|K\|_2^2 &= \frac{1}{2\pi}\int_{-\pi}^{\pi} |K^*(e^{i\theta})K(e^{i\theta})|d\theta \\ &\le \frac{1}{2\pi}\int_{-\pi}^{\pi} Trace\Big(K^*(e^{i\theta})K(e^{i\theta})\Big)d\theta \\ &= \frac{1}{2\pi} Trace\left(\int_{-\pi}^{\pi} K^*(e^{i\theta})K(e^{i\theta})d\theta\right) \\ &\le n\left|\frac{1}{2\pi}\int_{-\pi}^{\pi} K^*(e^{i\theta})K(e^{i\theta})d\theta\right|, \end{aligned}$$

as $|G|^2 \le Trace(G^*G) \le n|G|^2$ for any $G \in \mathbb{C}^{n\times n}$. The last bound coincides with $n\|\mathbf{K}\|^2$ because, by Parseval's Theorem,

$$\|\mathbf{K}h\|_{\ell_+^2}^2 = \frac{1}{2\pi}\int_{-\pi}^{\pi} h^*K^*(e^{i\theta})K(e^{i\theta})hd\theta = h^*\left(\frac{1}{2\pi}\int_{-\pi}^{\pi} K^*(e^{i\theta})K(e^{i\theta})d\theta\right)h \tag{89}$$

for an $h \in \mathbb{C}^n$. The Lemma follows. □

For any function $y = f(V)$, the notation Δy denotes $f(V_2) - f(V_1)$. Denote $\mathbf{\Gamma}_v$ by $\mathbf{\Gamma}$ and recall that $\rho = \|\mathbf{\Gamma}_v\| + \delta$.

Lemma 7.2. *The following inequalities hold.*

$$\begin{aligned}
\|\mathbf{G}\| &\le \rho\sqrt{1+\alpha} &(90)\\
\|\Delta(\rho^{-1}\mathbf{\Gamma})\| &\le \rho_2^{-1}(1+\sqrt{\alpha_1})\|\Delta\mathbf{\Gamma}\| &(91)\\
\|\Delta(\rho^{-1}\mathbf{\Gamma}^*\mathbf{\Gamma})\| &\le (\alpha_1^{\frac{1}{2}} + \alpha_2^{\frac{1}{2}} + \alpha_1^{\frac{1}{2}}\alpha_2^{\frac{1}{2}})\|\Delta\mathbf{\Gamma}\| &(92)\\
\|\Delta(\rho\mathbf{R})\| &\le [\rho_1\rho_2(1-\sqrt{\alpha_1})(1-\sqrt{\alpha_2})]^{-1}\|\Delta\mathbf{\Gamma}\| &(93)\\
\|\Delta(\rho\mathbf{T}_+\mathbf{\Gamma}\mathbf{R})\| &\le \frac{1+\sqrt{\alpha_1\alpha_2}}{(1-\alpha_1)(1-\alpha_2)}\left(\frac{1+\sqrt{\alpha_1}}{\rho_2}\right)\|\Delta\mathbf{\Gamma}\| &(94)
\end{aligned}$$

Proof: Inequality (90). Write $\mathbf{A} = \rho^{-2}\mathbf{\Gamma}^*\mathbf{\Gamma}$ and note that $\|\mathbf{A}\| \le \alpha < 1$. We can write,

$$(\mathbf{I} - \rho^{-2}\mathbf{\Gamma}^*\mathbf{\Gamma})^{-1} = (1-\alpha^2)^{-1}(\mathbf{I}+\mathbf{K})$$

where $\|\mathbf{K}\| \le \alpha$. (The bound on $\mathbf{K}$ follows from the observation that $\mathbf{K} = (\mathbf{A} - \alpha^2\mathbf{I})(\mathbf{I}-\mathbf{A})^{-1}$ and the observation that $(\mathbf{I}-\mathbf{A})$ is symmetric positive definite). Since projection reduces norm,

$$\mathbf{\Pi}_{\mathbb{C}^n}(\mathbf{I} - \rho^{-2}\mathbf{\Gamma}^*\mathbf{\Gamma})^{-1}|\mathbb{C}^n = (1-\alpha^2)^{-1}(\mathbf{I}_{\mathbb{C}^n} + \mathbf{K}_0)$$

where $\|\mathbf{K}_0\| \le \|\mathbf{K}\| \le \alpha$. Therefore,

$$\rho^{-1}\mathbf{G} = [\mathbf{\Pi}_{\mathbb{C}^n}(\mathbf{I}-\mathbf{A})^{-1}|\mathbb{C}^n]^{-\frac{1}{2}} = (1-\alpha^2)^{\frac{1}{2}}(\mathbf{I}_{\mathbb{C}^n} + \mathbf{K}_0)^{-\frac{1}{2}}.$$

As $\mathbf{K}_0$ is a contraction, $\|(\mathbf{I}_{\mathbb{C}^n} + \mathbf{K}_0)^{-\frac{1}{2}}\| \le (1-\alpha)^{-\frac{1}{2}}$ by elementary Banach algebra, and (90) holds.

Inequalities (91) and (92). Note that

$$|\rho_2 - \rho_1| \le |\mu_2 - \mu_1| \le \|\Delta\mathbf{\Gamma}\|. \tag{95}$$

Therefore,

$$\|\rho_2^{-1}\mathbf{\Gamma}_2 - \rho_1^{-1}\mathbf{\Gamma}_1\| \le |\rho_2^{-1} - \rho_1^{-1}|\|\mathbf{\Gamma}_1\| + \rho_2^{-1}\|\mathbf{\Gamma}_2 - \mathbf{\Gamma}_1\| \le \rho_2^{-1}(\sqrt{\alpha_1}+1)\|\Delta\mathbf{\Gamma}\|$$

which proves (91). Also,

$$\begin{aligned}
\|\rho_2^{-1}\mathbf{\Gamma}_2^*\mathbf{\Gamma}_2 - \rho_1^{-1}\mathbf{\Gamma}_1^*\mathbf{\Gamma}_1\| &\le \|\rho_2^{-1}\mathbf{\Gamma}_2^*(\mathbf{\Gamma}_2 - \mathbf{\Gamma}_1)\| + \|(\rho_2^{-1}\mathbf{\Gamma}_2^* - \rho_1^{-1}\mathbf{\Gamma}_1^*)\mathbf{\Gamma}_1\|\\
&\le \sqrt{\alpha_2}\|\Delta\mathbf{\Gamma}\| + \rho_1^{-1}(1+\sqrt{\alpha_2})\|\Delta\mathbf{\Gamma}\| \cdot \|\mathbf{\Gamma}_1\|
\end{aligned}$$

by (91), which implies (92).

Inequality (93). For $i = 1, 2$, $\rho_i \mathbf{R}_i = \rho_i^{-1}(\mathbf{I} - \rho_i^{-2}\mathbf{\Gamma}_i^*\mathbf{\Gamma}_i)^{-1} = \rho_i^{-1}(\mathbf{I} - \mathbf{A}_i)^{-1}$, where $\mathbf{A}_i := \rho_i^{-2}\mathbf{\Gamma}_i^*\mathbf{\Gamma}_i$, $\|\mathbf{A}_i\| \le \alpha_i$. Moreover,

$$\begin{aligned}
&\|\rho_2^{-1}(\mathbf{I}-\mathbf{A}_2)^{-1} - \rho_1^{-1}(\mathbf{I}-\mathbf{A}_1)^{-1}\| \\
= \;& (\rho_1\rho_2)^{-1}\|(\mathbf{I}-\mathbf{A}_2)^{-1}\{\rho_1(\mathbf{I}-\mathbf{A}_1) - \rho_2(\mathbf{I}-\mathbf{A}_2)\}(\mathbf{I}-\mathbf{A}_1)^{-1}\| \\
\le \;& (\rho_1\rho_2)^{-1}(1-\alpha_2)^{-1}\|\{(\rho_2-\rho_1)\mathbf{I} + \rho_1\mathbf{A}_1 - \rho_2\mathbf{A}_2\}\|(1-\alpha_1)^{-1} \\
\le \;& (\rho_1\rho_2)^{-1}(1-\alpha_1)^{-1}(1-\alpha_2)^{-1}(1+\sqrt{\alpha_1}+\sqrt{\alpha_2}+\sqrt{\alpha_1\alpha_2})\|\Delta\mathbf{\Gamma}\|
\end{aligned}$$

by (92) and as $|\rho_2 - \rho_1| \le \|\Delta\Gamma\|$. Therefore (93) holds.

Inequality (94). Let $\mathbf{X}_i := \mathbf{\Gamma}_i/\rho_i$, $i = 1, 2$, ($\|\mathbf{X}_i\| \le \sqrt{\alpha_i}$). We have,

$$\begin{aligned}
&\|\Delta(\rho\mathbf{T}_+\mathbf{\Gamma}\mathbf{R})\| \\
= \;& \|\rho_2\mathbf{\Gamma}_2\mathbf{R}_2 - \rho_1\mathbf{\Gamma}_1\mathbf{R}_1\| \\
= \;& \|\mathbf{X}_2(\mathbf{I}-\mathbf{X}_2^*\mathbf{X}_2)^{-1} - \mathbf{X}_1(\mathbf{I}-\mathbf{X}_1^*\mathbf{X}_1)^{-1}\| \\
= \;& \|(\mathbf{X}_2-\mathbf{X}_1)(\mathbf{I}-\mathbf{X}_2^*\mathbf{X}_2)^{-1} + \mathbf{X}_1[(\mathbf{I}-\mathbf{X}_2^*\mathbf{X}_2)^{-1} - (\mathbf{I}-\mathbf{X}_1^*\mathbf{X}_1)^{-1}]\| \\
\le \;& \|(\mathbf{X}_2-\mathbf{X}_1)\|\cdot\|(\mathbf{I}-\mathbf{X}_2^*\mathbf{X}_2)^{-1}\| \\
& +\|\mathbf{X}_1\|\cdot\|(\mathbf{X}_2^*\mathbf{X}_2 - \mathbf{X}_1^*\mathbf{X}_1)\|\cdot\|(\mathbf{I}-\mathbf{X}_1^*\mathbf{X}_1)^{-1}\|\|(\mathbf{I}-\mathbf{X}_2^*\mathbf{X}_2)^{-1}\| \\
\le \;& (1-\alpha_2)^{-1}\|\mathbf{X}_2-\mathbf{X}_1\| \\
& +\sqrt{\alpha_1}(1-\alpha_1)^{-1}(1-\alpha_2)^{-1}\|(\mathbf{X}_2^*\mathbf{X}_2 - \mathbf{X}_1^*\mathbf{X}_1)\|
\end{aligned}$$

and as

$$\begin{aligned}
\|\mathbf{X}_2^*\mathbf{X}_2 - \mathbf{X}_1^*\mathbf{X}_1\| &\le \|\mathbf{X}_2^*(\mathbf{X}_2-\mathbf{X}_1) + (\mathbf{X}_2^* - \mathbf{X}_1^*)\mathbf{X}_1\| \\
&\le (\sqrt{\alpha_2}+\sqrt{\alpha_1})\|\mathbf{X}_2-\mathbf{X}_1\|,
\end{aligned}$$

we obtain the bound

$$\begin{aligned}
\le \;& \left\{(1-\alpha_1) + \sqrt{\alpha_1}(\sqrt{\alpha_1}+\sqrt{\alpha_2})\right\}(1-\alpha_1)^{-1}(1-\alpha_2)^{-1}\|\mathbf{X}_2-\mathbf{X}_1\| \\
\le \;& (1+\sqrt{\alpha_1\alpha_2})((1-\alpha_1)(1-\alpha_2))^{-1}(1+\sqrt{\alpha_1})\rho_2^{-1}\|\Delta\mathbf{\Gamma}\|
\end{aligned}$$

(by (91),) which is (94). □

Proof of Theorem 7.1: Let $\tilde{Q}_i := (Q_-)_i G_i^{-1}$, $\tilde{P}_i := (P_+)_i G_i^{-1}$, $i = 1, 2$, where G_i denotes the constant matrix in L^∞ such that $G_i h = \mathcal{F}[\mathbf{G}_i h]$, $\mathbf{G}_i$ being the operator in (85). Then $S_i = \rho_i (Q_-)_i (P_+)_i^{-1} = \rho_i \tilde{Q}_i \tilde{P}_i^{-1}$, and

$$\|\Delta S\|_2 \;=\; \|\rho_2\tilde{Q}_2\tilde{P}_2^{-1} - \rho_1\tilde{Q}_1\tilde{P}_1^{-1}\|_2$$

$$
\begin{aligned}
&\le \|\rho_2\tilde{Q}_2 - \rho_1\tilde{Q}_1\|_2 \|\tilde{P}_2^{-1}\|_\infty + \|\rho_1\tilde{Q}_1(\tilde{P}_2^{-1} - \tilde{P}_1^{-1})\|_2 \\
&\le \left\{\|\Delta(\rho\tilde{Q})\|_2 + \rho_1\|\tilde{Q}_1\tilde{P}_1^{-1}\|_\infty \|\tilde{P}_2 - \tilde{P}_1\|_2\right\}\|\tilde{P}_2^{-1}\|_\infty .
\end{aligned}
$$

The suboptimality of $\rho_1(Q_-)_1(P_+)_1^{-1}$ implies that $\|\tilde{Q}_1\tilde{P}_1^{-1}\|_\infty \le 1$. Also, (86) implies that $\|(P_+)_2^{-1}\|_\infty \le 1$, from which $\|\tilde{P}_2^{-1}\| \le \| \le_2 (P_+)_2^{-1}\| \le \|G_2\|_\infty \le \rho_2\sqrt{1+\alpha_2}$ by (90). Therefore

$$
\begin{aligned}
\|\Delta S\|_2 &\le \sqrt{1+\alpha_2}\left\{\rho_2\|\Delta(\rho\tilde{Q})\|_2 + \rho_1\rho_2\|\Delta\tilde{P}\|_2\right\} \\
&\le n^{\frac{1}{2}}\sqrt{1+\alpha_2}\left\{\rho_2\|\Delta(\rho\tilde{\mathbf{Q}})\| + \rho_1\rho_2\|\Delta\tilde{\mathbf{P}}\|\right\}
\end{aligned}
$$

where $\tilde{\mathbf{P}}_i := \mathbf{P}_i\mathbf{G}_i^{-1} = \rho_i\mathbf{R}_i$, $\tilde{\mathbf{Q}}_i := \mathbf{Q}_i\mathbf{G}_i^{-1} = \mathbf{T}_+\mathbf{\Gamma}_i\mathbf{R}_i$. Now (93) and (94) yield

$$
\|\Delta\mathbf{S}\|_2 \le n^{\frac{1}{2}}\sqrt{1+\alpha}\left\{\frac{1}{(1-\sqrt{\alpha})^2} + \frac{(1+\alpha)(1+\sqrt{\alpha})}{(1-\alpha)^2}\right\}\|\Delta\mathbf{\Gamma}\|
$$

which implies (83), as $\|\Delta\mathbf{\Gamma}\| \le \|\Delta\mathbf{V}\|$. As $1-\sqrt{\alpha_i} = \delta(\mu_i+\delta)^{-1}$ and $\alpha < 1$, (84) follows. □

S will be called the *central δ-suboptimal interpolant* of (V', U) if U^*S is such an interpolant of (U^*V', I), where U is inner.

Corollary 7.1. *The central δ-suboptimal interpolant S of (V', U) depends Lipschitz continuously ($L^\infty \to L^2$) on the data, i.e.,*

$$
\|\Delta S\|_2 \le \gamma_{U^*V'}\|\Delta V'\|_\infty + \{\rho_1 + \gamma_{U^*V'}\|V_1'\|_\infty\}\|\Delta U\|_\infty
$$

where $\gamma_{U^*V'}$ is defined by (83), with $U^*V' = V$.

Proof: Let S_i be the central δ-suboptimal interpolants of (V_i', U_i). The corresponding interpolants of $(U_i^*V_i', I)$ are $U_i^*S_i$. Now,

$$
\begin{aligned}
\|S_2 - S_1\|_2 &= \|U_2U_2^*S_2 - U_1U_1^*S_1\|_2 \\
&\le \|U_2\|_\infty\|U_2^*S_2 - U_1^*S_1\|_2 + \|U_2 - U_1\|_2\|U_1^*S_1\|_\infty \\
&\le \|U_2^*S_2 - U_1^*S_1\|_2 + \rho_1\|\Delta U\|_2,
\end{aligned}
$$

as $\|U_i\|_\infty \le 1$, and $\|U_1^*S_1\| \le \rho_1$ by the AAK construction. By Theorem 7.1,

$$
\|U_2^*S_2 - U_1^*S_1\|_2 \le \gamma_{U^*V'}\|U_2^*V_2' - U_1^*V_1'\|_\infty
$$

$$\le \gamma_{U*V'}\left\{\|U_2^*\|_\infty\|V_2'-V_1'\|_\infty+\|U_2^*-U_1^*\|_\infty\|V_1'\|_\infty\right\}$$
$$\le \gamma_{U*V'}\left\{\|\Delta V'\|_\infty+\|\Delta U\|_\infty\|V_1'\|_\infty\right\}$$

Therefore,

$$\|\Delta S\|_2 \le \gamma_{U*V'}\|\Delta V'\|_\infty + \left[\rho_1+\gamma_{U*V'}\|V_1'\|_\infty\right]\|\Delta U\|_\infty$$

which proves the Corollary. □

7.3 Application to Sensitivity Design

We return to the problem posed in Section 4.1, of selecting a sensivity $\mathbf{S}^\ell$ subject to the local δ-suboptimality constraint (65) and Lipschitz condition (56), and resume using the terminology of Section 4. Recall that $(\mathbf{W},\mathbf{U})$ are in $\underline{\mathbf{E}}_\sigma$, and therefore must be in $\overline{\mathbf{E}}_{\sigma_0}$ for some $\sigma_0>\sigma$. For each $t\in\mathbf{Z}$, define

$$V_t(\cdot) := \widehat{\mathbf{U}}_t^*(\sigma_0(\cdot))\widehat{\mathbf{W}}_t(\sigma_0(\cdot)) \tag{96}$$

in L^∞. Let S_t^o be the δ-suboptimal AAK interpolant of $(\mathbf{V}_t,I)$ defined in (81) and $\widehat{\mathbf{S}}_t^\ell := \widehat{\mathbf{U}}_t S_t^o$. Then $\mathbf{S}^\ell$ locally interpolates $(\mathbf{W},\mathbf{U})$ in $\overline{\mathbf{E}}_{\sigma_0}$. As $\mathbf{U}$ is locally unitary, $\mathbf{S}^\ell$ satisfies

$$\mu_{\sigma_0}(\widehat{\mathbf{S}}^\ell) \le \mu_{\sigma_0}(\widehat{\mathbf{S}}^\ell_{opt})+\delta$$

($\mathbf{S}^\ell_{opt}$ defined in Section 4.1) which implies (65). Call $\mathbf{S}^\ell$ the *locally central δ-suboptimal sensitivity in* $\overline{\mathbf{E}}_{\sigma_0}$. Let γ_{V_t} be the Lipschitz constant defined in Theorem 7.1, Equation (83), with V_t identified as in (96). Write $\lambda_w := \sup_t \gamma_{V_t}$. Corollary 7.1 now immediately gives:

Corollary 7.2. *The locally central δ-suboptimal sensitivity $\mathbf{S}^\ell\in\overline{\mathbf{E}}_{\sigma_0}$ that locally interpolates $(\mathbf{W},\mathbf{U})$ (and which was introduced in Section 4.1) depends Lipschitz continuously ($L^\infty\to L^2$) on the data, i.e.*

$$\|\Delta(\widehat{\mathbf{S}}_t^\ell)\|_{H^2_{\sigma_0}} \le \lambda_W\|\Delta\widehat{\mathbf{W}}_t\|_{H^\infty_{\sigma_0}}+\{\mu+\delta+\lambda_w\mu_{\sigma_0}(\mathbf{W})\}\|\Delta\widehat{\mathbf{U}}_t\|_{H^\infty_{\sigma_0}} \tag{97}$$

(c.f. the Lipschitz condition (56)).

8 CONCLUDING REMARKS

The main point here is that by introducing notions of local spectrum and global spectrum it is possible to unify a lot of apparently diverse frozen-time results. Applications to stability analysis are typified by Corollary 3.1.

Although a particular approach to freezing was used, the key features of the double algebra structure, including the local product and ∇ operator, hold more generally.

Double algebras provide a natural mathematical framework for the frozen-time analysis of feedback systems. They suggest a simple approach to the approximate optimization of such systems.

One of the open questions of control theory has been: Can one provide a paradigm of adaptive feedback which does not depend on the structure or parameterization of the controller? Frozen-time optimization appears capable of providing such a paradigm, at least for the elementary case of slowly time-varying systems.

9 APPENDIX I. THE ALGEBRAS $\mathbb{E}_\sigma$ AND $\underline{\mathbb{E}}_\sigma$

Proposition A1.1. *The spaces $\mathbf{E}_\sigma$ and $\underline{\mathbf{E}}_\sigma$ under the norm $\|\cdot\|_{(\sigma)}$ and either (a) the global product, or (b) the local product, is a normed algebra, and $\mathbf{E}_\sigma$ is a Banach algebra.*

Proof: (a). Under the global product, for any $\mathbf{K}, \mathbf{M}$ in $\mathbf{E}_\sigma$ it will be shown that $\mathbf{KM} \in \mathbf{E}_\sigma$ and

$$\|\mathbf{KM}\|_{(\sigma)} \leq \|\mathbf{K}\|_{(\sigma)} \|\mathbf{M}\|_{(\sigma)} \tag{98}$$

which implies that $\mathbf{E}_\sigma$ is a normed algebra for each $\sigma \geq 1$. It then follows, from the representation of any $\mathbf{K} \in \mathbf{E}_\sigma$ by a Volterra sum with kernels in the complete space ℓ^1_σ of bounded ℓ^1_σ-norm, that the normed algebra $\mathbf{E}_\sigma$ is complete and therefore a Banach algebra. If $\mathbf{K}, \mathbf{M}$ are actually in the space $\underline{\mathbf{E}}_\sigma$, they are in some common $\mathbf{E}_{\sigma_0}$, $\sigma_0 > \sigma$, which has been shown to be an algebra, and therefore $\mathbf{KM} \in \mathbf{E}_{\sigma_0}$, which implies $\mathbf{KM} \in \underline{\mathbf{E}}_\sigma$, i.e. $\underline{\mathbf{E}}_\sigma$ is a normed algebra under the global product.

To prove (98), let $\mathbf{F} := \mathbf{KM}$, and denote the kernels of $\mathbf{F}, \mathbf{K}, \mathbf{M}$ by f, k, m. As k, m are in ℓ^1_σ, the following bounds and changes of order of summation are valid.

$$
\begin{aligned}
(\mathbf{KM}u)(t) &= \sum_{\eta=-\infty}^{\infty} k(t,\eta) \sum_{\theta=-\infty}^{\infty} m(\eta,\theta)u(\theta) \\
&= \sum_{\theta=-\infty}^{\infty} \left(\sum_{\eta=-\infty}^{\infty} k(t,\eta)m(\eta,\theta) \right) u(\theta), \\
\sum_{\theta=-\infty}^{\infty} |f(t,\theta)|\sigma^{(t-\theta)} &= \sum_{\theta=-\infty}^{\infty} \left| \sum_{\eta=-\infty}^{\infty} k(t,\eta)m(\eta,\theta) \right| \sigma^{(t-\theta)} \\
&\le \sum_{\eta=-\infty}^{\infty} |k(t,\eta)\sigma^{(t-\eta)}| \sum_{\theta=-\infty}^{\infty} |m(\eta,\theta)\sigma^{(\eta-\theta)}| \\
&\le \sum_{\eta=-\infty}^{\infty} |k(t,\eta)\sigma^{(t-\eta)}| \quad \|\mathbf{M}\|_{(\sigma)}. \qquad (99)
\end{aligned}
$$

After taking $\sup_{t\in\mathbf{Z}}$ of (99) we conclude that

$$\sup_t \|f(t,t-(\cdot))\|_{\ell^1_\sigma} \le \|\mathbf{K}\|_{(\sigma)} \|\mathbf{M}\|_{(\sigma)}$$

which implies that $\mathbf{KM} \in \mathbf{E}_\sigma$ and (98) holds. □

(b) Under the local product, for $\mathbf{K}, \mathbf{M}$ in $\mathbf{E}_\sigma$ and $t \in \mathbf{Z}$, $\mathbf{K}_t$ and $\mathbf{M}_t$ are in $\mathbf{E}_\sigma$ and

$$(\mathbf{K} \otimes \mathbf{M})_t = \mathbf{K}_t \mathbf{M}_t.$$

By Part (a), $\mathbf{K}_t\mathbf{M}_t$ is in $\mathbf{E}_\sigma$ and satisfies $\|\mathbf{K}_t\mathbf{M}_t\|_{(\sigma)} \le \|\mathbf{K}_t\|_{(\sigma)}\|\mathbf{M}_t\|_{(\sigma)}$. As $\mathbf{F}_t := \mathbf{K}_t\mathbf{M}_t$ is time-invariant, its kernel f_t satisfies

$$\|f_t\|_{\ell^1_\sigma} = \|\mathbf{F}_t\|_{(\sigma)} \le \|\mathbf{K}\|_{(\sigma)}\|\mathbf{M}\|_{(\sigma)},$$

and since this holds for all t,

$$\|\mathbf{K} \otimes \mathbf{M}\|_{(\sigma)} \le \|\mathbf{K}\|_{(\sigma)}\|\mathbf{M}\|_{(\sigma)},$$

i.e. $\mathbf{E}_\sigma$ is a normed algebra under $\otimes$. If $\mathbf{K}, \mathbf{M}$ are actually in $\underline{\mathbf{E}}_\sigma$, then so is $\mathbf{K} \otimes \mathbf{M}$, by the reasoning of part (a), so that $\underline{\mathbf{E}}_\sigma$ is a normed algebra. □

Proposition A1.2. *For $\mathbf{K} \in \mathbf{E}_{\sigma_0}$ with kernel k, and $\sigma < \sigma_0$,*

$$
\begin{aligned}
\|\mathbf{K}\|_{(\sigma)} &\le \kappa_{(\sigma_0/\sigma)}\sqrt{n}\mu_{\sigma_0}(\widehat{\mathbf{K}}), \qquad (100) \\
d_\sigma(\mathbf{K}) &\le \sigma\kappa_{(\sigma_0/\sigma)}\sqrt{n}\partial_{\sigma_0}(\widehat{\mathbf{K}}). \qquad (101)
\end{aligned}
$$

Proof: First, we will show that for $\sigma_1 \ge 1$,

$$\|k_t\|_{\ell^2_{\sigma_1}} \le \sqrt{n}\|\widehat{\mathbf{K}}_t\|_{L^2_{\sigma_1}}. \qquad (102)$$

Indeed, by the matrix inequalities $|G|^2 \leq Trace(G^*G) \leq n|G|^2$,

$$\begin{aligned}
\|k_t\|^2_{\ell^2_{\sigma_1}} &= \sum_{\tau=0}^{\infty}(|k(t,t-\tau)\sigma_1^\tau|)^2 \\
&\leq \sum_{\tau=0}^{\infty} Trace(k^*(t,t-\tau)k(t,t-\tau))\sigma_1^2 \\
&= \frac{1}{2\pi}\int_{-\pi}^{\pi} Trace(\widehat{\mathbf{K}}_t^*(\sigma_1 e^{i\theta})\widehat{\mathbf{K}}_t(\sigma_1 e^{i\theta})) \\
&\qquad \text{by Parseval's Theorem,} \\
&\leq n\|\widehat{\mathbf{K}}_t\|_{L^2_{\sigma_1}},
\end{aligned}$$

which proves (102).

For $t \in \mathbf{Z}$

$$\begin{aligned}
\|k_t\|_{\ell^1_\sigma} &= \sum_{r=0}^{\infty} |k(t,t-r)\sigma^r| \\
&\leq (\sum_{r=0}^{\infty} |k(t,t-r)\sigma_0^r|^2 \sum_{r=0}^{\infty} |\sigma/\sigma_0|^{2r})^{\frac{1}{2}} \\
&\qquad \text{by Schwartz's Inequality} \\
&= \kappa_{(\sigma_0/\sigma)}\sqrt{n}\|\widehat{\mathbf{K}}_t\|_{L^2_{\sigma_0}} \quad \text{by (102)} \\
&\leq \kappa_{(\sigma_0/\sigma)}\mu_{\sigma_0}(\widehat{\mathbf{K}})
\end{aligned}$$

which implies (100). Equation (101) follows from the inequalities

$$\begin{aligned}
\|\mathbf{KT} - \mathbf{TK}\|_{(\sigma)} &\leq \sigma \sup_t \|k(t, 1+(\cdot)) - k(t-1,\cdot)\|_{\ell^1_\sigma} \\
&\leq \sigma\kappa_{(\sigma_0/\sigma)}\sqrt{n}\partial_{\sigma_0}(\widehat{\mathbf{K}}) \qquad \text{by (100).}
\end{aligned}$$

□

Proof of Proposition 3.1: If $\mathbf{K} \in \underline{\mathbf{E}}_\sigma$, then for some $\sigma_0 > \sigma$, $\|\mathbf{K}\|_{(\sigma_0)} < \infty$. Therefore, for each $t \in \mathbf{Z}$, $\widehat{\mathbf{K}}_t \in H^\infty_{\sigma_0}$ satisfy

$$\|\widehat{\mathbf{K}}_t\|_{H^\infty_{\sigma_0}} \leq \|k_t\|_{\ell^1_{\sigma_0}} \leq \|\mathbf{K}\|_{(\sigma_0)},$$

where k_t is the kernel of $\mathbf{K}_t$, which implies that $\mu_{\sigma_0}(\widehat{\mathbf{K}}) \leq \|\mathbf{K}\|_{(\sigma_0)}$ and that $\mathbf{K} \in \underline{\widehat{\mathbf{E}}}_\sigma$.

Conversely, if $\mathbf{K} \in \underline{\widehat{\mathbf{E}}}_\sigma$ then for some $\sigma_0 > \sigma$, $\mu_{\sigma_0}(\widehat{\mathbf{K}}) < \infty$. For any σ_1, $\sigma < \sigma_1 < \sigma_0$, $\|\mathbf{K}\|_{(\sigma_1)} \leq const.\mu_{\sigma_0}(\widehat{\mathbf{K}})$ by (100), and $\mathbf{K} \in \mathbf{E}_{\sigma_1}$, which implies that $\mathbf{K} \in \underline{\mathbf{E}}_\sigma$.

This proves that the spaces $\underline{\mathbf{E}}_\sigma$, $\underline{\hat{\mathbf{E}}}_\sigma$ are identical. As $\underline{\mathbf{E}}_\sigma$ is a normed double algebra by Proposition A1.1, so is $\underline{\hat{\mathbf{E}}}_\sigma$. □

10 APPENDIX II. FREQUENCY-DOMAIN INEQUALITIES

Before proving (31 - 39) certain subsidiary inequalities will be established; for $\mathbf{K} \in \mathbf{E}_\sigma$, $u \in \ell^\infty$, $t \in \mathbf{Z}$,

$$\begin{aligned}
\|\mathbf{K}\|_{a(\sigma;t)} &\leq \|\hat{\mathbf{K}}\|_{H^\infty_\sigma} \quad \text{if } \mathbf{K} \text{ is shift invariant;} && (103)\\
\|\mathbf{K}\Pi_t u\|_{\ell^2_\sigma} &\leq \kappa_\sigma \sigma^t \|\hat{\mathbf{K}}\|_{H^\infty_\sigma} \|u\|_{a(\sigma)} \\
&\qquad \text{if } \mathbf{K} \text{ is shift invariant;} && (104)\\
\|\Pi_t(\mathbf{K} - \mathbf{K}_t)u\|_{\ell^2_\sigma} &\leq \kappa_\sigma \kappa_\sigma^{(p)} \sigma^t \partial_\sigma^{(p)}(\hat{\mathbf{K}}) \|u\|_{a(\sigma)}. && (105)
\end{aligned}$$

Also, it will be shown that the expressions $\mathbf{G}\nabla\mathbf{K}$ and $\mathbf{GK} - (\mathbf{G} \otimes \mathbf{K})_t$ truncated by Π_t can be represented by operator series

$$\begin{aligned}
\Pi_t[\mathbf{GK} - (\mathbf{G}\otimes\mathbf{K})_t] &= -\Pi_t \mathbf{G}_t \sum_{\tau=-\infty}^{t} \Pi_{\tau-1}(\mathbf{K}_\tau - \mathbf{K}_{\tau-1}) \\
&\quad - \sum_{\tau=-\infty}^{t} \Pi_{\tau-1}(\mathbf{G}_\tau - \mathbf{G}_{\tau-1})\mathbf{K}; && (106)\\
\Pi_t[\mathbf{G}\nabla\mathbf{K}] &= \Pi_t[\mathbf{GK} - (\mathbf{G}\otimes\mathbf{K})_t] \\
&\quad + \sum_{\tau=-\infty}^{t} \Pi_{\tau-1}(\mathbf{G}_\tau\mathbf{K}_\tau - \mathbf{G}_{\tau-1}\mathbf{K}_{\tau-1}); \\
& && (107)\\
(\Delta\Pi_t)[\mathbf{GK} - (\mathbf{G}\otimes\mathbf{K})_t] &= (\Delta\Pi_t)(\mathbf{G}\nabla\mathbf{K}) \\
&= -(\Delta\Pi_t)\mathbf{G}_t \sum_{\tau=-\infty}^{t} \Pi_{\tau-1}(\mathbf{K}_\tau - \mathbf{K}_{\tau-1}); \\
& && (108)
\end{aligned}$$

where $(\Delta\Pi_t) := \Pi_t - \Pi_{t-1}$, and the series of operators are weakly (ℓ^1) convergent.

Proof of Inequalities:

Inequalities (103) and (104). For $t \in \mathbf{Z}$, $u \in A$, and k the kernel of $\mathbf{K}$,

$$(\Pi_t \mathbf{K} u)(\tau)\sigma^\tau = (\Pi_t \mathbf{K}(\Pi_t u))(\tau)\sigma^\tau \quad \text{(as } \mathbf{K} \text{ is causal)}$$

$$= \sum_{\eta=-\infty}^{t} [k(\tau-\eta)\sigma^{(\tau-\eta)}][(\Pi_t u)(\eta)\sigma^{\eta}]$$

as **K** is shift-invariant. This is a convolution of functions whose transforms $\widehat{\mathbf{K}}, \widehat{\mathbf{Y}}$ are in $L_\sigma^\infty (= L^\infty$ of the circle $|z| = \sigma)$, where $\widehat{\mathbf{K}} := \mathcal{F}(k(\cdot))$, and $\widehat{\mathbf{Y}} := \mathcal{F}((\Pi_t u)(\cdot))$.

By Parseval's Theorem,

$$\begin{aligned}
\|\Pi_t \mathbf{K} u(\cdot)\|_{\ell_\sigma^2} &\leq \|\mathbf{K}\Pi_t u(\cdot)\sigma^{(\cdot)}\|_{\ell^2} \\
&= \{\frac{1}{2\pi}\int_0^{2\pi} |\widehat{\mathbf{K}}(\sigma e^{j\theta})\widehat{\mathbf{Y}}(\sigma e^{j\theta})|^2 d\theta\}^{\frac{1}{2}} \\
&\leq \|\widehat{\mathbf{K}}\|_{H_\sigma^\infty} \|\widehat{\mathbf{Y}}\|_{L_\sigma^2} \\
&= \|\widehat{\mathbf{K}}\|_{H_\sigma^\infty} \|\Pi_t u\|_{\ell_\sigma^2} \\
&\leq \kappa_\sigma \|\widehat{\mathbf{K}}\|_{H_\sigma^\infty} \|u\|_{a(\sigma)} \sigma^t
\end{aligned} \tag{109}$$

giving (103). As

$$\|\mathbf{K}\Pi_t u\|_{\ell_\sigma^2} \leq \|\widehat{\mathbf{K}}\|_{H_\sigma^\infty} \|\Pi_t u\|_{\ell_\sigma^2}$$

and $\|\Pi_t u\|_{\ell_\sigma^2} \leq \kappa_\sigma \sigma^t \|u\|_{a(\sigma)}$, (104) holds. □

Inequality (105). The cases $p = \infty$ and $2 \leq p < \infty$ will be treated by separate approaches. Consider $p = \infty$ first. $\Pi_t(\mathbf{K} - \mathbf{K}_t)$ can be resolved into a sum and then summed by parts,

$$\begin{aligned}
\Pi_t(\mathbf{K} - \mathbf{K}_t) &= \sum_{\tau=-\infty}^{t} \Delta\Pi_\tau(\mathbf{K} - \mathbf{K}_t) \\
&= \lim_{\tau\to-\infty} \Pi_\tau(\mathbf{K}_\tau - \mathbf{K}_t) - \sum_{\tau=-\infty}^{t} \Pi_{\tau-1}(\mathbf{K}_\tau - \mathbf{K}_{\tau-1})
\end{aligned}$$

where lim denotes a limit in the sense of weak (ℓ^1) operator convergence. Such a limit implies that $\lim_{t\to-\infty} \|\Pi_t(\mathbf{K}_t - \mathbf{K}_\tau)u\|_{\ell_\sigma^2} = 0$ for $u \in A$, i.e., the limit is null (a fact which will also be used in the remaining inequality proofs).Therefore, for $u \in A$,

$$\begin{aligned}
\|\Pi_t(\mathbf{K} - \mathbf{K}_t)u\|_{\ell_\sigma^2} &\leq \sum_{\tau=-\infty}^{t} \|\Pi_{\tau-1}(\mathbf{K}_\tau - \mathbf{K}_{\tau-1})u\|_{\ell_\sigma^2} \\
&\leq \kappa_\sigma \sum_{\tau=-\infty}^{t} \sigma^{\tau-1}\|\mathbf{K}_\tau - \mathbf{K}_{\tau-1}\|_{H_\sigma^\infty}\|u\|_{a(\sigma)}
\end{aligned}$$

by (104), which yields (105) for $p = \infty$.

Next, suppose $2 \le p < \infty$. Denote $(\mathbf{K} - \mathbf{K}_t)u$ by y, and the kernel of $\mathbf{K}_\tau$ by k_τ; as $k_\tau \in \ell_\sigma^1$, and $u \in \ell^\infty$, we may write

$$\begin{aligned}
|y(\tau)| &= \left| \sum_{\varsigma=-\infty}^{\tau} [k_\tau(\tau,\varsigma) - k_t(\tau,\varsigma)] u(\varsigma) \right| \\
&\le \left\{ \sum_{\varsigma=-\infty}^{\tau} |[k_\tau(\tau,\varsigma) - k_t(\tau,\varsigma)]\sigma^{(\tau-\varsigma)}|^2 \sum_{\varsigma'=-\infty}^{\tau} |u(\varsigma')\sigma^{-(\tau-\varsigma')}|^2 \right\}^{\frac{1}{2}} \\
&\qquad \text{(By Schwartz's Inequality, as } k_\tau, k_t \text{ are in } \ell_\sigma^2\text{),} \\
&\le \sqrt{n} \|\widehat{\mathbf{K}}_\tau - \widehat{\mathbf{K}}_t\|_{H_\sigma^2} \kappa_\sigma \|u\|_{a(\sigma)} \quad \text{by (102)} \\
&\le \kappa_\sigma \sqrt{n} \partial_\sigma^{(p)}(\mathbf{K}) |t - \tau| \|u\|_{a(\sigma)}.
\end{aligned}$$

Now multiply both sides of the inequality by σ^τ and sum, $(\sum_{\tau=-\infty}^{t} (\cdot)^2)^{\frac{1}{2}}$ to get (105) for $2 \le p < \infty$.

□

Series Expansions (106)-(108). We have the identities

$$\begin{aligned}
&\Pi_t[\mathbf{GK} - (\mathbf{G} \otimes \mathbf{K})_t] \\
&= \Pi_t[(\mathbf{G} - \mathbf{G}_t)\mathbf{K} + \mathbf{G}_t(\mathbf{K} - \mathbf{K}_t)] \quad (\text{as } (\mathbf{G} \otimes \mathbf{K})_t = \mathbf{G}_t\mathbf{K}_t) \\
&= \sum_{\tau=-\infty}^{t} (\Delta\Pi_\tau)(\mathbf{G} - \mathbf{G}_t)\mathbf{K} + \mathbf{\Pi}_t\mathbf{G}_t \sum_{\tau=-\infty}^{t} (\Delta\Pi_\tau)(\mathbf{K} - \mathbf{K}_t)
\end{aligned}$$

where Π_τ has been resolved into $\Sigma(\Delta\Pi_\tau)$. Now, for any $\mathbf{F} \in \mathbb{B}$, $(\Delta\Pi_\tau)\mathbf{F} = (\Delta\Pi_\tau)\mathbf{F}_\tau$, so $\mathbf{G}$ and $\mathbf{K}$ can be replaced by $\mathbf{G}_\tau, \mathbf{K}_\tau$ in the sums, which can be summed by parts to give (106), after the observation that $\Pi_\tau(\mathbf{G}_\tau - \mathbf{G}_t)$ and $\Pi_\tau(\mathbf{K}_\tau - \mathbf{K}_t)$ both weakly (ℓ^1) operator converge to 0 as $\tau \to -\infty$.

By definition of ∇,

$$\Pi_t(\mathbf{G}\nabla\mathbf{K}) = \Pi_t[(\mathbf{GK} - (\mathbf{G} \otimes \mathbf{K})_t) - (\mathbf{G} \otimes \mathbf{K} - (\mathbf{G} \otimes \mathbf{K})_t)]. \tag{110}$$

Resolution followed by partial summation gives

$$\begin{aligned}
&\Pi_t[\mathbf{G} \otimes \mathbf{K} - (\mathbf{G} \otimes \mathbf{K})_t] \\
&= \sum_{\tau=-\infty}^{t} (\Delta\Pi_\tau)(\mathbf{G}_\tau\mathbf{K}_\tau - \mathbf{G}_t\mathbf{K}_t) \\
&= -\sum_{\tau=-\infty}^{t} \Pi_{\tau-1}(\mathbf{G}_\tau\mathbf{K}_\tau - \mathbf{G}_{\tau-1}\mathbf{K}_{\tau-1}).
\end{aligned} \tag{111}$$

Equation (111) applied to (110) proves (107).

Equation (108) is obtained by multiplying (106), (107) by $(\Delta\Pi_t)$, observing that $(\Delta\Pi_t)\sum_{\tau=-\infty}^{t}\Pi_{\tau-1} = 0$, and equating the parts that do not vanish. □

Inequalities (31) - (33). Inequality (31) is the same as (103). To prove (32), begin with

$$\|\Pi_t \mathbf{K} u\|_{\ell^2_\sigma} \leq \|\Pi_t(\mathbf{K} - \mathbf{K}_t)u\|_{\ell^2_\sigma} + \|\Pi_t \mathbf{K}_t u\|_{\ell^2_\sigma}.$$

By (105), the first norm on the right is

$$\leq \kappa_\sigma \kappa_\sigma^{(p)} \sigma^t \partial_\sigma^{(p)}(\widehat{\mathbf{K}})\|u\|_{a(\sigma)},$$

and by the shift invariance of $\mathbf{K}_t$ and (104), the second norm is

$$\leq \kappa_\sigma \sigma^t \|\widehat{\mathbf{K}}_t\|_{H^\infty_\sigma}\|u\|_{a(\sigma)}.$$

Division by $\kappa_\sigma \sigma^t$ gives (32).

To prove (33), note that

$$\begin{aligned}
|(\mathbf{K}u)(t)|^2 &= \left|\sum_{\tau=-\infty}^{t} k(t,\tau)u(\tau)\right|^2 \\
&\leq \sum_{\tau=-\infty}^{t}\left|k(t,\tau)\sigma^{(t-\tau)}\right|^2 \sum_{\tau'=-\infty}^{t}\left|u(\tau')\sigma^{-(t-\tau')}\right|^2 \\
&\qquad \text{(by Schwartz's Inequality)} \\
&\leq n\left(\kappa_\sigma \left\|\widehat{\mathbf{K}}_t\right\|_{H^2_\sigma} \|u\|_{a(\sigma)}\right)^2
\end{aligned}$$

by (102). Therefore,

$$\|\mathbf{K}u\|_{a(\sigma;t)} \leq \|\mathbf{K}u\|_{\ell^\infty} \leq \sup_t \kappa_\sigma \sqrt{n}\|\widehat{\mathbf{K}}_t\|_{H^2_\sigma}\|u\|_{a(\sigma)}$$

which implies (33), as $\mu_\sigma^{(2)} \leq \mu_\sigma^{(p)}(\cdot) \leq \mu_\sigma^{(\infty)}(\cdot)$. □

Inequality (34). From the series expansion (106), and inequality (104), we get

$$\begin{aligned}
\|\Pi_t[\mathbf{GK} &- (\mathbf{G}\otimes\mathbf{K})_t]u\|_{\ell^2_\sigma} \\
&\leq \kappa_\sigma \|\widehat{\mathbf{G}}_t\|_{H^\infty_\sigma} \sum_{\tau=-\infty}^{t} \sigma^{(\tau-1)}\|(\mathbf{K}_\tau - \mathbf{K}_{\tau-1})u\|_{a(\sigma)} \\
&\quad + \kappa_\sigma \sum_{\tau=-\infty}^{t} \sigma^{(\tau-1)}\|(\mathbf{G}_\tau - \mathbf{G}_{\tau-1})\mathbf{K}u\|_{a(\sigma)}.
\end{aligned}$$

The series are summed, and the inequalities $\|\widehat{\mathbf{G}}_t\|_{H_\sigma^\infty} \le \mu_\sigma(\widehat{\mathbf{G}})$, and $\|(\mathbf{G}_\tau - \mathbf{G}_{\tau-1})\mathbf{K}u\|_{a(\sigma)} \le \partial_\sigma(\widehat{\mathbf{G}})\|\mathbf{K}u\|_{a(\sigma)}$ used, to obtain the following bound:

$$\begin{aligned}&\|\Pi_t[\mathbf{GK} - (\mathbf{G}\otimes\mathbf{K})_t]u\|_{\ell_\sigma^2}\\ \le\;& \kappa_\sigma\sigma^t(\sigma-1)^{-1}\{\mu_\sigma(\widehat{\mathbf{G}})\partial_\sigma(\widehat{\mathbf{K}}) + \partial_\sigma(\widehat{\mathbf{G}})\|\mathbf{K}\|_{a(\sigma)}\}\|u\|_{a(\sigma)}. \quad (112)\end{aligned}$$

The bound (32) on $\|\mathbf{K}\|_{a(\sigma)}$ gives

$$\begin{aligned}\|\Pi_t[\mathbf{GK} - (\mathbf{G}\otimes\mathbf{K})_t]u\|_{\ell_\sigma^2} \le\;& \kappa_\sigma\sigma^t(\sigma-1)^{-1}\{\mu_\sigma(\widehat{\mathbf{G}})\partial_\sigma(\widehat{\mathbf{K}})\\ &+\partial_\sigma(\widehat{\mathbf{G}})[\mu_\sigma(\widehat{\mathbf{K}}) + (\sigma-1)^{-1}\partial_\sigma(\widehat{\mathbf{K}})]\}.\end{aligned} \quad (113)$$

Now, by the triangle inequality,

$$\begin{aligned}\|\Pi_t(\mathbf{GK}+\mathbf{F})u\|_{\ell_\sigma^2} \le\;& \|\Pi_t(\mathbf{G}\otimes\mathbf{K}+\mathbf{F})_t u\|_{\ell_\sigma^2}\\ &+\|\Pi_t(\mathbf{GK} - (\mathbf{G}\otimes\mathbf{K})_t)u\|_{\ell_\sigma^2} + \|\Pi_t(\mathbf{F}-\mathbf{F}_t)u\|_{\ell_\sigma^2}.\end{aligned}$$

On the right-hand side, the first norm is bounded by $\kappa_\sigma\sigma^t\mu_\sigma(\widehat{\mathbf{G}\otimes\mathbf{K}} + \widehat{\mathbf{F}})\|u\|_{a(\sigma)}$ using (104); the second norm is bounded using (113); and the third using (105). After noting that $\|\cdot\|_{a(\sigma;t)} = \kappa_\sigma^{-1}\sigma^{-t}\|\Pi_t(\cdot)\|_{\ell_\sigma^2}$, (34) is obtained. □

Inequality (35).

$$\begin{aligned}&\|\mathbf{S}\|_{a(\sigma;t)} - \mu_\sigma(\widehat{\mathbf{S}}_t^\ell)\\ \le\;& \|\mathbf{S}-\mathbf{S}^\ell\|_{a(\sigma;t)} + \|\mathbf{S}^\ell - \mathbf{S}_t^\ell\|_{a(\sigma;t)} + \|\mathbf{S}_t^\ell\|_{a(\sigma;t)} - \mu_\sigma(\mathbf{S}_t^\ell)\\ \le\;& \|\mathbf{G}\nabla\mathbf{K}\|_{a(\sigma;t)} + \kappa_\sigma^{(p)}\partial_\sigma^{(p)}(\widehat{\mathbf{S}}^\ell)\end{aligned}$$

using $\mathbf{S}-\mathbf{S}^\ell = \mathbf{G}\nabla\mathbf{F}$, (105), (103) and (37), and giving (35). □

Inequality (36). By definition of ∇ and the triangle inequality,

$$\begin{aligned}\|\Pi_t(\mathbf{G}\nabla\mathbf{K})u\|_{\ell_\sigma^2} =\;& \|\Pi_t(\mathbf{GK} - \mathbf{G}\otimes\mathbf{K})u\|_{\ell_\sigma^2}\\ \le\;& \|\Pi_t(\mathbf{GK} - (\mathbf{G}\otimes\mathbf{K})_t)u\|_{\ell_\sigma^2}\\ &+\|\Pi_t(\mathbf{G}\otimes\mathbf{K} - (\mathbf{G}\otimes\mathbf{K})_t)u\|_{\ell_\sigma^2}.\end{aligned}$$

The last two norms are bounded using (112) and (105), to get (36). □

Inequality (37). Consider the case $p=\infty$ first. Since $\|\cdot\|_{a(\sigma)} \le \|\cdot\|_{\ell^\infty} = \sup_t \sigma^{-t}\,\|(\Delta\Pi_t)(\cdot)\|_{\ell_\sigma^2}$, we obtain

$$\|(\mathbf{G}\nabla\mathbf{K})u\|_{a(\sigma)} \le \sup_t \sigma^{-t}\|(\Delta\Pi_t)(\mathbf{G}\nabla\mathbf{K})u\|_{\ell_\sigma^2}. \quad (114)$$

But by (108),

$$\begin{aligned}\|(\Delta\Pi_t)(\mathbf{G}\nabla\mathbf{K})u\|_{\ell^2_\sigma} &= \|(\Delta\Pi_t)\mathbf{G}_t\sum_{\tau=-\infty}^{t}\Pi_{\tau-1}(\mathbf{K}_\tau-\mathbf{K}_{\tau-1})u\|_{\ell^2_\sigma} \\ &\leq \kappa_\sigma\sigma^t(\sigma-1)^{-1}\mu_\sigma(\widehat{\mathbf{G}})\partial_\sigma(\widehat{\mathbf{K}})\|u\|_{a(\sigma)} \qquad (115)\end{aligned}$$

by (103), (104), and the fact that $\|\widehat{\mathbf{G}}_t\|_{H^\infty_\sigma} \leq \mu_\sigma(\widehat{\mathbf{G}})$. Inequality (37) follows from (114), (115), for $p=\infty$.

For the case $2 \leq p < \infty$, let $y := (\mathbf{G}\nabla\mathbf{K})u$. We obtain

$$\begin{aligned}|y(t)| &= |(\mathbf{G}\mathbf{K}-\mathbf{G}\otimes\mathbf{K})u(t)| \\ &= |\mathbf{G}_t(\mathbf{K}-\mathbf{K}_t)u(t)| \\ &\leq \sigma^{-t}\|\Pi_t\mathbf{G}_t(\mathbf{K}-\mathbf{K}_t)u\|_{\ell^2_\sigma} \\ &\leq \mu_\sigma(\widehat{\mathbf{G}}_t)\kappa_\sigma\kappa_\sigma^{(p)}\partial_\sigma^{(p)}(\widehat{\mathbf{K}})\|u\|_{a(\sigma)} \qquad (116)\end{aligned}$$

by (104) and the shift invariance of $\mathbf{G}_t$, and (105). Therefore, (116) is a bound on $\|y\|_{\ell^\infty} \geq \|y\|_{a(\sigma)}$, and (37) is true for $2 \leq p < \infty$. □

Inequalities (38, 39). These two inequalities are implied by the following ones, which hold for $t \in \mathbf{Z}$,

$$\begin{aligned}\|(\widehat{\mathbf{G}}\otimes\widehat{\mathbf{K}})_t-(\widehat{\mathbf{G}}\otimes\widehat{\mathbf{K}})_{t-1}\|_{L^p_\sigma} &\leq \mu_\sigma^{(p)}(\widehat{\mathbf{G}}_t(\widehat{\mathbf{K}}_t-\widehat{\mathbf{K}}_{t-1})) \\ &\quad +\mu_\sigma^{(p)}((\widehat{\mathbf{G}}_t-\widehat{\mathbf{G}}_{t-1})\widehat{\mathbf{K}}_{t-1}); \\ \|\widehat{\mathbf{G}}_t^{\text{-1}}-\widehat{\mathbf{G}}_{t-1}^{\text{-1}}\|_{L^p_\sigma} &= \|\widehat{\mathbf{G}}_t^{-1}(\widehat{\mathbf{G}}_t-\widehat{\mathbf{G}}_{t-1})\widehat{\mathbf{G}}_{t-1}^{-1}\|_{L^p_\sigma} \\ &\leq \|\widehat{\mathbf{G}}_t^{-1}\|_{L^\infty_\sigma}\|(\widehat{\mathbf{G}}_t-\widehat{\mathbf{G}}_{t-1})\|_{L^p_\sigma}\|\widehat{\mathbf{G}}_{t-1}^{-1}\|_{L^\infty_\sigma}.\end{aligned}$$

□

Inequality (40). By the triangle inequality,

$$\|\mathbf{S}\|_{a(\sigma;t)} \leq \|\mathbf{S}^\ell\|_{a(\sigma;t)} + \|\mathbf{S}-\mathbf{S}^\ell\|_{a(\sigma;t)}.$$

Now

$$\begin{aligned}\|\mathbf{S}^\ell\|_{a(\sigma;t)} &\leq \mu_\sigma(\widehat{\mathbf{S}}_t^\ell)+\kappa_\sigma^{(p)}\partial_\sigma^{(p)}(\widehat{\mathbf{S}}^\ell) \quad by\ (32); \\ \mu_\sigma(\widehat{\mathbf{S}}_t^\ell) &\leq \mu_1(\widehat{\mathbf{S}}_t^\ell)\nu_{\sigma_0}(\widehat{\mathbf{S}}_t^\ell)^{\left(\frac{\ell n\,\sigma}{\ell n\,\sigma_0}\right)}\end{aligned}$$

by the radial growth condition; and $(\mathbf{S}-\mathbf{S}^\ell) = \mathbf{G}\nabla\mathbf{K}$; therefore $\mathbf{S}$ has the upper bound

$$\|\mathbf{S}\|_{a(\sigma;t)} \leq \mu_1(\widehat{\mathbf{S}}_t^\ell)\nu_{\sigma_0}(\widehat{\mathbf{S}}_t^\ell)^{\left(\frac{\ell n\,\sigma}{\ell n\,\sigma_0}\right)} + \kappa_\sigma^{(p)}\partial_\sigma^{(p)}(\widehat{\mathbf{S}}^\ell) + \|\mathbf{G}\nabla\mathbf{K}\|_{a(\sigma)}. \qquad (117)$$

We will show that $\mathbf{S}$ also has the lower bound

$$\|\mathbf{S}\|_{a(\sigma;t)} \geq \mu_1(\widehat{\mathbf{S}}_t^\ell) - \kappa_\sigma^{(p)} \partial_\sigma^{(p)}(\widehat{\mathbf{S}}^\ell) - \|\mathbf{G}\nabla\mathbf{K}\|_{a(\sigma)}, \tag{118}$$

which together with (117) implies (40).

To prove (118) observe that from the definition of $\mu_1(\cdot)$ there are sets of exponential inputs $u \in \ell^\infty(-\infty,\infty)$, $u(t) := \alpha \exp(j\theta t)$, for which $\|\widehat{\mathbf{S}}_t^\ell u\|_{\ell^\infty} = \mu_1(\widehat{\mathbf{S}}_t^\ell)\|u\|_{\ell^\infty}$. For imaginary exponentials the $\|\cdot\|_{a(\sigma)}$ and ℓ^∞ norms coincide, and there exists an exponential u for which

$$\|\Pi_t \mathbf{S}_t^\ell u\|_{a(\sigma;t)} \geq \mu_1(\widehat{\mathbf{S}}_t^\ell)\|u\|_{a(\sigma)}. \tag{119}$$

From the definition of $\|\cdot\|_{a(\sigma;t)}$, we now get

$$\begin{aligned}
\|\mathbf{S}u\|_{a(\sigma;t)} &\geq \kappa_\sigma^{-1}\sigma^t \|\mathbf{\Pi}_t \mathbf{S}u\|_{\ell_\sigma^2} \\
&\geq \kappa_\sigma^{-1}\sigma^t \{\|\mathbf{\Pi}_t \mathbf{S}_t^\ell u\|_{\ell_\sigma^2} - \|\mathbf{\Pi}_t(\mathbf{S}^\ell - \mathbf{S}_t^\ell)u\|_{\ell_\sigma^2} \\
&\qquad - \|\mathbf{\Pi}_t(\mathbf{S} - \mathbf{S}^\ell)u\|_{\ell_\sigma^2}\} \quad \text{(triangle inequality)} \\
&\geq \{\mu_1(\widehat{\mathbf{S}}_t^\ell) - \kappa_\sigma^{(p)}\partial_\sigma^{(p)}(\widehat{\mathbf{S}}^\ell) - \|\mathbf{G}\nabla\mathbf{K}\|_{a(\sigma)}\}\|u\|_{a(\sigma)}
\end{aligned}$$

by (119), (105) and as $(\mathbf{S} - \mathbf{S}^\ell) = \mathbf{G}\nabla\mathbf{K}$. This proves (118). □

Proof of Corollary 3.1: Write $\mathbf{F}_\lambda := (\lambda\mathbf{I} - \mathbf{G} \otimes \mathbf{K})$. If $\lambda \in \mathrm{Res}_{\mathbf{L}\overline{\mathbf{E}}_{\sigma};\gamma}(\mathbf{G} \otimes \mathbf{K})$ then $\mu_\sigma(\mathbf{F}_\lambda^{(-1)}) \leq \gamma$ by definition of $\mathrm{Res}_{(\cdot;\gamma)}$. As $\mathbf{K}$ is shift-invariant, $\mathbf{GK} = \mathbf{G} \otimes \mathbf{K}$. By Inversion Lemma 2.2, the resolvent conclusions will be true if $\|\mathbf{F}_\lambda^{(-1)} \nabla \mathbf{F}\|_{a(\sigma)} < 1$, for then $(\lambda\mathbf{I} - \mathbf{GK})^{-1} \in \mathbf{B}$. Let us evaluate this.

$$\begin{aligned}
&\|\mathbf{F}_\lambda^{(-1)} \nabla \mathbf{F}_\lambda\|_{a(\sigma)} \\
=\ &\|\mathbf{F}_\lambda^{(-1)} \nabla \mathbf{G} \otimes \mathbf{K}\|_{a(\sigma)} \quad \text{(as } \mathbf{K} \text{ and } \lambda\mathbf{I} \text{ are shift-invariant)} \\
\leq\ &(\sigma - 1)^{-1}\{\mu_\sigma(\widehat{\mathbf{F}}_\lambda^{(-1)})\partial_\sigma(\widehat{\mathbf{G}} \otimes \widehat{\mathbf{K}}) + \partial_\sigma(\widehat{\mathbf{F}}_\lambda^{(-1)})\ \|\mathbf{G} \otimes \mathbf{K}\|_{a(\sigma)}\}
\end{aligned} \tag{120}$$

as $\partial_\sigma(\mathbf{F}_\lambda^{(-1)} \otimes \mathbf{F}_\lambda) = 0$. The terms appearing in (120) can be bounded as follows.

$$\begin{aligned}
\mu_\sigma(\widehat{\mathbf{F}}_\lambda^{(-1)}) &\leq \gamma, \\
\partial_\sigma(\widehat{\mathbf{G}} \otimes \widehat{\mathbf{K}}) &\leq \partial_\sigma(\widehat{\mathbf{G}})\mu_\sigma(\widehat{\mathbf{K}}) \quad \text{(as } \mathbf{K} \text{ is shift-invariant)}, \\
\partial_\sigma(\widehat{\mathbf{F}}_\lambda^{(-1)}) &\leq \gamma^2\partial_\sigma(\widehat{\mathbf{F}}_\lambda) \leq \gamma^2\partial_\sigma(\widehat{\mathbf{G}})\mu_\sigma(\widehat{\mathbf{K}}), \\
\|\mathbf{G} \otimes \mathbf{K}\|_{a(\sigma)} &= \|\mathbf{GK}\|_{a(\sigma)} \leq \|\mathbf{G}\|_{a(\sigma)}\|\mathbf{K}\|_{a(\sigma)} \leq \mu_\sigma(\widehat{\mathbf{K}})
\end{aligned}$$

as $\mathbf{G}$ has no memory, $\|\mathbf{G}\|_{a(\sigma)} = \mu_\sigma(\widehat{\mathbf{G}}) = 1$, and $\mathbf{K}$ is shift invariant

Therefore the resolvent conclusions are true if

$$(\sigma - 1)^{-1}\{\gamma\mu_{\sigma}(\hat{\mathbf{K}}) + (\gamma\mu_{\sigma}(\hat{\mathbf{K}}))^2\}\partial_{\sigma}(\mathbf{G}) < 1,$$

which implies the Corollary. □

References

[1] V.M. Adamjan, D.Z. Arov, and M.G. Krein, Infinite Hankel Blocks Matrices and Related Extension Problems, *AMA Translations*, Vol.111, 1978, pp. 133-156.

[2] W. Arveson, Interpolation problems in nest algebras, *J. Functional Anal.*, (20), 1975, pp.208-233.

[3] J.A. Ball, C. Foias, J.W. Helton and A. Tannenbaum, On a local nonlinear commutant lifting theorem, preprint.

[4] H.M.J. Cantalloube, C.E. Nahum, and P.E. Caines, Robust adaptive control: a direct factorization approach, to be published.

[5] C.A. Desoer, Slowly Varying Discrete System $x_{i+1} = A_i x_i$, *Electronics Letters*, (6) 1970, pp. 339-40.

[6] M. Dahleh and M.A. Dahleh, On slowly time-varying systems, MIT Technical Report LIDS-P-1852, February 1989.

[7] W.L. Duren, *Theory of H^p Spaces.* New York: Academic, 1970.

[8] C.A. Desoer, R.W. Liu, J. Murray, and R. Saeks, Feedback system design: the fractional representation approach, *IEEE Trans. Auto. Const.,* vol. AC-25, 1980, pp.399-412.

[9] C.A. Desoer and M. Vidyasagar, Feedback systems: input-output properties, Academic Press, 1975.

[10] A. Feintuch and B.A. Francis, Uniformly optimal control of linear systems. *Automatica*, 21(5): 563-574, 1985.

[11] B.A. Francis, A course in H^∞ control theory, *Lecture Notes in Control and Information Sciences*, Vol.88, Springer-Verlag: New York, 1985.

[12] B.A. Francis and M. Vidyasagar, Algebraic and topological aspects of the servo problem for lumped systems, Dep. Eng. Appl. Sci., Yale Univ., New Haven, CT, S & IS Rep 8003, 1980.

[13] M. Freedman and G. Zames, Logarithmic variation criteria for the stability of systems with time-varying gains, *SIAM J. Control* Vol.6, No.3, 1968, pp.487-507.

[14] T.T. Georgiou, A.M. Pascoal and P.P. Khargonekar, On the robust stabilizability of uncertain linear time-invariant plants using nonlinear time-varying controllers, *Automatica*, 23(5): pp. 617-624, September 1987.

[15] P.A. Ioannou and K.S. Tsakalis, Time and frequency domain uncertainty bounds in robust adaptive control, preprint.

[16] P.P. Khargonekar and K. Poolla, On polynomial matrix fraction representations for linear time-varying systems, *Linear Algebra Appl.*, 80, pp. 1-37, 1986.

[17] P.P. Khargonekar and K. Poolla, Uniformly optimal control of linear time-invariant plants: Nonlinear time-varying controllers, *Systems and Control Letters*, Vol. 6, No. 5, pp. 303-308, 1986.

[18] M.C. Smith, Well-posedness of $\mathbf{H}^\infty$ optimal control problems, *SIAM J. Control and Optimization* Vol. 28, No. 2, pp. 342-358, March 1990.

[19] M.S. Verma, Robust stability of linear feedback systems under time-varying and nonlinear perturbations in the plant, preprint.

[20] M. Vidyasagar, H. Schneider, and B.A. Francis, Algebraic and topological aspects of feedback stabilization, *IEEE Trans. Automat. Contr.,* vol. AC-27, pp.880-894, Aug.1982.

[21] L.Y. Wang, Adaptive H^∞ Optimization, Ph.D Dissertation, 1989.

[22] L.Y. Wang, Lipschitz continuity of inner-outer factorization, *Systems and Control Letters* 16, pp. 281-287, 1991.

[23] L.Y. Wang and G. Zames, H^∞ optimization and slowly time-varying systems, *Proc. 26, Conf. Dec. Contr.*, Dec. 1987, 81-83.

[24] Ibid., Slowly time-varying systems and H^∞ Optimization, Proc. 8th IFAC Symp. on Identif. and Param. Est., Beijing, 1988, (1), pp. 492-495.

[25] Ibid., Local-global double algebras for slow H^∞ adaptation, *MTNS conference*, Amsterdam, Nethelands, June 22-28, 1989.

[26] Ibid., Local-global double algebras for slow H^∞ adaptation, *Proc. of IEEE 28th CDC conference*, Tampa, Florida, U.S.A., Dec. 13-15, 1989.

[27] Ibid., Local-global double algebras for slow H^∞ adaptation: The case of ℓ^2 disturbances, to appear in IMA J. of Control and Optimization.

[28] Ibid., Local-global double algebras for slow H^∞ adaptation, Part II: Optimization for stable plants, *IEEE Trans. Automat. Contr.* Vol. 36, No. 2, 143-151, 1991.

[29] G. Zames, Feedback and optimal sensitivity: model reference transformations, multiplicative seminorms, and approximate inverses, *IEEE Trans. Automat. Contr.,* vol. AC-26, pp.301-320, 1981.

[30] Ibid., On the metric complexity of causal linear systems, ε-entropy and ε-dimension for continuous time, *IEEE Trans. Automat. Contr.,* vol. AC-24, pp.222-230, 1979.

[31] G. Zames, Nonlinear time-varying feedback systems: Conditions for L^∞ boundedness derived using conic operators on exponentially weighted spaces, *Proc. 3rd Allerton Conference on Circuit and System Theory*, University of Illinois, Urbana, pp.460-471, 1965.

[32] Ibid., Input-output stability of nonlinear time-varying feedback systems, *IEEE trans. Automat. Contr.,* vol. AC-11, pp.228-238, 465-477, 1966.

[33] G. Zames and L.Y. Wang, Local-global double algebras for slow H^∞ adaptation, Part I: Inversion and stability, *IEEE Trans. Automat. Contr.* Vol. 36, No. 2, pp.130-142, 1991.

[34] G.Zames and L.Y. Wang, What is an adaptive-learning system?, *Proc. 1990 IEEE CDC Conference.*

ROBUST CONTROL TECHNIQUES FOR SYSTEMS WITH STRUCTURED STATE SPACE UNCERTAINTY

KENNETH M. SOBEL
WANGLING YU

The City College of New York
Department of Electrical Engineering
New York, NY 10031

I. INTRODUCTION

Robust controller analysis and design for linear systems has been studied using many different approaches. Doyle and Stein [1] model the uncertainty in the frequency domain by using H^{∞} norm bounds. An alternative approach using interval polynomials is based upon a result due to Kharitonov [2] which assumes that the coefficients of the characteristic polynomial exhibit independent variations and the stability region is the open left half complex plane. Much research has been done to relax these assumptions among which is the work of Petersen [3] who extends the Kharitonov approach to regions other than the open left half of the complex plane. Another result which greatly reduces the computational complexity is the so-called Edge Theorem proposed by Bartlett

CONTROL AND DYNAMIC SYSTEMS, VOL. 51

et. al. [4]. This theorem states that given a polytope in the space of the coefficients of the characteristic polynomial, then the stability of the whole polytope is equivalent to the stability of its edges. Yet another approach is that of finding maximal perturbation bounds or stability radii. This approach attempts to find the largest set of perturbations for which the roots of the characteristic polynomial remain in some stability region. Fu and Barmish [5] present a closed form description for the maximal perturbation bound of Hurwitz stable interval polynomials.

Ackermann et. al. [6] state that less progress has been made on the related important problem of the stability of state space systems. Patel and Toda [7] have presented a bound for the robust stability of an uncertain state space system. Later, Yedavalli [8] provided an improved stability robustness bound for structured perturbations. Yedavalli and Liang [9] introduced a method of state transformation to reduce the conservatism in these methods which utilize the solution to a matrix Lyapunov equation.

Another approach for state space systems, which is based upon the Gronwall lemma, has been proposed by Chen and Wong [10] for linear continuous time systems with norm bounded state space uncertainty. Sobel et. al. [11] extended the Gronwall lemma approach to include the structure of the uncertainty. This result ensures stability robustness to time varying structured state space uncertainty provided that the eigenvalues of the nominal linear time invariant (LTI) system lie to the left of a vertical line in the complex plane. The location of this vertical line is determined by a norm involving the structure of the uncertainty and the nominal closed loop eigenvector matrix. Therefore, this robustness result is especially well suited to the design of control systems using eigenstructure assignment. The robustness condition can be satisfied either by moving the nominal eigenvalues to the left in the complex plane or by changing the nominal eigenvectors. Later, Sobel and Yu [12] proposed a robust eigenstructure assignment design method by optimizing the sufficient condition for robust stability with an application to flight control.

A sufficient condition for the performance robustness of uncertain LTI systems has been proposed by Juang et. al. [13] by using a Lyapunov approach. The closed loop eigenvalues are guaranteed to lie within chosen performance regions when the LTI system is subjected to time invariant structured state space uncertainty. However, this approach requires the solution of a Lyapunov matrix equation involving complex matrices.

Yu and Sobel [14] used a lemma of Juang et. al. [13] to obtain a sufficient condition for performance robustness by using the Gronwall lemma approach. The closed loop eigenvalues are guaranteed to lie within chosen performance regions when the LTI system is subjected to time invariant structured state space uncertainty. If the performance region is chosen as the entire left half plane, then the new result reduces to the stability robustness condition described by Sobel et. al. [11]. The required computations include the nominal eigenvalues/eigenvectors and a matrix norm involving the uncertainty structure and the closed loop eigenvector matrix.

The problem of the conservatism of the new sufficient condition for performance robustness based upon the Gronwall lemma has also been addressed. Sobel et. al. [11] utilized a diagonal weighting matrix D whose real positive entries were chosen by using Perron weightings in order to reduce the conservatism. Later, Yu and Sobel [14] introduced an additional matrix Q which can further reduce the conservatism. Although, a simple closed form solution for the optimal matrix Q is not yet known, Yu and Sobel [14] proposed a choice based upon obtaining a unitary Q from the singular value decomposition of the product of the nominal closed loop eigenvector matrix and the weighting matrix D.

Yu and Sobel [15] proposed a robust eigenstructure assignment design method which optimizes either the sufficient condition for stability or performance robustness while constraining the dominant eigenvalues to lie within chosen performance regions in the complex plane. This constrained optimization problem was solved by using the sequential unconstrained minimization technique with a quadratic extended interior penalty function [16]. An example was presented which illustrates the design of a robust eigenstructure assignment controller for a pitch pointing/vertical translation maneuver of the AFTI F-16 aircraft. Several designs were shown which include (1) the non-robust design of Sobel and Shapiro [17], (2) a robust design using a state transformation to reduce conservatism in the sufficient condition, (3) a robust design which includes explicit constraints on certain eigenvector entries to illustrate the tradeoff between robustness and nominal performance, (4) a robust design which utilizes both a state transformation and a unitary weighting matrix to reduce conservatism, and (5) a performance robust design which ensures that the closed loop eigenvalues lie within a chosen settling time/damping ratio region for all uncertainty. Time responses of the different designs to unit step commands in both flight path angle and pitch attitude were shown which illustrate the penalty in nominal performance that results when using the robust design method.

This contribution begins with a statement of the problem formulation in Section II. Next, the stability robustness conditions for linear time varying structured state space uncertainty are described in Section III. The performance robustness results for linear time invariant uncertainty are discussed in Section IV. A robust eigenstructure assignment design method and its application to the design of a pitch pointing/vertical translation flight control law for the AFTI F-16 aircraft are shown in Section V. Finally, the conclusions are described in Section VI. All proofs are contained in the appendix which is designated as Section VII.

II. PROBLEM FORMULATION

Consider a nominal linear time invariant multi-input multi-output system described by

$$\dot{x} = Ax + Bu \tag{1}$$

$$y = Cx \tag{2}$$

where $x \in \mathbb{R}^n$ is the state vector, $u \in \mathbb{R}^m$ is the input vector, $y \in \mathbb{R}^r$ is the output vector, and A, B, C, are constant matrices.

Suppose that the nominal system is subject to linear time varying uncertainties in the entries of A and B described by $\Delta A(t)$ and $\Delta B(t)$, respectively. We shall assume that the entries of $\Delta A(t)$ and $\Delta B(t)$ are continuous functions of time. Then, the system with uncertainty is given by

$$\dot{x} = Ax + Bu + \Delta A(t)x + \Delta B(t)u \tag{3}$$

$$y = Cx \tag{4}$$

Further, suppose that bounds are available on the absolute values of the maximum variations in the elements of $\Delta A(t)$ and $\Delta B(t)$. That is,

$$|\Delta a_{ij}(t)| \le (a_{ij})_{max}; \quad i=1,...,n; \quad j=1,...,n \tag{5}$$

$$|\Delta b_{ij}(t)| \le (b_{ij})_{max}; \quad i=1,...,n; \quad j=1,...,m \tag{6}$$

Define $\Delta A^+(t)$ and $\Delta B^+(t)$ as the matrices obtained by replacing the entries of $\Delta A(t)$ and $\Delta B(t)$ by their absolute values. Also, define A_{max} and B_{max} as the matrices with entries $(a_{ij})_{max}$ and $(b_{ij})_{max}$, respectively. Then,

$$\{\Delta A(t): \Delta A^+(t) \le A_{max}\} \tag{7}$$

and

$$\{\Delta B(t): \Delta B^+(t) \le B_{max}\} \tag{8}$$

where "≤" is applied element by element to matrices and $A_{max} \in \mathbb{R}_+^{n\times n}$, $B_{max} \in \mathbb{R}_+^{n\times m}$ where $\mathbb{R}_+$ is the set of non-negative numbers.

Consider the constant gain output feedback control law described by

$$u(t) = Fy(t) \tag{9}$$

Then, the nominal closed loop system is given by

$$\dot{x}(t) = (A + BFC)x(t) \tag{10}$$

and the uncertain closed loop system is given by

$$\dot{x}(t) = (A + BFC)x(t) + [\Delta A(t) + \Delta B(t)FC]x(t) \tag{11}$$

Stability Robustness Problem: Given a feedback gain matrix $F \in \mathbb{R}^{m\times r}$ such that the nominal closed loop system exhibits desirable dynamic performance, determine if the uncertain closed loop system is asymptotically stable for all $\Delta A(t)$ and $\Delta B(t)$ described by Eqs.(7)-(8).

Performance Robustness Problem: A feedback gain matrix $F \in \mathbb{R}^{m\times r}$ is chosen such that all of the eigenvalues of the nominal closed loop system are inside a region R. Determine if all of the eigenvalues of the uncertain closed loop system are inside the region R for all time invariant ΔA and ΔB described by Eqs.(7)-(8).

III. STABILITY ROBUSTNESS

A. STABILITY THEOREMS

First, we present a preliminary lemma which will be needed later in the proof of Theorem 1.

Lemma 1

If the uncertain closed loop system is described by Eq.(11) with uncertainty which satisfies Eqs.(7)-(8) then

$$\|(MDQ)^{-1}[\Delta A(t)+\Delta B(t)FC]MDQ\|_2 \le$$

$$\|[(MDQ)^{-1}]^+[A_{max} + B_{max}(FC)^+](MDQ)^+\|_2 \tag{12}$$

Next, we present the theorem which describes the sufficient condition for robust stability.

Theorem 1: Stability Robustness Sufficient Condition

Let M be a modal matrix of (A+BFC), let D be a diagonal matrix with positive real entries, and let Q be a nonsingular matrix. Suppose F is such that the nominal closed loop system described by Eq.(10) is asymptotically stable with a non-defective modal matrix. Then, the uncertain closed loop system given by Eq.(11) is asymptotically stable for all $\Delta A(t)$ and $\Delta B(t)$ described by Eqs.(7)-(8) if

$$\alpha > \kappa_2(Q) \cdot \| [(MDQ)^{-1}]^+ [A_{max} + B_{max}(FC)^+](MDQ)^+ \|_2 \tag{13}$$

where

$$\alpha = -\max_i \, \mathrm{Re}[\lambda_i(A+BFC)]$$

and where $\kappa_2(Q) = \|Q^{-1}\|_2 \cdot \|Q\|_2$ is the 2-norm condition number of the matrix Q.

Theorem 1 presents a sufficient condition for robust stability in terms of the nominal closed loop system. Robust stability is ensured provided that the nominal closed loop eigenvalues lie to the left of a vertical line in the complex plane which is determined by a norm involving the structure of the uncertainty and the nominal closed loop modal matrix.

Corollary 1: Full State Feedback

Suppose $u=Fx$ with $F \in \mathbb{R}^{m\times n}$ such that the nominal closed loop system described by Eq.(10), with $C=I$, is asymptotically stable with a non-defective modal matrix. Then the uncertain closed loop system given by Eq.(11), with $C=I$, is asymptotically stable for all $\Delta A(t)$ and $\Delta B(t)$ described by Eqs.(7)-(8) if

$$\alpha > \kappa_2(Q) \cdot \| [(MDQ)^{-1}]^+ [A_{max} + B_{max}\{V\Sigma^{-1}U_0^T(M\Lambda M^{-1} - A)\}^+](MDQ)^+ \|_2 \tag{14}$$

where

$$B=[U_0 \quad U_1]\begin{bmatrix} \Sigma V^T \\ 0 \end{bmatrix}=U_0\Sigma V^T \tag{15}$$

is a singular value decomposition of B, $U=[U_0 \ U_1]$ is the matrix of left singular vectors, Σ is a diagonal matrix containing the singular values, V is the matrix of right singular vectors, and Λ is a diagonal matrix containing the eigenvalues of A+BF.

Corollary 2: Output Feedback

Suppose u=Fy with $F \in \mathbb{R}^{m\times r}$ is such that the nominal closed loop system described by Eq.(10) is asymptotically stable with a non-defective modal matrix. Let Λ_r be the (r×r) diagonal matrix whose entries are the assignable closed loop eigenvalues and let M_r be the (n×r) matrix whose columns are the corresponding achievable eigenvectors. Let

$$B=[U_0 \ U_1]\begin{bmatrix} \Sigma V^T \\ 0 \end{bmatrix} \tag{16}$$

and

$$CM_r=[U_{r0} \ U_{r1}]\begin{bmatrix} \Sigma_r V_r^T \\ 0 \end{bmatrix} \tag{17}$$

be the singular value decompositions of B and CM_r, respectively. Then, the uncertain closed loop system given by Eq.(11) is asymptotically stable for all $\Delta A(t)$ and $\Delta B(t)$ described by Eqs.(7)-(8) if

$$\alpha > \kappa_2(Q)\cdot$$

$$\|[(MDQ)^{-1}]^+[A_{max}+B_{max}\{V\Sigma^{-1}U_0^T(M_r\Lambda_r-AM_r)V_r\Sigma_r^{-1}U_{r0}^T C\}^+(MDQ)^+\|_2 \tag{18}$$

We observe from Corollary 1 that in the case of full state feedback, the norm on the right-hand side of the sufficient condition is in terms of the known matrices A_{max}, B_{max}, V, Σ, and U_0 and the nominal closed loop modal matrix. Thus, we can try to satisfy the sufficient condition by leaving the nominal eigenvalues at their desired locations and reducing the right-hand side by changing the nominal eigenvectors within the allowable subspaces. Alternatively, we could move the nominal eigenvalues further to the left in the complex plane.

We observe from Corollary 2 that in the case of output feedback, the norm on the right-hand side of the sufficient condition is in terms of both M and M_r. Recall that M is the nominal closed loop modal matrix while M_r is the matrix whose columns are the r eigenvectors we can partially assign. For the output feedback problem, we can also minimize the right-hand side of the sufficient condition. However, on each iteration we would obtain a new M_r by searching the r allowable eigenvector subspaces for the eigenvectors that we can partially assign. Then we would compute the eigenvectors of (A + BFC) in order to determine the modal matrix M. Thus, the output feedback control problem requires some additional computation as compared to the full state feedback problem.

B. CHOICE OF MATRICES D AND Q

The conservatism inherent in the sufficient condition of Theorem 1 can be reduced by the use of a diagonal weighting. First, we consider the optimal choice of the diagonal real positive matrix D in Eq.(13) when Q is the identity matrix. Thus, we seek D_{opt} which yields the infimum of

$$\|D^{-1}(M^{-1})^{+}[A_{max} + B_{max}(FC)^{+}]M^{+}D\|_{2} \tag{19}$$

or equivalently by the D-weighted 2-norm

$$\|(M^{-1})^{+}[A_{max} + B_{max}(FC)^{+}]M^{+}\|_{2D} \tag{20}$$

We now state a lemma that gives the infimum of the norm in Eq.(20). This result is attributed to Stoer and Witzgall [18].

Lemma 2 [18]

Define the Perron eigenvalue of the non-negative matrix A^{+}, denoted by $\pi(A^{+})$, to be the real non-negative eigenvalue $\lambda_{max}\geq 0$ such that $\lambda_{max}\geq|\lambda_i|$ for all eigenvalues (i=1,...,n) of A^{+}. Define the right and left eigenvectors corresponding to λ_{max} to be the right and left Perron eigenvectors, denoted by x and y, respectively. Normalize the Perron eigenvectors such that $\pi x=A^{+}x$ and $\pi y^{T}=y^{T}A^{+}$ with $\|x\|_{\infty}=\|y\|_{\infty}=1$.
Then,

$$\inf_{D}\|A^{+}\|_{2D}=\pi(A^{+}) \tag{21}$$

where the matrix D_{opt} which yields the infimum of $\|D^{-1}A^{+}D\|_{2}$ is given by

$$D_{opt}=\text{diag}\{[y(1)/x(1)]^{1/2},...,[y(n)/x(n)]^{1/2}\} \tag{22}$$

We observe that the matrix inside the norm in Eq.(20) is a non-negative matrix. Therefore, our sufficient condition becomes

$$\alpha > \inf_{D} \| (M^{-1})^{+} [A_{max} + B_{max} (FC)^{+}] M^{+} \|_{2D} \tag{23}$$

or, by using lemma 2

$$\alpha > \pi[(M^{-1})^{+} [A_{max} + B_{max} (FC)^{+}] M^{+}] \tag{24}$$

By taking the infimum on the right-hand side of Eq.(23), we have placed the vertical line in the complex plane as close to the imaginary axis as possible. Recall that the sufficient condition for robust stability requires that the nominal eigenvalues lie to the left of this vertical line. In this sense, we have reduced the conservatism of the sufficient condition.

Conjecture 1

A choice for the matrix Q which will reduce the norm in Eq.(13) is given by

$$Q = UV^{T} \tag{25}$$

where $U\Sigma V^{T}$ is the singular value decomposition of the matrix $(MD)^{T}$.

An optimal choice for the matrix Q is not yet available. However, the choice of Eq.(25) results in $\kappa_2(Q)$ equal to its minimum value of unity because U and V are unitary due to the property of the singular value decomposition. Furthermore, the choice of Eq.(25) solves the following problem [19]:

$$\min \|MDQ-I\|_F \quad \text{subject to } Q^{T}Q=I \tag{26}$$

where $\|\cdot\|_F$ denotes the Frobenius norm. This provides some motivation for the conjecture that the choice of Q in Eq.(25) will reduce the value of the norm in Eq.(13).

The proposed method is to first compute D_{opt} using Eq.(22) and then compute Q using Eq.(25). Then, these matrices are used to compute the norm in the stability robustness condition given by Eq.(13).

IV. PERFORMANCE ROBUSTNESS

A. PERFORMANCE ROBUSTNESS SUFFICIENT CONDITION

In this section, we present a theorem which describes a sufficient condition for robust performance in terms of the eigenstructure of the nominal closed loop system. Robust performance is ensured provided that two functions of the eigenvalues of the nominal closed loop system lie to the left of a vertical line in the complex plane. The phrase "performance robustness", as used here, means that the eigenvalues of the uncertain closed loop system with time invariant uncertainty ΔA and ΔB, lie inside the region R which is shown in Figure 1. This region ensures that a system with a dominant pair of complex conjugate eigenvalues has a damping ratio $\zeta \geq \zeta_{min}$ and a settling time $t_s \leq (t_s)_{max}$.

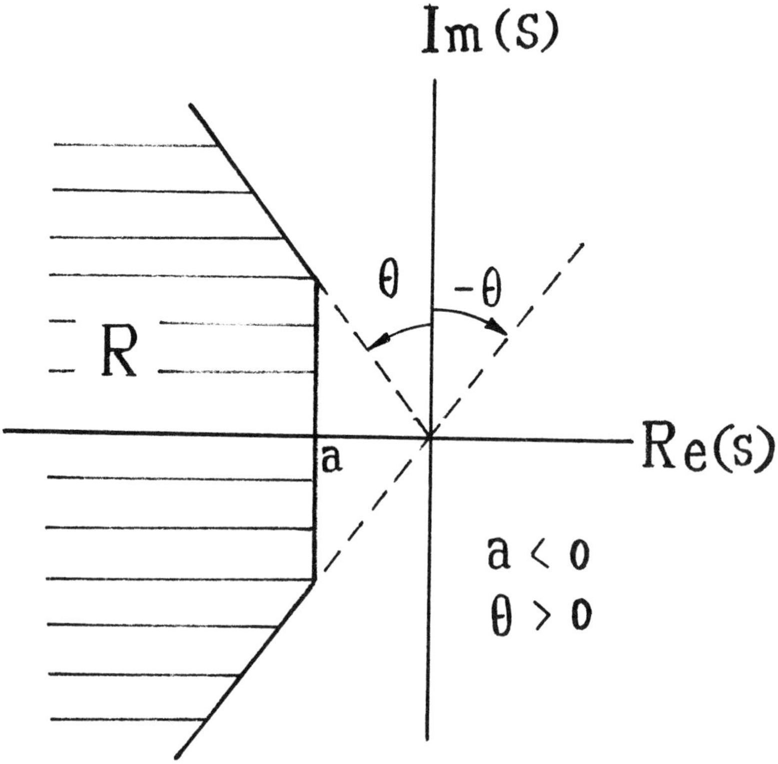

Fig. 1. Settling time/damping ratio region.

Theorem 2: Performance Robustness Sufficient Condition

Let the region R be the settling time/damping ratio region which is shown in Figure 1. Suppose F is such that the nominal closed loop system described by Eq.(10) has a non-defective modal matrix with eigenvalues all of which lie in the region R. Let M be a modal matrix of A+BFC, let D be a diagonal matrix with positive real entries, and let Q be a nonsingular matrix. Then, the eigenvalues of the uncertain closed loop system given by Eq.(11) will be in the region R for all time invariant ΔA and ΔB described by Eqs.(7)-(8) if

$$\min[\alpha_1,\alpha_2]>\kappa_2(Q)\cdot\|[(MDQ)^{-1}]^{+}[A_{max} + B_{max}(FC)^{+}][MDQ]^{+}\|_2 \qquad (27)$$

where

$$\alpha_1 = -\max_i \,[\mathrm{Re}\,\lambda_i] + a$$

and

$$\alpha_2 = -\max_i \,\mathrm{Re}[e^{-j\theta}\,\lambda_i]$$

where

λ_i is the i-th eigenvalue of A+BFC,

and where

$\kappa_2(Q) = \|Q^{-1}\|_2\cdot\|Q\|_2$ is the 2-norm condition number of the matrix Q,

and where the quantities a and θ are defined in Figure 1.

B. RESULTS NEEDED FOR THE PROOF OF THEOREM 2

In this section we present several results which are required in the proof of theorem 2. This section may be skipped by the reader without loss of continuity.

First, we present two definitions which relate the so-called transformed uncertain system to the uncertain system which is described by Eq.(11).

Definition 1

The transformed uncertain system is given by

$$\dot{\tilde{x}}(t)=e^{-j\theta}(A+BFC-aI)\tilde{x}(t)+e^{-j\theta}(\Delta A+\Delta BFC)\tilde{x}(t) \tag{28}$$

Definition 2

The nominal closed loop system matrix A_c, the uncertainty matrix ΔA_c, the transformed closed loop system matrix $\tilde{A}_c$, and the transformed uncertainty matrix $\Delta\tilde{A}_c$ are defined by

$$A_c = A + BFC$$

$$\Delta A_c = \Delta A + \Delta BFC$$

$$\tilde{A}_c=e^{-j\theta}(A_c-aI)$$

$$\Delta\tilde{A}_c = e^{-j\theta}A_c$$

Then, the uncertain closed loop system is given by

$$\dot{x}(t) = A_c x(t) + \Delta A_c x(t) \tag{29}$$

and the transformed uncertain closed loop system is given by

$$\dot{\tilde{x}}(t) = \tilde{A}_c\tilde{x}(t) + \Delta\tilde{A}_c\tilde{x}(t) \tag{30}$$

Next, we propose a lemma which shows that the transformed uncertain system is asymptotically stable if and only if the eigenvalues of the uncertain system of Eq.(29) are in the so-called H region. This H region was first considered by Juang et. al. [13] who used a line to divide the complex plane into two open half planes which are denoted by H and $\bar{H}$. This line intersects the real axis at the point "a" and makes an angle θ with respect to the positive imaginary axis. We shall denote a specific H region by $H(a,\theta)$.

Lemma 3

The eigenvalues of Eq.(29) are in the H region for all time invariant $\Delta A, \Delta B$ described by Eqs.(7)-(8) if and only if the eigenvalues of Eq.(30) are in the open left half complex plane for all time invariant $\Delta A, \Delta B$ described by Eqs.(7)-(8).

Now suppose that the real matrix A_c has a non-defective modal matrix. Then, the next lemma shows that the modal matrix of A_c will also diagonalize the complex matrix $\tilde{A}_c$.

Lemma 4

Suppose that the matrix A_c has a non-defective modal matrix. Let M be a modal matrix of A_c such that $\Lambda = M^{-1}A_cM$ is a diagonal matrix with the eigenvalues of A_c on the diagonal of Λ. Then, $\tilde{\Lambda} = M^{-1}\tilde{A}_cM$ is a diagonal matrix with the eigenvalues of $\tilde{A}_c$ on the diagonal of $\tilde{\Lambda}$.

The next lemma establishes an upper bound on $\|M^{-1}\Delta\tilde{A}_cM\|_2$. We observe that the upper bound of the norm of the complex matrix $M^{-1}\Delta\tilde{A}_cM$ is given by the norm of a real matrix.

Lemma 5

Let $\tilde{A}_c$ be given by definition 2, M is the modal matrix of A_c, and A_{max}, B_{max} are given by Eqs.(7)-(8). Then,

$$\|M^{-1}\Delta\tilde{A}_cM\|_2 \leq \|(M^{-1})^+[A_{max} + B_{max}(FC)^+]M^+\|_2 \tag{31}$$

Theorem 3: H Region Sufficient Condition

Suppose F is such that the nominal closed loop system described by Eq.(10) has a non-defective modal matrix with eigenvalues all of which lie in some $H(a,\theta)$ region. Let M be a modal matrix of A_c, let D be a diagonal matrix with positive real entries, and let Q be a nonsingular matrix. Then, the eigenvalues of the uncertain closed loop system given by Eq.(11) will be in the $H(a,\theta)$ region for all time invariant ΔA and ΔB described by Eqs.(7)-(8) if

$$\alpha > \kappa_2(Q)\cdot\|[(MDQ)^{-1}]^{+}[A_{max}+B_{max}(FC)^{+}][MDQ]^{+}\|_2 \tag{32}$$

where

$$\alpha = -\max_i \mathrm{Re}[\lambda_i(\tilde{A}_c)]$$

$$= -\max_i \mathrm{Re}[e^{-j\theta}(\lambda_i - a)]$$

and where $\kappa_2(Q) = \|Q^{-1}\|_2\cdot\|Q\|_2$ is the 2-norm condition number of the matrix Q.

Corollary 3: Multiple H Regions

Let the region R be the intersection of $H_k(a_k,\theta_k)$ for $k=1,2,\ldots,k_1$. Suppose F is such that the nominal closed loop system described by Eq.(10) has a non-defective modal matrix with eigenvalues which are all inside the region R. Let the matrices M and Q be defined as in theorem 3. Then, the eigenvalues of the uncertain closed loop system given by Eq.(11) will be in the region R for all time invariant ΔA and ΔB described by Eqs.(7)-(8) if

$$\min_k \alpha_k > \kappa_2(Q)\cdot\|[(MDQ)^{-1}]^+[A_{max}+B_{max}(FC)^+][MDQ]^+\|_2 \qquad (33)$$

where

$$\alpha_k = -\max_i \mathrm{Re}[e^{-j\theta_k}(\lambda_i - a_k)]$$

Outline of the Proof of Theorem 2

The proof of theorem 2 uses corollary 3 together with the observation that the region R in Figure 1 is the intersection of $H_1(a_1,0)$, $H_2(0,\theta_2)$, and $H_3(0,-\theta_2)$ with the definitions $a=a_1$ and $\theta=\theta_2$.

V. ROBUST EIGENSTRUCTURE ASSIGNMENT FLIGHT CONTROL DESIGN

A. EIGENSTRUCTURE ASSIGNMENT

Consider the linear time invariant system described by Eqs.(1)-(2). We shall assume that the matrices B and C have full rank. With these assumptions, the constant gain output feedback problem using eigenstructure assignment can be stated as follows: Given a set of desired eigenvalues $\{\lambda_i^d\}$, $i=1,\ldots,r$ and a corresponding set of desired eigenvectors $\{v_i^d\}$, $i=1,\ldots,r$ find a real $m\times r$ matrix F such that the eigenvalues of A+BFC contain $\{\lambda_i^d\}$ as a subset and the corresponding eigenvectors of A+BFC are close to the respective members of the set $\{v_i^d\}$.

Srinathkumar [20] has shown that for the system described by Eqs.(1)-(2), max(m,r) closed loop eigenvalues can be assigned and max(m,r) eigenvectors can be partially assigned with min(m,r) entries in each eigenvector arbitrarily chosen using constant gain output feedback.

In general, we may desire to exercise some control over more than min(m,r) entries in a particular eigenvector. Therefore, we shall now discuss the problem of first characterizing desired eigenvectors v_i^d that can be assigned as closed loop eigenvectors and of then determining the best possible set of achievable eigenvectors in case a desired eigenvector v_i^d is not achievable.

These problems were considered by Andry et. al. [21] who have shown the need for the eigenvector v_i to be in the subspace spanned by the columns of $(\lambda_i I-A)^{-1}B$. An alternative representation is described by Kautsky et. al. [22] who show that the subspace in which the eigenvector v_i must reside is also given by the nullspace of $U_1^T(\lambda_i I-A)$. The matrix U_1 is obtained from the singular value decomposition of B which is given by

$$B = \begin{bmatrix} U_0 & U_1 \end{bmatrix} \begin{bmatrix} \Sigma V^T \\ 0 \end{bmatrix} \tag{34}$$

In general, however, a desired eigenvector v_i^d will not reside in the prescribed subspace and, hence, cannot be achieved. Instead, a "best possible" choice for an achievable eigenvector is made. It is suggested by Andry et. al. [21] that this "best possible" eigenvector is the projection of v_i^d onto the subspace which is the nullspace of $U_1^T(\lambda_i I-A)$. We remark that the desired eigenvectors v_i^d are chosen based upon mode decoupling specifications. By using an orthogonal projection to obtain the achievable eigenvectors v_i^a, we are minimizing the 2-norm error between a desired eigenvector and its corresponding achievable eigenvector. It is important to note that the concept of stability robustness was not considered in the design approach proposed by Andry et. al. [21].

B. SEQUENTIAL UNCONSTRAINED MINIMIZATION TECHNIQUE

The constrained optimization to be considered [16] is to find a vector p* of design variables that minimizes either the function

$$J_1(p) = \pi\ [(M^{-1})^+[A_{max} + B_{max}(FC)^+]M^+] - \alpha \tag{35}$$

or the function

$$J_2(p)= \kappa_2(Q)\cdot\|[(MDQ)^{-1}]^+[A_{max}+B_{max}(FC)^+][MDQ]^+\|_2-\min(\alpha_1,\alpha_2) \tag{36}$$

subject to the constraints

$$h_i(p) \geq 0;\ i=1,\ldots,s \tag{37}$$

The vector p of design variables will contain a subset of the real and imaginary parts of the closed loop eigenvalues and a subset of the eigenvector parameters. To define the phrase "eigenvector parameters", we let L_i be a matrix whose columns form a basis for the nullspace of $U_1^T(\lambda_i I-A)$. Then, the i-th achievable eigenvector v_i^a is given by

$$v_i^a = L_i z_i;\ i=1,2,\ldots,n \tag{38}$$

where the vector z_i contains the free design parameters. Thus, we define the vectors z_i, i=1,2,...,n to be the "eigenvector parameters". The constraints $h_i(p)$ will be chosen to constrain some of the closed loop eigenvalues to lie in desired regions in the complex plane. Constraints may also be placed on certain entries of the closed loop eigenvectors in order to obtain a desired amount of mode decoupling. We remark that eigenvector constraints may be necessary to improve mode decoupling because a subset of the eigenvector parameters is included in the parameter vector p. Thus, the optimization will change this subset of eigenvector parameters without attempting to minimize the 2-norm difference between a desired eigenvector and its corresponding achievable eigenvector.

It is assumed that at least one of the constraints of Eq.(37) is critical at the minimum p*, that is, $h_i(p^*) = 0$ for some i. This constrained optimization problem may be transformed into a series of unconstrained minimization problems by introducing a penalty function associated with the constraints, and the transformed problem can be solved by the sequential unconstrained minimization technique [16]. The resulting transformed problem is to find the minimum of the function $P_k(\varepsilon)$ as ε goes to zero where

$$P_k(\varepsilon) = J_k(p) + \varepsilon \sum_{i=1}^{s} f_i(p) \quad ;k=1,2 \tag{39}$$

with $J_k(p)$ defined by either Eq.(35) or Eq.(36) and $f_i(p)$ defined by

$$f_i(p) = 1/h_i(p) \qquad \text{if } h_i(p) > 0 \tag{40}$$

The term $\varepsilon f_i(p)$ represents the penalty associated with the i-th constraint, and is an interior penalty function in the sense that it is defined only if p is inside the feasible design domain. With p^{ε} denoting the point in the design space where $P_k(\varepsilon)$ attains its minimum value for a given value of ε, it may be shown [16] that as ε goes to zero

$$\min P_k(\varepsilon) \to J_k(p^*)$$

$$p^{\varepsilon} \to p^*$$

Next, a quadratic extended interior penalty function [16] is introduced to permit the use of design points that are outside the feasible domain. The definition of the f_i in Eq.(40) for the quadratic extended penalty function is [16]

$$f_i = \begin{cases} 1/h_i & \text{if } h_i \geq h_0 \\ (1/h_0)[(h_i/h_0)^2 - 3(h_i/h_0) + 3] & \text{if } h_i \leq h_0 \end{cases} \tag{41}$$

where

$$h_0 = \sqrt{\varepsilon} \tag{42}$$

C. DESIGN OF A PITCH POINTING/VERTICAL TRANSLATION FLIGHT CONTROL LAW FOR THE AFTI F-16 AIRCRAFT

1. Problem Description and Non-Robust Design

Consider the LTI model of the AFTI F-16 aircraft described by

$$\begin{bmatrix} \dot{\gamma} \\ \dot{q} \\ \dot{\alpha} \\ \dot{\delta}_e \\ \dot{\delta}_f \end{bmatrix} = \begin{bmatrix} 0 & 0.00665 & 1.3411 & 0.16897 & 0.25183 \\ 0 & -0.86939 & 43.223 & -17.251 & -1.5766 \\ 0 & 0.99335 & -1.3411 & -0.16897 & -0.25183 \\ 0 & 0 & 0 & -20.0 & 0 \\ 0 & 0 & 0 & 0 & -20.0 \end{bmatrix} \begin{bmatrix} \gamma \\ q \\ \alpha \\ \delta_e \\ \delta_f \end{bmatrix}$$

$$+ \begin{bmatrix} 0 & 0 \\ 0 & 0 \\ 0 & 0 \\ 20 & 0 \\ 0 & 20 \end{bmatrix} \begin{bmatrix} \delta_{e_c} \\ \delta_{f_c} \end{bmatrix} \tag{43}$$

A pitch pointing/vertical translation controller was proposed by Sobel and Shapiro [17] to decouple the pitch attitude and flight path responses. The zero entries in the short period eigenvectors are for pitch pointing (θ command with no coupling to γ) while the zero entry in the gamma mode eigenvector is for vertical translation (γ command with no coupling to θ or q). The desired eigenvectors are shown in Table I where we observe the zero entries which are chosen to obtain the required decoupling. The feedback gains described by Sobel and Shapiro [17] are for the outputs $y^T = [q,\ n_{zp},\ \gamma,\ \delta_e,\ \delta_f]$. However, for simplicity, we will use full state feedback in our new design. The full state feedback gain matrix is shown in Table II and it is obtained by applying a transformation to the feedback gain matrix described by Sobel and Shapiro [17]. The achievable eigenvectors are shown in Table III where we observe that the two desired zero entries in the short period eigenvectors have been exactly achieved and the desired zero entry in the gamma mode eigenvector is very small. Thus, we expect excellent decoupling in both the pitch pointing and vertical translation responses.

Table I. Desired eigenvectors

Short Period		Gamma Mode	Actuator Mode	Actuator Mode	
0	0	1	x	x	γ
1	x	0	x	x	q
x	1	x	x	x	α
x	x	x	1	x	δ_e
x	x	x	x	1	δ_f

Table II. Closed loop eigenvalues and control gains

Closed Loop Eigenvalues	Feedforward Gains		Feedback Gains (u=-Fx)				
Non-Robust Design:							
	γ_c	θ_c	γ	q	α	δ_e	δ_f
$-5.60 \pm j4.19$	-.375	-2.87	-3.25	-.891	-7.112	.526	.0840
-1.00	4.12	1.98	6.101	.898	10.02	-.420	-.102
-19.0							
-19.5							
Robust Design:							
	γ_c	θ_c	γ	q	α	δ_e	δ_f
$-3.62 \pm j5.18$	-.992	-1.62	-2.61	-.581	-5.31	.288	.0555
-2.66	10.93	-6.47	4.47	.0642	1.11	.0131	.0204
-19.0							
-19.5							
Robust Design with Eigenvector Constraints:							
	γ_c	θ_c	γ	q	α	δ_e	δ_f
$-3.62 \pm j5.14$	-1.11	-2.01	-3.12	-.606	-5.72	.291	.0543
-3.00	12.37	-1.83	10.54	.361	5.97	-.0239	.0342
-19.0							
-19.5							
Robust Design with Matrix Q:							
	γ_c	θ_c	γ	q	α	δ_e	δ_f
$-3.61 \pm j5.18$	-.890	-1.68	-2.57	-.552	-5.29	.259	.0540
-2.40	9.90	-5.76	4.14	-.243	.925	.331	.0363
-19.0							
-19.5							
Performance Robust Design:							
	γ_c	θ_c	γ	q	α	δ_e	δ_f
$-3.61 \pm j5.18$	-1.01	-1.68	-2.69	-.557	-5.31	.262	.0529
-2.71	11.14	-5.67	5.47	-.183	1.23	.290	.0476
-19.0							
-19.5							

Table III. Closed loop eigenvectors (normalized $\|x\|_{\infty}=1$)

	short period		gamma mode	actuator mode	actuator mode
Non-	.0000	.0000	.3089	-.0054	-.0135
Robust	.5954	-.7679	.0001	1.0000	.0343
Design	-.1339	.0369	-.3090	-.0472	.0118
	-.4502	-.2629	-.8656	.9317	-.0249
	1.0000	0.0000	1.0000	.0081	1.0000
Robust	-.0434	.0301	-.2668	-.0052	-.0137
Design	1.0000	.0000	-.0038	1.0000	.0631
	-.0472	-.1598	.2682	-.0475	.0104
	.0215	-.7004	.5802	.9316	.0029
	.2153	-.0044	1.0000	-.0060	1.0000
Robust	-.0200	.0077	-.1911	-.0053	-.0137
Design	1.0000	.0000	.0010	1.0000	.0599
Eigenvector	-.0716	-.1378	.1908	-.0474	.0106
Constraint	-.0685	-.7063	.3868	.9312	-.0002
	.5301	.6895	1.0000	-.0018	1.0000
Robust	-.0418	.0119	-.3711	-.0053	-.0137
Design	1.0000	.0000	.0008	1.0000	.0587
with	-.0486	-.1419	.3708	-.0473	.0106
matrix Q	-.0177	-.6719	.8377	.9325	-.0013
	.5985	.1743	1.0000	.0018	1.0000
Performance	-.0421	.0119	-.2528	-.0053	-.0137
Robust	1.0000	0.0000	-.0026	1.0000	.0599
Design	-.0485	-.1420	.2537	-.0473	.0106
	-.0180	-.6714	.5440	.9374	-.0001
	.6023	.1711	1.0000	.0027	1.0000

Finally, feedforward gains based upon Broussard and O'Brien's [23] command generator tracker are computed to obtain zero steady state error to a step command. Thus, the control law is given by

$$u = [\Omega_{22} + F\Omega_{12}]u_c - Fy \tag{44}$$

where

$$\begin{bmatrix} \Omega_{11} & \Omega_{12} \\ \Omega_{21} & \Omega_{22} \end{bmatrix} = \begin{bmatrix} A & B \\ H & 0 \end{bmatrix}^{-1}$$

and

$$u_c = \begin{bmatrix} \gamma_c \\ \theta_c \end{bmatrix}$$

and where the matrix H is chosen so that

$$\begin{bmatrix} \gamma \\ \theta \end{bmatrix} = Hx$$

The control law described by Eq.(44) consists of a feedforward part to achieve zero steady state error to a step command u_c and a feedback part to achieve the desired transient response.

The vertical translation and pitch pointing responses are shown in Figure 2. As expected, the decoupling is excellent for both cases. However, the design of Sobel and Shapiro [17] does not consider stability robustness when the aircraft is subject to linear time varying structured state space uncertainty.

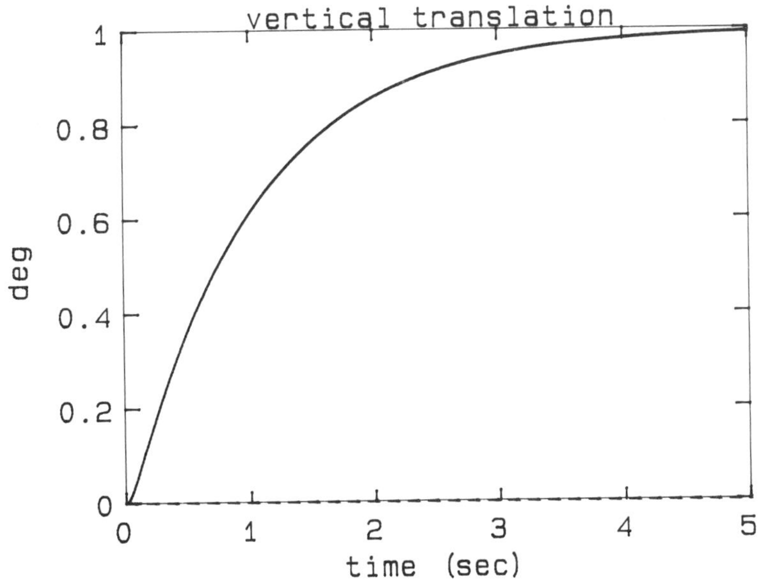

Fig. 2a. Non robust design: vertical translation response

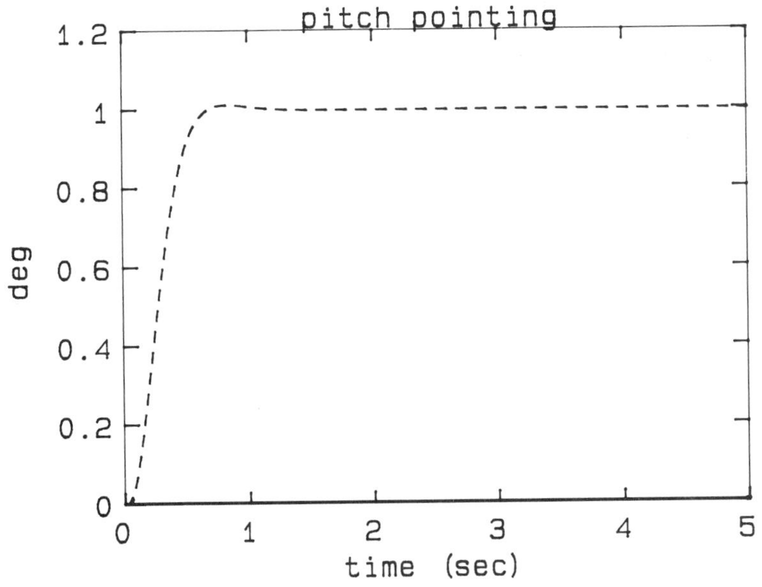

Fig. 2b. Non robust design: pitch pointing response

2. Robust Eigenstructure Assignment Design

In this section, we consider a robust eigenstructure assignment pitch pointing/vertical translation control law. First, to reduce the conservatism in Eq. (24), we use a constant similarity transformation to obtain the transformed state given by

$$\bar{x}(t) = [\theta,\ q,\ \alpha,\ \delta_e,\ \delta_f]^T \tag{45}$$

The utilization of such a transformation is valid because the stability of a linear time varying system is preserved when subjected to a constant similarity transformation [24]. The use of such a similarity transformation to reduce conservatism was used by Yedavalli and Liang [9] in connection with a Lyapunov approach to robustness. Yedavalli and Liang [9] indicate that there does not yet exist a systematic method for choosing the similarity transformation. We choose the transformation for our example to reduce the magnitude of the entries in the uncertainty matrix A_{max} with the conjecture that this will reduce the quantity on the right hand side of Eq. (24). In particular, our choice of the transformation yields a transformed system with a matrix A_{max} whose first row is zero.

In the transformed coordinates, the equation $\dot{\theta} = q$ is a physical relationship without uncertain parameters. Thus, the first row of A_{max} will contain all zeros. The matrices A_{max} and B_{max} are chosen to be

$$A_{max} = \begin{bmatrix} 0 & 0 & 0 & 0 & 0 \\ 0 & .0869 & 1.08 & .4313 & .158 \\ 0 & 0 & .1341 & .0169 & .0252 \\ 0 & 0 & 0 & 0.1 & 0 \\ 0 & 0 & 0 & 0 & 0.1 \end{bmatrix} \tag{46}$$

$$B_{max} = \begin{bmatrix} 0 & 0 \\ 0 & 0 \\ 0 & 0 \\ 0.1 & 0 \\ 0 & 0.1 \end{bmatrix} \tag{47}$$

This choice of A_{max} and B_{max} is for illustrative purposes. The chosen A_{max} and B_{max} correspond to maximum uncertainty of 10% in a_{22}, a_{25}, a_{33}, a_{34}, and a_{35}; 2.5% in a_{23} and a_{24}; 0.5% in the actuator parameters a_{44}, a_{55}, b_{41}, and b_{52}.

We emphasize that the transformed system is only used for computation of Eq.(24), while the original system of Eq.(43) is used for eigenvalue/eigenvector selection and gain computation. For the design proposed by Sobel and Shapiro [17], $\alpha = 1.0$ while the right hand side of Eq.(24) equals 4.28 which indicates that the sufficient condition for robust stability is not satisfied. We use a constrained optimization to solve

$$\min \lambda_{max} [(M^{-1})^{+}[A_{max} + B_{max}(FC)^{+}]M^{+}] - \alpha \tag{48}$$

while constraining some of the eigenvalues and eigenvectors. First, we choose to assign the actuator eigenvalues and eigenvectors to the same values as in the design presented by Sobel and Shapiro [17]. Then, an optimization is performed over a subset of the short period eigenvector parameters, short period eigenvalues, and the gamma mode eigenvalue. The term "eigenvector parameters" is the vector z_i which is defined in Eq.(38). This choice of optimization parameters is chosen by examining the gradient at the initial point corresponding to the design proposed by Sobel and Shapiro [17]. This gradient is given by

$$g = [\text{Re } z_{sp}(1), \text{Re } z_{sp}(2), \text{Im } z_{sp}(1), \text{Im } z_{sp}(2), \text{Re } \lambda_{sp}, \text{Im } \lambda_{sp}, z_{\gamma}(1), z_{\gamma}(2), \lambda_{\gamma}, z_{\delta_e}(1), z_{\delta_e}(2), \lambda_{\delta_e}, z_{\delta_f}(1), z_{\delta_f}(2), \lambda_{\delta_f}]^T \tag{49}$$

with numerical values given by

$$g = [-1.75, -0.389 \times 10^{-7}, -0.767, 0.393 \times 10^{-6}, -0.659, -0.461, -0.121, -0.0525, 1.47, 0.390, 0.109, 0.140, -0.0455, 0.181, 0.00567]^T \tag{50}$$

We choose to optimize over the parameters which correspond to $|g(i)| > 0.46$. The parameter vector for the optimization is chosen to be

$$p = [\text{Re } z_{sp}(1), \text{Im } z_{sp}(1), \text{Re } \lambda_{sp}, \text{Im } \lambda_{sp}, \lambda_{\gamma}]^T \tag{51}$$

Thus, during the optimization, the first entry (both real and imaginary parts) of the vector z_{sp} for the short period mode is allowed to change, the short period eigenvalues are allowed to change subject to the constraints $-7.6 \le Re[\lambda_{sp}] \le -3.6$, $3.2 \le Im[\lambda_{sp}] \le 5.2$, and the gamma mode eigenvalue is allowed to change subject to the constraint $-2.65 \le \lambda_{\gamma} \le -0.5$. Each time λ_{γ} is changed, its eigenvector is computed as the orthogonal projection of the desired gamma mode eigenvector onto the current achievable gamma mode subspace. In this way, the zero entry in the gamma mode eigenvector might be achieved in which case the vertical translation decoupling may be nearly the same as in the design shown by Sobel and Shapiro [17].

The constrained optimization is performed as a sequence of ten unconstrained optimizations by using IMSL subroutine ZXMIN [25] which implements a variable metric method. The first unconstrained optimization is initialized at the design shown by Sobel and Shapiro [17] with the variable ε in Eq.(39) set equal to 0.1. The variable ε is reduced by a factor of ten each time a new unconstrained minimization is performed.

When the optimization is complete $\alpha = 2.65$ and the right hand side of Eq.(24) equals 2.57; thus, the sufficient condition for robust stability is satisfied. The closed loop eigenvectors are shown in Table III from which we observe that the zero entry in the gamma mode eigenvector has been achieved. This suggests that the vertical translation decoupling will be the same as in the design shown by Sobel and Shapiro [17]. However, the entries in the short period eigenvectors which were zero in the design shown by Sobel and Shapiro [17] are now non-zero. This suggests that the pitch pointing decoupling will not be as good as in the design of Sobel and Shapiro [17]. The closed loop eigenvalues and feedback gains are shown in Table II. Observe that the short period eigenvalues have moved to the boundary of the constraint region corresponding to the largest settling time and smallest damping ratio. The gamma mode eigenvalue has moved to the leftmost point in its constraint interval which is to be expected since λ_γ appears explicitly in Eq.(24). Also observe that the magnitude of the feedback gains has been reduced. The vertical translation and pitch pointing responses are shown in Figure 3. As expected, the vertical translation decoupling is virtually identical to the design of Sobel and Shapiro [17] while the pitch pointing decoupling has been degraded.

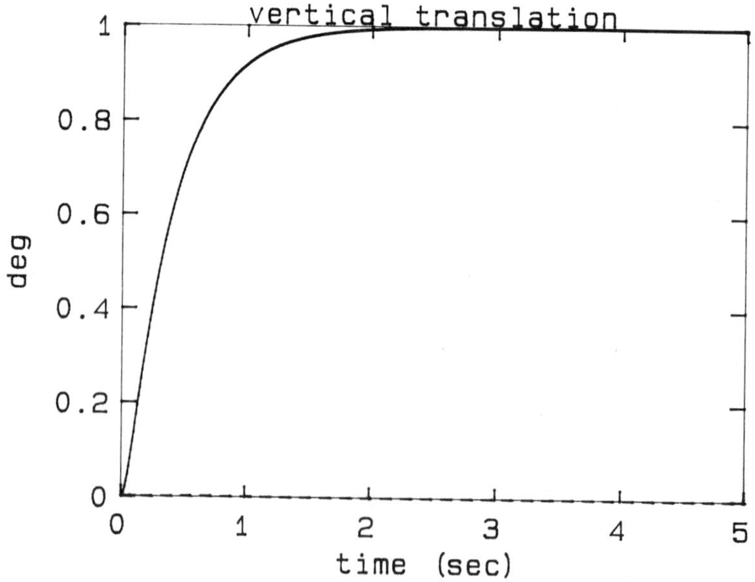

Fig. 3a. Robust design: vertical translation response

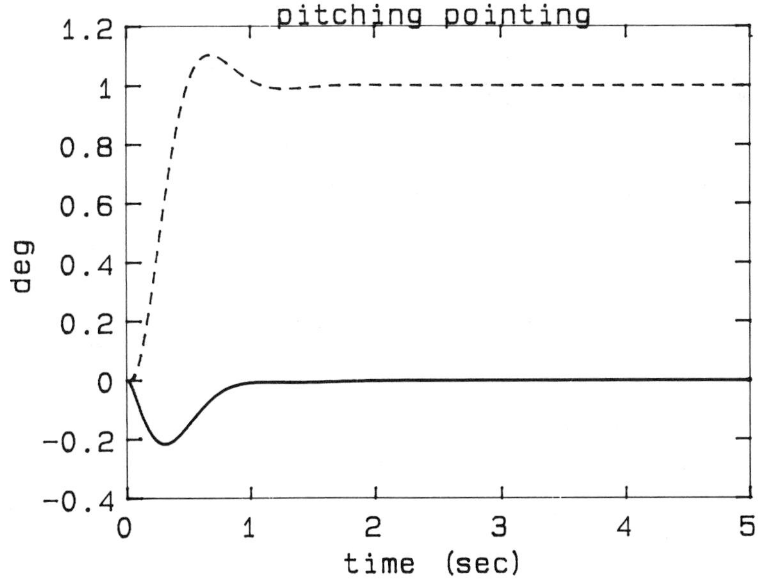

Fig. 3b. Robust design: pitch pointing response

The first column of Table IV shows the maximum absolute value of gamma for the pitch pointing responses from which we can observe the degradation in the pitch pointing decoupling. Table IV also shows that the new robust eigenstructure assignment design exhibits a significantly improved minimum singular value of the return difference matrix at the inputs and an improved condition number of the closed loop modal matrix as compared to the design of Sobel and Shapiro [17]. We also observe that the minimum singular value of the return difference matrix at the outputs has shown a slight improvement although it still remains small. We conclude that for our example, the robustness optimization yields a significant improvement in the multivariable stability margins at the aircraft inputs while having little effect on the margins at the aircraft outputs. Nevertheless, robust stability is guaranteed for the structured state space uncertainty which is represented by the bounds described by the matrices A_{max} and B_{max}.

Table IV. Comparison of Robustness Measures

	Max $\lvert\gamma\rvert$ $\theta_c = 1$	Min $\underline{\sigma}(I+FG)$	Min $\underline{\sigma}(I+GF)$	cond(M)
Non-Robust Design	6.66×10^{-4}	0.1983	0.0491	56.00
Robust Design	0.2171	0.5021	0.0563	27.30
Robust Design Eigenvector Constraints	0.0787	0.2802	0.0408	36.80
Robust Design with matrix Q	0.1617	0.4244	0.0516	27.11
Performance Robust Design	0.1596	0.4143	0.0504	27.84

3. Robust Design with Eigenvector Constraints

In this section, we place a constraint on the first entry of the complex conjugate short period eigenvectors.

This will allow us to place additional emphasis on the mode decoupling which is required for the pitch pointing maneuver. We remark that in the non-robust design we achieved mode decoupling by choosing each achievable eigenvector v_i^a to be close to its corresponding desired eigenvector v_i^d in a 2-norm sense. However, in the robust design we optimize over the short period eigenvector parameters instead of choosing v_{sp}^a to be close to v_{sp}^d in a 2-norm sense. Hence, the need for additional eigenvector constraints when the designer chooses to emphasize mode decoupling. The constraints are chosen to be $|\mathrm{Re}\ v_{sp}(1)| \le 0.01$ and $|\mathrm{Im}\ v_{sp}(1)| \le 0.01$ where the eigenvectors are normalized to unit length in the 2-norm sense. A solution could not be obtained for the eigenvalue constraints which were used in the previous design. Therefore, the gamma mode eigenvalue constraint is relaxed to become $-3.0 \le \lambda_\gamma \le -0.5$ which will have the effect of moving the gamma mode eigenvalue farther into the left half of the complex plane. The new closed loop eigenvectors are shown in Table III where we observe that the real and imaginary parts of the first entry of the short period eigenvectors are significantly smaller than in the previous design. Thus, we should expect that the pitch pointing response will be improved. The vertical translation and pitch pointing responses are shown in Figure 4. We observe that the coupling between γ and θ has been reduced by approximately 50% as compared to the previous design. However, we observe from Table II that the feedback gains are larger in magnitude. Also, from Table IV we observe that we no longer have a significant improvement in the minimum singular value of the return difference matrix at the inputs. Therefore, we observe a tradeoff between stability robustness and pitch pointing performance.

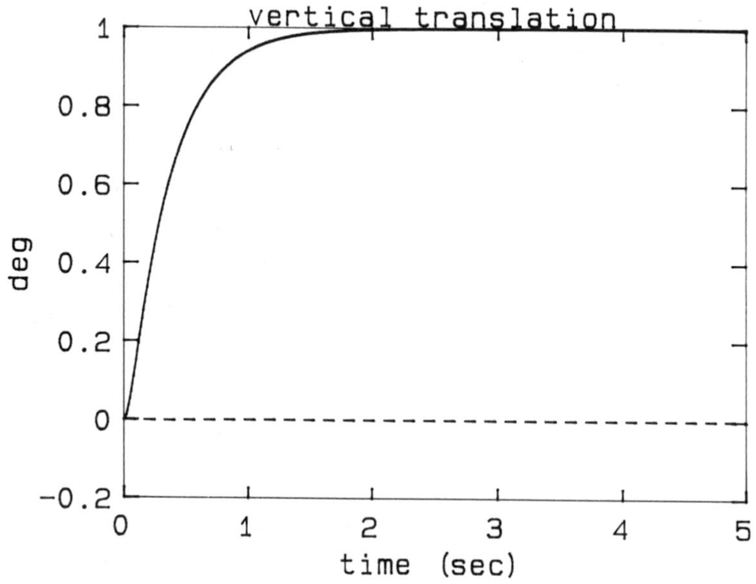

Fig. 4a. Robust design with eigenvector constraints: vertical translation response

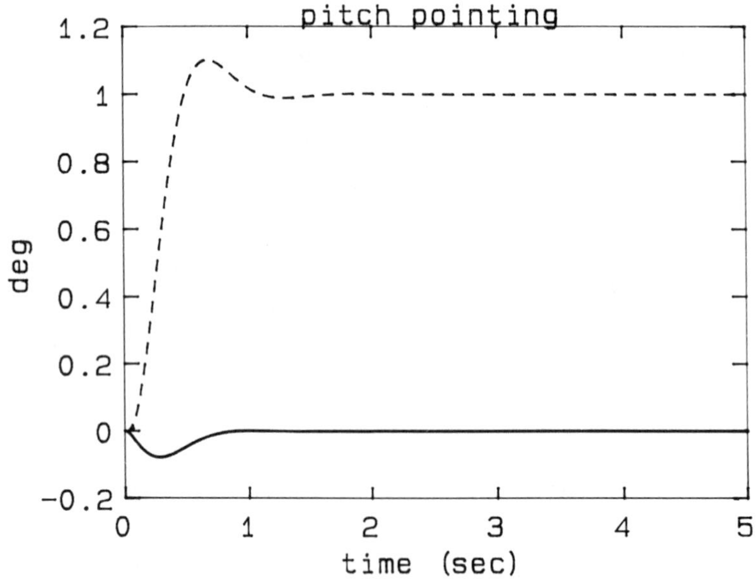

Fig. 4b. Robust design with eigenvector constraints: pitch pointing response

4. Robust Design with the Matrix Q

In this section, we remove the additional short period eigenvector constraints, but we introduce a unitary weighting matrix Q in order to reduce the conservatism in the robust stability condition. In this case, the sufficient condition is given by Eq.(13) which is repeated below:

$$\alpha > \kappa_2(Q) \cdot \| [(MDQ)^{-1}]^+ [A_{max} + B_{max}(FC)^+][MDQ]^+ \|_2 \tag{52}$$

where

$$\alpha = -\max_i \text{Re}\ \lambda_i(A + BFC)$$

and $\kappa_2(Q) = \|Q^{-1}\|_2 \cdot \|Q\|_2$ is the 2-norm condition number of the matrix Q.

The computation of the matrix D is based upon Eq.(22) of lemma 2. This will yield the optimal matrix D when Q is the identity matrix. Then, the matrix Q is computed by using Eq.(25) of conjecture 1. Thus, we utilize both a real, positive, diagonal weighting matrix D and a unitary weighting matrix Q.

Due to the use of the matrix Q, we can tighten the constraint on the gamma eigenvalue to $-2.4 \le \lambda_\gamma \le -0.5$. Thus, we obtain a solution with $\lambda_\gamma = -2.4$ which is to be compared with the previous solution of $\lambda_\gamma = -2.65$. So we do not need to move the gamma mode eigenvalue as far left in the complex plane as before. The vertical translation and pitch pointing responses are shown in Figure 5. The vertical translation response exhibits slightly less coupling than the first robust design of Figure 3 which did not utilize either the unitary matrix Q or explicit eigenvector constraints on $|\text{Re}\ v_{sp}(1)|$ and $|\text{Im}\ v_{sp}(1)|$.

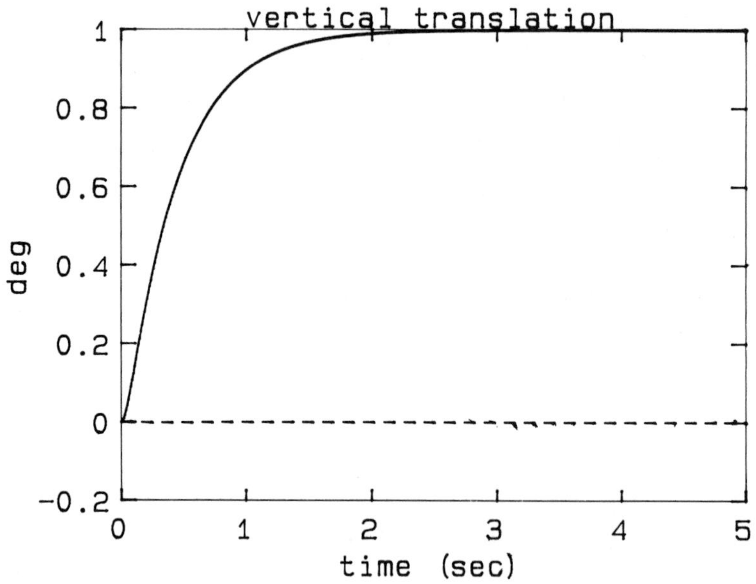

Fig. 5a. Robust design with matrix Q: vertical translation response

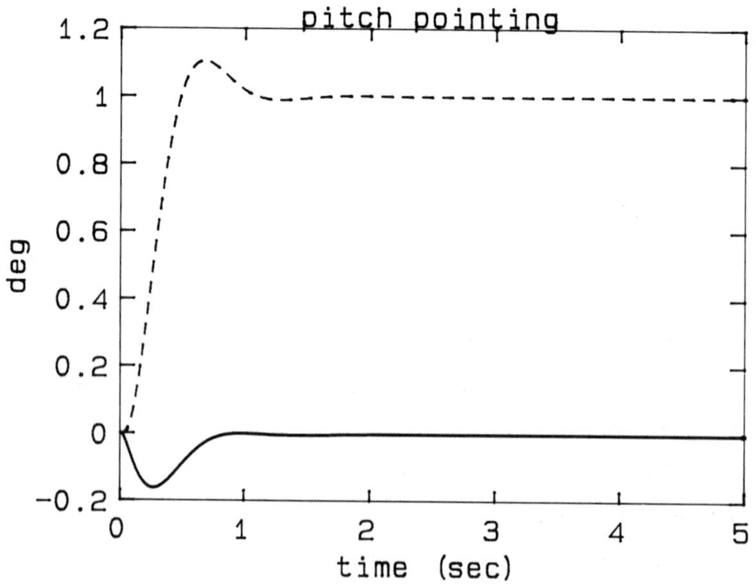

Fig. 5b. Robust design with matrix Q: pitch pointing response

5. Performance Robust Design

Finally, we design a controller by optimizing the performance robustness condition of Eq.(27). We choose a = -0.25 and θ = 8.63 deg which corresponds to a region in Figure 1 with $\zeta \geq 0.15$ and $t_s \leq 4$ sec. The quantity to be minimized is given by

$$\kappa_2(Q)\cdot\|[(MDQ)^{-1}]^+[A_{max}+B_{max}(FC)^+][MDQ]^+\|_2-\min(\alpha_1,\alpha_2) \qquad (53)$$

where

$$\alpha_1 = -\max_i\ [\mathrm{Re}\ \lambda_i] - 0.25$$

$$\alpha_2 = -\max\ [\mathrm{Re}\ e^{-j(8.63)(\pi/180)}\lambda_i]$$

For the design of Sobel and Shapiro [17], we find that $\min(\alpha_1, \alpha_2) = 0.75$ while the right hand side of Eq.(27) equals 4.546. Thus, the sufficient condition for performance robustness is not satisfied. The constraint for the gamma mode eigenvalue is chosen to be $-2.70 \leq \lambda_\gamma \leq -0.5$. After optimizing Eq.(53) we find that $\min(\alpha_1, \alpha_2) = 2.46$ while the right hand side of Eq.(27) equals 2.393. Thus, the sufficient condition for performance robustness is satisfied. The vertical translation and pitch pointing responses are shown in Figure 6. We observe that these responses are almost identical to the robust stability responses which were shown in Figure 5. However, we now guarantee that the closed loop eigenvalues remain in the performance region for all time invariant uncertainty which satisfies the A_{max} and B_{max} bounds. We observe from Table IV that this design has the smallest coupling of all the robust designs which do not use constraints on the first entry of the complex conjugate short period eigenvectors.

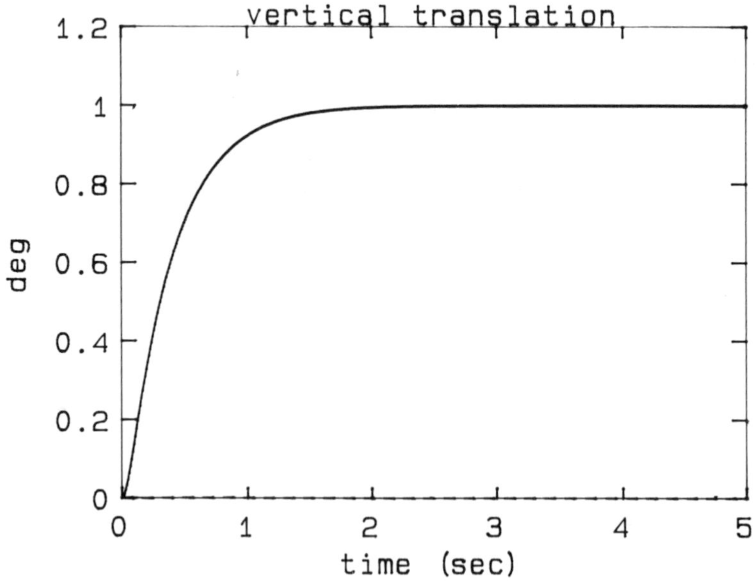

Fig. 6a. Performance robust design: vertical translation response

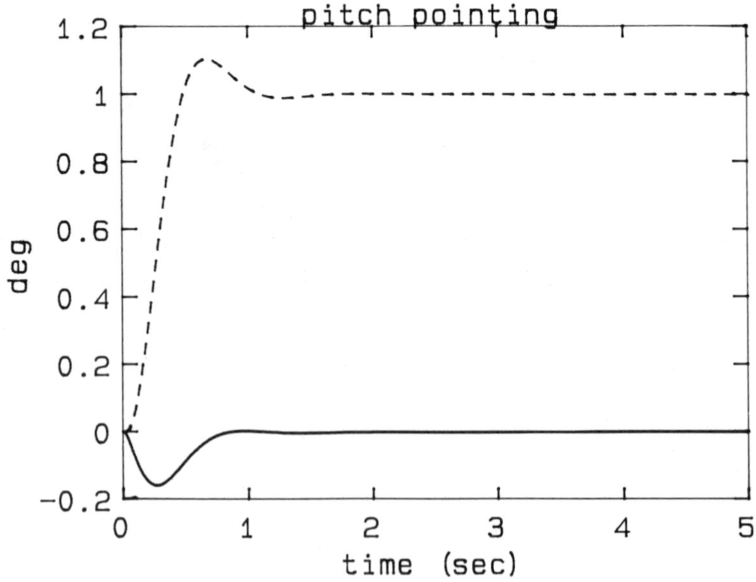

Fig. 6b. Performance robust design: pitch pointing response

VI. CONCLUSIONS

A new sufficient condition has been presented for the stability of a linear time invariant system subjected to time varying structured state space uncertainty. Robust stability is ensured provided that the nominal closed loop eigenvalues lie to the left of a vertical line in the complex plane which is determined by a norm involving the structure of the uncertainty and the nominal closed loop eigenvector matrix.

The sufficient condition was extended to the performance robustness of a linear time invariant system subjected to time invariant structured state space uncertainty. The performance robustness sufficient condition ensures that the eigenvalues of the uncertain closed loop system lie within a settling time/damping ratio region. The conservatism of the robustness conditions is addressed by introducing a real positive diagonal weighting matrix together with a unitary weighting matrix.

An example was presented which illustrates the design of a robust eigenstructure assignment controller for a pitch pointing/vertical translation maneuver of the AFTI F-16 aircraft. This example illustrates the tradeoff between robustness and nominal pitch pointing performance especially in the coupling between the pitch attitude command and the flight path angle transient response. For the particular example considered here, the robust eigenstructure assignment design exhibits both an improved minimum singular value of the return difference matrix at the inputs and an improved condition number of the closed loop modal matrix.

VII. APPENDIX: PROOF OF LEMMAS, THEOREMS, AND COROLLARIES

Let $\| \ \|$ denote the vector or matrix 2-norm.

A. Proof of Lemma 1

Use the result given by Kouvaritakis and Latchman [26] that for any matrix $A \in \mathbb{C}^{n \times m}$

$$\|A\| \leq \|A^+\| \tag{A1}$$

Then

$$\|(MDQ)^{-1}[\Delta A(t)+\Delta B(t)FC]MDQ\| \leq$$
$$\|\{(MDQ)^{-1}[\Delta A(t)+\Delta B(t)FC]MDQ\}^+\| \tag{A2}$$

$$= \sup \frac{\|\{(MDQ)^{-1}[\Delta A(t)+\Delta B(t)FC]MDQ\}^+ x\|}{\|x\|} \tag{A3}$$

$$\leq \sup \frac{\|\{(MDQ)^{-1}[\Delta A(t)+\Delta B(t)FC]MDQ\}^+ x^+\|}{\|x^+\|} \tag{A4}$$

$$\leq \sup \frac{\|[(MDQ)^{-1}]^+[\Delta A(t)+\Delta B(t)FC]^+ (MDQ)^+ x^+\|}{\|x^+\|} \tag{A5}$$

$$\leq \sup \frac{\|[(MDQ)^{-1}]^+[A_{max}+B_{max}(FC)^+](MDQ)^+ x^+\|}{\|x^+\|} \tag{A6}$$

$$= \|[(MDQ)^{-1}]^+[A_{max} + B_{max}(FC)^+](MDQ)^+\| \tag{A7}$$

B. Proof of Theorem 1

$$\dot{x}(t)=(A+BFC)x(t)+[\Delta A(t)+\Delta B(t)FC]x(t) \tag{A8}$$

Define

$$A_c=A+BFC$$

$$\Delta A_c(t)=\Delta A(t)+\Delta B(t)FC$$

$$(\Delta A_c)^+=A_{max}+B_{max}(FC)^+$$

Let M be a modal matrix of A+BFC. Use the similarity transformation MDQ where D is real, positive, diagonal and where Q is nonsingular to obtain

$$x(t)=MDQz(t) \tag{A9}$$

Then

$$\|x(t)\|\le\|MDQ\|\cdot\|z(t)\| \tag{A10}$$

Since $\|MDQ\|<\infty$, it follows that

$$\|z(t)\| \to 0 \text{ implies } \|x(t)\| \to 0 \tag{A11}$$

So it is sufficient to consider $\|z(t)\|$.

Apply the similarity transformation to (A8) and note that MD is also a modal matrix for (A+BFC) to obtain

$$\dot{z}(t)=Q^{-1}D^{-1}M^{-1}A_cMDQz(t)+Q^{-1}D^{-1}M^{-1}\Delta A_c(t)MDQz(t) \tag{A12}$$

which has a solution described by

$$z(t)=\exp(Q^{-1}\Lambda Qt)z(0)+$$

$$\int_0^t \exp[Q^{-1}\Lambda(t-\tau)Q]Q^{-1}D^{-1}M^{-1}\Delta A_c(\tau)MDQz(\tau)d\tau \qquad (A13)$$

where Λ is a diagonal matrix with entries $\lambda_i(A+BFC)$. Taking norms in Eq.(A13) and using $\exp[Q^{-1}\Lambda Qt)] = Q^{-1}\exp(\Lambda t)Q$ yields

$$\|z(t)\|\le\|Q^{-1}\|\cdot\|\exp(\Lambda t)\|\cdot\|Q\|\cdot\|z(0)\|+$$

$$\int_0^t \|Q^{-1}\|\cdot\|\exp[\Lambda(t-\tau)]\|\cdot\|Q\|\cdot\|Q^{-1}D^{-1}M^{-1}\Delta A_c(\tau)MDQ\|\cdot\|z(\tau)\|d\tau \qquad (A14)$$

Use $\|\exp(\Lambda t)\|\le\exp(-\alpha t)$ where $\alpha=-\max_i \text{Re}[\lambda_i(A+BFC)]$

and use $\kappa_2(Q)=\|Q\|\cdot\|Q^{-1}\|$ and $Q^{-1}D^{-1}M^{-1} = (MDQ)^{-1}$ to obtain

$$\|z(t)\| \le \kappa_2(Q)\cdot\exp(-\alpha t)\cdot\|z(0)\| +$$

$$\int_0^t \kappa_2(Q)\cdot\exp[-\alpha(t-\tau)]\cdot\|(MDQ)^{-1}\Delta A_c(\tau)MDQ\|\cdot\|z(\tau)\|d\tau \qquad (A15)$$

Multiply both sides by $\exp(\alpha t)$ to obtain

$$\|z(t)\|\exp(\alpha t) \le \kappa_2(Q)\cdot\|z(0)\| +$$

$$\int_0^t \kappa_2(Q)\cdot\|(MDQ)^{-1}\Delta A_c(\tau)MDQ\|\cdot\exp(\alpha\tau)\cdot\|z(\tau)\|d\tau \qquad (A16)$$

Use the Gronwall lemma, which is described in many textbooks among which is [27], to obtain

$$\|z(t)\|\cdot\exp(\alpha t) \le$$

$$\kappa_2(Q)\cdot\|z(0)\|\cdot\exp\left[\int_0^t \kappa_2(Q)\cdot\|(MDQ)^{-1}\Delta A_c(\tau)MDQ\|d\tau\right] \quad (A17)$$

Use lemma 1 to obtain

$$\|z(t)\|\cdot\exp(\alpha t) \le$$

$$\kappa_2(Q)\cdot\|z(0)\|\cdot\exp[\kappa_2(Q)\cdot\|[(MDQ)^{-1}]^+(\Delta A_c)^+(MDQ)^+\|\cdot t] \quad (A18)$$

or

$$\|z(t)\| \le$$

$$\kappa_2(Q)\cdot\|z(0)\|\cdot\exp[(-\alpha+\kappa_2(Q)\cdot\|[(MDQ)^{-1}]^+(\Delta A_c)^+(MDQ)^+\|)t] \quad (A19)$$

Thus, $\|z(t)\| \to 0$ if

$$\alpha > \kappa_2(Q)\cdot\|[(MDQ)^{-1}]^+(\Delta A_c)^+(MDQ)^+\| \quad (A20)$$

C. Proof of Corollary 1

Substitute the following equation for the feedback gain as described by Kautsky et. al. [22] into Eq.(13).

$$F=V\Sigma^{-1}U_0^T(M\Lambda M^{-1} - A) \quad (A21)$$

Then, the result follows from theorem 1.

D. Proof of Corollary 2

The r assignable eigenvalues and their corresponding eigenvectors satisfy the equations described by

$$(A+BFC)v_i = \lambda_i v_i; \quad i=1,2,\ldots,r \quad (A22)$$

Combining the r equations from Eq.(A22) yields

$$(A+BFC)M_r = M_r\Lambda_r \tag{A23}$$

Rearrange the matrices in Eq.(A23) to obtain

$$BFCM_r = M_r\Lambda_r - AM_r \tag{A24}$$

Substitute the singular value decompositions of B and CM_r into Eq.(A24) obtain

$$[U_o \ U_1]\begin{bmatrix}\Sigma V^T \\ 0\end{bmatrix} F[U_{ro} \quad U_{r1}]\begin{bmatrix}\Sigma_r V_r^T \\ 0\end{bmatrix} = M_r\Lambda_r - AM_r \tag{A25a}$$

$$U_o\Sigma V^T F U_{ro}\Sigma_r V_r^T = M_r\Lambda_r - AM_r \tag{A25b}$$

Taking the required inverses and recalling that U_o, U_{ro}, V, and V_r are unitary, we obtain

$$F = V\Sigma^{-1}U_o^T(M_r\Lambda_r - AM_r)V_r\Sigma_r^{-1}U_{ro}^T \tag{A26}$$

Recall that Eq.(13) from theorem 1 is given by

$$\alpha > \kappa_2(Q)\cdot\|[(MDQ)^{-1}]^+[A_{max} + B_{max}(FC)^+](MDQ)^+\| \tag{A27}$$

Finally, substituting Eq.(A26) into Eq.(A27) yields the desired result.

E. Proof of Lemma 3

The uncertain closed loop system may be written as

$$\dot{x}(t) = (A_c + \Delta A_c)x(t) \tag{A28}$$

and the transformed uncertain closed loop system may be written as

$$\dot{\tilde{x}}(t) = e^{-j\theta}(A_c + \Delta A_c - aI)\tilde{x}(t) \tag{A29}$$

Then, the result follows by using lemma 1 of reference 13 with $E=A_c+\Delta A_c$.

F. Proof of Lemma 4

Consider the transformed uncertain closed loop system which is described by

$$\dot{\tilde{x}}(t) = \tilde{A}_c \tilde{x}(t) + \Delta\tilde{A}_c \tilde{x}(t) \qquad \text{(A30)}$$

where

$$\tilde{A}_c = e^{-j\theta}(A_c - aI) \text{ and } \Delta\tilde{A}_c = e^{-j\theta}\Delta A_c.$$

Let λ_i be the i-th eigenvalue of A_c and let $\tilde{\lambda}_i$ be the corresponding eigenvalue of $\tilde{A}_c$.

Then, $\det[\tilde{\lambda}_i I - \tilde{A}_c] = 0$

$$\Rightarrow \det[\tilde{\lambda}_i I - e^{-j\theta}(A_c - aI)] = 0$$

$$\Rightarrow \det[(\tilde{\lambda}_i I + e^{-j\theta}aI) - e^{-j\theta}A_c] = 0$$

$$\Rightarrow \det[e^{-j\theta}I] \det[e^{j\theta}(\tilde{\lambda}_i + e^{-j\theta}a)I - A_c] = 0$$

$$\Rightarrow \lambda_i = e^{j\theta}\,\tilde{\lambda}_i + a$$

$$\Rightarrow \tilde{\lambda}_i = e^{-j\theta}\,(\lambda_i - a)$$

$$\Rightarrow \tilde{\Lambda} = e^{-j\theta}(\Lambda - aI)$$

Let $A_c = M\Lambda M^{-1}$. Since $\tilde{A}_c = e^{-j\theta}(A_c - aI)$, it follows that

$$\begin{aligned}\tilde{A}_c &= e^{-j\theta}(M\Lambda M^{-1} - aI)\\ &= e^{-j\theta}M(\Lambda - aI)M^{-1}\\ &= M[e^{-j\theta}(\Lambda - aI)]M^{-1}\\ &= M\tilde{\Lambda}M^{-1}\end{aligned}$$

Therefore, $\tilde{A}_c = M\tilde{\Lambda}M^{-1}$ and $\tilde{\Lambda} = M^{-1}\tilde{A}_c M$.

G. Proof of Lemma 5

Use the result given by Kouvaritakis and Latchman [26] that for any matrix $A \in \mathbb{C}^{n \times m}$

$$\|A\| \le \|A^{+}\| \tag{A31}$$

Thus,

$$\|M^{-1} \Delta\tilde{A}_c M\|_2 \le \|[M^{-1} \Delta\tilde{A}_c M]^{+}\| \tag{A32}$$

$$= \sup \frac{\|[M^{-1} \Delta\tilde{A}_c M]^{+} x\|}{\|x\|} \tag{A33}$$

$$\le \sup \frac{\|[M^{-1} \Delta\tilde{A}_c M]^{+} x^{+}\|}{\|x^{+}\|} \tag{A34}$$

$$\le \sup \frac{\|[M^{-1} e^{-j\theta} \Delta A_c M]^{+} x^{+}\|}{\|x^{+}\|} \tag{A35}$$

$$\le \sup \frac{\|\{M^{-1} e^{-j\theta} [\Delta A + \Delta BFC] M\}^{+} x^{+}\|}{\|x^{+}\|} \tag{A36}$$

$$\le \sup \frac{|e^{-j\theta}| \cdot \|\{(M^{-1})^{+} [\Delta A + \Delta BFC]^{+} M^{+}\} x^{+}\|}{\|x^{+}\|} \tag{A37}$$

$$\leq \sup \frac{\|(M^{-1})^{+}[A_{max} + B_{max}(FC)^{+}]M^{+} x^{+}\|}{\|x^{+}\|} \quad \text{(A38)}$$

$$= \|(M^{-1})^{+} [A_{max} + B_{max}(FC)^{+}] M^{+}\| \quad \text{(A39)}$$

H. Proof of Theorem 3

The transformed uncertain closed loop system is described by

$$\dot{\tilde{x}}(t) = \tilde{A}_c\tilde{x}(t) + \Delta\tilde{A}_c\tilde{x}(t) \quad \text{(A40)}$$

Let $\tilde{x}(t)=MDQz(t)$ where M and MD are both modal matrices of A_c.

Then

$$\|\tilde{x}(t)\| \leq \|MDQ\| \cdot \|z(t)\| \quad \text{(A41)}$$

Since $\|MDQ\|<\infty$, it follows that

$$\|z(t)\| \to 0 \text{ implies } \|\tilde{x}(t)\| \to 0 \quad \text{(A42)}$$

So it is sufficient to consider $\|z(t)\|$.

Apply the similarity transformation to (A40) to obtain

$$\dot{z}(t)=Q^{-1}D^{-1}M^{-1}\tilde{A}_cMDQz(t)+Q^{-1}D^{-1}M^{-1}\Delta\tilde{A}_c(t)MDQz(t) \quad \text{(A43)}$$

Use lemma 4 to obtain

$$z(t) = Q^{-1}\tilde{\Lambda}Qz(t) + Q^{-1}D^{-1}M^{-1}\Delta\tilde{A}_cMDQz(t) \quad \text{(A44)}$$

which has a solution described by

$$z(t)=\exp(Q^{-1}\tilde{\Lambda}Qt)z(0)+\int_0^t \exp[Q^{-1}\tilde{\Lambda}(t-\tau)Q]Q^{-1}D^{-1}M^{-1}\Delta\tilde{A}_c(\tau)MDQz(\tau)d\tau \qquad (A45)$$

Taking norms in Eq.(A45) and using $\exp[Q^{-1}\tilde{\Lambda}Qt] = Q^{-1}\exp(\tilde{\Lambda}t)Q$ yields

$$\|z(t)\|\le\|Q^{-1}\|\cdot\|\exp(\tilde{\Lambda}t)\|\cdot\|Q\|\cdot\|z(0)\|+\int_0^t\|Q^{-1}\|\cdot\|\exp[\tilde{\Lambda}(t-\tau)]\|\cdot\|Q\|\cdot\|Q^{-1}D^{-1}M^{-1}\Delta\tilde{A}_c(\tau)MDQ\|\cdot\|z(\tau)\|d\tau \qquad (A46)$$

Use $\|\exp(\tilde{\Lambda}t)\|\le\exp(-\alpha t)$

where

$$\alpha = -\max_i \mathrm{Re}[\lambda_i(\tilde{A}_c)] = -\max_i \mathrm{Re}[e^{-j\theta}(\lambda_i - a)]$$

and use

$$\kappa_2(Q)=\|Q\|\cdot\|Q^{-1}\| \text{ and } Q^{-1}D^{-1}M^{-1} = (MDQ)^{-1}$$

to obtain

$$\|z(t)\| \le \kappa_2(Q)\cdot\exp(-\alpha t)\cdot\|z(0)\| + \int_0^t\kappa_2(Q)\cdot\exp[-\alpha(t-\tau)]\cdot\|(MDQ)^{-1}\Delta\tilde{A}_c(\tau)MDQ\|\cdot\|z(\tau)\|d\tau \qquad (A47)$$

Multiply both sides by $\exp(\alpha t)$ to obtain

$$\|z(t)\|\exp(\alpha t) \le \kappa_2(Q)\cdot\|z(0)\| + \int_0^t \kappa_2(Q)\cdot\|(MDQ)^{-1}\Delta\tilde{A}_c(\tau)MDQ\|\cdot\exp(\alpha\tau)\cdot\|z(\tau)\|d\tau \tag{A48}$$

Use the Gronwall lemma, which is described in many textbooks among which is [27], to obtain

$$\|z(t)\|\cdot\exp(\alpha t) \le \kappa_2(Q)\cdot\|z(0)\|\cdot\exp\left[\int_0^t \kappa_2(Q)\cdot\|(MDQ)^{-1}\Delta\tilde{A}_c(\tau)MDQ\|d\tau\right] \tag{A49}$$

Use lemma 5 to obtain

$$\|z(t)\|\cdot\exp(\alpha t) \le \kappa_2(Q)\cdot\|z(0)\|\cdot\exp[\kappa_2(Q)\cdot\|[(MDQ)^{-1}]^+[A_{max}+B_{max}(FC)^+](MDQ)^+\|\cdot t] \tag{A50}$$

or

$$\|z(t)\| \le \kappa_2(Q)\|z(0)\|\exp[(-\alpha+\kappa_2(Q)\|[(MDQ)^{-1}]^+[A_{max}+B_{max}(FC)^+](MDQ)^+\|)t] \tag{A51}$$

Thus, $\|z(t)\| \to 0$ if

$$\alpha > \kappa_2(Q)\cdot\|[(MDQ)^{-1}]^+[A_{max} + B_{max}(FC)^+](MDQ)^+\| \tag{A52}$$

Then, z(t) is asymptotically stable

$\Rightarrow \tilde{x}(t)$ is asymptotically stable

$\Rightarrow \tilde{A}_c + \Delta\tilde{A}_c$ has all its eigenvalues in the open left half complex plane. Then, by using lemma 3, it follows that $A_c+\Delta A_c$ has all its eigenvalues in the H region.

I. Proof of Corollary 3

Using theorem 3, it follows that the eigenvalues of the uncertain closed loop system described by Eq.(11) are in H_k if

$$\alpha_k > \kappa_2(Q)\cdot\|[(MDQ)^{-1}]^+[A_{max} + B_{max}(FC)^+](MDQ)^+\| \qquad \text{(A53)}$$

The eigenvalues of the uncertain closed loop system described by Eq.(11) are in the region R if they are simultaneously in each H_k; $k=1,2,...,k_1$. This will be true if Eq.(33) is satisfied.

J. Proof of Theorem 2

From corollary 3 we have

$$\min[\alpha_1, \alpha_2, \alpha_3] > \kappa_2(Q)\cdot\|[(MDQ)^{-1}]^+[A_{max} + B_{max}(FC)^+](MDQ)^+\| \qquad \text{(A54)}$$

where

$$\alpha_k = -\max_i \mathrm{Re}[e^{-j\theta_k}(\lambda_i - a_k)]$$

Thus,

$$\alpha_1 = -\max_i \mathrm{Re}[e^{-j0}(\lambda_i - a_1)]$$

$$= -\max_i \mathrm{Re}[\lambda_i - a_1]$$

$$= -\max_i [\mathrm{Re}\ \lambda_i] + a_1$$

Also,

$$\alpha_2 = -\max_i \mathrm{Re}[e^{-j\theta_2}(\lambda_i - 0)]$$

$$= -\max_i \mathrm{Re}[e^{-j\theta_2}\ \lambda_i]$$

$$= -\max_i [\mathrm{Re}(\lambda_i)\cos\theta + \mathrm{Im}(\lambda_i)\sin\theta]$$

and

$$\alpha_3 = -\max_i \mathrm{Re}[e^{j\theta_2}(\lambda_i - 0)]$$

$$= -\max_i \mathrm{Re}[e^{j\theta_2}\ \lambda_i]$$

$$= -\max_i [\mathrm{Re}(\lambda_i)\cos\theta - \mathrm{Im}(\lambda_i)\sin\theta]$$

Since λ_i are the eigenvalues of the real matrix A_c, then the λ_i are either real or they occur in complex conjugate pairs. If the λ_i are real, then $\mathrm{Im}(\lambda_i)=0$ and $\alpha_2=\alpha_3$. Consider the complex pair $\lambda_i = a_i+jb_i$ and $\lambda_{i+1} = \bar{\lambda}_i = a_i-jb_i$. In this case,

$$\alpha_2 = -\max_i \{a_i\cos\theta + b_i\sin\theta,\ a_i\cos\theta - b_i\sin\theta\} \tag{A55}$$

$$\alpha_3 = -\max_i \{a_i\cos\theta - b_i\sin\theta,\ a_i\cos\theta + b_i\sin\theta\} \tag{A56}$$

So it follows that $\alpha_2=\alpha_3$. In general, α_2 and α_3 are the maximum of sets with the same members which appear in a different order. Thus, the maximum of the two sets is equal and $\alpha_2=\alpha_3$.

ACKNOWLEDGMENT

This work was supported by the Air Force Office of Scientific Research/AFSC, United States Air Force under contract F49620-88-C-0053 and by computer resources provided by the City University of New York/University Computer Center.

REFERENCES

1. J.C. Doyle and G. Stein, "Multivariable Feedback Design: Concepts for Classical/Modern Synthesis", *IEEE Transactions on Automatic Control* **AC-26**, pp. 4-16 (1981).

2. V.L. Kharitonov, "Asymptotic Stability of an Equilibrium Position of a Family of Systems of Linear Differential Equations", *Differentsyalnye Uravnenya* **14**(11), pp.1483-1485 (1978).

3. I.R. Petersen, "A Class of Stability Regions for Which a Kharitonov Like Theorem Holds", *Proceedings of the 26th IEEE Conference on Decision and Control*, Los Angeles CA, pp. 440-444 (1987).

4. A.C. Bartlett, C.V. Hollot, and H. Lin, "Root Locations of an Entire Polytope of Polynomials: It Suffices To Check the Edges", *Mathematics of Control Signals and Systems* **1**, pp. 61-71 (1988).

5. M. Fu and B.R. Barmish, "Maximal Unidirectional Perturbation Bounds for Stability of Polynomials and Matrices", *Systems and Control Letters* **11**, 173-179 (1988).

6. J. Ackermann, B.D.O. Anderson, C.V. Hollot, and P.P. Khargonekar, "Panel Discussion: New Trends in Robustness Analysis", *Proceedings of the 28th IEEE Conference on Decision and Control*, Tampa FL, pp. 2278-2279 (1989).

7. R.V. Patel and M. Toda, "Quantitative Measures of Robustness for Multivariable Systems", *Proceedings of the Joint Automatic Control Conference*, San Francisco CA, TP8-A (1980).

8. R.K. Yedavalli, "Perturbation Bounds for Robust Stability in Linear State Space Models with Structured Uncertainty", *International Journal of Control* **42**, pp. 1507-1517 (1985).

9. R.K. Yedavalli and Z. Liang, "Reduced Conservatism in Stability Robustness Bounds by State Transformation," *IEEE Transactions on Automatic Control* **31**, pp.863-866 (1986).

10. B.S. Chen and C.C. Wong, "Robust Linear Controller Design: Time Domain Approach", *IEEE Transactions on Automatic Control* **32**, pp. 161-164 (1987).

11. K.M. Sobel, S.S. Banda and H.H. Yeh, "Robust Control for Linear Systems with Structured State Space Uncertainty", *International Journal of Control* **50**, pp. 1991-2004 (1989).

12. K.M. Sobel and W. Yu, "Flight Control Application of Eigenstructure Assignment with Optimization of Robustness to Structured State Space Uncertainty", *Proceedings of the 28th IEEE Conference on Decision and Control*, Tampa FL, pp. 1705-1707 (1989).

13. Y.T. Juang, Z.C. Hong, and Y.T. Wang, "Robustness of Pole Assignment in a Specified Region", *IEEE Transactions on Automatic Control* **AC-34**, 758-760 (1989).

14. W. Yu and K.M. Sobel, "Performance Robustness for LTI Systems with Structured State Space Uncertainty", *Proceedings of the 1990 American Control Conference*, San Diego CA, pp. 2039-2042 (1990).

15. W. Yu and K.M. Sobel, "Robust Eigenstructure Assignment with Structured State Space Uncertainty", *Journal of Guidance, Control, and Dynamics* **14**(3), pp. 621-628 (1991).

16. R.T. Haftka and J.H. Starnes Jr., "Applications of a Quadratic Extended Interior Penalty Function for Structural Optimization", *AIAA Journal* **14**, pp. 718-724 (1976).

17. K.M. Sobel and E.Y. Shapiro, "A Design Methodology for Pitch Pointing Flight Control Systems," *Journal of Guidance, Control, and Dynamics* **8**(2), pp 181-187 (1985).

18. J. Stoer and C. Witzgall, "Transformations by Diagonal Matrices in a Normed Space", *Numerical Mathematics* **4**, pp. 458-471 (1964).

19. G.H. Golub and C.F. Van Loan, *Matrix Computations*, The Johns Hopkins University Press, Baltimore, MD (1983).

20. S. Srinathkumar, "Eigenvalue Eigenvector Assignment Using Output Feedback", *IEEE Transactions on Automatic Control* **23**, pp. 79-81 (1978).

21. A.N. Andry Jr., E.Y. Shapiro, and J. Chung, "Eigenstructure Assignment for Linear Systems," *IEEE Transactions on Aerospace and Electronic Systems* **AES-19**(5), pp.711-729 (1983).

22. J. Kautsky, N.J. Nichols, and P. Van Dooren, "Robust Pole Assignment in Linear State Feedback," *International Journal of Control* **41**(5), pp.1129-1159 (1985).

23. J.R. Broussard and M.J. O'Brien, "Feedforward Control to Track the Output of a Forced Model", *IEEE Transactions on Automatic Control* **25**, pp. 851-854 (1980).

24. C.T. Chen, *Introduction to Linear System Theory*, Holt, Rinehart, and Winston, Inc., New York, NY (1970).

25. IMSL Reference Manual, IMSL Inc., Houston, Texas, Release 9.1

26. B. Kouvaritakis and H. Latchman, "Singular Value and Eigenvalue Techniques in the Analysis of Systems with Structured Perturbations", *International Journal of Control* **41**, pp. 1381-1412 (1985).

27. M. Vidyasagar, *Nonlinear Systems Analysis*, Prentice-Hall, Englewood Cliffs, NJ (1978).

INDEX

D

I

Q

R

S

T